D1628630

Bacterial Signaling

Edited by
Reinhard Krämer and Kirsten Jung

Related Titles

G. Krauss

Biochemistry of Signal Transduction and Regulation

2008

ISBN: 978-3-527-31397-6

W. Schumann

Dynamics of the Bacterial Chromosome

Structure and Function

2006

ISBN: 978-3-527-30496-7

A. Mendez-Vilas (Ed.)

Modern Multidisciplinary Applied Microbiology

Exploiting Microbes and Their Interactions

2006

ISBN: 978-3-527-31611-3

G. Kahl

The Dictionary of Genomics, Transcriptomics and Proteomics

2009

ISBN: 978-3-527-32073-8

I. Ahmad, F. Aqil (Eds.)

New Strategies Combating Bacterial Infection

2009

ISBN: 978-3-527-32206-0

B.H.A. Rehm (Ed.)

Pseudomonas

Model Organism, Pathogen, Cell Factory

2008

ISBN: 978-3-527-31914-5

S.H.E. Kaufmann, P. van Helden, E. Rubin, W.J. Britton (Eds.)

Handbook of Tuberculosis

2008

ISBN: 978-3-527-31683-0

Bacterial Signaling

Edited by
Reinhard Krämer and Kirsten Jung

WILEY-VCH Verlag GmbH & Co. KGaA

The Editors

Prof. Dr. Reinhard Krämer
University of Cologne
Insitute of Biochemistry
Zülpicher Strasse 47
50674 Cologne
Germany

Prof. Dr. Kirsten Jung
Ludwig-Maximilians-Universität
Department of Biology – Microbiology
Grosshaderner Strasse 2–4
82152 Planegg-Martinsried
Germany

All books published by Wiley-VCH are carefully produced. Nevertheless, authors, editors, and publisher do not warrant the information contained in these books, including this book, to be free of errors. Readers are advised to keep in mind that statements, data, illustrations, procedural details or other items may inadvertently be inaccurate.

Library of Congress Card No.: applied for

British Library Cataloguing-in-Publication Data
A catalogue record for this book is available from the British Library.

Bibliographic information published by the Deutsche Nationalbibliothek
The Deutsche Nationalbibliothek lists this publication in the Deutsche Nationalbibliografie; detailed bibliographic data are available on the Internet at http://dnb.d-nb.de

© 2010 WILEY-VCH Verlag GmbH & Co. KGaA, Weinheim

All rights reserved (including those of translation into other languages). No part of this book may be reproduced in any form – by photoprinting, microfilm, or any other means – nor transmitted or translated into a machine language without written permission from the publishers. Registered names, trademarks, etc. used in this book, even when not specifically marked as such, are not to be considered unprotected by law.

Printed in the Federal Republic of Germany
Printed on acid-free paper

Cover Design Adam Design, Weinheim
Typesetting Thomson Digital, Noida, India
Printing Strauss GmbH, Mörlenbach
Binding Litges & Dopf Buchbinderei GmbH, Heppenheim

ISBN: 978-3-527–32365-4

Contents

Preface *XIX*
List of Contributors *XXI*

Part I Intercellular Communication *1*

Introduction *3*
Claudia Anetzberger and Kirsten Jung

1 Cell–Cell Communication and Biofilm Formation in Gram-Positive Bacteria *7*
Christine Heilmann and Friedrich Götz
1.1 Introduction *7*
1.2 Staphylococcal Infections and Biofilms *8*
1.3 Molecular Basis of Biofilm Formation in Staphylococci *8*
1.3.1 Attachment to Abiotic Surfaces *8*
1.3.2 Attachment to Biotic Surfaces *10*
1.3.3 Accumulation Process *11*
1.3.3.1 Polysaccharide-Associated Biofilm Accumulation *11*
1.3.3.2 Extracellular DNA *12*
1.3.3.3 Protein-Associated Biofilm Accumulation *12*
1.3.4 Biofilm Escape Factors *12*
1.4 QS in Staphylococcal Biofilms *13*
1.4.1 agr QS Locus *13*
1.4.2 luxS/AI-2 System *17*
References *17*

2 Cell–Cell Communication in Biofilms of Gram-Negative Bacteria *23*
Claudio Aguilar, Aurelien Carlier, Kathrin Riedel, and Leo Eberl
2.1 Introduction *23*
2.2 QS in Gram-Negative Bacteria *23*

2.3	QS and Biofilm Formation 24
2.3.1	When in the Biofilm Cycle does QS Play a Role? 27
2.4	QS-Regulated Factors Involved in Biofilm Formation 29
2.4.1	EPSs 29
2.4.2	Biosurfactants 30
2.4.3	DNA Release 31
2.4.4	Physiology: Dissimilatory Nitrate Reduction in *P. aeruginosa* 32
2.5	QS as a Target for the Eradication of Biofilms 32
2.6	Interspecies Signaling in Mixed Biofilms 34
2.7	Conclusions 34
	References 35

3	**Cell Interactions Guide the Swarming and Fruiting Body Development of Myxobacteria** 41
	Dale Kaiser
3.1	Introduction 41
3.2	Motility of Myxobacteria 41
3.3	Pilus Engine 42
3.4	Slime Secretion Engine 43
3.5	Swarming of Myxobacteria 44
3.6	Regulating Reversals 45
3.7	Fruiting Body Development 46
3.8	C-Signal and Fruiting Body Morphogenesis 48
3.9	Managing the Reversal Frequency 48
3.10	C-Signal Control of Gene Expression 49
	References 51

4	**Communication Between Rhizobia and Plants** 57
	Michael Göttfert
4.1	Introduction 57
4.2	Nodulation (*nod*) Genes are Induced by Flavonoids and are Under Positive and Negative Regulation 59
4.3	Activation of the *nod* Genes Results in the Synthesis and Export of Lipo-Chito-Oligosaccharide Signal Molecules 61
4.4	Rhizobia use Secreted Proteins as Effector Molecules 64
4.5	Microarray Studies Help in Elucidating the Flavonoid Stimulons 65
4.6	*nod* Genes as Accessory Components of the Rhizobial Core Genome 66
4.7	Conclusions and Outlook 66
	References 67

5	**Communication Between Pathogens and Eukaryotic Cells** *75*
	Jürgen Heesemann
5.1	Introduction *75*
5.2	Long-Distance Communication *77*
5.2.1	Language of Pathogen-Associated Molecular Patterns *77*
5.2.1.1	TLRs *78*
5.2.1.2	Cytosolic PAMP Receptors *80*
5.2.1.3	PAMPs as Chemoattractants *81*
5.2.2	Language of Hormones *83*
5.2.3	Extracellular Bacterial Toxins in Pathogen–Host Cell Communication *86*
5.2.3.1	Superantigens *86*
5.2.3.2	Cholera Toxin *86*
5.2.3.3	Bordetella Modulins *87*
5.2.3.4	Helicobacter *87*
5.2.3.5	VacA *87*
5.2.3.6	*Clostridium difficile* Toxins *87*
5.3	Short-Distance Communication *88*
5.3.1	Bacterial Adhesins and Host Cell Receptors *88*
5.4	Conclusions *91*
	References *91*

6	**Identification of Bacterial Autoinducers – *Methods Chapter*** *95*
	Agnes Fekete, Michael Rothballer, Anton Hartmann, and Philippe Schmitt-Kopplin
6.1	Introduction *95*
6.2	Biosensors *98*
6.2.1	Biosensor Construction *98*
6.2.2	AI Screening with Biosensors *98*
6.3	Sample Preparation Prior to Analysis *99*
6.3.1	Liquid–Liquid Extraction *99*
6.3.2	Principles of Liquid Chromatography *99*
6.3.2.1	SPE *100*
6.3.2.2	TLC *101*
6.3.2.3	(Semi)Preparative Liquid Chromatography *101*
6.4	Techniques for the Structural Analysis of AIs *102*
6.4.1	Mass Spectrometry *102*
6.4.2	NMR Spectroscopy *103*
6.5	Techniques for the Quantification of AIs *103*
6.5.1	Principles of the Analysis Methods *103*
6.5.2	Quantification Methods of the Known AIs *104*
6.5.2.1	Analysis of AHL-Based QS Signals *105*
6.5.2.2	Analysis of HAQ-Based QS Signals *106*
6.6	Conclusions and Future Perspectives *107*
	References *107*

Part II Transmembrane Signaling 113

Introduction 115
Reinhard Krämer

7 Outer Membrane Signaling in Gram-Negative Bacteria 117
Volkmar Braun
7.1 Introduction 117
7.2 A Sophisticated Mechanism: A Signaling Cascade Across the Outer Membrane in Transcriptional Regulation of the Ferric Citrate Transport Genes 117
7.3 Transfer of the Signal Across the Cytoplasmic Membrane 121
7.4 Signal Transfer into the Cytoplasm 121
7.5 FecI is an ECF Sigma Factor 122
7.6 Mechanism of Ferric Citrate Transcription Regulation 123
7.7 Transcription Regulation of the Fec Type in *Pseudomonas putida* 123
7.8 Transcription Regulation of the Fec Type in *Pseudomonas aeruginosa* 124
7.9 Transcriptional Regulation of the Fec Type in *Bordetella* 126
7.10 ECF Signaling in *Serratia marcescens* 127
7.11 ECF Signaling in *Ralstonia solanacearum* 127
7.12 Signaling in Outer Membrane Transport 127
7.13 Assumed Outer Membrane Signaling 128
7.14 Conclusions 128
References 130

8 Stimulus Perception and Signaling in Histidine Kinases 135
Ralf Heermann and Kirsten Jung
8.1 Introduction 135
8.2 Histidine Kinase Family 135
8.2.1 Basic Structure of Histidine Kinases 136
8.2.2 Specifics of Histidine Kinases in Comparison to Serine/Threonine/Tyrosine Kinases 138
8.3 Stimulus Perception and Signaling by Histidine Kinases 139
8.3.1 Chemical Stimuli 139
8.3.2 Physical Stimuli 145
8.4 Accessory Proteins of Histidine Kinases 148
8.5 Conclusions and Outlook 151
References 152

9 Chemotaxis and Receptor Localization 163
Victor Sourjik
9.1 Introduction 163
9.2 Architecture of the Sensory Complex 165
9.2.1 Structure and Function of Chemoreceptors 165

9.2.2	Protein Interactions in the Ternary Complex	*166*
9.3	Clustering of Sensory Complexes	*167*
9.3.1	Chemoreceptor Clusters	*167*
9.3.2	Cluster Assembly and Positioning	*168*
9.3.3	Cluster Positioning	*168*
9.3.4	Cluster Stability	*169*
9.4	Role of Clustering in Signal Processing	*170*
9.4.1	Role of Protein Localization	*170*
9.4.2	Signal Amplification	*170*
9.4.3	Allosteric Models and the Role of the Methylation System in High Sensitivity	*170*
9.4.4	Signal Integration	*172*
9.4.5	Adaptational Assistance Neighborhoods	*173*
9.5	Conclusions and Outlook	*173*
	References	*173*

10	**Photoreception and Signal Transduction**	*177*
	Sonja Brandt and Nicole Frankenberg-Dinkel	
10.1	Introduction	*177*
10.2	Bacterial Blue-Light Photoreceptors	*178*
10.2.1	Microbial Rhodopsins	*178*
10.2.2	Cryptochromes	*181*
10.2.3	Photoactive Yellow Proteins (Xanthopsins)	*183*
10.2.4	BLUF Domain Proteins	*183*
10.2.5	Phototropin-Like Microbial Photoreceptors	*184*
10.3	Red-Light Sensing – Phytochromes	*185*
10.3.1	Principle of Phytochrome Action	*185*
10.3.2	Domain Organization of Phytochromes	*186*
10.3.3	Cyanobacterial Phytochromes and Phytochrome-Like Proteins	*187*
10.3.4	Phytochromes in other Phototrophic Bacteria	*188*
10.3.5	Phytochromes in Heterotrophic Bacteria	*189*
10.4	Conclusions	*190*
	References	*190*

11	**Transmembrane Signaling**	*197*
	Melinda D. Baker and Matthew B. Neiditch	
11.1	Introduction	*197*
11.2	Transmembrane Receptor Domain Architecture	*198*
11.2.1	Transmembrane Histidine Kinase Domain Architecture	*198*
11.2.2	Chemoreceptor Domain Architecture	*201*
11.3	Structural Analysis of Transmembrane Signaling	*201*
11.3.1	CitA Transmembrane Signaling	*202*
11.3.2	LuxPQ Transmembrane Signaling	*203*
11.3.3	Chemotaxis Receptor Transmembrane Signaling	*204*

11.3.4	HAMP Linker Domain Structure *205*
11.4	Conclusions *206*
	References *207*

12 Sensory Transport Proteins *211*
Reinhard Krämer

12.1	Introduction *211*
12.2	Sensing of Transport Activity *212*
12.2.1	*Escherichia coli* Maltose System and the Global Regulator Mlc *213*
12.2.2	*E. coli* Uhp System *215*
12.2.3	Dicarboxylic Acid Uptake in *E. coli* and Rhizobia *216*
12.2.4	LysP/CadC System in *E. coli* *218*
12.2.5	Ammonium Signaling *218*
12.2.6	Further Transport Systems with Substrate Sensing Function *219*
12.3	Stress Sensing by Transport Proteins *219*
12.3.1	Mechanosensitive Channels *220*
12.3.2	Osmosensory Uptake Systems *221*
12.4	Conclusions and Perspective *223*
	References *224*

13 Regulated Intramembrane Proteolysis in Bacterial Transmembrane Signaling *229*
Thomas Wiegert

13.1	Introduction *229*
13.2	Bacterial I-CLiPs *231*
13.3	Regulation of ECF Sigma Factors by RIP *233*
13.3.1	Regulation of the *E. coli* σ^E-Dependent Envelope Stress Response *233*
13.3.2	σ^E Homologous Systems in Gram-Negative Pathogenic Bacteria *234*
13.3.3	Regulation of the *Bacillus subtilis* σ^W Regulon *235*
13.3.4	Possible Role of RIP in Regulation of other ECF Sigma Factors *236*
13.4	Regulation of ToxR-Like Transcriptional Regulators via RIP *236*
13.5	Involvement of RIP in Regulation of Bacterial Cell Division and Differentiation *237*
13.5.1	Involvement of RIP in Timing of Cell Division in *B. subtilis* *237*
13.5.2	Activation of the Alternative Sporulation Sigma Factor σ^K of *B. subtilis* *237*
13.5.3	Regulation of the Cell Polarity Determinant PodJ of *Caulobacter crescentus* *238*
13.6	Involvement of RIP in Cell–Cell Communication *239*
13.6.1	Production of Peptide Sex Pheromones in *Enterococcus faecalis* *239*
13.6.2	Rhomboid-Mediated QS in *Providencia stuartii* *240*
13.7	Conclusions *240*
	References *240*

14		**Protein Chemical and Electron Paramagnetic Resonance Spectroscopic Approaches to Monitor Membrane Protein Structure and Dynamics –** *Methods Chapter* *247*
		Daniel Hilger and Heinrich Jung
14.1		Introduction *247*
14.2		Cysteine Chemistry *248*
14.2.1		General Considerations *248*
14.2.2		Applications of Cysteine Chemistry *250*
14.2.2.1		Cysteine Accessibility Analyses *250*
14.2.2.2		Proximity Relationships in Proteins by Cysteine Cross-Linking *252*
14.3		Site-Directed Spin Labeling and EPR Spectroscopy *253*
14.3.1		Why EPR Spectroscopy? *253*
14.3.2		Site-Directed Spin Labeling of Proteins *253*
14.3.3		Information on Protein Structure and Dynamics Based on Spin Label Dynamics *255*
14.3.3.1		Example *256*
14.3.4		Information on Protein Structure and Dynamics Based on Spin Label Accessibility *256*
14.3.4.1		Example *257*
14.3.5		Polarity and Proticity in the Spin Label Microenvironment *257*
14.3.5.1		Example *258*
14.3.6		Intra- and Intermolecular Distances by Double Spin Labeling and Interspin Distance Measurements *258*
14.3.6.1		Example *258*
14.4		Conclusions *259*
		References *260*

Part III Intracellular Signaling *265*

Introduction *267*
Kirsten Jung, Michael Y. Galperin, and Reinhard Krämer

15	**Protein Domains Involved in Intracellular Signal Transduction** *269*
	Michael Y. Galperin
15.1	Introduction *269*
15.2	Computational Analysis of Signaling Domains *270*
15.3	Intracellular Sensory Domains *271*
15.3.1	PAS Domain *272*
15.3.2	GAF Domain *273*
15.3.3	BLUF Domain *273*
15.3.4	GCS Domain *274*
15.3.5	HNOB Domain *274*
15.3.6	Hr Domain *274*
15.3.7	KdpD Domain *275*
15.3.8	PHY Domain *275*

15.4	Intracellular Signal-Transducing and Output Domains 276
15.4.1	Two-Component Signal Transduction 276
15.4.2	Chemotaxis 276
15.4.3	Sugar: PTS 279
15.4.4	c-di-GMP-Mediated Signaling 279
15.4.5	Serine/threonine Protein Phosphorylation Signaling System 279
15.5	Diversity of Intracellular Signaling Pathways 280
	References 280

16	**Sensing of Oxygen by Bacteria** 289
	Gottfried Unden, Martin Müllner, and Florian Reinhart
16.1	Introduction 289
16.2	O_2 as a Signal 290
16.3	Direct O_2 Sensors 291
16.3.1	Heme B-Containing Sensors 291
16.3.1.1	FixL 291
16.3.1.2	Dos 294
16.3.1.3	HemAT 294
16.3.2	$[4Fe-4S]^{2+}$-Containing Sensors 294
16.3.2.1	FNR_{Ec} 294
16.3.2.2	FNR_{Bs} 296
16.3.2.3	NreB 296
16.3.2.4	WhiB3 296
16.3.2.5	$[4Fe-4S]^{2+}$ as a Universal Cofactor for O_2 Sensing 296
16.3.3	FAD-Containing Sensors 297
16.3.3.1	NifL 297
16.4	Indirect O_2 Sensors 298
16.4.1	Electron Transport-Linked Sensors 298
16.4.1.1	ArcB/ArcA 298
16.4.1.2	Aer 300
16.4.1.3	PrrB/PrrA and RegB/RegA 300
16.4.1.4	ResE/ResD and SrrA/SrrB 300
16.4.2	NADH-Linked Systems 301
16.4.2.1	Rex 301
	References 301

17	**Microbial Sensor Systems for Dihydrogen, Nitric Oxide, and Carbon Monoxide** 307
	Rainer Cramm and Bärbel Friedrich
17.1	Introduction 307
17.2	Sensing of Molecular Hydrogen 309
17.2.1	Environmental Signals that Direct Hydrogenase Control 310
17.2.2	Hydrogen-Activating Proteins in Nature 310
17.2.3	What Makes the H_2 Signaling Hydrogenase Different from the Energy-Providing Hydrogenase? 311

17.2.4	H$_2$ Signaling Cascade	*312*
17.2.5	H$_2$ Sensor Complex in Action	*314*
17.2.6	Concluding Remarks and Perspectives	*315*
17.3	Sensing of Nitric Oxide and Carbon Monoxide	*315*
17.3.1	Primary Sensors for NO	*316*
17.3.1.1	NorR-Type NO-Sensing Regulators	*317*
17.3.1.2	NsrR-Type NO Sensing Regulators	*318*
17.3.1.3	NO-Sensing Regulators Containing CAP Domains	*318*
17.3.2	Primary Sensors for CO	*319*
17.3.2.1	CooA – A CO Sensor of Anaerobic Carboxidotrophs	*320*
17.3.2.2	RcoM – A CO Sensor of Aerobic Carboxidotrophs	*321*
17.3.3	Hypothetical or Secondary Sensors Systems for NO and/or CO	*321*
17.3.3.1	Eukaryotic-Style NO Sensing in Prokaryotes	*321*
17.3.3.2	NO Sensing by Fur, SoxR, and OxyR	*322*
17.3.3.3	Detecting Multiple Diatomic Gases: Sensors Responding to O$_2$, CO, and NO	*323*
	References *324*	
18	**Signal Transduction by Trigger Enzymes: Bifunctional Enzymes and Transporters Controlling Gene Expression** *329*	
	Fabian M. Commichau and Jörg Stülke	
18.1	Introduction *329*	
18.2	Trigger Enzymes Active as DNA-Binding Transcription Factors	*332*
18.3	Trigger Enzymes Involved in Post-Transcriptional Regulation via Protein–RNA Interaction *333*	
18.4	Trigger Enzymes Controlling Gene Expression by Signal-Dependent Phosphorylation of Transcription Regulators *334*	
18.5	Trigger Enzymes Controlling the Activity of Transcription Factors by Protein–Protein Interactions *335*	
18.6	Evolution of Trigger Enzymes: From Enzymes via Trigger Enzymes to Regulators *338*	
	References *339*	
19	**Regulation of Carbohydrate Utilization by Phosphotransferase System-Mediated Protein Phosphorylation** *343*	
	Boris Görke and Birte Reichenbach	
19.1	Introduction *343*	
19.2	Unique Features of the Bacterial PTS *344*	
19.3	Phosphorylation of the IIA^{Glc} Subunit of the Glucose Transporter Triggers Global CCR in Enteric Bacteria *345*	
19.4	A Second Key Mechanism of CCR: Phosphorylation of IIA^{Glc} Controls Inducer Exclusion in Enteric Bacteria *347*	
19.5	Phosphorylation of Ser46 of HPr Triggers CCR in Low-GC Gram-Positive Bacteria *348*	

19.6	Phosphorylation of HPr by the Bifunctional Kinase/Phosphorylase Links CCR to the Metabolic State of the Cell in Gram-Positive Bacteria *349*	
19.7	HPr Controls Inducer Exclusion in Low-GC Gram-Positive Bacteria *349*	
19.8	Control of Transcription Regulators by EII *350*	
19.9	Catabolite Control of PRD-Containing Regulators by HPr(His~P)-Mediated Phosphorylation *351*	
19.10	PTS-Dependent Regulation of Chemotaxis *352*	
19.11	Regulatory Functions of Paralogous PTSs *352*	
	References *353*	
20	**cAMP Signaling in Prokaryotes** *357*	
	Knut Jahreis	
20.1	Introduction *357*	
20.2	CCR – A Short Historical Account *357*	
20.3	Regulation of Intracellular cAMP Levels: PTS as a Sensor and Signal Transduction System that Modulates AC Activity *358*	
20.4	Another Extension of the Simple Model: Catabolite Repression by Non-PTS Substrates: The PEP: Pyruvate Ratio is a Key Node in Carbon and Energy Metabolism *362*	
20.5	cAMP Excretion and Phosphodiesterase Activity *363*	
20.6	Function of the cAMP–CRP Complex *364*	
20.6.1	Transcriptional Regulation of the *crp* Gene *364*	
20.6.2	Properties of CRP *364*	
20.6.3	cAMP–CRP Complex-Dependent Promoter Activation and Repression *365*	
20.7	cAMP–CRP Modulon and the CFU "Carbohydrate Catabolism/Quest for Food" *366*	
20.8	Interactions with Other Regulatory Systems *367*	
20.8.1	Inducer Exclusion by Unphosphorylated EIIA^{Glc} *367*	
20.8.2	Interactions with Other Signaling Systems to Keep the Metabolic Balance: "Anticatabolite Repression" or Glucose Induction by Mlc *368*	
20.9	Mathematical and Computer-Assisted Modeling of Catabolite Repression *369*	
20.10	Conclusions *370*	
	References *370*	
21	**c-di-GMP Signaling** *377*	
	Christina Pesavento and Regine Hengge	
21.1	Introduction *377*	
21.2	Protein Domains Involved in c-di-GMP Signaling *377*	
21.2.1	Making and Breaking of c-di-GMP *377*	
21.2.2	Composite GGDEF, EAL, and HD-GYP Proteins *379*	

21.2.3	Recruitment of GGDEF and EAL Domains for c-di-GMP-Unrelated Functions *379*	
21.2.4	Regulation of DGC and PDE Activity and Expression *380*	
21.2.5	c-di-GMP-Binding Effectors *380*	
21.3	Signaling Specificity *381*	
21.4	c-di-GMP Signaling in *E. coli* *382*	
21.5	c-di-GMP signaling in *V. cholerae* *385*	
21.6	c-di-GMP Signaling in *C. crescentus* *387*	
21.7	Conclusions and Outlook *388*	
	References *389*	
22	**ppGpp Signaling** *395*	
	Rolf Wagner	
22.1	Introduction *395*	
22.2	Induction of the Effector (p)ppGpp Through Synthesis and Degradation *396*	
22.3	ppGpp – A *Bona Fide* Global Regulator *398*	
22.3.1	Transcriptional Profiling in Different Bacterial Strains *398*	
22.3.2	Lack of (p)ppGpp Signaling in Obligate Intracellular Bacterial Pathogens and Archaea *399*	
22.3.3	ppGpp in Plants *399*	
22.3.4	(p)ppGpp as a Mediator of Bacterial Social Behavior and Cell–Cell Signaling Mechanisms *400*	
22.3.5	(p)ppGpp Signaling in Virulence and Pathogen–Host Interaction *400*	
22.3.6	(p)ppGpp as a Regulator for Toxin–Antitoxin Systems in Bacterial Programmed Cell Death *401*	
22.3.7	Persister Cells and Enhanced Mutation Frequency *401*	
22.4	Effects on Macromolecular Synthesis *402*	
22.4.1	Role of ppGpp as an Inhibitor of Replication *402*	
22.4.2	Inhibition of Translation: Effect on Initiation Factor 2 *403*	
22.5	Regulation of Transcription: RNA Polymerase is the Target *403*	
22.5.1	Role of RNA Polymerase ω Subunit *404*	
22.5.2	Promoter-Specific Effects of (p)ppGpp *404*	
22.5.3	Rate-Limiting Step in ppGpp-Dependent Transcription Initiation *405*	
22.5.4	Different Mechanism of rRNA Regulation Between *E. coli* and *B. subtilis* *406*	
22.5.5	Involvement of Coregulators: RNA Polymerase Secondary Channel-Binding Proteins *406*	
22.5.6	Positive Stringent Control *407*	
22.5.7	Passive Regulation by Sigma Factor Competition – Direct Versus Indirect Effects *407*	
	References *408*	

23	**Sensory RNAs** *415*	
	Franz Narberhaus	
23.1	Introduction *415*	
23.2	RNA as a Regulatory Molecule *415*	
23.3	Riboswitches *416*	
23.4	RNA Thermometers *420*	
23.5	Conclusions *422*	
	References *423*	
24	**Signal Transduction by Serine/Threonine Protein Kinases in Bacteria** *427*	
	Michael Bott	
24.1	Introduction *427*	
24.2	Discovery and Distribution of STPKs in Prokaryotes *427*	
24.3	Serine/Threonine Phosphorylation versus Histidine/Aspartate Phosphorylation *428*	
24.4	Domain Architecture of STPKs *428*	
24.5	Structural Studies on STPKs *429*	
24.6	Signal Transduction by STPKs *432*	
24.7	Control of Gene Expression by PknB via the Activity of Sigma Factors *434*	
24.8	Control of Gene Expression by PknH via the Transcriptional Regulator EmbR *435*	
24.8.1	Discovery of EmbR in Mycobacteria and its Phosphorylation by PknH *435*	
24.8.2	Structure of EmbR *435*	
24.8.3	Effects of EmbR Phosphorylation by PknH *435*	
24.8.4	Model of Signal Transduction by PknH and EmbR *436*	
24.9	Direct Control of Enzyme Activities by STPKs *436*	
24.10	Indirect Control of Enzyme Activity by PknG and its Target Protein OdhI/GarA *438*	
24.10.1	Distribution of PknG *438*	
24.10.2	Structure of PknG *438*	
24.10.3	Evidence for a Role of PknG in the Pathogenicity of Mycobacteria *439*	
24.10.4	Studies on PknG and its Target Protein OdhI in *C. glutamicum* *439*	
24.10.5	Inhibition of 2-Oxoglutarate Dehydrogenase by Corynebacterial OdhI *440*	
24.10.6	Identification of the OdhI Homolog GarA in Mycobacteria *440*	
24.10.7	Identification of GarA as a Substrate of Mycobacterial PknB *440*	
24.10.8	Identification of GarA as Substrate of Mycobacterial PknG *441*	
24.10.9	Functions of GarA in Mycobacteria *441*	
24.10.10	Putative Mechanism of GarA/OdhI Function *442*	
24.10.11	Model of Signal Transduction by PknG and OdhI/GarA *443*	
24.11	Conclusions and Outlook *443*	
	References *444*	

25 Regulatory Proteolysis and Signal Transduction in Bacteria 449
Kürşad Turgay

25.1 Introduction 449
25.2 Hsp100/Clp and other AAA+ Protease Systems in Bacteria 450
25.3 Substrate Recognition and Adaptor Proteins 452
25.3.1 Substrate Recognition 452
25.3.2 Adaptor Proteins 452
25.4 Examples of Regulatory Proteolysis in *B. subtilis* 454
25.4.1 Competence Development and the Proteolytic Switch 454
25.4.2 Heat Shock Adaptation 455
25.5 Conclusions 456
References 457

26 Intracellular Signaling and Gene Target Analysis – *Methods Chapter* 463
Jörn Kalinowski

26.1 Introduction 463
26.2 Genome-Wide Expression Analysis 463
26.3 Finding Unknown Target Genes 465
26.3.1 Systematic Evolution of Ligands by Exponential Enrichment (SELEX) 465
26.3.2 Chromatin Immunoprecipitation (ChIP) 467
26.4 Analyzing Known Targets 468
26.4.1 DNA Affinity Chromatography (DAC) 469
26.4.2 Electrophoretic Mobility Shift Assay (EMSA) 469
26.5 Conclusions and Outlook 471
References 472

Index 473

Preface

Throughout their life, bacteria interact with their surroundings by exchanging information with other cells, by exploring optimal growth conditions, and by sensing and responding to environmental stress. Thus, the signaling network of bacteria is a complex and indispensable aspect of bacterial life. Therefore, it is not surprising that research in this field is highly dynamic, and novel and important phenomena and mechanisms related to bacterial signaling are continuously uncovered and elucidated.

The bacterial cell is surrounded by the cell envelope, which is the basis for the cell's shape and its physiological individuality. Signaling can thus be conceptually divided into processes that occur outside the cell, across the membrane between the interior and the exterior, and within the cytoplasmic compartment. Although a clear and systematic distinction on the basis of this approach is not possible in a strict way, we have grouped the individual contributions in this book according to the major site of their action, such as signaling between cells, across the membrane, and within the cytoplasm.

This book provides an overview of a large number of examples of signaling mechanisms related to core functions in bacterial life. Established principles of signaling, such as catabolite repression, chemotaxis, and histidine kinase/response regulator systems, are comprehensively described. The book also focuses on new principles and concepts, such as intramembrane proteolysis, trigger enzymes, and c-di-GMP signaling. Each of the three major sections is complemented by a special chapter that provides methodological insights into the analysis of important features of signaling mechanisms.

This book offers a comprehensive overview of the present state of art on the major facets of the bacterial signaling network. It is thought of as a valuable source of information for researchers and advanced students who wish to learn about one of the most exciting fields of modern microbiology.

Cologne and Munich, November 2009   *Reinhard Krämer and Kirsten Jung*

Bacterial Signaling. Edited by Reinhard Krämer and Kirsten Jung
Copyright © 2010 WILEY-VCH Verlag GmbH & Co. KGaA, Weinheim
ISBN: 978-3-527-32365-4

List of Contributors

Claudio Aguilar
University of Zurich
Institute of Plant Biology
Department of Microbiology
Zollikerstrasse 107
8008 Zurich
Switzerland

Claudia Anetzberger
Ludwig-Maximilians-Universität
München
Munich Center for Integrated Protein
Science (CiPSM) at the Department
of Biology I – Microbiology
Großhaderner Strasse 2–4
82152 Planegg-Martinsried
Germany

Melinda D. Baker
UMDNJ-New Jersey Medical School
Department of Microbiology and
Molecular Genetics
225 Warren Street
Newark, NJ 07103
USA

Michael Bott
Forschungszentrum Jülich
Institut für Biotechnologie
52425 Jülich
Germany

Sonja Brandt
Ruhr-Universität Bochum
Fakultät für Biologie und
Biotechnologie
Physiologie der Mikroorganismen
Universitätsstrasse 150
44780 Bochum
Germany

Volkmar Braun
Max-Planck-Institute for
Developmental Biology
Spemannstrasse 35
72076 Tübingen
Germany

Aurelien Carlier
University of Zurich
Institute of Plant Biology
Department of Microbiology
Zollikerstrasse 107
8008 Zurich
Switzerland

Fabian M. Commichau
Georg-August-University Göttingen
Department of General Microbiology
Griesebachstrasse 8
37077 Göttingen
Germany

Bacterial Signaling. Edited by Reinhard Krämer and Kirsten Jung
Copyright © 2010 WILEY-VCH Verlag GmbH & Co. KGaA, Weinheim
ISBN: 978-3-527-32365-4

Rainer Cramm
Humboldt University of Berlin
Institute of Biology–Microbiology
Chausseestrasse 117
10115 Berlin
Germany

Leo Eberl
University of Zurich
Institute of Plant Biology
Department of Microbiology
Zollikerstrasse 107
8008 Zurich
Switzerland

Agnes Fekete
Helmholtz Zentrum München
German Research Center for
Environmental Health
Institute of Ecological Chemistry
Ingolstädter Landstrasse 1
85764 Neuherberg
Germany

Nicole Frankenberg-Dinkel
Ruhr-Universität Bochum
Fakultät für Biologie und
Biotechnologie
Physiologie der Mikroorganismen
Universitätsstrasse 150
44780 Bochum
Germany

Bärbel Friedrich
Humboldt University of Berlin
Institute of Biology/Microbiology
Chausseestrasse 117
10115 Berlin
Germany

Michael Y. Galperin
National Institutes of Health
National Center for Biotechnology
Information
National Library of Medicine
8600 Rockville Pike
Bethesda, MD 20894
USA

Boris Görke
Universität Göttingen
Institut für Mikrobiologie und Genetik
Abteilung für Allgemeine Mikrobiologie
Grisebachstrasse 8
37077 Göttingen
Germany

Michael Göttfert
TU Dresden
Institut für Genetik
Helmholtzstraße 10
01069 Dresden
Germany

Friedrich Götz
University Hospital Münster
Institute of Medical Microbiology
Domagkstrasse 10
48149 Münster
Germany

Anton Hartmann
Helmholtz Zentrum München
German Research Center for
Environmental Health
Department of Microbe–Plant
Interactions
Ingolstädter Landstrasse 1
85764 Neuherberg
Germany

Ralf Heermann
Ludwig-Maximilians-Universität
München
Biozentrum – Bereich Mikrobiologie
Großhardener Strasse 2–4
82152 Planegg-Martinsried
Germany

Jürgen Heesemann
Ludwig-Maximilians-Universität
München
Max von Pettenkofer Institut
Pettenkoferstrasse 9a
80336 München
Germany

Christine Heilmann
University Hospital Münster
Institute of Medical Microbiology
Domagkstrasse 10
48149 Münster
Germany

Regine Hengge
Freie Universität Berlin
Institut für Biologie – Mikrobiologie
Königin-Luise-Strasse 12–16
14195 Berlin
Germany

Daniel Hilger
Ludwig-Maximilians-Universität
München
Department of Biology –
Microbiology
Großhaderner Strasse 2–4
82152 Planegg-Martinsried
Germany

Knut Jahreis
Universität Osnabrück
Fachbereich Biologie/Chemie
AG Genetik
Barbarastrasse 11
49069 Osnabrück
Germany

Heinrich Jung
Ludwig-Maximilians-Universität
München
Department of Biology –
Microbiology
Großhaderner Strasse 2–4
82152 Planegg-Martinsried
Germany

Kirsten Jung
Ludwig-Maximilians-Universität
München
Department of Biology –
Microbiology
Großhaderner Strasse 2–4
82152 Planegg-Martinsried
Germany

Dale Kaiser
Stanford University School of Medicine
Departments of Biochemistry and
Developmental Biology
279 Campus Drive
Stanford, CA 94305
USA

Jörn Kalinowski
Universität Bielefeld
Institut für Genomforschung und
Systembiologie
Universitätsstrasse 27
33615 Bielefeld
Germany

Reinhard Krämer
University of Cologne
Institute of Biochemistry
Zülpicher Strasse 47
50674 Köln
Germany

Martin Müllner
Universität Mainz
Institut für Mikrobiologie und
Weinforschung
Becherweg 15
55099 Mainz
Germany

Franz Narberhaus
Ruhr-Universität Bochum
Lehrstuhl für Biologie der
Mikroorganismen
Universitätstrasse 150
44780 Bochum
Germany

Matthew B. Neiditch
UMDNJ-New Jersey Medical School
Department of Microbiology and
Molecular Genetics
225 Warren Street
Newark, NJ 07103
USA

Christina Pesavento
Freie Universität Berlin
Institut für Biologie – Mikrobiologie
Königin-Luise-Strasse 12–16
14195 Berlin
Germany

Birte Reichenbach
Georg-August-University Göttingen
Department of General Microbiology
Grisebachstrasse 8
37077 Göttingen
Germany

Florian Reinhart
Universität Mainz
Institut für Mikrobiologie und
Weinforschung
Becherweg 15
55099 Mainz
Germany

Kathrin Riedel
University of Zurich
Institute of Plant Biology
Department of Microbiology
Zollikerstrasse 107
8008 Zurich
Switzerland

Michael Rothballer
Helmholtz Zentrum München
German Research Center for
Environmental Health
Department of Microbe–Plant
Interactions
Ingolstädter Landstrasse 1
85764 Neuherberg
Germany

Philippe Schmitt-Kopplin
Helmholtz Zentrum München
German Research Center for
Environmental Health
Institute of Ecological Chemistry
Ingolstädter Landstrasse 1
85764 Neuherberg
Germany

Victor Sourjik
Universität Heidelberg
Zentrum für Molekulare Biologie
Im Neuenheimer Feld 282
69120 Heidelberg
Germany

Jörg Stülke
Georg-August-University Göttingen
Department of General Microbiology
Grisebachstrasse 8
37077 Göttingen
Germany

Kürşad Turgay
Freie Universität Berlin
Institut für Biologie – Mikrobiologie
Königin-Luise-Strasse 12–16
14195 Berlin
Germany

Gottfried Unden
Universität Mainz
Institut für Mikrobiologie und Weinforschung
Becherweg 15
55099 Mainz
Germany

Rolf Wagner
Heinrich-Heine-Universität Düsseldorf
Institut für Physikalische Biologie
Universitätsstrasse 1
40225 Düsseldorf
Germany

Thomas Wiegert
University of Bayreuth
Institute of Genetics
Universitätsstrasse 30
95440 Bayreuth
Germany

Part I
Intercellular Communication

Introduction

Claudia Anetzberger and Kirsten Jung

For a long time bacteria were regarded as dump singled-cell living organisms that ensure their survival by adaptation to rapidly changing environmental conditions without communication. The concept of intercellular communication within bacterial populations originates from different discoveries in the 1960s and 1970s proposing the involvement of external factors excreted by bacteria themselves. In *Streptomyces griseus* aerial hyphae formation was found to be inducible by an old culture [1]; 10 years later this compound was identified as γ-butyrolactone (A-factor) [2]. In 1965, Tomasz assumed that an external factor, "a hormone-like cell product," was important for genetic competence in *Streptococcus pneumoniae* (formerly known as *Pneumococcus*) [3], which was later shown to be a modified peptide. Hasting *et al.* found that *Vibrio fischeri*, a bioluminescent marine bacterium, produced light at high cell density, but not in dilute suspensions [4]. Light production could be stimulated by the addition of cell-free culture fluid. They called the responsible component "autoinducer," which was later identified as an acyl-homoserine lactone [5]. At this time autoinduction was defined as an environmental sensing system that allows bacteria to observe their own population density by monitoring the concentration of these autoinducers. At low cell density autoinducers are available in a low concentration, while at high cell density these compounds accumulate to the critical concentration required for activation of certain genes. The term "quorum sensing" was introduced by Greenberg in 1994 as cell-density-linked, coordinated gene expression in populations that experience threshold signal concentrations to induce a synchronized population response [6]. "Quorum sensing" spread like wildfire, making its way into virtually every following paper on autoinduction [7], and the scientific community accepted that bacterial social behaviors described, for example, for *Myxococcus* (see Chapter 3), *Bacillus*, or *Vibrio* are the norm and not the exception in the bacterial world [8].

Processes controlled by quorum sensing are usually those that are unproductive when undertaken by an individual bacterium, but become effective when undertaken by a group. Thus, quorum sensing allows bacteria to behave like a multicellular organism. In addition to competence, bioluminescence and morphological differ-

Bacterial Signaling. Edited by Reinhard Krämer and Kirsten Jung
Copyright © 2010 WILEY-VCH Verlag GmbH & Co. KGaA, Weinheim
ISBN: 978-3-527-32365-4

entiation, quorum sensing controls virulence factor secretion, biofilm formation, and sporulation [9].

The chemical vocabulary used by bacteria for communication is constantly increasing as new molecules are discovered. In general, Gram-negative quorum sensing bacteria communicate with acyl-homoserine lactones [10] (see Chapter 2), while Gram-positive bacteria predominantly communicate with short peptides that often contain chemical modifications [11] (see Chapter 1). In addition, a family of compounds termed as autoinducer-2, derived from the common precursor, 4,5-dihydroxy-2,3 pentandione, has been found to be widespread in the bacterial world, and autoinducer-2 allows interspecies communication. Other molecules, including 3-hydroxypalmitic acid methyl ester [12], 4-hydroxy-2-alkyl-quinolones [13], 3-hydroxy-tri- and pentadecan-4-one [14, 15], cis-11-methyl-2-dodecenoic acid [16], and p-coumaroyl-homoserine lactone [17], are also important in bacterial cell–cell signaling.

It is now becoming increasingly apparent that there is also chemical communication between species from different domains of life. As examples, the communication between rhizobia and plants (Chapter 4) and between pathogenic bacteria and animal cells (Chapter 5) are discussed in more detail in this book. Finally, methods to quantitatively analyze known quorum sensing molecules as well as general strategies to identify and characterize unknown signal molecules are summarized in Chapter 6.

The bacterial language is still very primitive and we are convinced that there are many more bacterial signal molecules that need to be discovered. A better understanding of interorganismic chemical communication will open new possibilities to manipulate bacterial behavior, including the blockage of pathogens from multiplying or the promotion of the growth of beneficial bacteria.

References

1 Dondero, N.C. and Scotti, T. (1957) Excretion by streptomycetes of factors causing formation of aerial hyphae by old cultures. *J. Bacteriol.*, **73**, 584–585.

2 Khokhlov, A.S., Tovarova, I.I., Borisova, L.N., Pliner, S.A., Shevchenko, L.N., Kornitskaia, E.I. et al. (1967) The A-factor, responsible for streptomycin biosynthesis by mutant strains of *Actinomyces streptomycini*. *Dokl. Akad. Nauk. SSSR*, **177**, 232–235.

3 Tomasz, A. (1965) Control of the competent state in *Pneumococcus* by a hormone-like cell product: an example for a new type of regulatory mechanism in bacteria. *Nature*, **208**, 155–159.

4 Nealson, K.H., Platt, T., and Hastings, J.W. (1970) Cellular control of the synthesis and activity of the bacterial luminescent system. *J. Bacteriol.*, **104**, 313–322.

5 Eberhard, A., Burlingame, A.L., Eberhard, C., Kenyon, G.L., Nealson, K.H., and Oppenheimer, N.J. (1981) Structural identification of autoinducer of *Photobacterium fischeri* luciferase. *Biochemistry*, **20**, 2444–2449.

6 Fuqua, W.C., Winans, S.C., and Greenberg, E.P. (1994) Quorum sensing in bacteria: the LuxR–LuxI family of cell density-responsive transcriptional regulators. *J. Bacteriol.*, **176**, 269–275.

7 Turovskiy, Y., Kashtanov, D., Paskhover, B., and Chikindas, M.L. (2007) Quorum sensing: fact, fiction, and everything in between. *Adv. Appl. Microbiol.*, **62**, 191–234.

8 Shapiro, J.A. (1998) Thinking about bacterial populations as multicellular organisms. *Annu. Rev. Microbiol.*, **52**, 81–104.

9 Bassler, B.L. and Losick, R. (2006) Bacterially speaking. *Cell*, **125**, 237–246.

10 Visick, K.L. and Fuqua, C. (2005) Decoding microbial chatter: cell–cell communication in bacteria. *J. Bacteriol.*, **187**, 5507–5519.

11 Waters, C.M. and Bassler, B.L. (2005) Quorum sensing: cell-to-cell communication in bacteria. *Annu. Rev. Cell Dev. Biol.*, **21**, 319–346.

12 Flavier, A.B., Clough, S.J., Schell, M.A., and Denny, T.P. (1997) Identification of 3-hydroxypalmitic acid methyl ester as a novel autoregulator controlling virulence in *Ralstonia solanacearum*. *Mol. Microbiol.*, **26**, 251–259.

13 Dubern, J.F. and Diggle, S.P. (2008) Quorum sensing by 2-alkyl-4-quinolones in *Pseudomonas aeruginosa* and other bacterial species. *Mol. Biosyst.*, **4**, 882–888.

14 Higgins, D.A., Pomianek, M.E., Kraml, C.M., Taylor, R.K., Semmelhack, M.F., and Bassler, B.L. (2007) The major *Vibrio cholerae* autoinducer and its role in virulence factor production. *Nature*, **450**, 883–886.

15 Spirig, T., Tiaden, A., Kiefer, P., Buchrieser, C., Vorholt, J.A., and Hilbi, H. (2008) The *Legionella* autoinducer synthase LqsA produces an alpha-hydroxyketone signaling molecule. *J. Biol. Chem.*, **283**, 18113–18123.

16 Wang, L.H., He, Y., Gao, Y., Wu, J.E., Dong, Y.H., He, C. *et al.* (2004) A bacterial cell–cell communication signal with cross-kingdom structural analogues. *Mol. Microbiol.*, **51**, 903–912.

17 Schaefer, A.L., Greenberg, E.P., Oliver, C.M., Oda, Y., Huang, J.J., Bittan-Banin, G. *et al.* (2008) A new class of homoserine lactone quorum-sensing signals. *Nature*, **454**, 595–599.

1
Cell–Cell Communication and Biofilm Formation in Gram-Positive Bacteria
Christine Heilmann and Friedrich Götz

1.1
Introduction

It is now widely accepted that naturally, bacteria prefer to live in surface-associated communities called biofilms. In the biofilms, the bacteria are embedded in an extracellular polymeric matrix, and are protected against environmental stresses, antimicrobial treatment, and the host immune system. Biofilms have been implicated in a variety of human infections, such as endocarditis, osteomyelitis, chronic otitis media, foreign-body-associated infections, gastrointestinal ulcers, urinary tract infections, chronic lung infections in cystic fibrosis patients, caries, and periodontitis [1]. The causative agents of biofilm-associated infections are different Gram-positive species of *Staphylococcus*, *Streptococcus*, and *Enterococcus* as well as Gram-negative bacteria, such as *Pseudomonas aeruginosa*, *Escherichia coli*, and *Actinobacillus actinomycetemcomitans*.

Within the biofilm community, bacteria communicate with each other by using chemical signal molecules in response to population density in a process that is called quorum sensing (QS; reviewed in [2]). The cell–cell communication via QS involves the production, release, detection, and response to small hormone-like molecules termed pheromones or autoinducers (AIs). During growth, bacteria produce the AIs, which activate the QS system upon reaching a threshold concentration. Three different types of AIs are currently known: *N*-acyl-homoserine lactones that are mainly used by Gram-negative bacteria and secreted cyclic oligopeptides with a thiolactone structure that are preferred by Gram-positive bacteria. LuxS/AI-2 are produced by both Gram-negative and Gram-positive bacteria, and are believed to function in interspecies communication [2].

Various of physiological activities are regulated via QS in Gram-positive bacteria, including biofilm formation in staphylococci, streptococci, and enterococci, expression of virulence factors in staphylococci, development of competence in streptococci, sporulation in *Bacillus*, and antibiotic biosynthesis in *Lactococcus lactis* [2].

Among the Gram-positive bacteria, biofilm formation and QS has been most intensely studied with staphylococci. In contrast to many biofilms found in natural environments, where a biofilm usually consists of a multispecies microbial community, infections due to staphylococci mostly, but not always, are monospecific [3]. The most important staphylococcal species involved in biofilm-associated infections are *Staphylococcus epidermidis* (primarily causing foreign-body-associated infections) and *Staphylococcus aureus* (typically causing infections associated with colonization of the host tissue).

1.2
Staphylococcal Infections and Biofilms

Staphylococci are ubiquitous commensals of the skin and mucous membranes of humans and animals. In humans, *S. aureus* and the coagulase-negative *S. epidermidis* are among the most leading causes of nosocomial infections [4]. Infections due to *S. epidermidis* typically are more subacute or even chronic and require a predisposed or immunocompromised host, such as patients with indwelling medical devices (e.g., prosthetic heart valves and joints, artificial pacemakers, and intravascular catheters) [5]. In contrast, *S. aureus* causes more acute infections associated with the colonization of the host tissue, such as endocarditis and osteomyelitis, which may lead to sepsis. However, *S. aureus* is also a common cause of foreign-body-associated infections and, occasionally, *S. epidermidis* may cause native valve endocarditis.

The most critical pathogenicity factor in these infections is the colonization of abiotic or biotic surfaces by the formation of a three-dimensional structure called a biofilm. The presence of large adherent biofilms on explanted intravascular catheters has been demonstrated by scanning electron microscopy [6]. Microorganisms within a biofilm are protected against antimicrobial chemotherapy as well as against the immune system of the host.

To form a biofilm, staphylococci first attach either to host tissue or to the surface of a medical device, and then proliferate and accumulate into multilayered cell clusters, which are embedded in an amorphous extracellular material that mainly is composed of *N*-acetyl-glucosamine, cell wall teichoic acids, DNA, and host products [7–9]. A mature biofilm contains fluid-filled channels that ensure the delivery of nutrients and oxygen to bacterial cells located deeper in the biofilm [1]. From a mature biofilm, individual cells or cell aggregates can detach. Upon detachment from the biofilm, the bacteria may disseminate via the blood stream, which is thought to lead to metastatic infection and/or development of sepsis. In the following, the molecular mechanisms involved in staphylococcal biofilm formation and detachment are summarized (Figure 1.1).

1.3
Molecular Basis of Biofilm Formation in Staphylococci

1.3.1
Attachment to Abiotic Surfaces

Microbial adherence to biomaterials largely depends on the nature of the polymer material and on the cell surface characteristics of the bacteria. The initial interactions

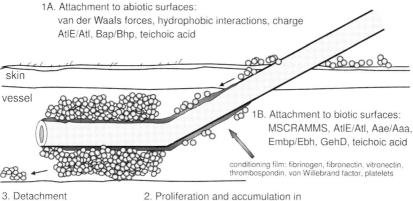

Figure 1.1 Model of different phases of staphylococcal biofilm formation and factors involved. Biofilms develop by initial attachment to surfaces, which may be abiotic (polymer surface) or biotic (polymer surface coated with extracellular matrix and plasma proteins or host tissue), and subsequent proliferation and accumulation into multilayered cell clusters, which requires intercellular adhesion. From a mature biofilm, cells or cell aggregates can detach and disseminate. The different phases and factors involved are indicated.

are believed to occur via nonspecific physicochemical forces such as charge, van der Waals forces, and hydrophobic interactions. The S. aureus colonization of abiotic surfaces depends on the charge of its teichoic acid. S. aureus teichoic acids are highly charged cell wall polymers, composed of alternating phosphate and ribitol (wall teichoic acids) or glycerol (lipoteichoic acids) groups, which are substituted with D-alanine and N-acetyl-glucosamine. A *dltA* mutant lacks D-alanine in its teichoic acid rendering it higher negatively charged. The *dltA* mutant has a biofilm-negative phenotype due to a decreased initial attachment to polystyrene or glass, which is hydrophobic or negatively charged, respectively [10].

Initial adherence has also been attributed to bacterial surface proteins. Using transposon mutagenesis, the autolysin AtlE of *S. epidermidis* O-47 was identified as a surface-associated component that mediates primary attachment of bacterial cells to a polystyrene surface [11]. The 148-kDa AtlE and the homologous autolysin Atl from *S. aureus* are proteolytically cleaved into two bacteriolytically active domains – an N-terminal amidase and a C-terminal glucosaminidase [11, 12]. In the central part of the proteins, there are three repetitive sequences, possibly involved in the adhesive function.

Another protein from *S. aureus*, the 239-kDa biofilm-associated protein Bap, is involved in attachment to a polystyrene surface and intercellular adhesion leading to biofilm formation [13]. The structural features of Bap correspond to those of other typical Gram-positive surface proteins, called MSCRAMMs (microbial surface components recognizing adhesive matrix molecules; see below). The clinical significance of Bap is not clear, because it is apparently present in only 5% of 350 bovine mastitis and absent in all human clinical *S. aureus* isolates tested so far. However, a gene encoding a Bap-homologous protein, the 258-kDa Bhp, is present in the human clinical strain *S. epidermidis* RP62A [14].

1.3.2
Attachment to Biotic Surfaces

Implanted material rapidly becomes coated with plasma and extracellular matrix proteins, such as fibronectin, fibrinogen, vitronectin, thrombospondin, bone sialoprotein, collagen, and von Willebrand factor, or platelets. Thus, all these host factors could serve as specific receptors for colonizing bacteria [15, 16]. Moreover, *S. aureus* is especially able to directly adhere to host tissue, such as the host epithelium or endothelium. Staphylococcal host-factor-binding proteins typically belong to the MSCRAMM family [17]. MSCRAMMs have a common overall organization including an N-terminal signal peptide, an exposed ligand-binding domain, a characteristic cell-wall-spanning region that often contains repetitive sequences, and a C-terminal LPXTG motif responsible for covalent cell wall anchorage. While the *S. aureus* genomes contain a larger number of genes encoding MSCRAMMs (at least 20), there are only 12 genes in the *S. epidermidis* RP62A genome [18]. MSCRAMMs can bind to one or more host extracellular matrix and plasma protein, and include in *S. aureus* fibronectin-binding proteins (FnBpA, FnBpB), fibrinogen-binding proteins (clumping factors ClfA and ClfB), a collagen-binding protein (Cna), a bone-sialoprotein-binding protein (Bbp), and a von Willebrand factor-binding protein A (Spa) [17, 19–22]. However, not all the ligands of all MSCRAMMs have yet been identified. Less data on host factor-binding MSCRAMMs of *S. epidermidis* are available. The fibrinogen-binding 119-kDa Fbe and the almost identical 97-kDa SdrG show significant similarity to the ClfA of *S. aureus* [23].

Staphylococcal surface-associated proteins that are anchored to the cell surface by different means (noncovalently) include the giant 1.1-mDa fibronectin-binding protein Ebh of *S. aureus* and the homologous Embp of *S. epidermidis* [24, 25], whose fibronectin-binding sites seem to be unrelated to those of the *S. aureus* FnBPs, autolysins, the collagen-binding GehD lipase in *S. epidermidis* [14], and the elastin-binding protein EbpS [26]. Further examples of noncovalently associated surface proteins of *S. aureus* are two proteins with a broad binding spectrum, the extracellular matrix and plasma-binding protein Emp and the extracellular adherence protein Eap [27, 28]. Aside from proteins, the cell wall teichoic acid is involved in the adherence of *S. epidermidis* to fibronectin [29].

The autolysin AtlE from *S. epidermidis* not only mediates primary attachment to a polystyrene surface (see Section 1.3.1), but also binds vitronectin [11]. By using a catheter-associated infection model, an *in vivo* role for AtlE was suggested [30]. Further multifunctional autolysin/adhesins include the Aae from *S. epidermidis* and the homologous Aaa from *S. aureus*. Aae and Aaa have bacteriolytic activity, and bind to fibrinogen, fibronectin, and vitronectin in a dose-dependent and saturable fashion and with high affinity [31, 32].

1.3.3
Accumulation Process

After successful attachment to a surface, bacteria proliferate and accumulate in multilayered cell clusters, which requires intercellular adhesion. Probably the same mechanisms are involved in biofilm accumulation on biotic and abiotic surfaces. Staphylococcal biofilm accumulation can be mediated by polysaccharide as well as protein factors.

1.3.3.1 Polysaccharide-Associated Biofilm Accumulation

Transposon mutants not able to accumulate in multilayered cell clusters lack a specific polysaccharide antigen referred to as polysaccharide intercellular adhesin (PIA) [33, 34], which was later also designated as poly-N-acetyl-glucosamine (PNAG) [33, 35]. Purification and structural analysis of PIA revealed that it is a linear β-1–6-linked N-acetyl-glucosaminoglycan with 15–20% of the N-acetyl-glucosaminyl residues being non-N-acetylated [8]. Thus, the designation as PNAG is certainly not correct.

PIA is also produced by S. aureus. It has been reported that the N-acetyl-glucosamine residues of PIA from S. aureus are completely succinylated, which led to its designation as poly-N-succinyl β1–6-glucosamine [36]. However, it is now clear that the succinyl groups were an artifact [35].

The partial deacetylation of 15–20% of the N-acetyl-glucosaminyl residues renders the polysaccharide positively charged, which determines its biological activity. Possibly, it functions as an intercellular adhesin by electrostatically attracting the negatively charged teichoic acid at the bacterial cell surface. The structure of PIA so far is unique. However, PIA-mediated biofilm formation might represent a common principle, because PIA-related structures have also been identified to play a role in the biofilm formation of other pathogenic bacteria, such as the Gram-negative E. coli and A. actinomycetemcomitans [37].

PIA is produced by the gene products encoded by the icaADBC operon. The icaADBC operon was first identified in S. epidermidis, and is also present in S. aureus and other staphylococcal species [34, 38]. The N-acetyl-glucosaminyltransferase activity is carried out by IcaA, which requires IcaD for full activity. With its transmembrane helices, IcaC very likely is an integral membrane protein that putatively transports the N-acetyl-glucosamine oligomers across the membrane [39]. IcaB is mainly found in the culture supernatant and deacetylates PIA [39, 40].

The importance of PIA as a pathogenicity factor has been confirmed in various foreign-body animal infection models with different S. epidermidis icaADBC mutants [30, 41]. However, in S. aureus conflicting results were obtained: PIA production did not increase the capacity to induce persistent infections in a tissue cage model [42]. A study investigating the pathogenic properties of S. epidermidis strains obtained from polymer-associated septicemic disease compared with saprophytic skin and mucosal isolates demonstrated a strong correlation of biofilm formation and presence of the ica gene cluster essentially associated with disease isolates [43].

1.3.3.2 Extracellular DNA

Another polymeric molecule, extracellular DNA, has been identified as an important component of the biofilm matrix of several bacterial species, such as *Streptococcus pneumoniae*, *P. aeruginosa*, and *Enterococcus faecalis* [44–46]. Although it does not seem to mediate biofilm accumulation by itself, it contributes to *S. aureus* biofilm development [47]. DNA is a negatively charged molecule that upon its release could interact with the positively charged extracellular polymer PIA, thus acting as an additional "glue."

1.3.3.3 Protein-Associated Biofilm Accumulation

Staphylococcal biofilm formation is not always polysaccharide-mediated. There are examples of infection-related biofilm-forming *S. epidermidis* strains that do not carry the *icaADBC* gene cluster [48]. In these strains, biofilm formation may be mediated by surface proteins. Surface proteins conferring biofilm accumulation include the 220-kDa accumulation-associated protein Aap from *S. epidermidis* and the homologous *S. aureus* surface protein G (SasG) [49, 50]. The function of Aap in the accumulation process was speculated to be the anchoring of PIA to the cell surface. However, recently, it was shown that Aap is able to mediate intercellular adhesion and biofilm accumulation in a completely PIA-independent background. Intercellular adhesion is mediated by a repeat domain B, which becomes active only after proteolytic cleavage of the N-terminal A domain [49]. Most recently, the B repeats of Aap (also known as G5 domains) were found to be zinc-dependent adhesion modules and a "zinc zipper" mechanism was suggested for G5 domain-based intercellular adhesion in Aap- or SasG-mediated biofilm accumulation [51]. Recently, transmission electron microscopy revealed that Aap has a fibrillar structure [52].

The biofilm-associated protein Bap mentioned above is involved in *S. aureus* adherence to a polystyrene surface, intercellular adhesion, and biofilm accumulation, [13]. The Bap-homologous protein Bhp may be involved in biofilm accumulation in *S. epidermidis* [14].

1.3.4
Biofilm Escape Factors

Biofilm detachment may lead to the dissemination of a staphylococcal infection, and thus to colonization of new sites and metastatic infection. Factors involved in biofilm detachment may include enzymatic activities that lead to the disintegration of the "glue." Depending on the nature of the substance that mediates the "stickiness", enzymatic activities like glycosyl hydrolases that would degrade PIA, proteases that would degrade protein components (such as Aap/SasG or Bap/Bhp), or nucleases that would degrade extracellular DNA, might be involved. Indeed, the Gram-negative periodontal pathogen *A. actinomycetemcomitans* produces dispersin B, which is a soluble glycosyl hydrolase that degrades the self-synthesized extracellular polysaccharide PGA. Like PIA, PGA is a linear polymer of β1–6-linked *N*-acetyl-glucosamine residues [37]. Dispersin B is also able to dissolve biofilms of clinical *S. epidermidis* strains by hydrolyzing the glycosidic linkages of PIA [37, 53]. However, the *S. aureus* and *S. epidermidis* genomes do not seem to encode analogous enzymatic activities.

Extracellular DNA has been shown to be an important component of the *S. aureus* biofilm matrix (see Section 1.3.3.2) [47]. Accordingly, the addition of DNase I inhibits biofilm formation of *S. aureus* and promotes the detachment of preformed *S. aureus* biofilms [54]. Therefore, it may be speculated that the activity of an extracellular *S. aureus* nuclease would also contribute to biofilm detachment. The expression of the *S. aureus* nuclease gene (*nuc*) is under control of the *agr* QS system (see Section 1.4.1) [55]. In contrast to *S. aureus*, DNase I only slightly inhibits biofilm formation in *S. epidermidis*, but does not promote the detachment of preformed biofilms. Thus, in *S. epidermidis* extracellular DNA seems to effect initial attachment to a surface, rather than biofilm accumulation and detachment [54].

Several studies indicate that the biofilm matrix of a significant proportion of biofilm-forming staphylococcal strains mainly contained teichoic acid and proteins, but not PIA [48, 56]. In this case, protease treatment disintegrated the biofilms, although sometimes only partially [48, 57]. At least in *S. aureus*, protease-mediated biofilm detachment is dependent on a functional *agr* QS system (see Section 1.4.1) [58].

Another strategy leading to biofilm detachment involves the production and release of small peptides called phenol-soluble modulins (PSMs). PSMs were first described as proinflammatory agents in *S. epidermidis* [59]. According to their length, the PSMs can be divided in two classes: α-type peptides have a length of approximately 20 amino acids and β-type peptides are 40–45 amino acids in length. PSMs are supposed to have a surfactant-like effect due to their amphipathic α-helical character, which might be responsible for their role in biofilm detachment [60]. The expression of the genes encoding the PSMs is under the control of the *agr* QS system (see Section 1.4.1) [61].

1.4
QS in Staphylococcal Biofilms

In staphylococci, two QS systems have been described so far: the accessory gene regulator (*agr*) system, which has been studied in great detail in *S. aureus* and is also present in other staphylococcal species [55], and the *luxS*/AI-2 system identified in *S. epidermidis* as well as in *S. aureus* [62, 63].

1.4.1
agr QS Locus

S. aureus uses a biphasic strategy to cause disease. At low cell density, the bacteria produce protein factors, such as the MSCRAMMS and other adhesins that promote attachment and biofilm accumulation. In contrast, at high cell density, the bacteria repress the genes encoding the colonization factors, and initiate secretion of a variety of toxins (such as α-toxin, δ-toxin, and toxic shock syndrome toxin-1) and enzymes (such as proteases, lipases, hyaluronidase, and nuclease) that are involved in tissue destruction and/or biofilm detachment probably required for dissemination of the infection and colonization of new sites.

The transcription of the virulence genes is regulated by a 514-nucleotide RNA molecule, termed RNAIII. RNAIII is a component of the global *agr* QS system that activates the transcription of genes encoding secreted toxins and enzymes and represses the transcription of genes encoding cell surface proteins (Figure 1.2) (reviewed in [55]). The *S. aureus agr* locus, approximately 3.5 kbp in size, consists of the genes *agrA*, *agrC*, *agrD*, and *agrB*, which are cotranscribed (RNAII), and the divergently transcribed gene for the regulatory RNAIII molecule, which also encodes the gene for the 26-amino-acid δ-toxin (*hld*). Transcription of RNAII is controlled by the P2 promoter and transcription of RNAIII is controlled by the P3 promoter. The autoinducing peptide (AIP) is a post-translationally modified cyclic peptide that

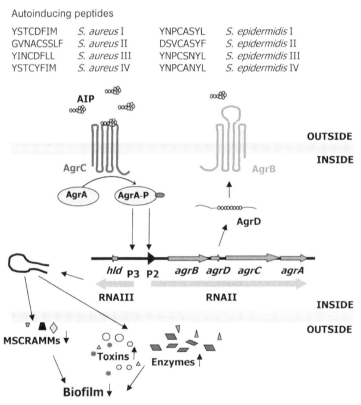

Figure 1.2 Model of the *Staphylococcus agr* QS system [55]. AIP is encoded by *agrD* and processed by *agrB*. The response regulator AgrA~P activates promotors P2 and P3 to transcribe RNAII encoding *agrBDCA* and the effector molecule RNAIII, respectively. RNAIII also contains the δ-toxin gene (*hld*). RNAIII inhibits the expression of the genes encoding MSCRAMMs, and stimulates the expression of genes encoding extracellular enzymes and toxins, and as a consequence downregulates biofilm formation. The amino acid sequences of autoinducing peptides of *S. aureus* and *S. epidermidis* specificity groups I–IV are listed.

contains a thiolactone ring structure and is encoded by *agrD*. The AgrB protein processes, modifies, and exports the AIP. The modification is a cyclic thiolactone bond between the central cysteine and the C-terminal carboxyl group. The proteins encoded by *agr*A and *agr*C constitute a classical two-component regulatory system. Binding of the AIP to AgrC leads to phosphorylation of AgrA. Phospho-AgrA then activates the promotor P3 and thus induces the expression of the regulatory RNAIII. Moreover, phospho-AgrA activates the promotor P2 leading to the autoinduction of the *agr* system. The *agr* system is induced when the AIP reaches a certain threshold concentration in the culture medium, which usually occurs in the late exponential growth phase. In *S. aureus*, four different classes of AIPs have been identified, each belonging to another specificity group. The AIP of one specificity group activates the respective homologous *agr* system, while inhibiting the heterologous *agr* systems [64]. The *agr* specificity groups can be correlated with different pathotypes (e.g., most menstrual toxic shock syndrome strains belong to *agr* group III) [55].

In *S. epidermidis* and also in other staphylococci, *agr* homologs and different *agr* specificity groups have been identified [55, 64]. DNA sequence analysis revealed a pronounced similarity between the *S. epidermidis* and *S. aureus agr* system. However, there is no striking sequence similarity between the AIPs of *S. epidermidis*, *S. aureus* or *S. lugdunensis* (hepta-, octa-, or nonapeptides) except for the central cysteine and its distance to the C-terminus, suggesting that these conserved structural features are necessary for the thiolactone formation.

The influence of the *agr* QS system on staphylococcal biofilm formation is multifaceted, as expected for a global regulator. Since in *S. aureus*, the *agr* system downregulates the expression of genes encoding colonization factors and upregulates the expression of genes encoding detachment factors, the *agr* system might influence several stages of biofilm formation. Generally, the *agr* system downregulates biofilm formation in both *S. aureus* and *S. epidermidis*: *agr* mutants of *S. aureus* and *S. epidermidis* form a more pronounced biofilm than their parental counterparts [65–68].

The *S. epidermidis agr* mutant showed an increased attachment to polystyrene and expression of the autolysin AtlE, which is involved the attachment phase [11, 68]. Moreover, the *agr* mutant revealed significantly enhanced binding to epithelial cells, suggesting that decreased *agr* activity promotes the colonization of *S. epidermidis*. These results could be confirmed by *in vivo* data – the *agr* mutant revealed a higher infectivity in a rabbit model of device-associated infection. Furthermore, it has been observed that nonfunctional *agr* variants occur at a higher rate among clinical infection strains associated with joint prostheses (36%) in comparison to strains isolated from healthy individuals (4.7%), suggesting an inactive *agr* enhances the success of *S. epidermidis* to cause polymer-associated infections [67].

Further comparison of the *S. epidermidis agr* mutant with its wild type revealed that it showed a significantly altered protein expression: the expression of surface-associated proteins was increased, whereas the expression of extracellular proteins, such as lipases and proteases, was decreased [66]. Accordingly, microarray transcriptional analysis of the *agr* mutant showed that the expression of lipases and proteases as well as that of PSMs is upregulated by the *agr* system [69]. Proteome analysis confirmed that these proteins were produced in a significantly lower amount in the *agr* mutant [70]. However, the same proteome analysis also indicated that the production of the autolysin AtlE was reduced in the *agr* mutant, which contradicts earlier findings (see above) [68]. The higher level of biofilm formation in the *S. epidermidis agr* mutant could not be explained by an enhanced expression of genes associated with biofilm accumulation, such as the *icaADBC* gene cluster or *aap* [69, 70].

Generally, *agr* transcription was significantly downregulated in *S. epidermidis* cells grown in a biofilm in comparison with planktonically grown bacteria as shown by genome microarray transcription analyzes [60]. More specifically, *agr* expression was restricted to the externally located regions of the biofilm, whereas no *agr* expression was detected in deeper, internally located biofilm layers. This suggested that *agr* might be involved in the biofilm detachment process [67].

Similar results were obtained with *S. aureus*. In a large collection of 105 clinical *S. aureus* isolates, a strong correlation between *agr* and biofilm formation has been found: 78% of *agr*-negative, but only 6% of *agr*-positive strains formed a biofilm [65]. In contrast to *S. epidermidis*, this effect did not correlate with an altered production of the autolysin Atl, because in the *agr* mutant the expression of *atl* was even slightly reduced. Furthermore, PIA production was unchanged und therefore is not under the control of *agr*. Rather, this effect might at least in part be due to an increased production of PSMs, because the surfactant-like structure of PSMs led to a decreased attachment of the bacterial cells to polystyrene [65]. Another study confirmed that *agr* repressed biofilm formation, but only under static growth conditions. In a flow cell, no significant differences in biofilm formation were observed with the wild-type and an *agr* mutant strain [71]. The same study also indicated that cells detaching from a biofilm revealed a highly activated *agr* system, while bacteria within the biofilm repressed the *agr* system, which is consistent with the observations made in *S. epidermidis*. Recently, it was reported that the repression of the *agr* system is required to form a biofilm and that the induction of the *agr* system in established biofilms promotes detachment, which at least in part depends on extracellular protease activity (see Section 1.3.4) [58].

The expression of extracellular enzymes and toxins seems to be regulated by *agr* in the same way in *S. epidermidis* and *S. aureus*. In contrast, different regulatory mechanisms seem to be involved in the regulation of the genes encoding colonization factors between *S. epidermidis* and *S. aureus*: while the *agr* system in *S. aureus* downregulates the MSCRAMMs, several cell surface proteins of *S. epidermidis* are expressed mainly in the stationary growth phase rather than in the exponential phase [72].

However, as shown in numerous reports, in *S. aureus* as well as in *S. epidermidis*, biofilm formation is significantly reduced by the *agr* QS system. At least in part, this may be explained by an increased biofilm detachment via the upregulation by *agr* of different genes that might be involved in biofilm detachment, such as nucleases, proteases, and PSMs [55, 58, 61]. Partially conflicting results sometimes obtained for the role of the *agr* QS system in staphylococcal biofilm formation may be explained by different growth conditions, such as static or under flow, different growth phases observed, different supply of nutrients, or strain differences [71].

1.4.2
luxS/AI-2 System

The *luxS* QS system has been identified in several Gram-positive and Gram-negative bacterial species, and affects biofilm formation not only in staphylococci, but also in *Streptococcus mutans, Actinomyces naeslundii*, and *Helicobacter pylori* [73, 74]. The *luxS* gene encodes the production of the autoinducer AI-2, which is a furanone derivative, in *S. epidermidis* as well as in *S. aureus*. [62, 63]. The production of AI-2 is growth-phase dependent with a peak production observed during exponential growth. The inactivation of the *luxS* gene in *S. epidermidis* had the same effect as the inactivation of the *agr* system: an *S. epidermidis luxS* mutant was able to form a thicker and stronger biofilm than its parental strain. Transcriptional analysis indicated that the *luxS* system repressed biofilm formation by downregulating the *icaADBC* expression. Accordingly, the production of PIA was elevated in the *luxS* mutant compared with the wild type [62]. This contrasts the effects of the *agr* system, which does not influence *icaADBC* transcription and PIA production. In a rat central venous catheter infection model, the *luxS* mutant turned out to be a more successful colonizer and had a higher capacity to cause infection [62]. However, a recent genome-wide gene expression study indicated that in *S. epidermidis*, mostly genes involved in metabolism, such as sugar, nucleotide, amino acid, and nitrogen, are under the control of the AI-2 [75]. Additionally, *luxS* controls virulence-associated genes encoding lipase and PSMs, suggesting that the stronger biofilm formation in the *luxS* mutant may at least partially be due to a decreased production of PSMs and thus a reduced detachment rate. Surprisingly, the *icaADBC* genes were not found to be differentially expressed in the *luxS* mutant, contradicting earlier findings [75].

In contrast to *S. epidermidis*, a role of the *luxS* system in *S. aureus* biofilm formation and expression of virulence-associated genes could not be detected. Instead, a role for *luxS* in metabolism was suggested [63]. Thus, there seem to exist important species-specific differences in *luxS*-dependent gene regulation among staphylococci. A contrasting effect of *luxS* on biofilm formation has also been observed with other bacterial species. While *luxS* represses biofilm formation in *S. mutans* and *H. pylori*, a *luxS* mutant of *Salmonella* was not able to develop a complete biofilm.

Taken together, the *luxS* QS system has a profound effect on biofilm formation and pathogenicity in *S. epidermidis*, but not in *S. aureus*. Thus, at least in *S. epidermidis* both known QS systems, *agr* and *luxS*, repress biofilm formation.

References

1 Costerton, J.W., Stewart, P.S., and Greenberg, E.P. (1999) Bacterial biofilms: a common cause of persistent infections. *Science*, **284**, 1318–1322.
2 Waters, C.M. and Bassler, B.L. (2005) Quorum sensing: cell-to-cell communication in bacteria. *Annu. Rev. Cell. Dev. Biol.*, **21**, 319–346.
3 von Eiff, C., Heilmann, C., Herrmann, M., and Peters, G. (1999) Basic aspects of the pathogenesis of staphylococcal polymer-associated infections. *Infection*, **27**, S7–S10.

4 Karlowsky, J.A., Jones, M.E., Draghi, D.C., Thornsberry, C., Sahm, D.F., and Volturo, G.A. (2004) Prevalence and antimicrobial susceptibilities of bacteria isolated from blood cultures of hospitalized patients in the United States in 2002. *Ann. Clin. Microbiol. Antimicrob.*, **3**, 7.

5 Götz, F. and Peters, G. (2000) Colonization of medical devices by coagulase-negative staphylococci, in *Infections Associated with Indwelling Medical Devices*, 3rd edn (eds F.A. Waldvogel and A.L. Bisno), ASM Press, Washington, DC, pp. 55–88.

6 Peters, G., Locci, R., and Pulverer, G. (1981) Microbial colonization of prosthetic devices. II. Scanning electron microscopy of naturally infected intravenous catheters. *Zentralbl. Bakteriol. Mikrobiol. Hyg.*, **173**, 293–299.

7 Baldassarri, L., Donnelli, G., Gelosia, A., Voglino, M.C., Simpson, A.W., and Christensen, G.D. (1996) Purification and characterization of the staphylococcal slime-associated antigen and its occurrence among *Staphylococcus epidermidis* clinical isolates. *Infect. Immun.*, **64**, 3410–3415.

8 Mack, D., Fischer, W., Krokotsch, A., Leopold, K., Hartmann, R., Egge, H., and Laufs, R. (1996) The intercellular adhesin involved in biofilm accumulation of *Staphylococcus epidermidis* is a linear beta-1,6-linked glucosaminoglycan: purification and structural analysis. *J. Bacteriol.*, **178**, 175–183.

9 Hussain, M., Wilcox, M.H., and White, P.J. (1993) The slime of coagulase-negative staphylococci: biochemistry and relation to adherence. *FEMS Microbiol. Rev.*, **10**, 191–207.

10 Gross, M., Cramton, S.E., Götz, F., and Peschel, A. (2001) Key role of teichoic acid net charge in *Staphylococcus aureus* colonization of artificial surfaces. *Infect. Immun.*, **69**, 3423–3426.

11 Heilmann, C., Hussain, M., Peters, G., and Götz, F. (1997) Evidence for autolysin-mediated primary attachment of *Staphylococcus epidermidis* to a polystyrene surface. *Mol. Microbiol.*, **24**, 1013–1024.

12 Biswas, R., Voggu, L., Simon, U.K., Hentschel, P., Thumm, G., and Götz, F. (2006) Activity of the major staphylococcal autolysin Atl. *FEMS Microbiol. Lett.*, **259**, 260–268.

13 Cucarella, C., Solano, C., Valle, J., Amorena, B., Lasa, I., and Penades, J.R. (2001) Bap, a *Staphylococcus aureus* surface protein involved in biofilm formation. *J. Bacteriol.*, **183**, 2888–2896.

14 Bowden, M.G., Visai, L., Longshaw, C.M., Holland, K.T., Speziale, P., and Höök, M. (2002) Is the GehD lipase from *Staphylococcus epidermidis* a collagen binding adhesin? *J. Biol. Chem.*, **277**, 43017–43023.

15 Herrmann, M., Lai, Q.J., Albrecht, R.M., Mosher, D.F., and Proctor, R.A. (1993) Adhesion of *Staphylococcus aureus* to surface-bound platelets: role of fibrinogen/fibrin and platelet integrins. *J. Infect. Dis.*, **167**, 312–322.

16 Herrmann, M., Hartleib, J., Kehrel, B., Montgomery, R.R., Sixma, J.J., and Peters, G. (1997) Interaction of von Willebrand factor with *Staphylococcus aureus*. *J. Infect. Dis.*, **176**, 984–991.

17 Patti, J.M., Allen, B.L., McGavin, M.J., and Höök, M. (1994) MSCRAMM-mediated adherence of microorganisms to host tissues. *Annu. Rev. Microbiol.*, **48**, 585–617.

18 Gill, S.R., Fouts, D.E., Archer, G.L., Mongodin, E.F., Deboy, R.T., Ravel, J., Paulsen, I.T., Kolonay, J.F., Brinkac, L., Beanan, M. et al. (2005) Insights on evolution of virulence and resistance from the complete genome analysis of an early methicillin-resistant *Staphylococcus aureus* strain and a biofilm-producing methicillin-resistant *Staphylococcus epidermidis* strain. *J. Bacteriol.*, **187**, 2426–2438.

19 Flock, J.I., Froman, G., Jonsson, K., Guss, B., Signas, C., Nilsson, B., Raucci, G., Höök, M., Wadstrom, T., and Lindberg, M. (1987) Cloning and expression of the gene for a fibronectin-binding protein from *Staphylococcus aureus*. *EMBO J.*, **6**, 2351–2357.

20 McDevitt, D., Francois, P., Vaudaux, P., and Foster, T.J. (1994) Molecular characterization of the clumping factor (fibrinogen receptor) of *Staphylococcus aureus*. *Mol. Microbiol.*, **11**, 237–248.

21 Tung, H., Guss, B., Hellman, U., Persson, L., Rubin, K., and Ryden, C. (2000) A bone

sialoprotein-binding protein from *Staphylococcus aureus*: a member of the staphylococcal Sdr family. *Biochem. J.*, **345**, 611–619.
22 Hartleib, J., Kohler, N., Dickinson, R.B., Chhatwal, G.S., Sixma, J.J., Hartford, O.M., Foster, T.J., Peters, G., Kehrel, B.E., and Herrmann, M. (2000) Protein A is the von Willebrand factor binding protein on *Staphylococcus aureus*. *Blood*, **96**, 2149–2156.
23 McCrea, K.W., Hartford, O., Davis, S., Eidhin, D.N., Lina, G., Speziale, P., Foster, T.J., and Höök, M. (2000) The serine–aspartate repeat (Sdr) protein family in *Staphylococcus epidermidis*. *Microbiology*, **146**, 1535–1546.
24 Clarke, S.R., Harris, L.G., Richards, R.G., and Foster, S.J. (2002) Analysis of Ebh, a 1.1-megadalton cell wall-associated fibronectin-binding protein of *Staphylococcus aureus*. *Infect. Immun.*, **70**, 6680–6687.
25 Williams, R.J., Henderson, B., Sharp, L.J., and Nair, S.P. (2002) Identification of a fibronectin-binding protein from *Staphylococcus epidermidis*. *Infect. Immun.*, **70**, 6805–6810.
26 Downer, R., Roche, F., Park, P.W., Mecham, R.P., and Foster, T.J. (2002) The elastin-binding protein of *Staphylococcus aureus* (EbpS) is expressed at the cell surface as an integral membrane protein and not as a cell wall-associated protein. *J. Biol. Chem.*, **277**, 243–250.
27 Hussain, M., Becker, K., von Eiff, C., Schrenzel, J., Peters, G., and Herrmann, M. (2001) Identification and characterization of a novel 38.5-kilodalton cell surface protein of *Staphylococcus aureus* with extended-spectrum binding activity for extracellular matrix and plasma proteins. *J. Bacteriol.*, **183**, 6778–6786.
28 McGavin, M.H., Krajewska-Pietrasik, D., Ryden, C., and Höök, M. (1993) Identification of a *Staphylococcus aureus* extracellular matrix-binding protein with broad specificity. *Infect. Immun.*, **61**, 2479–2485.
29 Hussain, M., Heilmann, C., Peters, G., and Herrmann, M. (2001) Teichoic acid enhances adhesion of *Staphylococcus epidermidis* to immobilized fibronectin. *Microb. Pathog.*, **31**, 261–270.
30 Rupp, M.E., Fey, P.D., Heilmann, C., and Götz, F. (2001) Characterization of the importance of *Staphylococcus epidermidis* autolysin and polysaccharide intercellular adhesin in the pathogenesis of intravascular catheter-associated infection in a rat model. *J. Infect. Dis.*, **183**, 1038–1042.
31 Heilmann, C., Hartleib, J., Hussain, M., and Peters, G. (2005) The multifunctional *Staphylococcus aureus* autolysin Aaa mediates adherence to immobilized fibrinogen and fibronectin. *Infect. Immun.*, **73**, 4793–4802.
32 Heilmann, C., Thumm, G., Chhatwal, G.S., Hartleib, J., Uekötter, A., and Peters, G. (2003) Identification and characterization of a novel autolysin (Aae) with adhesive properties from *Staphylococcus epidermidis*. *Microbiology*, **149**, 2769–2778.
33 Mack, D., Nedelmann, M., Krokotsch, A., Schwarzkopf, A., Heesemann, J., and Laufs, R. (1994) Characterization of transposon mutants of biofilm-producing *Staphylococcus epidermidis* impaired in the accumulative phase of biofilm production: genetic identification of a hexosamine-containing polysaccharide intercellular adhesin. *Infect. Immun.*, **62**, 3244–3253.
34 Heilmann, C., Schweitzer, O., Gerke, C., Vanittanakom, N., Mack, D., and Götz, F. (1996) Molecular basis of intercellular adhesion in the biofilm-forming *Staphylococcus epidermidis*. *Mol. Microbiol.*, **20**, 1083–1091.
35 Maira-Litran, T., Kropec, A., Abeygunawardana, C., Joyce, J., Mark, G., Goldmann, D.A., and Pier, G.B. (2002) Immunochemical properties of the staphylococcal poly-*N*-acetylglucosamine surface polysaccharide. *Infect. Immun.*, **70**, 4433–4440.
36 McKenney, D., Pouliot, K.L., Wang, Y., Murthy, V., Ulrich, M., Döring, G., Lee, J.C., Goldmann, D.A., and Pier, G.B. (1999) Broadly protective vaccine for *Staphylococcus aureus* based on an *in vivo*-expressed antigen. *Science*, **284**, 1523–1527.
37 Kaplan, J.B., Velliyagounder, K., Ragunath, C., Rohde, H., Mack, D., Knobloch, J.K.,

and Ramasubbu, N. (2004) Genes involved in the synthesis and degradation of matrix polysaccharide in *Actinobacillus actinomycetemcomitans* and *Actinobacillus pleuropneumoniae* biofilms. *J. Bacteriol.*, **186**, 8213–8220.

38 Cramton, S.E., Gerke, C., Schnell, N.F., Nichols, W.W., and Götz, F. (1999) The intercellular adhesion (*ica*) locus is present in *Staphylococcus aureus* and is required for biofilm formation. *Infect. Immun.*, **67**, 5427–5433.

39 Gerke, C., Kraft, A., Sussmuth, R., Schweitzer, O., and Götz, F. (1998) Characterization of the N-acetylglucosaminyltransferase activity involved in the biosynthesis of the *Staphylococcus epidermidis* polysaccharide intercellular adhesin. *J. Biol. Chem.*, **273**, 18586–18593.

40 Vuong, C., Kocianova, S., Voyich, J.M., Yao, Y., Fischer, E.R., DeLeo, F.R., and Otto, M. (2004) A crucial role for exopolysaccharide modification in bacterial biofilm formation, immune evasion, and virulence. *J. Biol. Chem.*, **279**, 54881–54886.

41 Rupp, M.E., Ulphani, J.S., Fey, P.D., and Mack, D. (1999) Characterization of *Staphylococcus epidermidis* polysaccharide intercellular adhesin/hemagglutinin in the pathogenesis of intravascular catheter-associated infection in a rat model. *Infect Immun*, **67**, 2656–2659.

42 Kristian, S.A., Golda, T., Ferracin, F., Cramton, S.E., Neumeister, B., Peschel, A., Götz, F., and Landmann, R. (2004) The ability of biofilm formation does not influence virulence of *Staphylococcus aureus* and host response in a mouse tissue cage infection model. *Microb. Pathog.*, **36**, 237–245.

43 Ziebuhr, W., Heilmann, C., Götz, F., Meyer, P., Wilms, K., Straube, E., and Hacker, J. (1997) Detection of the intercellular adhesion gene cluster (*ica*) and phase variation in *Staphylococcus epidermidis* blood culture strains and mucosal isolates. *Infect. Immun.*, **65**, 890–896.

44 Hall-Stoodley, L., Nistico, L., Sambanthamoorthy, K., Dice, B., Nguyen, D., Mershon, W.J., Johnson, C., Hu, F.Z., Stoodley, P., Ehrlich, G.D. *et al.* (2008) Characterization of biofilm matrix, degradation by DNase treatment and evidence of capsule downregulation in *Streptococcus pneumoniae* clinical isolates. *BMC Microbiol.*, **8**, 173.

45 Allesen-Holm, M., Barken, K.B., Yang, L., Klausen, M., Webb, J.S., Kjelleberg, S., Molin, S., Givskov, M., and Tolker-Nielsen, T. (2006) A characterization of DNA release in *Pseudomonas aeruginosa* cultures and biofilms. *Mol. Microbiol.*, **59**, 1114–1128.

46 Thomas, V.C., Thurlow, L.R., Boyle, D., and Hancock, L.E. (2008) Regulation of autolysis-dependent extracellular DNA release by *Enterococcus faecalis* extracellular proteases influences biofilm development. *J. Bacteriol.*, **190**, 5690–5698.

47 Rice, K.C., Mann, E.E., Endres, J.L., Weiss, E.C., Cassat, J.E., Smeltzer, M.S., and Bayles, K.W. (2007) The *cidA* murein hydrolase regulator contributes to DNA release and biofilm development in *Staphylococcus aureus*. *Proc. Natl. Acad. Sci. USA*, **104**, 8113–8118.

48 Rohde, H., Burandt, E.C., Siemssen, N., Frommelt, L., Burdelski, C., Wurster, S., Scherpe, S., Davies, A.P., Harris, L.G., Horstkotte, M.A. *et al.* (2007) Polysaccharide intercellular adhesin or protein factors in biofilm accumulation of *Staphylococcus epidermidis* and *Staphylococcus aureus* isolated from prosthetic hip and knee joint infections. *Biomaterials*, **28**, 1711–1720.

49 Rohde, H., Burdelski, C., Bartscht, K., Hussain, M., Buck, F., Horstkotte, M.A., Knobloch, J.K., Heilmann, C., Herrmann, M., and Mack, D. (2005) Induction of *Staphylococcus epidermidis* biofilm formation via proteolytic processing of the accumulation-associated protein by staphylococcal and host proteases. *Mol. Microbiol.*, **55**, 1883–1895.

50 Corrigan, R.M., Rigby, D., Handley, P., and Foster, T.J. (2007) The role of *Staphylococcus aureus* surface protein SasG in adherence and biofilm formation. *Microbiology*, **153**, 2435–2446.

51 Conrady, D.G., Brescia, C.C., Horii, K., Weiss, A.A., Hassett, D.J., and Herr, A.B. (2008) A zinc-dependent adhesion module is responsible for intercellular adhesion in

staphylococcal biofilms. *Proc. Natl. Acad. Sci. USA*, **105**, 19456–19461.

52 Banner, M.A., Cunniffe, J.G., Macintosh, R.L., Foster, T.J., Rohde, H., Mack, D., Hoyes, E., Derrick, J., Upton, M., and Handley, P.S. (2007) Localized tufts of fibrils on *Staphylococcus epidermidis* NCTC 11047 are comprised of the accumulation-associated protein. *J. Bacteriol.*, **189**, 2793–2804.

53 Itoh, Y., Wang, X., Hinnebusch, B.J., Preston, J.F. 3rd., and Romeo, T. (2005) Depolymerization of beta-1,6-*N*-acetyl-D-glucosamine disrupts the integrity of diverse bacterial biofilms. *J. Bacteriol.*, **187**, 382–387.

54 Izano, E.A., Amarante, M.A., Kher, W.B., and Kaplan, J.B. (2008) Differential roles of poly-*N*-acetylglucosamine surface polysaccharide and extracellular DNA in *Staphylococcus aureus* and *Staphylococcus epidermidis* biofilms. *Appl. Environ. Microbiol.*, **74**, 470–476.

55 Novick, R.P. (2006) Staphylococcal pathogenesis and pathogenicity factors: genetics and regulation, in *Gram-Positive Pathogens*, 2nd edn (eds V.A. Fischetti, J.J. Ferretti, D.A. Portnoy, J.I. Rood, and R.P. Novick) ASM Press, Washington, DC, pp. 496–516.

56 Kogan, G., Sadovskaya, I., Chaignon, P., Chokr, A., and Jabbouri, S. (2006) Biofilms of clinical strains of *Staphylococcus* that do not contain polysaccharide intercellular adhesin. *FEMS Microbiol. Lett.*, **255**, 11–16.

57 Chaignon, P., Sadovskaya, I., Ragunah, C., Ramasubbu, N., Kaplan, J.B., and Jabbouri, S. (2007) Susceptibility of staphylococcal biofilms to enzymatic treatments depends on their chemical composition. *Appl. Microbiol. Biotechnol.*, **75**, 125–132.

58 Boles, B.R. and Horswill, A.R. (2008) Agr-mediated dispersal of *Staphylococcus aureus* biofilms. *PLoS Pathog.*, **4**, e1000052.

59 Mehlin, C., Headley, C.M., and Klebanoff, S.J. (1999) An inflammatory polypeptide complex from *Staphylococcus epidermidis*: isolation and characterization. *J. Exp. Med.*, **189**, 907–918.

60 Yao, Y., Sturdevant, D.E., and Otto, M. (2005) Genomewide analysis of gene expression in *Staphylococcus epidermidis* biofilms: insights into the pathophysiology of *S. epidermidis* biofilms and the role of phenol-soluble modulins in formation of biofilms. *J. Infect. Dis.*, **191**, 289–298.

61 Vuong, C., Durr, M., Carmody, A.B., Peschel, A., Klebanoff, S.J., and Otto, M. (2004) Regulated expression of pathogen-associated molecular pattern molecules in *Staphylococcus epidermidis*: quorum-sensing determines pro-inflammatory capacity and production of phenol-soluble modulins. *Cell Microbiol.*, **6**, 753–759.

62 Xu, L., Li, H., Vuong, C., Vadyvaloo, V., Wang, J., Yao, Y., Otto, M., and Gao, Q. (2006) Role of the *luxS* quorum-sensing system in biofilm formation and virulence of *Staphylococcus epidermidis*. *Infect. Immun.*, **74**, 488–496.

63 Doherty, N., Holden, M.T., Qazi, S.N., Williams, P., and Winzer, K. (2006) Functional analysis of luxS in *Staphylococcus aureus* reveals a role in metabolism but not quorum sensing. *J. Bacteriol.*, **188**, 2885–2897.

64 Otto, M., Süssmuth, R., Vuong, C., Jung, G., and Götz, F. (1999) Inhibition of virulence factor expression in *Staphylococcus aureus* by the *Staphylococcus epidermidis agr* pheromone and derivatives. *FEBS Lett.*, **450**, 257–262.

65 Vuong, C., Saenz, H.L., Götz, F., and Otto, M. (2000) Impact of the agr quorum-sensing system on adherence to polystyrene in *Staphylococcus aureus*. *J. Infect. Dis.*, **182**, 1688–1693.

66 Vuong, C., Götz, F., and Otto, M. (2000) Construction and characterization of an *agr* deletion mutant of *Staphylococcus epidermidis*. *Infect. Immun.*, **68**, 1048–1053.

67 Vuong, C., Kocianova, S., Yao, Y., Carmody, A.B., and Otto, M. (2004) Increased colonization of indwelling medical devices by quorum-sensing mutants of *Staphylococcus epidermidis in vivo*. *J. Infect. Dis.*, **190**, 1498–1505.

68 Vuong, C., Gerke, C., Somerville, G.A., Fischer, E.R., and Otto, M. (2003) Quorum-sensing control of biofilm factors in *Staphylococcus epidermidis*. *J. Infect. Dis.*, **188**, 706–718.

69 Yao, Y., Vuong, C., Kocianova, S., Villaruz, A.E., Lai, Y., Sturdevant, D.E., and Otto, M.

(2006) Characterization of the *Staphylococcus epidermidis* accessory-gene regulator response: quorum-sensing regulation of resistance to human innate host defense. *J. Infect. Dis.*, **193**, 841–848.

70 Batzilla, C.F., Rachid, S., Engelmann, S., Hecker, M., Hacker, J., and Ziebuhr, W. (2006) Impact of the accessory gene regulatory system (Agr) on extracellular proteins, *codY* expression and amino acid metabolism in *Staphylococcus epidermidis*. *Proteomics*, **6**, 3602–3613.

71 Yarwood, J.M., Bartels, D.J., Volper, E.M., and Greenberg, E.P. (2004) Quorum sensing in *Staphylococcus aureus* biofilms. *J. Bacteriol.*, **186**, 1838–1850.

72 Bowden, M.G., Chen, W., Singvall, J., Xu, Y., Peacock, S.J., Valtulina, V., Speziale, P., and Höök, M. (2005) Identification and preliminary characterization of cell-wall-anchored proteins of *Staphylococcus epidermidis*. *Microbiology*, **151**, 1453–1464.

73 Merritt, J., Qi, F., Goodman, S.D., Anderson, M.H., and Shi, W. (2003) Mutation of *luxS* affects biofilm formation in *Streptococcus mutans*. *Infect. Immun.*, **71**, 1972–1979.

74 Rickard, A.H., Palmer, R.J. Jr., Blehert, D.S., Campagna, S.R., Semmelhack, M.F., Egland, P.G., Bassler, B.L., and Kolenbrander, P.E. (2006) Autoinducer 2: a concentration-dependent signal for mutualistic bacterial biofilm growth. *Mol. Microbiol.*, **60**, 1446–1456.

75 Li, M., Villaruz, A.E., Vadyvaloo, V., Sturdevant, D.E., and Otto, M. (2008) AI-2-dependent gene regulation in *Staphylococcus epidermidis*. *BMC Microbiol.*, **8**, 4.

2
Cell–Cell Communication in Biofilms of Gram-Negative Bacteria

Claudio Aguilar, Aurelien Carlier, Kathrin Riedel, and Leo Eberl

2.1
Introduction

Evidence has accumulated over the past few years suggesting that biofilm formation proceeds through an ordered series of steps [1, 2]. With most bacteria investigated so far, at least three developmental steps can be distinguished: (i) initial attachment of individual cells to a surface, (ii) aggregation of these cells to microcolonies, and (iii) differentiation of the microcolonies into a mature biofilm (i.e., the development of a typical three-dimensional biofilm architecture). Two mutually nonexclusive mechanisms underlying this developmental process are currently debated. Biofilm formation could follow a genetic program that directs the cells from the planktonic into the sessile mode of growth or it could just be the consequence of a series of metabolic adjustments of the cells to rapidly changing microenvironments and thus represents a self-organizing process.

Regardless of these ongoing debates there is one factor that has emerged to play an important role in the establishment and development of biofilms with a typical architecture: cell–cell communication, commonly referred to as quorum sensing (QS). In this chapter we summarize the knowledge on the role of QS in biofilm formation in different Gram-negative bacteria. Furthermore, QS as target for the eradication of biofilms is also discussed.

2.2
QS in Gram-Negative Bacteria

QS is a generic regulatory mechanism used by many bacteria to perceive and respond to the density of the population. QS enables bacterial cells to communicate with each other and to coordinate their activities, thus acting in a concerted manner similar to multicellular organisms. QS systems rely on small signal molecules. Although various signal molecules have been described, the two most thoroughly investigated classes are the *N*-acyl-homoserine lactones (AHLs), which are produced by many Gram-negative

Bacterial Signaling. Edited by Reinhard Krämer and Kirsten Jung
Copyright © 2010 WILEY-VCH Verlag GmbH & Co. KGaA, Weinheim
ISBN: 978-3-527-32365-4

bacteria, and small peptides, which are utilized by many Gram-positive species [3, 4]. More recently, a furanosyl borate diester (autoinducer-2 (AI-2)), which is synthesized by members of the LuxS protein family, was shown to be produced by various Gram-negative and Gram-positive organisms. The fact that this signaling system is widespread among diverse bacteria led to the suggestion that this signal molecule represents a universal bacterial language [5]. Several examples showing that AI-2- and peptide-based QS systems influence biofilm formation at different stages are available, and these have been reviewed recently [6, 7]. In this chapter primarily focus on the role of AHL-dependent signaling systems in biofilm formation, but we also include some examples of QS systems in Gram-negative bacteria that rely on other signal molecules (Table 2.1).

AHL-based systems typically rely on two proteins – an AHL synthase, usually a member of the LuxI family of proteins, and an AHL receptor protein belonging to the LuxR family of transcriptional regulators. At low population densities cells synthesize a basal level of AHL via the activity of the AHL synthase. With increasing cell density the signal molecule not only accumulates in the growth medium, but also inside the cells, as AHLs can diffuse across the cell wall. On reaching a critical threshold concentration, the signal molecule binds to its cognate receptor, which in turn activates or represses expression of target genes. To date, AHL-dependent QS circuits have been identified in more than 80 species of Gram-negative bacteria, in which they regulate a variety of functions including bioluminescence, plasmid conjugative transfer, synthesis of antibiotics and extracellular hydrolytic enzymes, motility, and production of virulence factors (reviewed in [8]). More recently, evidence has emerged that AHL signal molecules are also recognized by eukaryotes and induce specific responses often affecting the immune status of the organism [9].

2.3
QS and Biofilm Formation

The involvement of an AHL-based QS system in the regulation of biofilm formation was first reported for the opportunistic human pathogen *Pseudomonas aeruginosa* [10]. That study showed that a *lasI* mutant formed biofilms that were flat, densely packed, and undifferentiated when compared to the wild-type, which formed heterogeneous biofilms consisting of typical mushroom-shaped microcolonies that were separated by water channels. The mutant biofilm was also demonstrated to be much more sensitive to the biocide sodium dodecylsulfate than to the wild-type biofilm. The *las* system is one of two QS systems that have been identified in *P. aeruginosa*, and consists of the transcriptional activator LasR and the AHL synthase LasI, which directs the synthesis of N-3-oxo-dodecanoyl-homoserine lactone (3-oxo-C12-HSL). The second system, designated *rhl*, consists of RhlR and RhlI, which directs the synthesis of N-butanoyl-L-homoserine lactone (C4-HSL). The two systems do not operate independently as the *las* system positively regulates expression of both *rhlR* and *rhlI*. Thus, the two QS systems of *P. aeruginosa* are hierarchically arranged with the *las* system being on top of the signaling cascade (for a review, see [11]). Moreover, the QS circuitry is subject to various additional layers of regulation both at the

Table 2.1 QS-regulated factors that affect biofilm development in selected organisms.

Organism	Major QS molecule(s)	Factor(s) known to be regulated by QS	Biofilm of a QS mutant	References
Pseudomonas aeruginosa	C4-HSL; 3-oxo-C12-HSL	rhamnolipids, eDNA, NO detoxification, Pel and Psl EPSs	undifferentiated, sensitive to sodium dodecylsulfate	[22, 45, 47, 58, 59, 65–67]
Burkholderia cenocepacia	C6-HSL, C8-HSL, BDSF	unknown	defective in late stages	[23, 27]
Aeromonas hydrophila	C4-HSL	unknown	defective in late stages	[9]
Pseudomonas putida	3-oxo-C10-HSL, 3-oxo-C12-HSL	putisolvin I and II	structured biofilm (QS as biofilm repressor)	[25, 56]
Serratia marcescens	C4-HSL	EPS	thin and nonstructured	[26, 30]
Pantoea stewartii	3-oxo-C6-HSL	stewartan EPS	flat and uniform	[32, 42, 43]
Xanthomonas campestris	DSF	unknown	no development or no dispersal	[34, 35]
Xylella fastidiosa	DSF	unknown	no development	[17]

transcriptional and post-transcriptional level [11]. This highly sophisticated regulatory network, which integrates various environmental parameters, may explain why the influence of QS on biofilm structures was found to be highly dependent on experimental conditions. In two studies, in which slightly changed experimental settings were used, no difference between biofilms of the wild-type and those formed by signal-negative mutants were observed [12, 13]. In a further study, Purevdorj *et al.* [14] reported minor structural differences between wild-type and mutant biofilms – differences only apparent when particular hydrodynamic conditions were used for growing the biofilms. Even though many factors might contribute to the observed inconsistencies between the studies, one key factor appears to be the nutritional status of the cells. Most intriguingly in this context is the study of Klausen *et al.* [15], who showed that different carbon sources can lead to greatly changed biofilm structures. Biofilms grown on glucose were very heterogeneous, and contained discrete tower and mushroom structures separated by water channels and voids. In contrast, biofilms grown under identical conditions, but with citrate as the carbon source, appeared flat and uniform. Further evidence for the importance of the carbon source for biofilm architecture was obtained by O'Toole *et al.* [16], who showed that the Crc catabolite repression control protein, which plays a central role in the regulation of carbon metabolism, is involved in the regulation of biofilm formation.

In addition to the AHL signaling system, *P. aeruginosa* synthesizes a quinolone signal, 2-heptyl-3-hydroxy-4-quinolone (PQS), to regulate gene expression in a cell-density fashion [17]. PQS has been detected in the lungs of cystic fibrosis (CF) patients and it has been postulated that PQS production by *P. aeruginosa* may vary at different stages of lung colonization [18]. In this environment, *P. aeruginosa* forms a biofilm that is resistant to both antimicrobials and the host immune system. It has been shown that PQS promotes adherence and biofilm formation by increasing the production of PA-IL lectin [19]. PQS also appears to contribute to biofilm formation by inducing autolysis of cells and thus is involved in the release of DNA – a component of the extracellular matrix [20–22].

A role for AHL-mediated QS in biofilm formation has also been demonstrated for *Burkholderia cenocepacia* [23], *Aeromonas hydrophila* [24], *Pseudomonas putida* [25], and *Serratia marcescens* [26]. Like *P. aeruginosa*, AHL-negative mutants of *B. cenocepacia* and *A. hydrophila* showed defects in the late stages of biofilm development, and thus were unable to form biofilms with a wild-type architecture. Employing a quorum quenching approach (i.e., the enzymatic degradation of AHL signal molecules) it was shown that QS regulates biofilm formation not only in *B. cenocepacia*, but also in the large majority of strains from nine other *Burkholderia* species [27]. The biofilm formed by the *P. putida* strain IsoF on abiotic surfaces is very homogenous and elaborate structural elements are absent. However, QS mutants were shown to form structured biofilms similar to those formed by the wild-type strains of *B. cenocepacia* and *A. hydrophila*, indicating that in this organism QS represses the development of biofilm structures [25]. In a follow-up study the QS-regulated proteome was analyzed and the profiles of proteins expressed either in planktonic cultures or in biofilms formed by *P. putida* IsoF were compared [28]. It was found that, of the 30 proteins differentially expressed in both conditions, 17 were also controlled by QS. Moreover,

11 out of 23 surface-induced proteins appeared to be downregulated by QS. An important conclusion derived from this study is that the expression of QS-regulated proteins is strongly dependent on the lifestyle of this organism.

The opportunistic pathogen *S. marcescens* MG1 forms biofilms through a series of defined stages that leads to a highly porous biofilm composed of cell chains, filaments, and cell clusters [26, 29]. QS was shown to play important roles in the regulation of several stages of the biofilm lifecycle of this organism. A *swrI* mutant of this strain, in which the AHL synthase had been inactivated, formed only a thin and nonstructured biofilm – a defect that could be reversed to a wild-type biofilm by the addition of the cognate AHL. The products of two QS-controlled genes, *bsmA* and *bsmB*, were shown to be involved in adhesion and colonization on both biotic and abiotic surfaces and in late-stage biofilm development [30].

Plant-associated bacteria can have diverse ecological roles as pathogens, commensals, or mutualists for which the ability to form biofilms is likely an advantage, if not essential (reviewed in [31]). For example, in *Pantoea stewartii*, the EsaI/R QS system has been shown to be of crucial importance in virulence, controlling the expression of genes necessary for biofilm formation and subsequent colonization of the plant [32].

Xanthomonas campestris is a ubiquitous plant pathogen, the causal agent of black rot disease in cruciferous crops. In this organism, regulation of QS is mediated by a small diffusible molecule named diffusible signal factor (DSF). The structure of DSF has been recently identified as a *cis*-11-methyl-2-dodecenoic acid [33]. DSF was demonstrated to be involved in biofilm dispersal, since mutations affecting genes involved in DSF synthesis, sensing, or signal transduction resulted in the formation of cell aggregates [34]. Interestingly, a recent study by Torres *et al.* [35] presented evidence that, under certain growth conditions, DSF plays a role not only in biofilm dispersal but also in biofilm formation (see Section 2.4.1 for further details).

Xylella fastidiosa, which causes Pierce's disease of grapevine and other important plant diseases, is a xylem-limited bacterium that depends on insect vectors for transmission. *X. fastidiosa* produces a DSF molecule that is different from, although structurally related to, that of *X. campestris*. This signal molecule was shown to be required for biofilm formation in the insect vector, for vector transmission to plants, and for suppression of virulence to grape (for a recent review, see [36]).

Another DSF-like signal molecule, named diffusible signal factor from *Burkholderia cenocepacia* BDSF, has been recently identified in the culture supernatant of *B. cenocepacia* [37]. This molecule was shown to inhibit the hyphal growth of *Candida albicans* and although it was demonstrated that BDSF restores biofilm formation in *X. campestris*, its role in *B. cenocepacia* QS regulation of biofilm formation or dispersal remains to be elucidated.

2.3.1
When in the Biofilm Cycle does QS Play a Role?

Bacteria living in biofilms are embedded in a matrix composed of extracellular polysaccharides, proteins, and nucleic acids, and therefore cell densities are obviously very high in these communities. Moreover, the biofilm matrix may also be a

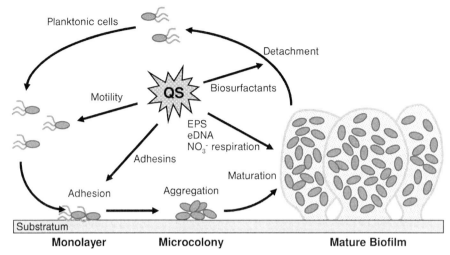

Figure 2.1 QS in the biofilm cycle. QS controls the expression of several functions important in the biofilm lifecycle of Gram-negative bacteria: motility and adhesins, which are important in the early steps of biofilm development; EPS and eDNA, which are crucial components of the extracellular matrix and nitrate respiration for biofilm maturation; and biosurfactants for dispersion of cells from the biofilm. See text for further details.

diffusion barrier for signal molecules, creating an environment that is ideal for the induction of QS.

QS systems regulate gene expression in response to the size of the population. This means that when a certain critical mass, the "quorum" has been attained, cells collectively induce the expression of particular phenotypic traits, which are not observed with isolated individual cells. Hence, QS is often considered an example of primitive multicellular behavior. In this light it is conceivable that QS plays an important role in triggering expression of genes required for particular steps during biofilm development (Figure 2.1). For example, in some bacteria different forms of motility or the production of adhesins are QS-regulated and this would affect the early events of biofilm formation (i.e., attachment to a surface and microcolony formation). In other bacteria, QS regulates the production of extrapolymeric substances (EPSs) and controls DNA release – factors which are important for biofilm maturation. Finally, QS appears to be also important for biofilm dispersal. For example, mutants of *Rhodobacter sphaeroides* and *Yersinia pseudotuberculosis* that were defective in AHL-mediated QS formed aggregates that could be dispersed by the addition of cognate AHLs, suggesting that escape from clumping was under QS control [38, 39].

At present it is unknown how QS can affect biofilm formation at different stages and thus different population sizes. However, one could imagine that several points contribute to this flexibility: (i) depending on the bacterial species or strain the minimal population size required to trigger QS (i.e., the quorum) may be different, (ii) QS systems can induce or repress gene expression and thus certain genes may be oppositely regulated depending on the organism, (iii) QS molecules can differ greatly with respect to their physical properties, most importantly in their diffusion rates, (iv) QS systems are often embedded in more complex regulatory networks and thus additional factors may influence target gene expression, and (v) QS molecules need to be turned over, likely by specific enzymes.

2.4
QS-Regulated Factors Involved in Biofilm Formation

In this section we will summarize the current knowledge on QS-regulated factors that affect the biofilm developmental cycle (see also Table 2.1).

2.4.1
EPSs

The biofilm matrix plays a critical role in initiating and maintaining the structure of the biofilm as well as protecting it from harmful environmental factors such as desiccation and predator grazing. The composition and properties of EPSs within the biofilm matrix vary widely among species and culture conditions [40]. However, a growing number of bacterial species are recognized to use QS to regulate EPS synthesis and degradation during biofilm development. For instance, the corn pathogen *P. stewartii* ssp. *stewartii* uses the EsaI/EsaR QS circuitry to regulate stewartan EPS production, necessary for biofilm formation, plant colonization, and virulence. Stewartan EPS is a well-characterized acidic polymer of glucose, galactose and glucuronic acid that enmeshes bacteria in mature biofilms [41, 42]. At low cell density and low AHL concentrations, the EsaR QS regulator represses transcription of the stewartan EPS biosynthetic genes and bacteria are competent for attachment to abiotic surfaces [32, 43]. *P. stewartii esaI* mutants are defective in AHL signal synthesis and only form flat, uniform biofilms lacking the structure of wild-type biofilms. Precise timing of EPS synthesis is essential, and *esaR* mutants that synthesize stewartan EPS constitutively are unable to attach securely to surfaces and only form fragile biofilms [32].

S. marcescens MG1 produces a calcofluor-binding EPS in a C4-HSL-dependent manner. Very little is known about the structure of the EPS produced by *S. marcescens*, but the calcofluor dye displays specificity towards cellulose- or chitin-like insoluble polymers containing β1–3 or β1–4 glycosidic linkages [44]. This type of EPS is thought to promote attachment of *S. marcescens* to abiotic surfaces in the first stages of biofilm formation [30]. This is in contrast to what is found in most other bacteria, where QS stimulates EPS production at later stages of biofilm development to provide structural support of the mature biofilm.

P. aeruginosa uses a similar strategy to regulate the EPS structural components of its biofilms. The biofilm matrix of *P. aeruginosa* PA14 contains at least three distinct EPS types. Pel and Psl EPS are thought to be most important for biofilm structure under standard laboratory conditions, with alginate playing only a minor role [45]. A defined *P. aeruginosa* PA14 mutant defective in Pel EPS production was shown to be severely impaired in the formation of air–liquid interface biofilms and failed to establish sturdy, mature biofilms on submerged abiotic surfaces [46]. Downregulation of Pel biosynthesis in the absence of a functional LasI/LasR QS regulatory network likely contributes, at least partially, to the frail biofilm phenotype observed for *las* mutants of *P. aeruginosa* PA14 [47]. Unfortunately, the structure of the glucose-rich, insoluble Pel EPS is unknown, albeit it appears to be distinct from cellulose and has also been shown to play a critical role in maintaining the structural integrity of biofilms of the nonpiliated *P. aeruginosa* PAK strain [48, 49].

Bacteria may use QS to trigger biofilm dispersal when the biofilm population reaches a critical size. At this point, the cells have to disintegrate the biofilm matrix to re-enter the planktonic mode of growth. Such QS-dependent matrix degradation is well documented for the Gram-negative plant pathogen *X. campestris* pv. *campestris* (*X. campestris*). *X. campestris* uses the DSF signal to sense population density and regulate formation and maturation of biofilms [35]. *X. campestris rpf* mutants defective in DSF synthesis or detection produce low levels of xanthan EPS – a necessary component of *X. campestris* biofilms [35, 50, 51]. Interestingly, the DSF signal is also involved in disintegration of bacterial aggregates in early stationary phase cultures, and possibly in the dispersal of bacterial cells from mature biofilms. Disruption of bacterial aggregates depends upon the DSF-activated secretion of an endo-β1–4-mannanase at high cell densities [34, 52]. Although the precise substrate of the secreted mannanase remains to be identified, the role of that substrate glycopolymer may be to provoke gelling of the EPS matrix by interacting with xanthan EPS [34, 53]. Digestion of one component of the EPS gel by the QS-regulated mannanase may cause disintegration of the biofilm matrix, releasing bacterial cells from the biofilm.

2.4.2
Biosurfactants

P. putida PCL1445 produces two novel lipodepsipeptides, putisolvin I and II, consisting of a C_6 lipid moiety and a 12-amino-acid peptide – both synthesized by the products of the gene cluster *pso* [54, 55]. When first studied in the context of liquid, shaken cultures, it was observed that the production of putisolvin started at the onset of stationary phase, suggesting that putisolvin synthesis is regulated by QS. In *P. putida*, the QS response is controlled by the genetic cluster *ppuI–rsaL–ppuR* and the product of *ppuI* is necessary for AHL biosynthesis [25]. It was demonstrated that the addition of 3-oxo-C12-HSL restores putisolvin production in a *ppuI* mutant background. This suggested that the *ppuI/ppuR* genes are responsible for both the production of AHLs and the regulation of putisolvin expression in *P. putida* [56]. Since putisolvin I and II have been shown not only to inhibit the formation but also to break down mature biofilms of various pseudomonads, including *Pseudomonas fluorescens* WCS365 and *P. aeruginosa* PA14 [55], it is very likely that QS-controlled expression of putisolvin is involved in the dispersal of cells from the mature biofilm. The dispersing properties of putisolvin may explain in part why the wild-type biofilm is flat and unstructured while QS-deficient mutants form structured biofilms with characteristic microcolonies and water-filled channels [25].

For *P. aeruginosa* it has been demonstrated that production of the extracellular rhamnolipid biosurfactants play an important role in the maintenance of a distinct biofilm architecture. Specifically, Davey *et al.* [57] demonstrated that rhamnolipids are not only required to keep the fluid-filled channels of the biofilm open, but also to prevent other bacteria from colonizing the void spaces and thus protects the biofilm from invaders. In agreement with this model, it has been shown that rhamnolipid synthesis preferentially occurs in the stalks rather than in the caps of mature mushroom structures [58]. Subsequent studies have presented evidence suggesting that rhamnolipids have in fact multiple roles in biofilm development [59]. In addition to maintaining the biofilm structure, rhamnolipids are necessary for microcolony formation (i.e., initial biofilm stages), as well as for dispersion of cells from the mature biofilm [60]. Thus, precise expression of rhamnolipids throughout biofilm development seems to be a crucial factor determining what role these biosurfactants will play in biofilm development. In *P. aeruginosa*, the expression of the rhamnolipid synthesis operon *rhlAB* is controlled by the C4-HSL-dependent RhlI/RhlR QS system [61–63]. A more detailed analysis showed that, in addition to QS, the stationary phase sigma factor RpoS is involved in the regulation of this operon [64]. Hence, QS in concert with other regulatory factors appears to ensure the fine tuned expression of rhamnolipids at particular sites and time points during biofilm development.

2.4.3
DNA Release

The extracellular DNA (eDNA) present in bacterial biofilms formed by *P. aeruginosa* has been suggested to have a role as a cell–cell interconnecting compound, necessary for the establishment of biofilms in flow chambers [22]. Treatment with DNase I was shown to prevent the formation of biofilms, suggesting that it could be beneficial as an early prophylactic measure to prevent the establishment of chronic *P. aeruginosa* infection of the CF lung by inhibiting biofilm formation. In *P. aeruginosa* biofilms grown in flow chambers, the eDNA was shown to be organized in distinct patterns throughout biofilm formation. Early in biofilm development, the highest concentration of eDNA was found on the surface of microcolonies, while in older biofilms it was found to be localized in discrete layers on the stalks and cups of the mushroom-like structures, typically formed by this bacterium [65]. The mechanism by which the DNA is exported to the extracellular matrix of the biofilm is still not fully understood. Evidence suggests that there is a basal level of eDNA generated through a QS-independent pathway. It was observed, however, that biofilms formed by QS-deficient mutants (*lasI* and *rhlI*) contained considerably less eDNA, suggesting that QS plays an active role in the generation of eDNA [65, 66].

2.4.4
Physiology: Dissimilatory Nitrate Reduction in P. aeruginosa

P. aeruginosa has been shown to form robust biofilms when grown under anaerobic conditions [67]. In the absence of oxygen P. aeruginosa can utilize NO_3^- and NO_2^- as terminal electron acceptors. A toxic byproduct of this denitrification process is nitric oxide (NO). Interestingly, in a recent study by Barraud et al. [68] it was shown that NO, when present at sublethal (nanomolar) concentrations, induces the dispersal of P. aeruginosa biofilms through the stimulation of both swimming and swarming motility. QS plays an important role in the detoxification of NO and its has been demonstrated that inactivation of *rhlR* leads to premature cell death due to the accumulation of toxic concentrations of NO [67]. In agreement with this observation, it was suggested that the two QS systems repress denitrification activity of P. aeruginosa [69].

2.5
QS as a Target for the Eradication of Biofilms

According to a public announcement of the US National Institutes of Health, over 80% of microbial infections in the human body are caused by biofilms [70], including lung infections in CF patients, wound infections in burn patients, otitis media, periodontitis, endocarditis, urinary tract infections, and infections caused by implants (reviewed in [71]). To make it even worse, bacterial biofilms display a tremendous capacity to resist traditional antimicrobial therapies such as antibiotics or biocides (reviewed in [72]). Despite the urgent need for innovative agents that efficiently eradicate bacterial biofilms, only few chemical compounds exhibit specific antibiofilm properties and none of them are currently used in clinical applications.

Given that AHL-mediated cell–cell communication acts as central regulator of biofilm formation and expression of virulence factors in many pathogens such as P. aeruginosa and Burkholderia species, QS appears to be a highly attractive target for battling these undesired bacteria [11, 73]. The major advantage of this therapeutic approach is that QS-interfering compounds would diminish the ability of a pathogen to infect its host while not affecting growth of the pathogen (and other beneficial microorganisms), thereby circumventing the development of resistance, as it is the case with classic antibiotics. Given that QS blockers inhibit the production of virulence factors the pathogens can no longer adapt to the host environment and consequently will be cleared by the host defenses. The QS cascade reveals several targets for inhibition, (i) AHL production, (ii) AHL stability, (iii) AHL receptor binding, and (iv) receptor DNA binding. To date, mainly strategies (ii) and (iii) have been followed, and are delineated in more detail in the following.

A number of Gram-positive bacteria, in particular numerous *Bacillus* [74, 75], *Arthrobacter* [76] and *Rhodococcus* [77] species, produce enzymes capable of degrading AHL molecules by hydrolyzing their lactone ring moiety – a process also referred to as quorum quenching. Transgenic plants expressing the *Bacillus* lactonase AiiA were highly resistant to colonization and infection by the Gram-negative AHL-producing plant pathogen *Erwinia carotovora* [75]. Moreover, it has been shown that recombinant *aiiA* expression in *P. aeruginosa* markedly reduces the production of QS-regulated virulence factors and impairs biofilm formation [78, 79]. Although enzymes are not commonly used as antimicrobial agents, they might have a potential in fighting biofilm-mediated lung infections, as enzymes can be administered by mobilization.

Probably the most promising strategy to block AHL-dependent gene expression is to interfere with the binding of AHL signal molecules to their cognate receptors. In fact, several eukaryotic organisms employ such a mechanism, as the production of AHL antagonists was demonstrated for fungi [80], the marine red alga *Delisea pulchra* [81], different higher plants such as pea, tomato, and rice [82], garlic [83], cinnamon [84], Florida medicinal plants [85], and marine animals (i.e., *Flustra foliacea* [86] and *Luffariella variabilis* [87]). Pure compounds capable of specifically blocking QS were isolated from *D. pulchra* and *F. foliacea*, and were identified as halogenated furanones and brominated alkaloids, respectively. Halogenated furanones were shown to displace radiolabeled AHL signal molecules from *E. coli* cells expressing the cognate receptor protein [88]. More recently, however, it was shown that the furanones rather induce the rapid degradation of the receptor protein [89].

While these molecules are not suited for treatment of infections in humans due to their high reactivity, they may serve as lead structures for the development of drugs with less side-effects. In fact, synthetic derivatives of certain halogenated furanones were shown to effectively antagonize QS and biofilm formation in *P. aeruginosa* [73, 90, 91]. Furthermore, combinatorial chemistry has been successfully used to identify compounds with antagonistic activity towards QS-regulated phenotypes and biofilm formation in *P. aeruginosa*. Recently, rational drug design was employed to develop an agent that specifically targets the QS system of *B. cenocepacia* [92]. The compound not only reduced biofilm formation of *B. cenocepacia*, but also attenuated the pathogenicity of the bacterium. However, the exact mode of action of the antimicrobial compound remains to be elucidated.

Although the compounds discussed above exhibit considerable potential to serve as lead structures for the design of novel anti-QS and antibiofilm therapies, to date no clinically useful therapy capable of eradicating bacterial biofilm infections has been developed. Even if such a compound were to be found, it remains to be seen whether the agent should be administered alone, hoping the host immune system is able to overwhelm the remaining bacteria, or whether it has to be applied synergistically with a conventional antibiotic to ensure the exhaustive elimination of the pathogen. The most promising strategy may be a prophylactic administration of QS inhibitors, which is supposed to protect the patient against the establishment of chronic infections. Finally, it should be noted that apart from being strongly desirable for clinical applications, biofilm-eradicating compounds could also be exploited to fight biofilms causing microbial corrosion and costly damage in various industrial settings, such as ship hulls, water pipelines, and cooling systems.

2.6
Interspecies Signaling in Mixed Biofilms

In nature, bacteria usually do not live solitarily, but rather assemble in polymicrobial biofilms where cells are located in close proximity to each other. Given that QS systems are widespread, it may not be surprising that the signal molecules produced by one species can affect the behavior of a coresident species employing a similar cell–cell signaling system (interspecies signaling; reviewed in [93]). Communication between different species occupying the same ecological niche and competing for common resources includes synergistic as well as antagonistic interactions; moreover, interspecies signaling might play an important role in influencing bacterial virulence and response to antimicrobial therapy [94]. In the case of AHL-mediated cross-talk it has been demonstrated that *P. aeruginosa* stimulates the expression of virulence factors of *B. cenocepacia* in mixed biofilms growing under laboratory conditions as well as in the lungs of infected mice [95]. This finding is of particular relevance since the lungs of patients affected with CF can be coinfected with both *P. aeruginosa* and *B. cenocepacia*, and the mixed consortium is generally more virulent than the individual strains [96]. DSF has also been implicated in interspecies signaling in mixed biofilms of *Stenotrophomonas maltophilia* and *P. aeruginosa* [93]. Specifically, it was demonstrated that DSF produced by *S. maltophilia* stimulates expression of proteins involved in stress tolerance and resistance to cationic antimicrobial peptides in *P. aeruginosa*.

Some signal molecules may also be used by bacteria to interact with biofilm formation of eukaryotic organisms. For example 3-oxo-C12-HSL, which is produced by *P. aeruginosa*, inhibits filamentation of *C. albicans* [97]. As a consequence the yeast is unable to form biofilms [98]. Likewise DSF and BDSF, signal molecules used by *X. campestris* and *B. cenocepacia*, are molecules known to inhibit filament formation of *C. albicans* [37].

2.7
Conclusions

Given the high cell densities in biofilms and the possible diffusion limitation of signal molecules within the biofilm matrix, it appears that QS is a particularly valuable regulatory mechanism for controlling the formation of surface-associated consortia. In fact, QS systems utilizing different types of signal molecules were shown to affect various stages of the biofilm lifecycle, particularly biofilm maturation and dispersal. Several QS-regulated factors have been identified that are important in these processes, including the production of the biofilm matrix, biosynthesis of biosurfactants, and metabolic processes. However, it is likely that many more QS-regulated functions will influence biofilm development. It will therefore be one of the major challenges in the future to further identify those QS-regulated structural genes that affect biofilm formation.

Although only a few examples of interspecies and interkingdom communication systems have been reported, this line of research is still in its infancy and additional work will be required to unravel the underlying mechanisms, as well as to understand its role in syntrophy and pathogenesis.

Finally, the inhibition of QS systems has to be considered an attractive option not only to inhibit the production of virulence factors, but also to prevent and eradicate biofilms of important pathogens.

Acknowledgments

Financial support for work on QS and biofilm formation from the Swiss National Fund (Projects 3100A0–104215 and -114215) is gratefully acknowledged.

References

1 Watnick, P. and Kolter, R. (2000) Biofilm, city of microbes. *J. Bacteriol.*, **182**, 2675–2679.

2 O'Toole, G., Kaplan, H., and Kolter, R. (2000) Biofilm formation as microbial development. *Annu. Rev. Microbiol.*, **54**, 49–79.

3 Fuqua, C., Winans, S., and Greenberg, E. (1996) Census and consensus in bacterial ecosystems: the LuxR–LuxI family of quorum-sensing transcriptional regulators. *Annu. Rev. Microbiol.*, **50**, 727–751.

4 Whitehead, N., Barnard, A., Slater, H., Simpson, N., and Salmond, G. (2001) Quorum-sensing in Gram-negative bacteria. *FEMS Microbiol. Rev.*, **25**, 365–404.

5 Xavier, K.B. and Bassler, B.L. (2003) LuxS quorum sensing: more than just a numbers game. *Curr. Opin. Microbiol.*, **6**, 191–197.

6 Hardie, K. and Heurlier, K. (2008) Establishing bacterial communities by 'word of mouth': LuxS and autoinducer 2 in biofilm development. *Nat. Rev. Microbiol.*, **6**, 635–643.

7 Otto, M. (2008) Staphylococcal biofilms. *Curr. Top. Microbiol. Immunol.*, **322**, 207–228.

8 Williams, P., Winzer, K., Chan, W.C., and Camara, M. (2007) Look who's talking: communication and quorum sensing in the bacterial world. *Philos. Trans. R. Soc. Lond. B Biol. Sci.*, **362**, 1119–1134.

9 Kravchenko, V., Kaufmann, G., Mathison, J., Scott, D., Katz, A., Grauer, D., Lehmann, M., Meijler, M., Janda, K., and Ulevitch, R. (2008) Modulation of gene expression via disruption of NF-kappaB signaling by a bacterial small molecule. *Science*, **321**, 259–263.

10 Davies, D.G., Parsek, M.R., Pearson, J.P., Iglewski, B.H., Costerton, J.W., and Greenberg, E.P. (1998) The involvement of cell-to-cell signals in the development of a bacterial biofilm. *Science*, **280**, 295–298.

11 Juhas, M., Eberl, L., and Tummler, B. (2005) Quorum sensing: the power of cooperation in the world of *Pseudomonas*. *Environ. Microbiol.*, **7**, 459–471.

12 Heydorn, A., Ersboll, B., Kato, J., Hentzer, M., Parsek, M., Tolker-Nielsen, T., Givskov, M., and Molin, S. (2002) Statistical analysis of *Pseudomonas aeruginosa* biofilm development: impact of mutations in genes involved in twitching motility, cell-to-cell signaling, and stationary-phase sigma factor expression. *Appl. Environ. Microbiol.*, **68**, 2008–2017.

13 Stoodley, P., Lewandowski, Z., Boyle, J., and Lappin-Scott, H. (1999) The formation of migratory ripples in a mixed species bacterial biofilm growing in turbulent flow. *Environ. Microbiol.*, **1**, 447–455.

14 Purevdorj, B., Costerton, J.W., and Stoodley, P. (2002) Influence of hydrodynamics and cell signaling on the structure and behavior of *Pseudomonas aeruginosa* biofilms. *Appl. Environ. Microbiol.*, **68**, 4457–4464.

15 Klausen, M., Heydorn, A., Ragas, P., Lambertsen, L., Aaes-Jorgensen, A., Molin, S., and Tolker-Nielsen, T. (2003) Biofilm formation by *Pseudomonas aeruginosa* wild type, flagella and type IV pili mutants. *Mol. Microbiol.*, **48**, 1511–1524.

16 O'Toole, G.A., Gibbs, K.A., Hager, P.W., Phibbs, P.V. Jr., and Kolter, R. (2000) The global carbon metabolism regulator Crc is a component of a signal transduction pathway required for biofilm development by *Pseudomonas aeruginosa*. *J. Bacteriol.*, **182**, 425–431.

17 Pesci, E.C., Milbank, J.B., Pearson, J.P., McKnight, S., Kende, A.S., Greenberg, E.P., and Iglewski, B.H. (1999) Quinolone signaling in the cell-to-cell communication system of *Pseudomonas aeruginosa*. *Proc. Natl. Acad. Sci. USA*, **96**, 11229–11234.

18 Guina, T., Purvine, S., Yi, E., Eng, J., Goodlett, D., Aebersold, R., and Miller, S. (2003) Quantitative proteomic analysis indicates increased synthesis of a quinolone by *Pseudomonas aeruginosa* isolates from cystic fibrosis airways. *Proc. Natl. Acad. Sci. USA*, **100**, 2771–2776.

19 Diggle, S., Winzer, K., Chhabra, S., Worrall, K., Camara, M., and Williams, P. (2003) The *Pseudomonas aeruginosa* quinolone signal molecule overcomes the cell density-dependency of the quorum sensing hierarchy, regulates *rhl*-dependent genes at the onset of stationary phase and can be produced in the absence of LasR. *Mol. Microbiol.*, **50**, 29–43.

20 D'Argenio, D., Calfee, M., Rainey, P., and Pesci, E. (2002) Autolysis and autoaggregation in *Pseudomonas aeruginosa* colony morphology mutants. *J. Bacteriol.*, **184**, 6481–6489.

21 Whitchurch, C., Beatson, S., Comolli, J., Jakobsen, T., Sargent, J., Bertrand, J., West, J., Klausen, M., Waite, L., Kang, P., Tolker-Nielsen, T., Mattick, J., and Engel, J. (2005) *Pseudomonas aeruginosa* fimL regulates multiple virulence functions by intersecting with Vfr-modulated pathways. *Mol. Microbiol.*, **55**, 1357–1378.

22 Whitchurch, C.B., Tolker-Nielsen, T., Ragas, P.C., and Mattick, J.S. (2002) Extracellular DNA required for bacterial biofilm formation. *Science*, **295**, 1487.

23 Huber, B., Riedel, K., Hentzer, M., Heydorn, A., Gotschlich, A., Givskov, M., Molin, S., and Eberl, L. (2001) The *cep* quorum-sensing system of *Burkholderia cepacia* H111 controls biofilm formation and swarming motility. *Microbiology*, **147**, 2517–2528.

24 Lynch, M.J., Swift, S., Kirke, D.F., Keevil, C.W., Dodd, C.E., and Williams, P. (2002) The regulation of biofilm development by quorum sensing in *Aeromonas hydrophila*. *Environ. Microbiol.*, **4**, 18–28.

25 Steidle, A., Allesen-Holm, M., Riedel, K., Berg, G., Givskov, M., Molin, S., and Eberl, L. (2002) Identification and characterization of an *N*-acylhomoserine lactone-dependent quorum-sensing system in *Pseudomonas putida* strain IsoF. *Appl. Environ. Microbiol.*, **68**, 6371–6382.

26 Labbate, M., Queck, S.Y., Koh, K.S., Rice, S.A., Givskov, M., and Kjelleberg, S. (2004) Quorum sensing-controlled biofilm development in *Serratia liquefaciens* MG1. *J. Bacteriol.*, **186**, 692–698.

27 Wopperer, J., Cardona, S., Huber, B., Jacobi, C., Valvano, M., and Eberl, L. (2006) A quorum-quenching approach to investigate the conservation of quorum-sensing-regulated functions within the *Burkholderia cepacia* complex. *Appl. Environ. Microbiol.*, **72**, 1579–1587.

28 Arevalo-Ferro, C., Reil, G., Gorg, A., Eberl, L., and Riedel, K. (2005) Biofilm formation of *Pseudomonas putida* IsoF: the role of quorum sensing as assessed by proteomics. *Syst. Appl. Microbiol.*, **28**, 87–114.

29 Rice, S., Koh, K., Queck, S., Labbate, M., Lam, K., and Kjelleberg, S. (2005) Biofilm formation and sloughing in *Serratia marcescens* are controlled by quorum sensing and nutrient cues. *J. Bacteriol.*, **187**, 3477–3485.

30 Labbate, M., Zhu, H., Thung, L., Bandara, R., Larsen, M., Willcox, M., Givskov, M.,

Rice, S., and Kjelleberg, S. (2007) Quorum-sensing regulation of adhesion in *Serratia marcescens* MG1 is surface dependent. *J. Bacteriol.*, **189**, 2702–2711.

31 Danhorn, T. and Fuqua, C. (2007) Biofilm formation by plant-associated bacteria. *Annu. Rev. Microbiol.*, **61**, 401–422.

32 Koutsoudis, M.D., Tsaltas, D., Minogue, T.D., and von Bodman, S.B. (2006) Quorum-sensing regulation governs bacterial adhesion, biofilm development, and host colonization in *Pantoea stewartii* subspecies *stewartii*. *Proc. Natl. Acad. Sci. USA*, **103**, 5983–5988.

33 Wang, L.H., He, Y., Gao, Y., Wu, J.E., Dong, Y.H., He, C., Wang, S.X., Weng, L.X., Xu, J.L., Tay, L., Fang, R., and Zhang, L. (2004) A bacterial cell–cell communication signal with cross-kingdom structural analogues. *Mol. Microbiol.*, **51**, 903–912.

34 Dow, J.M., Crossman, L., Findlay, K., He, Y.Q., Feng, J.X., and Tang, J.L. (2003) Biofilm dispersal in *Xanthomonas campestris* is controlled by cell–cell signaling and is required for full virulence to plants. *Proc. Natl. Acad. Sci. USA*, **100**, 10995–11000.

35 Torres, P., Malamud, F., Rigano, L., Russo, D., Marano, M., Castagnaro, A., Zorreguieta, A., Bouarab, K., Dow, J., and Vojnov, A. (2007) Controlled synthesis of the DSF cell–cell signal is required for biofilm formation and virulence in *Xanthomonas campestris*. *Environ. Microbiol.*, **9**, 2101–2109.

36 Chatterjee, S., Almeida, R.P., and Lindow, S. (2008) Living in two worlds: the plant and insect lifestyles of *Xylella fastidiosa*. *Annu. Rev. Phytopathol.*, **46**, 243–271.

37 Boon, C., Deng, Y., Wang, L., He, Y., Xu, J., Fan, Y., Pan, S., and Zhang, L. (2008) A novel DSF-like signal from *Burkholderia cenocepacia* interferes with *Candida albicans* morphological transition. *ISME J.*, **2**, 27–36.

38 Atkinson, S., Throup, J., Stewart, G., and Williams, P. (1999) A hierarchical quorum-sensing system in *Yersinia pseudotuberculosis* is involved in the regulation of motility and clumping. *Mol. Microbiol.*, **33**, 1267–1277.

39 Puskas, A., Greenberg, E., Kaplan, S., and Schaefer, A. (1997) A quorum-sensing system in the free-living photosynthetic bacterium *Rhodobacter sphaeroides*. *J. Bacteriol.*, **179**, 7530–7537.

40 Sutherland, I.W. (2001) The biofilm matrix – an immobilized but dynamic microbial environment. *Trends Microbiol.*, **9**, 222–227.

41 Braun, E. (1982) Ultrastructural investigation of resistant and susceptible maize inbreds infected with *Erwinia stewartii*. *Phytopathology*, **72**, 159–166.

42 Nimtz, M., Mort, A., Wray, V., Domke, T., Zhang, Y., Coplin, D., and Geider, K. (1996) Structure of stewartan, the capsular exopolysaccharide from the corn pathogen *Erwinia stewartii*. *Carbohydr. Res.*, **288**, 189–201.

43 von Bodman, S., Majerczak, D., and Coplin, D. (1998) A negative regulator mediates quorum-sensing control of exopolysaccharide production in *Pantoea stewartii* subsp. *stewartii*. *Proc. Natl. Acad. Sci. USA*, **95**, 7687–7692.

44 Rattee, I.D. and Brener, M.M. (1974) *The Physical Chemistry of Dye Adsorption*, Academic Press, New York.

45 Ryder, C., Byrd, M., and Wozniak, D.J. (2007) Role of polysaccharides in *Pseudomonas aeruginosa* biofilm development. *Curr. Opin. Microbiol.*, **10**, 644–648.

46 Friedman, L. and Kolter, R. (2004) Genes involved in matrix formation in *Pseudomonas aeruginosa* PA14 biofilms. *Mol. Microbiol.*, **51**, 675–690.

47 Sakuragi, Y. and Kolter, R. (2007) Quorum-sensing regulation of the biofilm matrix genes (*pel*) of *Pseudomonas aeruginosa*. *J. Bacteriol.*, **189**, 5383–5386.

48 Friedman, L. and Kolter, R. (2004) Two genetic loci produce distinct carbohydrate-rich structural components of the *Pseudomonas aeruginosa* biofilm matrix. *J. Bacteriol.*, **186**, 4457–4465.

49 Vasseur, P., Vallet-Gely, I., Soscia, C., Genin, S., and Filloux, A. (2005) The *pel* genes of the *Pseudomonas aeruginosa* PAK strain are involved at early and late stages of biofilm formation. *Microbiology*, **151**, 985–997.

50 Slater, H., Alvarez-Morales, A., Barber, C., Daniels, M., and Dow, J. (2000) A two-component system involving an HD-GYP

domain protein links cell–cell signalling to pathogenicity gene expression in *Xanthomonas campestris*. *Mol. Microbiol.*, **38**, 986–1003.

51 Tang, J., Liu, Y., Barber, C., Dow, J., Wootton, J., and Daniels, M. (1991) Genetic and molecular analysis of a cluster of *rpf* genes involved in positive regulation of synthesis of extracellular enzymes and polysaccharide in *Xanthomonas campestris* pathovar *campestris*. *Mol. Gen. Genet.*, **226**, 409–417.

52 Crossman, L. and Dow, J. (2004) Biofilm formation and dispersal in *Xanthomonas campestris*. *Microbes Infect.*, **6**, 623–629.

53 Morris, E., Rees, D., Young, G., Walkinshaw, M., and Darke, A. (1977) Order–disorder transition for a bacterial polysaccharide in solution. A role for polysaccharide conformation in recognition between *Xanthomonas* pathogen and its plant host. *J. Mol. Biol.*, **110**, 1–16.

54 Dubern, J.F., Lugtenberg, B.J., and Bloemberg, G.V. (2006) The *ppuI–rsaL–ppuR* quorum-sensing system regulates biofilm formation of *Pseudomonas putida* PCL1445 by controlling biosynthesis of the cyclic lipopeptides putisolvins I and II. *J. Bacteriol.*, **188**, 2898–2906.

55 Kuiper, I., Lagendijk, E., Pickford, R., Derrick, J., Lamers, G., Thomas-Oates, J., Lugtenberg, B., and Bloemberg, G. (2004) Characterization of two *Pseudomonas putida* lipopeptide biosurfactants, putisolvin I and II, which inhibit biofilm formation and break down existing biofilms. *Mol. Microbiol.*, **51**, 97–113.

56 Dubern, J., Lugtenberg, B., and Bloemberg, G. (2006) The *ppuI–rsaL–ppuR* quorum-sensing system regulates biofilm formation of *Pseudomonas putida* PCL1445 by controlling biosynthesis of the cyclic lipopeptides putisolvins I and II. *J. Bacteriol.*, **188**, 2898–2906.

57 Davey, M.E., Caiazza, N.C., and O'Toole, G.A. (2003) Rhamnolipid surfactant production affects biofilm architecture in *Pseudomonas aeruginosa* PAO1. *J. Bacteriol.*, **185**, 1027–1036.

58 Lequette, Y. and Greenberg, E.P. (2005) Timing and localization of rhamnolipid synthesis gene expression in *Pseudomonas aeruginosa* biofilms. *J. Bacteriol.*, **187**, 37–44.

59 Pamp, S.J. and Tolker-Nielsen, T. (2007) Multiple roles of biosurfactants in structural biofilm development by *Pseudomonas aeruginosa*. *J. Bacteriol.*, **189**, 2531–2539.

60 Boles, B.R., Thoendel, M., and Singh, P.K. (2005) Rhamnolipids mediate detachment of *Pseudomonas aeruginosa* from biofilms. *Mol. Microbiol.*, **57**, 1210–1223.

61 Ochsner, U., Fiechter, A., and Reiser, J. (1994) Isolation, characterization, and expression in *Escherichia coli* of the *Pseudomonas aeruginosa rhlAB* genes encoding a rhamnosyltransferase involved in rhamnolipid biosurfactant synthesis. *J. Biol. Chem.*, **269**, 19787–19795.

62 Ochsner, U. and Reiser, J. (1995) Autoinducer-mediated regulation of rhamnolipid biosurfactant synthesis in *Pseudomonas aeruginosa*. *Proc. Natl. Acad. Sci. USA*, **92**, 6424–6428.

63 Pearson, J., Pesci, E., and Iglewski, B. (1997) Roles of *Pseudomonas aeruginosa las* and *rhl* quorum-sensing systems in control of elastase and rhamnolipid biosynthesis genes. *J. Bacteriol.*, **179**, 5756–5767.

64 Medina, G., Juarez, K., Valderrama, B., and Soberon-Chavez, G. (2003) Mechanism of *Pseudomonas aeruginosa* RhlR transcriptional regulation of the rhlAB promoter. *J. Bacteriol.*, **185**, 5976–5983.

65 Allesen-Holm, M., Barken, K.B., Yang, L., Klausen, M., Webb, J.S., Kjelleberg, S., Molin, S., Givskov, M., and Tolker-Nielsen, T. (2006) A characterization of DNA release in *Pseudomonas aeruginosa* cultures and biofilms. *Mol. Microbiol.*, **59**, 1114–1128.

66 Barken, K., Pamp, S., Yang, L., Gjermansen, M., Bertrand, J., Klausen, M., Givskov, M., Whitchurch, C., Engel, J., and Tolker-Nielsen, T. (2008) Roles of type IV pili, flagellum-mediated motility and extracellular DNA in the formation of mature multicellular structures in *Pseudomonas aeruginosa* biofilms. *Environ. Microbiol.*, **10**, 2331–2343.

67 Yoon, S.S., Hennigan, R.F., Hilliard, G.M., Ochsner, U.A., Parvatiyar, K., Kamani, M.C., Allen, H.L., DeKievit, T.R., Gardner, P.R., Schwab, U., Rowe, J., Iglewski, B.,

McDermott, T., Mason, R., Wozniak, D., Hancock, R., Parsek, M., Noah, T., Boucher, R., and Hassett, D. (2002) *Pseudomonas aeruginosa* anaerobic respiration in biofilms: relationships to cystic fibrosis pathogenesis. *Dev. Cell*, **3**, 593–603.

68 Barraud, N., Hassett, D., Hwang, S., Rice, S., Kjelleberg, S., and Webb, J. (2006) Involvement of nitric oxide in biofilm dispersal of *Pseudomonas aeruginosa*. *J. Bacteriol.*, **188**, 7344–7353.

69 Toyofuku, M., Nomura, N., Fujii, T., Takaya, N., Maseda, H., Sawada, I., Nakajima, T., and Uchiyama, H. (2007) Quorum sensing regulates denitrification in *Pseudomonas aeruginosa* PAO1. *J. Bacteriol.*, **189**, 4969–4972.

70 Davies, D. (2003) Understanding biofilm resistance to antibacterial agents. *Nat. Rev. Drug Discov.*, **2**, 114–122.

71 Lynch, A.S. and Robertson, G.T. (2008) Bacterial and fungal biofilm infections. *Annu. Rev. Med.*, **59**, 415–428.

72 Anderson, G.G. and O'Toole, G.A. (2008) Innate and induced resistance mechanisms of bacterial biofilms. *Curr. Top Microbiol. Immunol.*, **322**, 85–105.

73 Hentzer, M. and Givskov, M. (2003) Pharmacological inhibition of quorum sensing for the treatment of chronic bacterial infections. *J. Clin. Invest.*, **112**, 1300–1307.

74 Dong, Y., Gusti, A., Zhang, Q., Xu, J., and Zhang, L. (2002) Identification of quorum-quenching *N*-acyl homoserine lactonases from *Bacillus* species. *Appl. Environ. Microbiol.* **68**, 1754–1759.

75 Dong, Y.H., Xu, J.L., Li, X.Z., and Zhang, L.H. (2000) AiiA, an enzyme that inactivates the acylhomoserine lactone quorum-sensing signal and attenuates the virulence of *Erwinia carotovora*. *Proc. Natl. Acad. Sci. USA*, **97**, 3526–3531.

76 Park, S., Lee, S., Oh, T., Oh, J., Koo, B., Yum, D., and Lee, J. (2003) AhlD, an *N*-acylhomoserine lactonase in *Arthrobacter* sp., and predicted homologues in other bacteria. *Microbiology*, **149**, 1541–1550.

77 Uroz, S., Oger, P., Chapelle, E., Adeline, M., Faure, D., and Dessaux, Y. (2008) A Rhodococcus *qsdA*-encoded enzyme defines a novel class of large-spectrum quorum-quenching lactonases. *Appl. Environ. Microbiol.*, **74**, 1357–1366.

78 Reimmann, C., Ginet, N., Michel, L., Keel, C., Michaux, P., Krishnapillai, V., Zala, M., Heurlier, K., Triandafillu, K., Harms, H., Défago, G., and Haas, D. (2002) Genetically programmed autoinducer destruction reduces virulence gene expression and swarming motility in *Pseudomonas aeruginosa* PAO1. *Microbiology*, **148**, 923–932.

79 Wang, Y., Dai, Y., Zhang, Y., Hu, Y., Yang, B., and Chen, S. (2007) Effects of quorum sensing autoinducer degradation gene on virulence and biofilm formation of *Pseudomonas aeruginosa*. *Sci. China C Life. Sci.*, **50**, 385–391.

80 Rasmussen, T., Skindersoe, M., Bjarnsholt, T., Phipps, R., Christensen, K., Jensen, P., Andersen, J., Koch, B., Larsen, T., Hentzer, M., Eberl, L., Hoiby, N., and Givskov, M. (2005) Identity and effects of quorum-sensing inhibitors produced by Penicillium species. *Microbiology*, **151**, 1325–1340.

81 Givskov, M., de Nys, R., Manefield, M., Gram, L., Maximilien, R., Eberl, L., Molin, S., Steinberg, P.D., and Kjelleberg, S. (1996) Eukaryotic interference with homoserine lactone-mediated prokaryotic signalling. *J. Bacteriol.*, **178**, 6618–6622.

82 Teplitski, M., Robinson, J., and Bauer, W. (2000) Plants secrete substances that mimic bacterial *N*-acyl homoserine lactone signal activities and affect population density-dependent behaviors in associated bacteria. *Mol. Plant-Microbe Interact.*, **13**, 637–648.

83 Bjarnsholt, T., Jensen, P. O., Rasmussen, T., Christophersen, L., Calum, H., Hentzer, M., Hougen, H., Rygaard, J., Moser, C., Eberl, L., Høiby, N., and Givskov, M. (2005) Garlic blocks quorum sensing and promotes rapid clearing of pulmonary *Pseudomonas aeruginosa* infections. *Microbiology*, **151**, 3873–3880.

84 Niu, C., Afre, S., and Gilbert, E.S. (2006) Subinhibitory concentrations of cinnamaldehyde interfere with quorum sensing. *Lett. Appl. Microbiol.*, **43**, 489–494.

85 Adonizio, A., Kong, K., and Mathee, K. (2008) Inhibition of quorum sensing-controlled virulence factor production in

Pseudomonas aeruginosa by South Florida plant extracts. *Antimicrob. Agents Chemother.*, **52**, 198–203.

86 Peters, L., Konig, G.M., Wright, A.D., Pukall, R., Stackebrandt, E., Eberl, L., and Riedel, K. (2003) Secondary metabolites of *Flustra foliacea* and their influence on bacteria. *Appl. Environ. Microbiol.*, **69**, 3469–3475.

87 Skindersoe, M.E., Ettinger-Epstein, P., Rasmussen, T.B., Bjarnsholt, T., de Nys, R., and Givskov, M. (2008) Quorum sensing antagonism from marine organisms. *Mar. Biotechnol. (NY)*, **10**, 56–63.

88 Manefield, M., de Nys, R., Kumar, N., Read, R., Givskov, M., Steinberg, P., and Kjelleberg, S. (1999) Evidence that halogenated furanones from *Delisea pulchra* inhibit acylated homoserine lactone (AHL)-mediated gene expression by displacing the AHL signal from its receptor protein. *Microbiology*, **145**, 283–291.

89 Manefield, M., Rasmussen, T.B., Henzter, M., Andersen, J.B., Steinberg, P., Kjelleberg, S., and Givskov, M. (2002) Halogenated furanones inhibit quorum sensing through accelerated LuxR turnover. *Microbiology*, **148**, 1119–1127.

90 Hentzer, M., Riedel, K., Rasmussen, T., Heydorn, A., Andersen, J., Parsek, M., Rice, S., Eberl, L., Molin, S., Høiby, N., Kjelleberg, S., and Givskov, M. (2002) Inhibition of quorum sensing in *Pseudomonas aeruginosa* biofilm bacteria by a halogenated furanone compound. *Microbiology*, **148**, 87–102.

91 Kim, C., Kim, J., Park, H., Park, H., Lee, J., Kim, C., and Yoon, J. (2008) Furanone derivatives as quorum-sensing antagonists of *Pseudomonas aeruginosa*. *Appl. Microbiol. Biotechnol.*, **80**, 37–47.

92 Riedel, K., Kothe, M., Kramer, B., Saeb, W., Gotschlich, A., Ammendola, A., and Eberl, L. (2006) Computer-aided design of agents that inhibit the *cep* quorum-sensing system of *Burkholderia cenocepacia*. *Antimicrob. Agents Chemother.*, **50**, 318–323.

93 Ryan, R.P. and Dow, J.M. (2008) Diffusible signals and interspecies communication in bacteria. *Microbiology*, **154**, 1845–1858.

94 Ahmer, B.M. (2004) Cell-to-cell signalling in *Escherichia coli* and *Salmonella enterica*. *Mol. Microbiol.*, **52**, 933–945.

95 Riedel, K., Hentzer, M., Geisenberger, O., Huber, B., Steidle, A., Wu, H., Høiby, N., Givskov, M., Molin, S., and Eberl, L. (2001) *N*-acylhomoserine-lactone-mediated communication between *Pseudomonas aeruginosa* and *Burkholderia cepacia* in mixed biofilms. *Microbiology*, **147**, 3249–3262.

96 Eberl, L. and Tummler, B. (2004) *Pseudomonas aeruginosa* and *Burkholderia cepacia* in cystic fibrosis: genome evolution, interactions and adaptation. *Int. J. Med. Microbiol.*, **294**, 123–131.

97 Hogan, D., Vik, A., and Kolter, R. (2004) A *Pseudomonas aeruginosa* quorum-sensing molecule influences *Candida albicans* morphology. *Mol. Microbiol.*, **54**, 1212–1223.

98 Lewis, R.E., Lo, H.J., Raad, I.I., and Kontoyiannis, D.P. (2002) Lack of catheter infection by the *efg1/efg1 cph1/cph1* double-null mutant, a *Candida albicans* strain that is defective in filamentous growth. *Antimicrob. Agents Chemother.*, **46**, 1153–1155.

3
Cell Interactions Guide the Swarming and Fruiting Body Development of Myxobacteria

Dale Kaiser

3.1
Introduction

Myxobacteria are voluntary multicellular organisms that are common in topsoil. Given a complete medium these Gram-negative bacteria grow as independent cells in liquid culture. However, when in contact with each other on a solid medium, they interact and grow as a swarm. The swarm spreads indefinitely outwards from the point of inoculation to gain access to nutrients. Cell interactions occur at every stage in the lifecycle (Figure 3.1) of *Myxococcus xanthus*. Interactions include the mechanical effects of collisions between cells, signal transmission by end-to-end contact between two cells, and signaling by diffusible small molecules. An inventory of *M. xanthus* genes reveals that it evolved from a δ-proteobacterial progenitor by a 4–5-Mb genome expansion. More than 1500 duplications specific to the myxobacterial lineage were identified, representing more than 15% of the total genes. Genes were not duplicated at random; rather, genes for cell–cell signaling, small-molecule sensing, and integrative control of transcription were amplified selectively [1]. Among δ-proteobacterial genomes that have been sequenced, only the myxobacteria have the capacity to develop multicellular fruiting bodies whose form is species specific.

3.2
Motility of Myxobacteria

M. xanthus cells are elongated, flexible rods with a 10 : 1 length: width ratio [2]. They lack flagella and are unable to swim. They can only move on surfaces and do so by gliding in the direction of their long axis [3]. Two molecular motors, retractile type IV pili at their leading end (S motility) and nozzles for secreting a slime gel at the trailing end (A motility), provide the adhesion and thrust necessary for moving on surfaces, including the surfaces of soil particles.

Bacterial Signaling. Edited by Reinhard Krämer and Kirsten Jung
Copyright © 2010 WILEY-VCH Verlag GmbH & Co. KGaA, Weinheim
ISBN: 978-3-527-32365-4

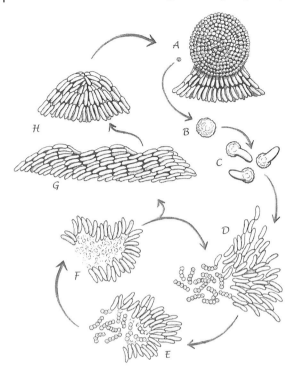

Figure 3.1 The lifecycle of M. xanthus. A swarm (a group of moving and interacting cells) can have either of two fates depending on their environment. The fruiting body (A) is a spherical structure of approximately 1×10^5 cells that have become stress-resistant spores (B). The fruiting body is small (0.1 mm high), sticky, and its spores are tightly packed. When a fruiting body receives nutrients, the individual spores germinate (C) and thousands of M. xanthus cells emerge together as an "instant" swarm (D). When prey is available (micrococci in the figure), the swarm becomes a predatory collective that surrounds the prey. Swarm cells feed by contacting, lysing and consuming the prey bacteria (E and F). Fruiting body development is advantageous given the collective hunting behavior. Nutrient-poor conditions elicit a unified starvation stress response. That response initiates a program that changes cell movement behavior, leading to aggregation. The movement behaviors include wave formation (G) and streaming into mounded aggregates (H), which become spherical (A). Spores differentiate within mounded and spherical aggregates. (Reprinted from [1] with permission of the publisher).

3.3
Pilus Engine

At the leading end of the cell are lengthy (each several μm long), very thin (6–8 nm), type IV pilus fibers. M. xanthus, Neisseria gonorrhoeae, Pseudomonas aeruginosa, Synechocystis PCC6803, and other Gram-negative bacteria share a common set of 10 pilus proteins and produce functionally similar pili [4, 5]. The pilus fibers are helical arrays of pilin (PilA in M. xanthus) monomers, whose sequence-conserved N-termini assembles a coiled-coil down the center of the fiber that provides its tensile strength [6]. Considered to lack a specialized tip structure and being helical, the *M. xanthus* pilus tip exposes the surfaces of several pilin monomers that would have bound other monomers within a fiber. These exposed surfaces attach to cells ahead, or more precisely, to multistranded polysaccharide "fibrils" that envelop clusters of adjacent cells with a network resembling a fisherman's net [7–9]. After firm attachment to fibrils, the cell pulls itself forward by retracting its pilus and storing the dissociated pilin monomers in the cytoplasmic membrane for reuse. The retraction motor is PilT, an AAA-ATPase, located in the inner membrane, whose structure has recently been solved [10], suggesting large domain movements. Currently, PilT is the strongest known molecular motor, capable of developing more than 100 pN of tension in the case of *N. gonorrhoeae* [11]. That tension is the reason the pilus tip must attach multivalently to a fibril, a bundle of polysaccharide chains, because such attachment provides enough binding strength to withstand 100 pN of tension without rupturing. A signal is thought to pass from the tip of a securely attached pilus down to the base, activating PilT retraction. Pili are too thin and flexible to push cells when they elongate; pushing against a solid merely causes the fiber to bend. The pilus fiber slides through PilQ, a gated channel in the outer membrane of *M. xanthus*, which is a slippery bushing for extension and retraction of the fiber. Tgl is an outer membrane lipoprotein assembly factor for PilQ [12, 13]. The mechanics of pulling, together with the observation that *M. xanthus* pili are unipolar [14], imply that pili are located at the leading end of the cell, and there is visual evidence from dye binding [15].

3.4
Slime Secretion Engine

Electron microscopy shows several filaments of slime emerging from a cell end [16]. Images of complete cells show that filaments emerge from one end only; the opposite ends do not extrude. Several hundred thick-walled pores are found at both cell ends. Unlike the type IV pili and the filaments of slime, the pores are found at both ends of the cell. Presumably the filaments of slime are emerging from a pore or a cluster of neighboring pores at one end only. Biochemical and genetic experiments [17] indicate that the thick-walled pores are secretion nozzles for the slime [18]. Differential interference contrast microscopy reveals a single filament at one pole in wild-type cells [17]. Apparently, the narrow filaments associate laterally to create a gel having the width of the whole cell. Slime secretion from the rear is demonstrably linked to cell movement by the time-lapse movies of Lars Jelsbak [19]. It is evident that, each time the cell moves, one of the slime trails lengthen.

Recently, Mignot *et al.* observed focal adhesions in the leading half of moving cells that were essential for A motility [20]. The authors speculated that an apicomplexan-like gliding force is injected at each site of adhesion, but did not describe the motor. Focal adhesions without force injection that are consistent with Mignot's observations could be used for propulsion by polar slime secretion, as described [19].

Significantly, the slime secretion engine is unipolar like the pilus engine and slime secretion pushes cells at the back, while pilus retraction pulls cells from the front. Thus, at any instant, the cell is structured to move in one particular direction – it has a dedicated head and a dedicated tail. The cell also has a template whose activity is revealed at cell division, when the two new poles created by the division septum gain engines. Those new engines, specified by the template, always complement the engines at their old poles. Both daughters are motile at birth. Polarity is conserved through cell division. Template activity is also revealed by the fact that the rate of swarm expansion of cells with both engines ($A^+ S^+$) is 50% greater than the sum of the swarm rates of two mutants, one with S engines only and the other with only the A engines [21]. The synergism between engines shows that cells always have pili at one end and active slime nozzles at the other – apparently the engines never oppose each other. Although the structure of the template remains to be found, its activity is clearly essential for swarming.

3.5
Swarming of Myxobacteria

By moving as an organized swarm, *M. xanthus* is able to spread rapidly over a solid surface like agar or a particle of organic matter in soil. When myxobacteria swarm, there is a net flow of cells from the center of the swarm, where most cells are borne out to the edge. By escaping from the high-cell-density center to the low-density edge, swarming reduces competition among cells for nutrients that are available from the substratum as well as for oxygen from the atmosphere. Competition is high in the swarm center where many layers of cells cover each other. Laboratory swarms of *M. xanthus* spread for several weeks at a constant rate of 2.4 mm, or about 340 cell lengths, per day. Swarming requires growth, even though growth accounts for only 10% of the swarm expansion rate, while cell movement accounts for 90% [21]. Although they are moving at high density, the synergism implies not only that cells stay out of each others' way, but that they facilitate each others' motion. Many observations indicate not only that growth is necessary for swarming, but that there is also a need for periodic reversal of gliding direction, which was unexpected. The logical need for reversal was recently demonstrated computationally [22]. To identify the cell behaviors that are important for the ability of *Myxococcus* to swarm, we have taken time-lapse movies of the cells at the edge of an expanding swarm. The movies show single cells reversing and the formation of regular multicellular structures. The movies revealed other qualities important for the ability to swarm – flexible cells that bend to resolve end-to-side collisions and the following of slime trails, on which cells can slip past one another [23]. The movies also implicate the attachment of pili to fibrils on cells ahead. Finally, the movies show a regular flow of cells among multicellular rafts of cells. The power of myxobacteria to build multicellular fruiting bodies appears to have arisen from their capacity to swarm.

3.6
Regulating Reversals

The intracellular regulatory network that triggers the coordinate reversal of gliding direction is sketched out in the right half of Figure 3.2. FrzCD and FrzE proteins encode a two-component chemosensory system [24]. In 1985, Blackhart and Zusman discovered that mutations in FrzCD and FrzE change the frequency of gliding reversals. These proteins are related by amino acid sequence, but are not identical to, the chemotaxis proteins of *E. coli* and *Salmonella* [25]. The methyl-accepting chemosensory protein, FrzCD, is a cytoplasmic protein in *M. xanthus*, not a membrane receptor as in *Escherichia coli* – providing evidence that FrzCD receives cytoplasmic, and not extracellular, signals. FrzE, the histidine protein kinase for this two-component system is signaled by methylated FrzCD. FrzE can autopho-

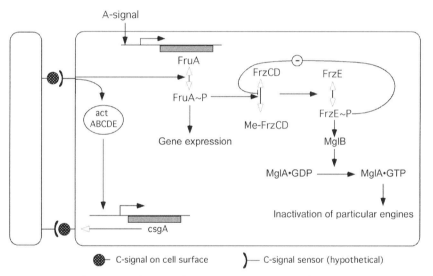

Figure 3.2 Regulatory C-signal circuit. The C-signal is a 17-kDa cell surface protein. Cells must make end-to-end contact to transmit the signal, as indicated. Reception of C-signal activates FruA by forming FruA~P. FruA~P drives the oscillation of MeFrzCD and FrzE~P. FrzE~P switches MglA·GDP to MglA·GTP that, in turn, inactivates old engines. C-signaling increases *csgA* transcription, directly or indirectly via the proteins of the *act* operon. Waves, aggregation, sporulation, and C-signal-dependent gene expression are induced by increasing levels of FruA~P. MXAN4885 is proposed to be the branch from reception of C-signal to FruA~P [32]. FruA is an orphan response regulator of the two-component signal transduction family containing a DNA-binding output domain that is thought to be activated by phosphorylation of a conserved aspartic acid in its receiver domain [62, 96]. Since C-signaling raises the level of FruA~P higher and higher as development proceeds, it activates the expression of more and more genes. Since developmentally regulated genes have different FruA (or FruA~P) requirements, governed by the affinities of their DNA-binding sites [97], gene expression in each cell would be determined directly by the number of C-signal molecules on its surface, and indirectly to its past movements and positions.

sphorylate [26]. Igoshin et al. [27] discovered that the levels of Me-FrzCD and FrzE~P oscillated out of phase with each other. They proposed that there is negative feedback from FrzE~P back onto the methylation of FrzCD, which would create an oscillator – the frizzilator. While seeking biochemical evidence for feedback, the frizzilator remains a precise working hypothesis. Wild-type cells reverse once every 8 min on average, but a FrzCD or a FrzE null mutant reverses with a 34-min period [28]. Pulses of FrzE~P from the frizzilator drive the MglAB switch to oscillate on/off with the 8-min period.

MglA is a small, Ras-like G-protein [29] that is placed in its active GTP-bound state by the Roadblock/LC7 MglB modulator [30]. Formerly, MglA mutants were said to "hyper reverse" [31], but MglA mutant reversals are not produced by the frizzilator, as in swarming; instead, they reflect statistical fluctuations in the number of active slime secretion engines. MglA mutants secrete slime from both poles as if they were trying to move forward and backward at the same time [17]. They are observed to move rapidly back and forth less than one-fifth of a cell length per stroke, which can be explained quantitatively by the statistics of the difference in the rate of slime secretion between two ends [32]. For swarming, once MglA protein has been switched to its GTP-bound state by a pulse of FrzE~P, it recognizes the currently active pili at the cytoplasmic face of the inner membrane for Tgl destruction and loss of the type IV pili [19]. Simultaneously, MglA·GTP switches slime secretion from its current pole to the opposite pole using RomR [33]. During such a regulated reversal, a cell simply stops, pauses for about a minute, then moves off in the opposite direction [34]. The strictly coordinated polarity of two different engines uses the template described above, which must be a structure that can be smoothly inverted. The phenotype of MglA mutants clearly shows that engine destruction is the process that is regulated while new engine production is like normal cell division. As an MglA deletion mutant is unable to degrade CglB, it secretes slime from both its ends [17].

3.7
Fruiting Body Development

Swarming requires concurrent growth, which it also enhances. When a swarming and growing population of M. xanthus senses the onset of starvation, it stops growing but continues to move. M. xanthus needs to detect starvation early because it must synthesize more than 30 new proteins from its reserves or by lysis of its siblings before it can sporulate [35]. The stringent response [36] and the quorum sensing A-signal are involved in computing the most appropriate population response to the quality of starvation that it faces. About 2 h after starvation, those cells that sense a likelihood that they will starve to death secrete a fixed, small amount of the A-signal [37, 38]. A-signal molecules were purified from medium conditioned by starving cells and six amino acids (Trp, Pro, Phe, Tyr, Leu, and Ile) were found to account for 95% of the A-signal activity in the conditioned medium [39]. All cells in the population take the A-signal up, but they respond to the signal only if its concentration is above threshold (a quorum). Cells respond by expressing a set of

A-signal-dependent genes that include FruA, CsgA, and other aggregation genes (Figure 3.2). This voting mechanism is thought to ensure that the decision is appropriate for the overall nutritional conditions.

Most of our (clearly incomplete) understanding of aggregation arises from fitting cell movement into a scheme of signaling and regulation. When there is an A-signal quorum, cells cease to flow smoothly past one another as they do when swarming, instead they form traffic jams and circulate around them. Traffic jams formed where cells had converged from three or more directions in the submerged cultures of the Welch type [40]. In submerged culture of the Kuner type [41], traffic jams arose on the dense edges of an initial facet pattern [40]. In the former, traffic jams have rounded due to their covering of circulating cells; the latter foci have not yet been covered. They were formed on apparently random edges of the facet pattern. In both culture types, cells accumulate around initially asymmetric traffic jams [40, 42]. Their asymmetry suggests that the foci were not formed by chemotaxis. Nevertheless taxis has often been invoked [43, 44], and many orthologs of genes employed for chemotaxis in *E. coli* are found in *M. xanthus* and *Stigmatella aurantiaca* [45], but the majority have been shown to regulate expression of specific genes [46]. There is evidence that cells slow down while circulating inside a nascent aggregate [47], which could be explained by an increase in frictional drag with increased cell density. Although this was not the authors' favorite, they mentioned it as a plausible alternative. There also is evidence that the reversal frequency decreases progressively throughout aggregation [34, 48, 49]. This differs from a chemotactic biased random walk. Movement continues until all the spores have formed; spores are nonmotile and accumulate in the center of the spherical fruiting body, while the moving rod cells occupy an outer domain [50]. Reversals are controlled by the frizzilator throughout development, which is being managed by the cell-bound C-signal (Figure 3.2).

Each of the 50 different species of myxobacteria build fruiting bodies that have a distinctive form. Their individual morphologies track their phylogeny perfectly [51]. Most species, including *Myxococcus stipitatis*, mount their spores on a stalk, which suggests that the main reason for building a fruiting body is myxospore dispersal [52]. Spore elevation should facilitate transport by small animals in soil; most probably they carry the spore-filled fruiting body cyst to a place where the animal finds food [53]. With pedestal stalks that are easier to form, *M. xanthus* fruiting bodies are remarkably uniform in size and shape [54, 55]. That uniformity reflects the reliability of morphogenetic C-signaling between adjacent cells. C-signal deficient mutants (*csgA*) were found to grow normally but neither to aggregate nor to sporulate [56–58]. The C-signal is a 17-kDa cell-surface-bound protein that signals when a pair of cells makes end-to-end contact with each other [59]. Side-by-side or end-by-side contacts apparently do not exchange C-signal for aggregation or sporulation [60]. Is the proposal that local cell-contact signaling around traffic jams sufficient to explain aggregation? A continuous three-dimensional computational simulation was carried out [42, 61]. The simulations qualitatively reproduced the experimentally observed stages of fruiting body formation: asymmetric initial aggregates (the traffic jams), linear streams, formation of hemispherical mounds whose centers have low cell density, and finally the induction of spores when C-signal passed a threshold. These simulations suggest that the C-signaling model may be sufficient; further experiments to establish whether the model is also necessary await plausible alternative models.

3.8
C-Signal and Fruiting Body Morphogenesis

The simple fact that gliding engines use cell contact with their substratum to develop force suggests how a flat, asymmetric traffic jam becomes a spherically symmetric fruiting body resting on a short stalk. As shown in Figure 3.2, C-signaling induces the phosphorylation of FruA – a developmentally important response regulator [62] controlled by the C-signal. Whenever cells engage in C-signaling, expression of the C-signal is increased by a positive feedback loop involving proteins of the *act* operon [63]. The circuit depicted in Figure 3.2 was built to account for the phenotype of gene knockout mutants for each of its elements. The simulations by Sozinova *et al.* [42] confirmed the positive feedback loop, which causes the cell surface level of C-signal to increase progressively with each episode of C-signaling. The five proteins of the act operon increase expression of the *csgA* gene [63]. They produce a 25-kDa protein that is secreted to the cell surface where it is cleaved to the active 17-kDa form by a cell surface protease [64–66]. This mechanism prevents a cell from C-signaling to itself. At the start of development, there are few C-signal molecules per cell. Cells making end-to-end contact respond to signal exchange by reversing their direction of gliding which creates the traveling waves [67]. One consequence of the positive feedback by Act is that the wave pattern is transient [63]. Each time C-signal is exchanged between cells in the crests of two colliding waves, the positive feedback increases expression of *csgA* and elevates the number of CsgA molecules on the pair of signaling cells.

Traveling waves are produced by the C-signal level found at the start of development. The low level of FruA$\sim$P, induced at the start, drives the frizzilator (Figure 3.2), producing a very regular 8-min period and sharp wave crests. The circuit oscillates because FrzE$\sim$P inhibits methylation of FrzCD (or stimulates demethylation of MeFrzCD). Waves start as broad and diffuse ridges. Then, because the frizzilators in a pair of cells that are signaling to each other synchronize, the waves sharpen [27].

3.9
Managing the Reversal Frequency

Due to the positive feedback, more signal is produced. Higher levels of C-signal induce higher levels of FruA$\sim$P [68–70], raising the cytoplasmic level of FruA$\sim$P (Figure 3.2) to a threshold – the streaming threshold. The existence of a threshold level of FruA$\sim$P that can stop further oscillation by driving the entire FrzE pool into its non-phosphorylated state follows from the kinetics of negative feedback within the frizzilator, as described by Igoshin *et al.* [27]. Without FrzE$\sim$P, the cells remain moving in the direction they had been moving since their last reversal. The population of moving cells makes a transition from frequent reversals that give traveling waves to the streaming into aggregation centers. This is observed when the waves fade away and simultaneously the aggregation centers enlarge [40]. This transition in cell behavior agrees with the tracks of individual cells that were observed by Jelsbak [49]. Unidirectional movement allows the cells to form streams, and a stream that happens to be heading toward a nascent aggregate helps to enlarge it [65]. To be clear, streams are not attracted by the aggregate; streams that happen to collide with a nascent aggregate are trapped by it. As there are many nascent aggregates, eventually all moving cells in an aggregation field find themselves trapped within one or another aggregate, as observed in the movie of Welch [40].

How is a stream trapped? When a stream of cells encounters a traffic jam, the stream is deflected and turns to glide over or around the jam. In effect, the stream of cells sees the traffic jam as a particle that blocks its way and each cell bends as it glides over or around the particle. The elastic wall of *M. xanthus* can temporarily adopt a bent shape, as it adheres to the particle surface [71]. Persistence of that shape traps the cell causing it to follow a circular orbit over the surface of the aggregate. The cells continue to stream in orbit, they continue to C-signal, and more positive feedback raises the C-signal level. Ultimately this population of cells arrives at a still higher threshold, the threshold for spore differentiation that is set by the *dev* operon, [72]. The sporulation threshold is also observed experimentally [63, 68, 69]. When rod cells differentiate into spherical spores, they lose the capacity to propel themselves [50]. Although spores can no longer glide, it is supposed that adjacent motile rod-shaped cells push them out of the way and into the low-cell-density center of the, now spherical aggregate [73, 74]. Although the movements of fruiting body morphogenesis can be explained qualitatively by C-signaling, more experiments are needed to reveal the molecular mechanics of the Act and Dev proteins, and to do so quantitatively.

3.10
C-Signal Control of Gene Expression

The level of expression of several hundred (among the 7400 total) genes in *M. xanthus* increases during fruiting body development. Dr. Mitchell Singer is analyzing the expression of all developmentally regulated genes using whole genome DNA microarrays at UC Davis. A sampling of genes whose expression increased more than threefold during development had been obtained with the transposable promoter probe, Tn5 *lac*, were investigated [75, 76]. The sample showed that all of the developmentally upregulated genes belonged to one of three classes: one class depended only on starvation, a second class depended on starvation and A-signal, and a third on starvation, A-signal, and C-signal [39, 77]. Also, MrpC, a transcription factor of the cAMP receptor family [78–80], is needed for the aggregation response to A-signal [81]. The upstream regions of several developmentally regulated genes in the Tn5 *lac* sample have been cloned and segmentally deleted to reveal their upstream activation and repression sites, at which transcription factors, including sigma factors might bind. That search has revealed a cascade of 13 signal-responsive transcription factors other than sigma factors [81]. Unlike *Bacillus subtilis* with its cascade of sigma factors, only σ^D changes in *M. xanthus* – it falls when (p)ppGpp increases [82].

In general it is thought that the synthesis of each developmental protein is regulated in time and in the proper amount to maximize the number of environmentally resistant spores produced by starving cells. To see how that task might be actually be accomplished with a cascade of transcription factors, consider that 53 genes in the *M. xanthus* genome encode enhancer binding proteins (EBPs). They are specialized transcription factors that interact with RNA polymerase at σ^{54} promoters. EBPs bind regulatory DNA sequences upstream or downstream from σ^{54} promoters that are called enhancers [83]. Starting with EBP deletion mutations, mutants for at least 20 EBPs have been shown to exhibit aberrant fruiting body development [46, 63, 78, 79, 84–90]. Many EBPs are part of signal transduction circuits that respond to environmental cues. EBPs have an almost universal domain organization, in which a central AAA-ATPase domain is responsible for ATP hydrolysis and interaction with σ^{54}, the C-terminal domain binds DNA, and at the N-terminus a sensory domain regulates the ATPase activity of the central domain. Thus, the promoter is opened for transcription only when some sensory condition is satisfied. The N-terminal sensory domains show the most variation from one EBP to another and two large groups of N-terminal sequences are recognized. One has a response regulator sensory domain suggesting that they are each part of a two-component system. The way that EBPs fit into the cascade of transcription factors has recently been analyzed by Kroos [81].

A second group of 12 EBPs in *M. xanthus* have a forkhead-associated (FHA) domain at their sensory N-termini. These EBPs share three specific amino acid motifs: G69–R70, S85–XX–H88, and N107–G108, and they are orthologs of the prototypical RAD53^{FHA1} from yeast (GenBank A39616). The FHA domain in RAD53 has been shown to recognize a phosphothreonine moiety and is thought to interact with a protein partner in a process regulated by reversible protein phosphorylation. In particular, the N-terminal FHA domain in the MXAN4885 deletion mutant strongly suggests that it interacts with an autophosphorylated serine/threonine protein kinase [81, 91, 92]. The experimental data suggest that the normal function of MXAN4885 EBP is to adjust the level of FruA~P to track the increasing number of C-signal molecules on the signal donor, with precision. This holds particular current interest because the *M. xanthus* genome encodes at least 99 different serine/threonine protein kinases [1], whose regulatory functions are just beginning to be understood [93]. Pérez *et al.* compare the number and properties of the serine/threonine protein kinases in the sequenced myxobacteria with other sequenced bacteria [94]. *M. xanthus* has 99 different serine/threonine protein kinases, *S. aurantiaca* has 194 [45], *Sorangium cellulosum* has 317 [95]. The numbers suggest that the myxobacteria are exploring the usefulness of specific serine/threonine protein kinases for the sensory coordination of its multicellular activities.

Both groups of EBPs allow the level of target gene expression to be adjusted with precision over a wide dynamic range by signal input to EBPs in the transcription factor cascade. The timing of gene expression is thus controlled either by a two-component system or what might be called a three-component system, which is an EBP that has a FHA domain at its N-terminus and therefore has at least one additional sensory input ahead of the first two-component sensor. The resulting three-component system integrates multiple sensory inputs to control gene expression. A major challenge is specifying the critical sensory inputs and understanding how they fit into the program that enables an elementary multicellular organism to survive.

References

1. Goldman, B.S., Nierman, W.C., Kaiser, D., Slater, S.C., Durkin, A.S., Eisen, J.A., Ronning, C.M. *et al.* (2006) Evolution of sensory complexity recorded in a myxobacterial genome. *Proc. Natl. Acad. Sci. USA*, **103**, 15200–15205.
2. Reichenbach, H. (1993) Biology of the myxobacteria: ecology and taxonomy, in *Myxobacteria II* (eds M. Dworkin and D. Kaiser), ASM Press, Washington, DC, pp. 13–62.
3. Burchard, R.P. (1981) Gliding motility of prokaryotes: ultrastructure, physiology, and genetics. *Annu. Rev. Microbiol.*, **35**, 497–529.
4. Nudleman, E. and Kaiser, D. (2004) Pulling together with type IV pili. *J. Mol. Microbiol. Biotechnol.*, **7**, 52–62.
5. Craig, L., Pique, M., and Tainer, J.A. (2004) Type IV pilus structure and pathogenicity. *Nat. Rev. Microbiol.*, **2**, 363–378.
6. Craig, L., Volkmann, N., Arvai, A., Pique, M., Yeager, M., Engleman, E., and Tainer, J.A. (2006) Type IV pilus structure by cryo-electron microscopy and crystallography: implications for pilus assembly and functions. *Mol. Cell*, **23**, 651–662.
7. Arnold, J.W. and Shimkets, L.J. (1988) Cell surface properties correlated with cohesion in *Myxococcus xanthus*. *J. Bacteriol.*, **170**, 5771–5777.
8. Behmlander, R.M. and Dworkin, M. (1994) Biochemical and structural analyses of the extracellular matrix fibrils of *Myxococcus xanthus*. *J. Bacteriol.*, **176**, 6295–6303.
9. Chang, B.Y. and Dworkin, M. (1994) Isolated fibrils rescue cohesion and development in the Dsp mutant of *Myxococcus xanthus*. *J. Bacteriol.*, **176**, 7190–7196.
10. Satyshur, K., Worzalla, G., Meyer, L., Heiniger, E., Aukema, K., Misic, A., and Forest, K.T. (2007) Crystal structures of the pilus retraction motor PilT suggests large domain movements and subunit cooperation drive motility. *Structure*, **15**, 363–376.
11. Maier, B., Potter, L., So, M., Seifert, H.S., and Sheetz, M.P. (2002) Single pilus motor forces exceed 100 pN. *Proc. Natl. Acad. Sci. USA*, **99**, 16012–16017.
12. Nudleman, E., Wall, D., and Kaiser, D. (2005) Cell-to-cell transfer of bacterial outer-membrane lipoproteins. *Science*, **309**, 125–127.
13. Nudleman, E., Wall, D., and Kaiser, D. (2006) Polar assembly of the type IV pilus secretin in *Myxococcus xanthus*. *Mol. Microbiol.*, **60**, 16–29.
14. Kaiser, A.D. (1979) Social gliding is correlated with the presence of pili in *Myxococcus xanthus*. *Proc. Natl. Acad. Sci. USA*, **76**, 5952–5956.
15. Mignot, T., Merlie, J.P., and Zusman, D. (2005) Regulated pole-to-pole oscillations of a bacterial gliding motility protein. *Science*, **310**, 855–857.
16. Wolgemuth, C., Hoiczyk, E., Kaiser, D., and Oster, G. (2002) How myxobacteria glide. *Curr. Biol.*, **12**, 369–377.
17. Yu, R. and Kaiser, D. (2007) Gliding motility and polarized slime secretion. *Mol. Microbiol.*, **63**, 454–467.
18. Raetz, C.R.H. and Whitfield, C. (2002) Lipopolysaccharide endotoxins. *Annu. Rev. Biochem.*, **71**, 635–700.
19. Kaiser, D. (2008) *Myxococcus* – from single cell polarity to complex multi-cellular patterns. *Annu. Rev. Genet.*, **42**, 109–130.

20 Mignot, T., Shaevitz, J., Hartzell, P., and Zusman, D. (2007) Evidence that focal adhesions power bacterial gliding motility. *Science*, **315**, 853–856.

21 Kaiser, A.D. and Crosby, C. (1983) Cell movement and its coordination in swarms of *Myxococcus xanthus*. *Cell Motil.*, **3**, 227–245.

22 Wu, Y., Kaiser, D., Jiang, Y., and Alber, M. (2009) Periodic reversal of direction allows myxobacteria to swarm. *Proc. Natl. Acad. Sci. USA*, **106**, 1222–1227.

23 Wu, Y., Kaiser, D., Jiang, Y., and Alber, M. (2007) Social interactions in myxobacterial swarming. *PLoS Comp. Biol.*, **31**, e253.

24 Ward, M.J., Lew, H., Treuner-Lange, A., and Zusman, D.R. (1998) Regulation of motility behavior in *Myxococcus xanthus* may require an extracytoplasmic-function sigma factor. *J. Bacteriol.*, **180**, 5668–5675.

25 Blackhart, B.D. and Zusman, D. (1985) Frizzy genes of *Myxococcus xanthus* are involved in control of frequency of reversal of gliding motility. *Proc. Natl. Acad. Sci. USA*, **82**, 8767–8770.

26 Acuna, G., Shi, W., Trudeau, K., and Zusman, D. (1995) The cheA and cheY domains of *Myxococcus xanthus* FrzE function independently *in vitro* as an autokinase and a phosphate acceptor, respectively. *FEBS Lett.*, **358**, 31–33.

27 Igoshin, O., Goldbetter, A., Kaiser, D., and Oster, G. (2004) A biochemical oscillator explains the developmental progression of myxobacteria. *Proc. Natl. Acad. Sci. USA*, **101**, 15760–15765.

28 Bustamante, V.H., Martínez-Flores, I., Vlamakis, H.C., and Zusman, D. (2004) Analysis of the Frz signal transduction system of *Myxococcus xanthus* shows the importance of the conserved C-terminal region of the cytoplasmic chemoreceptor FrzCD in sensing signals. *Mol. Microbiol.*, **53**, 1501–1513.

29 Hartzell, P.L. (1997) Complementation of sporulation and motility defects in a prokaryote by a eukaryotic GTPase. *Proc. Natl. Acad. Sci. USA*, **94**, 9881–9886.

30 Koonin, E. and Aravind, L. (2001) Dynein light chains of the roadblock/LC7 group belong to an ancient protein superfamily implicated in NTPase regulation. *Curr. Biol.*, **10**, R774–776.

31 Spormann, A.M. and Kaiser, D. (1999) Gliding mutants of *Myxococcus xanthus* with high reversal frequencies and small displacements. *J. Bacteriol.*, **181**, 2593–2601.

32 Kaiser, D. (2008) Reversing *Myxococcus xanthus* polarity, in *Myxobacteria: Multicellularity and Differentiation* (ed. D.E. Whitworth), ASM Press, Washington, DC, pp. 93–102.

33 Leonardy, S., Freymark, G., Hebener, S., Ellehauge, E., and Søgaard-Andersen, L. (2007) Coupling of protein localization and cell movement by a dynamically localized response regulator in *Myxococcus xanthus*. *EMBO J.*, **26**, 4433–4444.

34 Jelsbak, L. and Søgaard-Andersen, L. (1999) The cell-surface associated C-signal induces behavioral changes in individual *M. xanthus* cells during fruiting body morphogenesis. *Proc. Natl. Acad. Sci. USA*, **96**, 5031–5036.

35 Nariya, H. and Inouye, M. (2008) MazF, an mRNA interferase, mediates programmed cell death during multicellular *Myxococcus* development. *Cell*, **132**, 55–66.

36 Singer, M. and Kaiser, D. (1995) Ectopic production of guanosine penta- and tetra-phosphate can initiate early developmental gene expression in *Myxococcus xanthus*. *Genes Dev.*, **9**, 1633–1644.

37 Kaplan, H.B. and Plamann, L. (1996) A *Myxococcus xanthus* cell density-sensing system required for multicellular development. *FEMS Microbiol. Lett.*, **139**, 89–95.

38 Plamann, L. and Kaplan, H.B. (1999) Cell-density sensing during early development in *Myxococcus xanthus*, in *Cell–Cell Signaling in Bacteria* (eds G.M. Dunny and S.C. Winans), ASM Press, Washington, DC, pp. 67–82.

39 Kuspa, A., Kroos, L., and Kaiser, D. (1986) Intercellular signaling is required for developmental gene expression in *Myxococcus xanthus*. *Dev. Biol.*, **117**, 267–276.

40 Kaiser, D. and Welch, R. (2004) Dynamics of fruiting body morphogenesis. *J. Bacteriol.*, **186**, 919–927.

41 Kuner, J. and Kaiser, D. (1982) Fruiting body morphogenesis in submerged cultures of *Myxococcus xanthus*. *J. Bacteriol.*, **151**, 458–461.

42 Sozinova, O., Jang, Y., Kaiser, D., and Alber, M.S. (2005) Three-dimensional model of myxobacterial aggregation by contact-mediated interaction. *Proc. Natl. Acad. Sci. USA*, **102**, 11308–11312.

43 Lev, M. (1954) Demonstration of a diffusible fruiting factor in Myxobacteria. *Nature*, **173**, 501.

44 McVittie, A. and Zahler, S.A. (1962) Chemotaxis in *Myxococcus*. *Nature*, **194**, 1299–1300.

45 Ronning, C.M. and Nierman, W.C. (2008) The genomes of *Myxococcus xanthus* and *Stigmatella aurantiaca*, in *Myxobacteria: Multicellularity and Differentiation* (ed. D.E. Whitworth), ASM Press, Washington, DC, pp. 285–298.

46 Kirby, J.R. and Zusman, D.R. (2003) Chemosensory regulation of developmental gene expression in *Myxococcus xanthus*. *Proc. Natl. Acad. Sci. USA*, **100**, 2008–2013.

47 Sliusarenko, O., Zusman, D.R., and Oster, G. (2007) Aggregation during fruiting body formation in *Myxococcus xanthus* is driven by reducing cell movement. *J. Bacteriol.*, **189**, 611–619.

48 Jelsbak, L. and Søgaard-Andersen, L. (2000) Pattern formation: fruiting body morphogenesis in *Myxococcus xanthus*. *Curr. Opin. Microbiol.*, **3**, 637–642.

49 Jelsbak, L. and Søgaard-Andersen, L. (2002) Pattern formation by a cell-surface associated morphogen in *M. xanthus*. *Proc. Natl. Acad. Sci. USA*, **99**, 2032–2037.

50 Sager, B. and Kaiser, D. (1993) Spatial restriction of cellular differentiation. *Genes Dev.*, **7**, 1645–1653.

51 Sproer, C., Reichenbach, H., and Stackebrandt, E. (1999) Correlation between morphological and phylogenetic classification of myxobacteria. *Int. J. Syst. Bacteriol.*, **49**, 1255–1262.

52 Bonner, J.T. (1982) Evolutionary strategies and developmental constraints in the cellular slime molds. *Am. Nat.*, **119**, 530–552.

53 Reichenbach, H. (1984) Myxobacteria: a most peculiar group of social prokaryotes, in *Myxobacteria* (ed. E. Rosenberg), Springer, New York, pp. 1–50.

54 Velicer, G., Kroos, L., and Lenski, R.E. (1998) Loss of social behaviors by *Myxococcus xanthus* during evolution in an unstructured habitat. *Proc. Natl. Acad. Sci. USA*, **95**, 12376–12380.

55 Velicer, G.J. (2003) Social strife in the microbial world. *Trends Microbiol.*, **11**, 330–337.

56 Hagen, D.C., Bretscher, A.P., and Kaiser, D. (1978) Synergism between morphogenetic mutants of *Myxococcus xanthus*. *Dev. Biol.*, **64**, 284–296.

57 Shimkets, L.J., Gill, R.E., and Kaiser, D. (1983) Developmental cell interactions in *Myxococcus xanthus* and the *spoC* locus. *Proc. Natl. Acad. Sci. USA*, **80**, 1406–1410.

58 Kim, S.K. and Kaiser, D. (1990) Purification and properties of *Myxococcus xanthus* C-factor, an intercellular signaling protein. *Proc. Natl. Acad. Sci. USA*, **87**, 3635–3639.

59 Kim, S.K. and Kaiser, D. (1990) Cell alignment required in differentiation of *Myxococcus xanthus*. *Science*, **249**, 926–928.

60 Sager, B. and Kaiser, D. (1994) Intercellular C-signaling and the traveling waves of *Myxococcus*. *Genes Dev.*, **8**, 2793–2804.

61 Sozinova, O., Jang, Y., Kaiser, D., and Alber, M. (2006) A three-dimensional model of myxobacterial fruiting body formation. *Proc. Natl. Acad. Sci. USA*, **103**, 17255–17259.

62 Ellehauge, E., Norregaard-Madsen, M., and Søgaard-Andersen, L. (1998) The FruA signal transduction protein provides a checkpoint for the temporal coordination of intercellular signals in *M. xanthus* development. *Mol. Microbiol.*, **30**, 807–813.

63 Gronewold, T.M.A. and Kaiser, D. (2001) The *act* operon controls the level and time of C-signal production for *M. xanthus* development. *Mol. Microbiol.*, **40**, 744–756.

64 Lobedanz, S. and Søgaard-Andersen, L. (2003) Identification of the C-signal, a contact-dependent morphogen coordinating multiple developmental responses in *Myxococcus xanthus*. *Genes Dev.*, **17**, 2151–2161.

65 Søgaard-Andersen, L. (2008) Contact-dependent signaling in *Myxococcus xanthus*: the function of the C-signal in fruiting body morphogenesis, in *Myxobacteria, Multicellularity and Differentiation* (ed. D.E. Whitworth), ASM Press, Washington, DC, pp. 77–91.

66 Rolbetski, A., Ammon, M., Jakovljevic, V., Konovalova, A., and Søgaard-Andersen, L. (2008) Regulated secretion of a protease activity activates intercellular signaling during fruiting body formation in M. xanthus. Dev. Cell, **15**, 627–634.

67 Welch, R. and Kaiser, D. (2001) Cell behavior in traveling wave patterns of myxobacteria. Proc. Natl. Acad. Sci. USA, **98**, 14907–14912.

68 Kim, S.K. and Kaiser, D. (1991) C-factor has distinct aggregation and sporulation thresholds during Myxococcus development. J. Bacteriol., **173**, 1722–1728.

69 Li, S., Lee, B.U., and Shimkets, L. (1992) csgA expression entrains Myxococcus xanthus development. Genes Dev., **6**, 401–410.

70 Kruse, T., Lobendanz, S., Bertheleson, N.M.S., and Søgaard-Andersen, L. (2001) C-signal: a cell surface-associated morphogen that induces and coordinates multicellular fruiting body morphogenesis and sporulation in M. xanthus. Mol. Microbiol., **40**, 156–168.

71 Wolgemuth, C. (2005) Force and flexibility of flailing myxobacteria. Biophys. J., **89**, 945–950.

72 Thöny-Meyer, L. and Kaiser, D. (1993) devRS, an autoregulated and essential genetic locus for fruiting body development in Myxococcus xanthus. J. Bacteriol., **175**, 7450–7462.

73 Sager, B. and Kaiser, D. (1993) Two cell-density domains within the Myxococcus xanthus fruiting body. Proc. Natl. Acad. Sci. USA, **90**, 3690–3694.

74 Lux, R., Li, Y., Lu, A., and Shi, W. (2005) Detailed three-dimensional analysis of structural features of Myxococcus xanthus fruiting bodies using confocal laser scanning microscopy. Biofilms, **1**, 293–303.

75 Kroos, L. and Kaiser, D. (1984) Construction of Tn5 lac, a transposon that fuses lacZ expression to exogenous promoters, and its introduction into Myxococcus xanthus. Proc. Natl. Acad. Sci. USA, **81**, 5816–5820.

76 Kroos, L., Kuspa, A., and Kaiser, D. (1986) A global analysis of developmentally regulated genes in Myxococcus xanthus. Dev. Biol., **117**, 252–266.

77 Kroos, L. and Kaiser, D. (1987) Expression of many developmentally regulated genes in Myxococcus depends on a sequence of cell interactions. Genes Dev., **1**, 840–854.

78 Sun, H. and Shi, W. (2001) Analyses of mrp genes during Myxococcus xanthus development. J. Bacteriol., **183**, 6733–6739.

79 Sun, H. and Shi, W. (2001) Genetic studies of mrp, a locus essential for cellular aggregation and sporulation of Myxococcus xanthus. J. Bacteriol., **183**, 4786–4795.

80 Higgs, P.I., Jagadeesan, S., Mann, P., and Zusman, D.R. (2008) EspA, an orphan hybrid histidine protein kinase, regulates the timing of expression of key developmental proteins of Myxococcus xanthus. J. Bacteriol., **190**, 4416–4426.

81 Kroos, L. (2007) The Bacillus and Myxococcus developmental networks and their transcriptional regulators. Annu. Rev. Genet., **41**, 13–39.

82 Viswanathan, P. and Kroos, L. (2006) Role of sigma^D in regulating genes and signals during Myxococcus xanthus development. J. Bacteriol., **188**, 3246–3256.

83 Buck, M., Gallegos, M.T., Studholme, D.J., Guo, Y., and Gralla, J.D. (2000) The bacterial enhancer-dependent sigma^{54} (sigma^N) transcription factor. J. Bacteriol., **182**, 4129–4136.

84 Jakobsen, J.S., Jelsbak, L., Jelsbak, L., Welch, R., Cummings, C., Goldman, B., Stark, E. et al. (2004) Sigma^{54} enhancer binding proteins and Myxococcus xanthus fruiting body development. J. Bacteriol., **186**, 4361–4368.

85 Gorski, L. and Kaiser, D. (1998) Targetted mutagenesis of sigma-54 activator proteins in Myxococcus xanthus. J. Bacteriol., **180**, 5896–5905.

86 Guo, D., Wu, Y., and Kaplan, H.B. (2000) Identification and characterization of genes required for early Myxococcus xanthus developmental gene expression. J. Bacteriol., **182**, 4564–4571.

87 Hager, E., Tse, H., and Gill, R.E. (2001) Identification and characterization of spdR mutations that bypass the BsgA protease-dependent regulation of developmental gene expression in Myxococcus xanthus. Mol. Microbiol., **39**, 765–780.

88 Caberoy, N.B., Welch, R.D., Jakobsen, J.S., Slater, S.C., and Garza, A.G. (2003) Global mutational analysis of NtrC-like activators in Myxococcus xanthus: identifying

activator mutants defective for motility and fruiting body development. *J. Bacteriol.*, **185**, 6083–6094.

89 Diodati, M., Ossa, F., Caberoy, N.B., Jose, I.R., Hiraiwa, W., Igo, M.M., Singer, M. *et al.* (2006) Nla18, a key regulatory protein required for normal growth and development of *Myxococcus xanthus*. *J. Bacteriol.*, **188**, 1733–1743.

90 Ossa, F., Diodati, M., Caberoy, N.B., Giglio, K., Edmunds, M., Singer, M., and Garza, A.G. (2007) The *Myxococcus xanthus* Nla4 protein is important for expression of stringent response-associated genes, ppGpp accumulation, and fruiting body development. *J. Bacteriol.*, **189**, 8474–8483.

91 Jelsbak, L., Givskov, M., and Kaiser, D. (2005) Enhancer-binding proteins with a forkhead-associated domain and the sigma54 regulon in *Myxococcus xanthus* fruiting body development. *Proc. Natl. Acad. Sci. USA*, **102**, 3010–3015.

92 Kroos, L. (2005) Eukaryotic-like signaling and gene regulation in a prokaryote that undergoes multicellular development. *Proc. Natl. Acad. Sci. USA*, **102**, 2681–2682.

93 Inouye, S., Jain, R., Ueki, T., Nariya, H., Xu, C., Hsu, M., Fernandez-Luque, B.A. *et al.* (2000) A large family of eukaryotic-like protein Ser/Thr kinases of *Myxococcus xanthus*, a developmental bacterium. *Microb. Comp. Genomics*, **5**, 103–120.

94 Pérez, J., Castañeda-García, A., Jenke-Kodama, H., Müller, R., and Muñoz-Dorado, J. (2008) Eukaryotic-like protein kinases in the prokaryotes and the myxobacterial kinome. *Proc. Natl. Acad. Sci. USA*, **105**, 15950–15955.

95 Schneiker, S., Perlova, O., Alici, A., Altmeyer, M.O., Bartels, D., Bekel, T., Beyer, S. *et al.* (2007) Complete sequence of the largest known bacterial genome from the myxobacterium *Sorangium cellulosum*. *Nat. Biotechnol.*, **25**, 1281–1289.

96 Ogawa, M., Fujitani, S., Mao, X., Inouye, S., and Komano, T. (1996) FruA, a putative transcription factor essential for the development of *Myxococcus xanthus*. *Mol. Microbiol.*, **22**, 757–767.

97 Viswanathan, P., Ueki, T., Inouye, S., and Kroos, L. (2007) Combinatorial regulation of genes essential for *Myxococcus xanthus* development involves an upstream response regulator and a downstream LysR-type regulator. *Proc. Natl. Acad. Sci. USA*, **104**, 7969–7974.

4
Communication Between Rhizobia and Plants
Michael Göttfert

4.1
Introduction

Rhizobia are Gram-negative soil bacteria able to enter symbiosis with many members of the family Leguminosae. Within this symbiosis bacteria fix atmospheric nitrogen. The fixed nitrogen is released into the plant cytosol and metabolized. The amount of nitrogen supplied can be as high as 300 kg N/ha/year [1]. The ability of the plant to grow without nitrogen fertilizer makes this symbiosis particularly attractive for agriculture. In Brazil, one of the world biggest soybean producers, most of the crop obtains nitrogen from biological nitrogen fixation [2]. Nitrogenase, the enzyme complex responsible for reduction of atmospheric nitrogen, is very oxygen-sensitive [3, 4]. Therefore, nitrogen-fixing aerobic bacteria had to develop strategies to protect the enzyme. On the other hand, nitrogen fixation is very energy demanding – at least 16 mol ATP are needed to reduce 1 mol molecular nitrogen. Rhizobia generate this ATP by respiration. Therefore, oxygen has to be supplied at a low but constant concentration. Root nodules are ideally equipped to meet both requirements. The cortex cells that surround the infected nodule tissue form a barrier against high oxygen concentrations. Within the infected cells leghemoglobin, a myoglobin-like compound, which reversibly binds oxygen, is present at high concentration. This ensures a constant supply of oxygen at a concentration of about 20 nM [4].

In the fully developed nodule bacteria, which are now called bacteroids, are surrounded by a plant-derived membrane (peribacteroid membrane) and located in the plant cytosol. In exchange for the released ammonia, bacteroids obtain all nutrients from the plant. The intimate contact of bacterium and plant, and the mutual dependency, requires that both organisms have to make sure that they find the proper partner. For this they have to communicate. This chapter summarizes our present knowledge of the signal exchange between legumes and rhizobia with emphasis on the bacterial side.

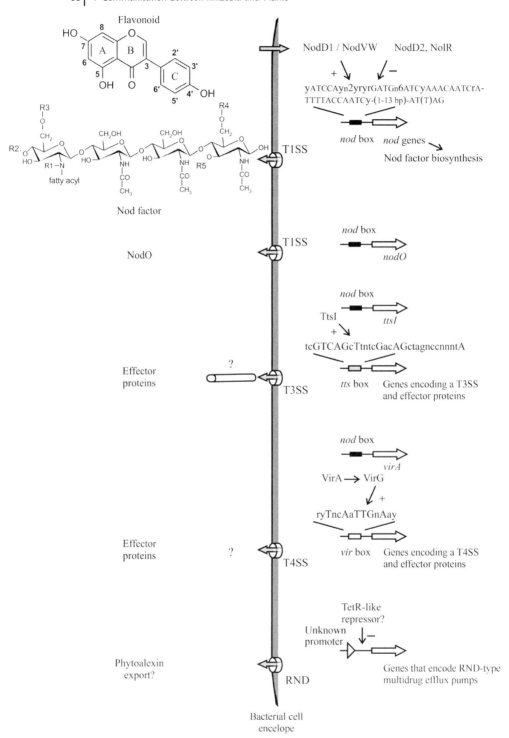

Figure 4.1 (see caption on page 59)

4.2
Nodulation (*nod*) Genes are Induced by Flavonoids and are Under Positive and Negative Regulation

In general, rhizobia infect their host plant in the region of developing root hairs. The rhizosphere is a nutrient-rich environment due to root exudate from the plant. This attracts many bacterial species including symbionts and pathogens. Root exudate also contains flavonoids. Flavonoids are secondary metabolites of the phenylpropanoid and acetate–malonate pathway with a large structural variability and diverse functions [5–7]. The basic structure is a 15-carbon skeleton forming two phenyl rings (A and B) connected by a C-ring (Figure 4.1). Flavonoids differ, for example, in ring position and hydroxylation pattern, and legumes differ in the flavonoid cocktail they release. Table 4.1 lists a few flavonoids that are released from roots or germinating seeds. Rhizobia perceive these flavonoids as specific signal. Luteolin (5,7,3′,4′-tetrahydroxyflavone), isolated from alfalfa seeds, acts as potent signal for *Sinorhizobium meliloti*, the symbiont of alfalfa. Genistein (5,7,4′-trihydroxyisoflavone; Figure 4.1), released by soybean roots, is recognized by *Bradyrhizobium japonicum*, the suitable symbiont. The presence of a symbiont may lead to an increased synthesis and release of flavonoids from the plant [8]. However, root exudate of a host plant may also contain flavonoids that lead to repression of *nod* genes. Alfalfa, for example, releases coumestrol, which is an anti-inducer of *nod* gene expression in *S. meliloti* [9].

It is still not completely understood how flavonoids meet the matching regulatory proteins in rhizobia. No uptake system has been described for flavonoids. Flavonoids that are known as signal molecules have a low solubility in water (low micromolar range) and might accumulate in the cytoplasmic membrane. This was shown for 5,7,4′-trihydroxyflavanone, which is an inducer for *nod* genes in *Rhizobium leguminosarum* bv. *viciae* [10]. For this strain, it was suggested that the cytoplasmic membrane is the site where naringenin binds to NodD [11]. However, it is unclear how NodD would be sequestered to the membrane. Because flavonoids act as inducers at very low concentrations in the nanomolar range, their concentration in the cytoplasm might reach a concentration that is high enough for induction. Thus, further research is needed to locate the (in)active NodD in the cell.

Figure 4.1 Flavonoid-inducible genes in the rhizobia–legume interaction. Almost all rhizobia encode *nod* genes that are inducible by flavonoids. A characteristic motif in promoter regions of *nod* genes is the *nod* box. *nod* box motifs were also found upstream of regulatory genes that activate protein secretion systems. Microarray experiments indicate that flavonoids also induce the expression of RND efflux systems. See text for further details. Open arrows symbolize one or more genes, that might be organized in several operons. Less well-conserved nucleotides in the indicated motifs are shown in lower case. Examples of Nod factor modifications (R1–R5) are given in Table 4.3. Question marks indicate that it is unknown how secreted effector proteins enter plant cells. In the case of rhizobial T3SSs extracellular pili structures were found. Abbreviations: T1SS, type I secretion system (see text for other abbreviations); nucleotide abbreviations: r, purine; y, pyrimidine; n, any nucleotide. + positive regulation; − negative regulation.

Table 4.1 Representative rhizobial strains, host plants, and released flavonoids.

Species/genome or replicon size (kb)[a]	Host[b]	Flavonoid present in root or seed exudate[c]	Reference
A. caulinodans/5369	Sesbania rostrata	7,4'-dihydroxyflavanone	[92, 93]
B. japonicum/9105	soybean	5,7,4'-trihydroxyisoflavone; 7,4'-dihydroxyisoflavone	[94, 95]
Bradyrhizobium (Arachis) sp.	peanut	5,7,4'-trihydroxyisoflavone	[96]
M. loti/7036,351,208	Lotus japonicus	—	[60]
Rhizobium sp. strain NGR234	cowpea	5,7,4'-trihydroxyisoflavone; 7,4'-dihydroxyisoflavone	[97]
R. etli CFN 42/4381,642, 505,371,250,194,184	common bean	5,7,4'-trihydroxyflavanone; 5,7,4'-trihydroxyisoflavone	[98, 99]
R. leguminosarum bv. viciae/5057,870,684,488, 352,151,147	vetch	3,5,7,3'-tetrahydroxy-4'-methoxyflavanone; 7,3'-dihydroxy-4'-methoxyflavanone	[100, 101]
R. leguminosarum bv. trifolii	clover	7,4'-dihydroxyflavone; 7,4'-dihydroxy-3'-methoxyflavone	[102]
S. fredii	soybean	5,7,4'-trihydroxyisoflavone	[95]
S. meliloti/3654,1683,1354	alfalfa	5,7,3',4'-tetrahydroxyflavone; 4,4'-dihydroxy-2'-methoxychalcone; 7,4'-dihydroxyflavone	[103–105]

a) Genome sizes were taken from http://bacteria.kazusa.or.jp/rhizobase. The sequenced A. caulinodans and B. japonicum genomes consist of one circular chromosome; other species may contain a variable set of plasmids.
b) Host ranges of rhizobia are very different. Rhizobium sp. NGR234 is able to infect legumes from at least 112 genera[106]; A. caulinodans, in contrast, has a narrow host range ([92] and further references therein).
c) Only selected flavonoids are listed. In cases where the flavonoid inducer from the host plant is unknown, it is possible to introduce a characterized nodD gene into the rhizobial strain and to use the matching flavonoid to study Nod factor production.

NodD is a typical regulator of the LysR family [12]. The N-terminal 65-amino-acid residues, which contain a helix–turn–helix DNA-binding domain, are well conserved within this family. For signal recognition, two central domains critical for flavonoid-binding are predicted [12, 13]. The NodD copies of different bacteria have a similar length of about 314 amino acids and in general share an amino acid identity of at least 60%. Despite this strong conservation, individual NodD proteins accept different flavonoids as inducers [14, 15]. A flavonoid that activates NodD from one rhizobial strain may block the activity of a NodD protein from a different strain. So far – based on the amino acid sequence of NodD – it is not possible to predict which flavonoid might act as inducer.

NodD binds to the *nod* box as a tetramer and protects a DNA region from about −20 to −75 with respect to the transcriptional start site [16, 17]. Binding of NodD takes place even in the absence of inducer. However, for S. meliloti it was shown that luteolin increases the binding affinity [18, 19]. The *nod* box consists of three highly conserved regions that encompass almost 50 nucleotides and a fourth more variable region at the 3′-end (Figure 4.1). In *B. japonicum* and *S. meliloti* transcription starts about 9–12 nucleotides downstream of this last conserved element [20, 21]. The *nod* box is present in the promoter region of all genes that are essential for Nod factor biosynthesis. This ensures that these genes are coordinately expressed.

Although NodD was found in all investigated strains other transcriptional activators might be important. In *B. japonicum*, a *nodD* deletion derivative is still nodulation competent. In this mutant, *nod* genes are activated by the two-component regulatory system NodVW. Mutants defective in NodW are no longer able to infect the three host plants *Vigna unguiculata*, *Vigna radiata*, and *Macroptilium atropurpureum* [22]. NodV encodes a sensor protein with a highly hydrophobic N-terminal domain. Therefore, NodV is likely to be membrane-anchored. NodW contains the receiver and output domains that are typical for this regulator family [23]. In the presence of genistein, NodV phosphorylates NodW [24]. It is still an open question how NodV senses genistein and if other flavonoids may also act as signal molecules for NodV. Likewise, the DNA target site for NodW is unknown. Electrophoretic mobility shift assays that would prove binding to the *nod* box promoter failed. It is also unclear how NodW and NodD might interact at the same promoter.

Most rhizobial species characterized so far contain several nonidentical *nodD* genes. In this case, the individual proteins respond differently to flavonoids. They may act as transcriptional activators or as repressors. In *B. japonicum*, NodD2 was characterized as a repressor [25]. In *S. meliloti*, all three NodD proteins are activators [26]; negative regulation is exerted by NolR [27]. NolR is a member of the ArsR subfamily. ArsR acts as a metal-regulated homodimeric repressor, which dissociates from DNA in the presence of metal ions [28]. In *S. meliloti*, NolR is a global regulator, which is present in several other but not all rhizobial strains [27]. In the case of NolR-regulated nodulation genes, the binding site of NolR is overlapping with the *nod* box promoter [29].

Even though flavonoids are the best characterized and probably the most important inducers of *nod* genes, other molecules might also affect *nod* gene expression. Trigonelline and stachydrine, which are present in alfalfa seed rinses, activate NodD2 of *S. meliloti* [30]. Aldonic acids, present in rood exudates of *Lotus albus*, have the capacity to activate *nod* genes [31]. Jasmonic acid and methyl jasmonate trigger Nod factor synthesis in soybean [32]. In the latter study it was shown that genistein and jasmonates act synergistically, which is quite interesting also because of the very different structures of the inducer compounds. The underlying mechanism of this synergism is not yet understood.

4.3
Activation of the *nod* Genes Results in the Synthesis and Export of Lipo-Chito-Oligosaccharide Signal Molecules

Nodulation genes are essential for the synthesis of Nod factors (Table 4.2). Although each rhizobial species secretes its own set of Nod factors, they share a common basic structure (Figure 4.1) [33]. They consist of a backbone of three to six β1–4-*N*-acetyl-

Table 4.2 Functions of selected nodulation genes.

Gene[a]	Function of encoded protein
nodA[b]	fatty acyl transferase; together with NodB attachment of the fatty acyl chain to the sugar backbone
nodB[b]	deacetylase; see nodA
nodC[b]	N-acetyl-glucosaminyl transferase; synthesis of the Nod factor backbone
nodD[b]	LysR-type regulator; activation of nod genes; if a strain encodes several NodD proteins, they may also act as repressors
nodE	β-ketoacyl synthase; together with NodF involved in synthesis of the fatty acyl chain; present, for example, in S. meliloti and R. leguminosarum
nodF	acyl carrier protein; see nodE
nodG	3-oxoacyl-acyl carrier protein reductase; involved in the synthesis of the fatty acyl chain; present, for example, in S. meliloti
nodH	sulfotransferase; sulfurylation of the Nod factor in combination with NodP and NodQ (R4)[c]; present, for example, in S. meliloti
nodI[b]	ABC transporter, ABC; together with NodJ secretion of the Nod factor
nodJ[b]	ABC transporter, membrane protein; see nodI
nodL	O-acetylation of the terminal nonreducing glucosaminyl residue (R3)[c]; present, for example, in R. leguminosarum
nodM	glucosamine synthetase; involved in backbone synthesis; the housekeeping glucosamine synthetase probably compensates for the lack of nodM in some rhizobial strains
nodO	Ca^{2+} binding; secreted ion channel forming protein; present, for example, in R. leguminosarum
nodP	ATP sulfurylase; see nodH
nodQ	APS kinase; see nodH
nodS	methyl transferase (R1)[c]; present, for example, in A. caulinodans
nodU	6-O-carbamoylation of the terminal nonreducing glucosaminyl residue (R3)[c]; present, for example, in A. caulinodans
nodV	two-component regulator family, sensor protein; present in B. japonicum
nodW	two-component regulator family, regulator protein; present in B. japonicum
nodX	6-O-acetylation at the reducing N-acetyl-glucosaminyl residue (R4)[c]; present in R. leguminosarum bv. viciae strain TOM
nodZ	fucosyl transferase; fucosylation of the terminal reducing glucosaminyl residue (R4)[c]; present, for example, in B. japonicum
nolK	GDP-fucose synthesis; present, for example, in A. caulinodans
nolL	O-acetylation of the fucose residue at the terminal reducing glucosaminyl residue (R4)[c]; present, for example, in Rhizobium sp. NGR234
nolO	carbamoylation at the nonreducing terminus (R2)[c]; present, for example, in Rhizobium sp. NGR234
nolR	repressor of nod genes; present, for example, in S. meliloti
noeC	involved in arabinosylation (R5)[c]; present in A. caulinodans

a) The list of nodulation genes is not complete. The genes initially received the designation nod followed by an alphabetic character; thereafter, nodZ, nol, and noe followed by an alphabetic character were used.
b) These genes were found in all investigated strains.
c) R1–R5 refer to the modified position in the Nod factor as depicted in Figure 4.1.

4.3 Activation of the nod Genes Results in the Synthesis

Table 4.3 Examples of Nod factor structures[a].

Strain	Fatty acyl chain	R1	R2	R3	R4	R5	Reference
A. caulinodans ORS571	C16:0, C18:0, C18:1	Me	H	Cb, H	Fuc, Ara, H	Ara, H	[107, 108]
B. japonicum USDA110	C18:1	H	H	H	2-O-MeFuc	H	[109]
M. loti E1R	C18:0, C18:1	Me	Cb	H	4-O-AcFuc	H	[110]
R. etli CE3	C18:0, C18:1	Me	Cb	H	4-O-AcFuc	H	[111]
R. galegae HAMBI1207	C18:1, C18:2, C18:3, C20:2, C20:3	H	H	Cb, H	H	H	[112]
R. leguminosarum bv. viciae RBL5560	C18:4	H	H	Ac	H	H	[113]
S. meliloti RCR2011	C16:2	H	H	Ac, H	S	H	[51, 114]

a) A more comprehensive list of Nod factor structures is given in [33]. In general, a single species produces multiple Nod factors.
Abbreviations: Ac, acetyl; Ara, arabinosyl; Cb, carbamoyl; Fuc, fucosyl; Me, methyl; S, sulfate.
R1–R5 refer to Figure 4.1.

glucosaminyl residues, which is synthesized by NodC. A fatty acyl chain is always attached to the nonreducing residue of this oligosaccharide. This attachment is the concerted action of NodA and NodB. Strains differ in the attached fatty acyl chain and a single strain may also use different fatty acyl residues (Table 4.3). The *B. japonicum* Nod factor contains vaccenic acid (C18:1). *S. meliloti* synthesizes a Nod factor with an unsaturated C16:2 fatty acid. For the synthesis of these "nonstandard" fatty acids, *S. meliloti* uses the two genes *nodF* and *nodE* (Table 4.2). Modifications at other positions also require the presence of specialized *nod* genes. These decorations influence the host range of the corresponding bacterium. A *nodH* mutant of *S. meliloti* produces a Nod factor that is not sulfurylated [34]. As a consequence, the mutant is no longer able to nodulate its natural host alfalfa, but it gains the ability to induce (inefficient) nodules on vetch.

The Nod factor has to be secreted, which is the function of NodI and NodJ [35–37]. NodJ has the characteristics of an integral membrane protein with at least six transmembrane helices. NodI belongs to the family of ATP-binding cassette (ABC) transporter proteins. ABC transporters form a very versatile transporter family (type I secretion system), members of which might be involved in protein export, high-affinity uptake systems, or multidrug resistance [38]. Owing to the central function of the *nodIJ* genes, they have been identified in all characterized Nod factor-producing strains. However, Nod factors can be also detected at a very low level in the supernatant of a *nodJ* mutant [37], indicating that alternative transporters might export the signal as well. Most probably, Nod factors are required only early in infection. This is supported by the finding that most *nod* genes are not expressed in mature nodules [39, 40]. Why a few *nod* genes are expressed in some symbiosis, such as *nodA* of *Azorhizobium caulinodans* [41] or *nolY, nolZ,* and *noeI* of *B. japonicum* [42, 43], is not understood.

Once the Nod factors are secreted, they have to be recognized by the plant. The most likely candidates for Nod factor perception are LysM (*lysin motif*) domain-containing receptor-like kinases, such as NFR1 and NFR5 of *Lotus japonicus* [44], and NFP of *Medicago truncatula* [45, 46]. The N-terminal ends of these proteins contain a signal peptide followed by LysM domains. The LysM domain most likely recognizes *N*-acetyl-glucosamine-containing molecules [47]. The LysM domains are followed by a transmembrane motif, which anchors the protein in the cytoplasmic membrane. The C-terminal part has similarity to serine/threonine kinases. Mutants affected in these genes do not respond to inoculation or Nod factor treatment [44, 45]. In recent years a number of plant genes have been identified that are involved in root nodule development [48]. Interestingly, there is an overlap with the signaling cascade induced by mycorrhizal infections. Since mycorrhiza form the more ancient plant–microbe interaction, it is generally believed that the Nod factor signaling cascade derives from the mycorrhizal pathway.

4.4
Rhizobia use Secreted Proteins as Effector Molecules

NodO of *R. leguminosarum* bv. *viciae* was the first described secreted protein that affects symbiosis [49]. The gene is preceded by a *nod* box and inducible by naringenin [49]. Interestingly, in nodulation assays it can compensate for the absence of the *nodFE* genes [50], which are required for the synthesis of the unsaturated C16:2 and C16:3 acyl chains [51]. NodO has the ability to form ion channels in membranes [52]. If this activity affects Nod factor perception or the signaling cascade further downstream is unclear. NodO is one of several proteins in *R. leguminosarum* that is secreted by a type I secretion system [53]. Although the secreted proteins are dispensable for symbiosis, they might have auxiliary functions that are relevant under the competitive conditions in the field.

A type IV secretion system (T4SS) has been described in *Mesorhizobium loti* strain R7A [54]. Protein candidates to be transported by the T4SS are known. Mutants that contain insertions within the corresponding genes exhibit a nodulation delay [54]. The T4SS is regulated by the two-component regulatory system VirA and VirG [55]. The *virA* gene, which encodes the sensor component, is preceded by a *nod* box and inducible by flavonoids. VirA probably interacts with the regulator VirG, which then activates promoters that contain a *vir* box. Since no flavonoid inducer is known for *M. loti*, the authors introduced the *nodD* gene of *R. leguminosarum* bv. *viciae* into strain R7A. This resulted in a naringenin-inducible activation of the described signal cascade.

Another, very versatile protein transport machinery is the type III secretion system (T3SS). T3SSs were first identified in pathogenic bacteria [56–58]. They are used to transport effector proteins directly into the plant cell. T3SSs can be easily recognized by their conserved core components, which are involved in the formation of the transport machinery. The transported effector proteins are, however, very diverse, reflecting their different functions. Within the *Rhizobiaceae*, T3SSs have been identified in *B. japonicum* [59], *B. elkanii* (Okazaki *et al.*, unpublished), *M. loti* strain MAFF303099 [60],

Rhizobium sp. NGR237 [61], and *S. fredii* strains USDA257 and HH103 [62, 63]. The regulation of the corresponding genes seems to be very similar in all studied strains. One of the genes, *ttsI*, is preceded by a *nod* box. Hence, transcription of *ttsI* is inducible by flavonoids [59, 61, 64]. By sequence similarity, TtsI belongs to the two-component regulatory family. Its target site is the *tts* box – a conserved sequence motif upstream of genes involved in type III secretion [65]. The *tts* box has a tripartite composition and is located about 10 nucleotides upstream of the transcriptional start site [66]. This regulation links the expression of the T3SS with the expression of nodulation genes. Therefore, it is likely that the T3SS is active already very early in infection. Data obtained with a *nopB–uidA* reporter gene fusion – NopB is a conserved component of rhizobial T3SSs – indicate that the T3SS is active during nodule development in *Macroptilium atropurpureum* and soybean ([66] and unpublished results). It was also shown that several genes of the type III secretion cluster are expressed in fully developed nodules of *M. atropurpureum*. In contrast, microarray data revealed that in soybean the expression of these genes significantly drops between 2 and 3 weeks after inoculation [42, 43]. How this downregulation is executed is unknown.

The influence of the T3SS on symbiosis is very variable and depends on the host investigated. Mutations may be beneficial or detrimental to symbiosis or have no effect. On *Pachyrhizus tuberosus*, for example, a type III secretion mutant of NGR234 exhibited an increased nodulation in comparison with the wild-type, whereas the same mutant had a reduced nodulation capacity on *Tephrosia vogelii* [61]. How effector proteins are transported into the plant cell is not understood. However, it is likely that NopA, NopB, and NopX, which are associated with extracellular pili structures, are involved in this [62, 67–69].

Little is known about the function of the putative effector proteins. GunA2 and Pgl are secreted proteins of *B. japonicum* ([70] and Hempel *et al.* unpublished). GunA2 has cellulase activity [71]; Pgl has similarity to polygalacturonases. Thus, both proteins might be involved in cell wall degradation or modification. However, no phenotype could be associated with corresponding mutants [71]. NopT of *Rhizobium* sp. NGR234 belongs to the peptidase C58 superfamily [72]. Transient expression in tobacco elicits a hypersensitive reaction. NopP and NopL are phosphorylated by plant kinases, and might modulate plant signal transduction pathways [73–75].

Although the above examples clearly show that secreted proteins are often important determinants of the rhizobia–legume interaction, further research is needed to understand their activity. In general three questions need to be answered: (i) to which plant cell compartment are the secreted proteins addressed, (ii) with which proteins or other molecules do the effector proteins interact, and (iii) what is their mode of action?

4.5
Microarray Studies Help in Elucidating the Flavonoid Stimulons

The impact of flavonoids on *nod* gene expression and protein secretion is well studied. In addition, flavonoids like genistein may have a negative effect on some bacteria [76, 77]. *B. japonicum* is sensitive to the phytoalexin glyceollin, which is produced by soybean [78]. Resistance to this phytoalexin can be induced by genistein; however, the mechanism of this resistance is not known. Flavonoids also stimulate a chemotactic response in rhizobia. Therefore, it is obvious that other "targets" for flavonoids exist. Within the recent years several rhizobial genomes have been sequenced (Table 4.1). This permits a genome-wide search for flavonoid-inducible genes. For *S. meliloti* it was reported that a total of 27 genes are upregulated by luteolin. This included the known nodulation genes and 12 genes that were induced independent of *nodD1* [79]. In *B. japonicum*, more than 100 genes were found to be induced by genistein, including all nodulation genes and all genes of a flagellar cluster [80]. Except for the *nod* genes, little is known about regulatory proteins that are involved in transcriptional activation. Several genes of *B. japonicum* that are induced by genistein are likely to encode multidrug efflux pumps of the resistance–nodulation–cell division (RND) family. Although data are not available, it is tempting to speculate that they might be involved in phytoalexin resistance and/or give the bacteria a competitive advantage in the rhizosphere. It is unknown how these genes are regulated. One possibility is the involvement of TetR-like regulators. The *acrAB* operon, for example, which in *Escherichia coli* encodes one of the best-studied RND members, is regulated by AcrR – a TetR-type regulator [81].

4.6
nod Genes as Accessory Components of the Rhizobial Core Genome

The taxonomic tree based on 16S rDNA sequences indicates that rhizobia are polyphyletic [82–84] and for many rhizobial strains closely related nonsymbiotic bacteria are known. The capacity to nodulate legumes even extends beyond the α-subclass of Proteobacteria and includes *Burkholderia* species that are members of the β-subclass [85, 86]. In this case, comparison of NodA sequences suggests that nodulation genes originate from α-rhizobia [87], indicating that horizontal gene transfer leads to the dissemination of symbiotic genes. Horizontal gene transfer in the field has been indeed demonstrated [88, 89]. Therefore, nodulation genes and many other genes that are often located on plasmids can be considered as accessory components of the genome, which evolve more rapidly than the core genome [90].

Although it was shown many times that *nod* genes are essential for the infection of legumes there are exceptions. The photosynthetic bradyrhizobia BTAi1 and ORS278, which induce nodules on *Aeschynomene sensitiva*, have nitrogen fixation genes but are devoid of *nod* genes [91]. It will be interesting to unravel the mechanism how these bacteria trigger nodule organogenesis on *Aeschynomene*.

4.7
Conclusions and Outlook

The last 30 years of research has resulted in a profound knowledge about how rhizobia use nodulation and nitrogen fixation genes to carry out their symbiotic task.

However, it also became clear that many other factors like cell surface structures and secreted proteins are crucial determinants of host specificity and symbiotic efficiency. Rhizobia rely on a number of exchangeable and variable building blocks whose components and functions are not well understood. The availability of a number of rhizobial genome sequences and technological advancements in transcriptome, proteome, and metabolome analyzes will help in elucidating their role.

References

1 Vance, C.P. (1998) Legume symbiotic nitrogen fixation: agronomic aspects, in *The Rhizobiaceae* (eds H. Spaink, A. Kondorosi, and P.J.J. Hooykaas), Kluwer, Dordrecht, pp. 509–530.
2 Alves, B.J.R., Boddey, R.M., and Urquiaga, S. (2003) The success of BNF in soybean in Brazil. *Plant Soil*, **252**, 1–9.
3 Fischer, H.M. (1994) Genetic regulation of nitrogen fixation in rhizobia. *Microbiol. Rev.*, **58**, 352–386.
4 Fischer, H.M. (1996) Environmental regulation of rhizobial symbiotic nitrogen fixation genes. *Trends Microbiol.*, **4**, 317–320.
5 Harborne, J.B. and Williams, C.A. (2000) Advances in flavonoid research since 1992. *Phytochemistry*, **55**, 481–504.
6 Iwashina, T. (2003) Flavonoid function and activity to plants and other organisms. *Biol. Sci. Space*, **17**, 24–44.
7 Cooper, J.E. (2004) Multiple responses of rhizobia to flavonoids during legume root infection. *Adv. Bot. Res.*, **41**, 1–62.
8 Schlaman, H.R.M., Phillips, D.A., and Kondorosi, E. (1998) Genetic organization and transcriptional regulation of rhizobial nodulation genes, in *The Rhizobiaceae* (eds H. Spaink, A. Kondorosi, and P.J.J. Hooykaas), Kluwer, Dordrecht, pp. 361–386.
9 Zuanazzi, J.A.-S. (1998) Production of *Sinorhizobium meliloti nod* gene activator and repressor flavonoids from *Medicago sativa* roots. *Mol. Plant-Microbe Interact.*, **11**, 784–794.
10 Recourt, K., van Brussel, A.A., Driessen, A.J., and Lugtenberg, B.J. (1989) Accumulation of a *nod* gene inducer, the flavonoid naringenin, in the cytoplasmic membrane of *Rhizobium leguminosarum* biovar *viciae* is caused by the pH-dependent hydrophobicity of naringenin. *J. Bacteriol.*, **171**, 4370–4377.
11 Schlaman, H.R., Spaink, H.P., Okker, R.J., and Lugtenberg, B.J. (1989) Subcellular localization of the *nodD* gene product in *Rhizobium leguminosarum*. *J. Bacteriol.*, **171**, 4686–4693.
12 Schell, M.A. (1993) Molecular biology of the LysR family of transcriptional regulators. *Annu. Rev. Microbiol.*, **47**, 597–626.
13 Györgypal, Z. and Kondorosi, A. (1991) Homology of the ligand-binding regions of *Rhizobium* symbiotic regulatory protein NodD and vertebrate nuclear receptors. *Mol. Gen. Genet.*, **226**, 337–340.
14 Györgypal, Z., Kondorosi, E., and Kondorosi, A. (1991) Diverse signal sensitivity of NodD protein homologs from narrow and broad host range rhizobia. *Mol. Plant-Microbe. Interact.*, **4**, 356–364.
15 Spaink, H.P., Wijffelman, C.A., Pees, E., Okker, R.-J.H., and Lugtenberg, B.-J.J. (1987) *Rhizobium* nodulation gene *nodD* as a determinant of host specificity. *Nature*, **328**, 337–340.
16 Feng, J., Li, Q., Hu, H.L., Chen, X.C., and Hong, G.F. (2003) Inactivation of the *nod* box distal half-site allows tetrameric NodD to activate *nodA* transcription in an inducer-independent manner. *Nucleic Acids Res.*, **31**, 3143–3156.
17 Fisher, R.F. and Long, S.R. (1989) DNA footprint analysis of the transcriptional activator proteins NodD1 and NodD3 on inducible *nod* gene promoters. *J. Bacteriol.*, **171**, 5492–5502.
18 Yeh, K.C., Peck, M.C., and Long, S.R. (2002) Luteolin and GroESL modulate

in vitro activity of NodD. *J. Bacteriol.*, **184**, 525–530.

19 Peck, M.C., Fisher, R.F., and Long, S.R. (2006) Diverse flavonoids stimulate NodD1 binding to *nod* gene promoters in *Sinorhizobium meliloti*. *J. Bacteriol.*, **188**, 5417–5427.

20 Fisher, R.F., Egelhoff, T.T., Mulligan, J.T., and Long, S.R. (1988) Specific binding of proteins from *Rhizobium meliloti* cell-free extracts containing NodD to DNA sequences upstream of inducible nodulation genes. *Genes Dev.*, **2**, 282–293.

21 Wang, S.P. and Stacey, G. (1991) Studies of the *Bradyrhizobium japonicum nodD1* promoter: a repeated structure for the *nod* box. *J. Bacteriol.*, **173**, 3356–3365.

22 Göttfert, M., Grob, P., and Hennecke, H. (1990) Proposed regulatory pathway encoded by the *nodV* and *nodW* genes, determinants of host specificity in *Bradyrhizobium japonicum*. *Proc. Natl. Acad. Sci. USA*, **87**, 2680–2684.

23 Laub, M.T. and Goulian, M. (2007) Specificity in two-component signal transduction pathways. *Annu. Rev. Genet.*, **41**, 121–145.

24 Loh, J., Garcia, M., and Stacey, G. (1997) NodV and NodW, a second flavonoid recognition system regulating *nod* gene expression in *Bradyrhizobium japonicum*. *J. Bacteriol.*, **179**, 3013–3020.

25 Loh, J. and Stacey, G. (2003) Nodulation gene regulation in *Bradyrhizobium japonicum*: a unique integration of global regulatory circuits. *Appl. Environ. Microbiol.*, **69**, 10–17.

26 Maillet, F., Debellé, F., and Dénarié, J. (1990) Role of the *nodD* and *syrM* genes in the activation of the regulatory gene *nodD3*, and of the common and host-specific *nod* genes of *Rhizobium meliloti*. *Mol. Microbiol.*, **4**, 1975–1984.

27 Chen, H., Gao, K., Kondorosi, E., Kondorosi, A., and Rolfe, B.G. (2005) Functional genomic analysis of global regulator NolR in *Sinorhizobium meliloti*. *Mol. Plant-Microbe. Interact.*, **18**, 1340–1352.

28 Busenlehner, L.S., Pennella, M.A., and Giedroc, D.P. (2003) The SmtB/ArsR family of metalloregulatory transcriptional repressors: structural insights into prokaryotic metal resistance. *FEMS Microbiol. Rev.*, **27**, 131–143.

29 Kondorosi, E., Gyuris, J., Schmidt, J., John, M., Duda, E., Hoffmann, B., Schell, J., and Kondorosi, A. (1989) Positive and negative control of *nod* gene expression in *Rhizobium meliloti* is required for optimal nodulation. *EMBO J.*, **8**, 1331–1340.

30 Phillips, D.A., Joseph, C.M., and Maxwell, C.A. (1992) Trigonelline and stachydrine released from alfalfa seeds activate NodD2 protein in *Rhizobium meliloti*. *Plant Physiol.*, **99**, 1526–1531.

31 Gagnon, H. and Ibrahim, R.K. (1998) Aldonic acids: a novel family of *nod* gene inducers of *Mesorhizobium loti*, *Rhizobium lupini*, and *Sinorhizobium meliloti*. *Mol. Plant-Microbe. Interact.*, **11**, 988–998.

32 Mabood, F., Souleimanov, A., Khan, W., and Smith, D.L. (2006) Jasmonates induce Nod factor production by *Bradyrhizobium japonicum*. *Plant Physiol. Biochem.*, **44**, 759–765.

33 D'Haeze, W. and Holsters, M. (2002) Nod factor structures, responses, and perception during initiation of nodule development. *Glycobiology*, **12**, 79R–105R.

34 Roche, P., Debellé, F., Maillet, F., Lerouge, P., Faucher, C., Truchet, G., Dénarié, J., and Promé, J.-C. (1991) Molecular basis of symbiotic host specificity in *Rhizobium meliloti*: *nodH* and *nodPQ* genes encode the sulfation of lipo-oligosaccharide signals. *Cell*, **67**, 1131–1143.

35 Mckay, I.A. and Djordjevic, M.A. (1993) Production and excretion of Nod metabolites by *Rhizobium leguminosarum* bv. *trifolii* are disrupted by the same environmental factors that reduce nodulation in the field. *Appl. Environ. Microbiol.*, **59**, 3385–3392.

36 Spaink, H.P., Wijfjes, A.H., and Lugtenberg, B.J. (1995) *Rhizobium* NodI and NodJ proteins play a role in the efficiency of secretion of lipochitin oligosaccharides. *J. Bacteriol.*, **177**, 6276–6281.

37 Fernández-López, M., D'Haeze, W., Mergaert, P., Verplancke, C., Promé, J.-C., Van Montagu, M., and Holsters, M.

(1996) Role of *nodI* and *nodJ* in lipo-chitooligosaccharide secretion in *Azorhizobium caulinodans* and *Escherichia coli*. *Mol. Microbiol.*, **20**, 993–1000.

38 Fath, M.J. and Kolter, R. (1993) ABC transporters: bacterial exporters. *Microbiol. Rev.*, **57**, 995–1017.

39 Sharma, S.B. and Signer, E.R. (1990) Temporal and spatial regulation of the symbiotic genes of *Rhizobium meliloti* in planta revealed by transposon Tn5-*gusA*. *Genes Dev.*, **4**, 344–356.

40 Schlaman, H.R., Horvath, B., Vijgenboom, E., Okker, R.J., and Lugtenberg, B.J. (1991) Suppression of nodulation gene expression in bacteroids of *Rhizobium leguminosarum* biovar *viciae*. *J. Bacteriol.*, **173**, 4277–4287.

41 Gao, M., D'Haeze, W., De Rycke, R., and Holsters, M. (2001) Dual control of the *nodA* operon of *Azorhizobium caulinodans* ORS571 by a *nod* box and a NifA-sigma^{54}-type promoter. *Mol. Genet. Genomics*, **265**, 1050–1059.

42 Chang, W.S., Franck, W.L., Cytryn, E., Jeong, S., Joshi, T., Emerich, D.W., Sadowsky, M.J., Xu, D., and Stacey, G. (2007) An oligonucleotide microarray resource for transcriptional profiling of *Bradyrhizobium japonicum*. *Mol. Plant-Microbe Interact.*, **20**, 1298–1307.

43 Pessi, G., Ahrens, C.H., Rehrauer, H., Lindemann, A., Hauser, F., Fischer, H.-M., and Hennecke, H. (2007) Genome-wide transcript analysis of *Bradyrhizobium japonicum* bacteroids in soybean root nodules. *Mol. Plant-Microbe Interact.*, **20**, 1353–1363.

44 Radutoiu, S., Madsen, L.H., Madsen, E.B., Felle, H.H., Umehara, Y., Grønlund, M., Sato, S., Nakamura, Y., Tabata, S., Sandal, N. *et al.* (2003) Plant recognition of symbiotic bacteria requires two LysM receptor-like kinases. *Nature*, **425**, 585–592.

45 Amor, B.B., Shaw, S.L., Oldroyd, G.E., Maillet, F., Penmetsa, R.V., Cook, D., Long, S.R., Dénarié, J., and Gough, C. (2003) The *NFP* locus of *Medicago truncatula* controls an early step of Nod factor signal transduction upstream of a rapid calcium flux and root hair deformation. *Plant J.*, **34**, 495–506.

46 Arrighi, J.F., Barre, A., Ben, A.B., Bersoult, A., Soriano, L.C., Mirabella, R., de Carvalho-Niebel, F., Journet, E.P., Ghérardi, M., Huguet, T. *et al.* (2006) The *Medicago truncatula* lysine motif-receptor-like kinase gene family includes NFP and new nodule-expressed genes. *Plant Physiol.*, **142**, 265–279.

47 Buist, G., Steen, A., Kok, J., and Kuipers, O.P. (2008) LysM, a widely distributed protein motif for binding to (peptido) glycans. *Mol. Microbiol.*, **68**, 838–847.

48 Oldroyd, G.E. and Downie, J.A. (2008) Coordinating nodule morphogenesis with rhizobial infection in legumes. *Annu. Rev. Plant Biol.*, **59**, 519–546.

49 de Maagd, R.A., Wijfjes, A.H., Spaink, H.P., Ruiz-Sainz, J.E., Wijffelman, C.A., Okker, R.J., and Lugtenberg, B.J. (1989) *nodO*, a new *nod* gene of the *Rhizobium leguminosarum* biovar *viciae* sym plasmid pRL1JI, encodes a secreted protein. *J. Bacteriol.*, **171**, 6764–6770.

50 Downie, J.A. and Surin, B.P. (1990) Either of two *nod* gene loci can complement the nodulation defect of a *nod* deletion mutant of *Rhizobium leguminosarum* bv *viciae*. *Mol. Gen. Genet.*, **222**, 81–86.

51 Demont, N., Debellé, F., Aurelle, H., Dénarié, J., and Promé, J.C. (1993) Role of the *Rhizobium meliloti nodF* and *nodE* genes in the biosynthesis of lipo-oligo-saccharidic nodulation factors. *J. Biol. Chem*, **268**, 20134–20142.

52 Sutton, J.M., Lea, E.J., and Downie, J.A. (1994) The nodulation-signaling protein NodO from *Rhizobium leguminosarum* biovar *viciae* forms ion channels in membranes. *Proc. Natl. Acad. Sci. USA*, **91**, 9990–9994.

53 Krehenbrink, M. and Downie, J.A. (2008) Identification of protein secretion systems and novel secreted proteins in *Rhizobium leguminosarum* bv. *viciae*. *BMC Genomics*, **9**, 55.

54 Hubber, A., Vergunst, A.C., Sullivan, J.T., Hooykaas, P.J., and Ronson, C.W. (2004) Symbiotic phenotypes and translocated effector proteins of the *Mesorhizobium loti* strain R7A VirB/D4 type IV secretion system. *Mol. Microbiol.*, **54**, 561–574.

55 Hubber, A.M., Sullivan, J.T., and Ronson, C.W. (2007) Symbiosis-induced cascade

regulation of the *Mesorhizobium loti* R7A VirB/D4 type IV secretion system. *Mol. Plant-Microbe Interact.*, **20**, 255–261.

56 Büttner, D. and Bonas, U. (2006) Who comes first? How plant pathogenic bacteria orchestrate type III secretion. *Curr. Opin. Microbiol.*, **9**, 193–200.

57 Hueck, C.J. (1998) Type III protein secretion systems in bacterial pathogens of animals and plants. *Microbiol. Mol. Biol. Rev.*, **62**, 379–433.

58 Mota, L.J., Sorg, I., and Cornelis, G.R. (2005) Type III secretion: the bacteria-eukaryotic cell express. *FEMS Microbiol. Lett.*, **252**, 1–10.

59 Krause, A., Doerfel, A., and Göttfert, M. (2002) Mutational and transcriptional analysis of the type III secretion system of *Bradyrhizobium japonicum*. *Mol. Plant-Microbe Interact.*, **15**, 1228–1235.

60 Kaneko, T., Nakamura, Y., Sato, S., Asamizu, E., Kato, T., Sasamoto, S., Watanabe, A., Idesawa, K., Ishikawa, A., Kawashima, K. *et al.* (2000) Complete genome structure of the nitrogen-fixing symbiotic bacterium *Mesorhizobium loti*. *DNA Res.*, **7**, 331–338.

61 Viprey, V., Del Greco, A., Golinowski, W., Broughton, W.J., and Perret, X. (1998) Symbiotic implications of type III protein secretion machinery in *Rhizobium*. *Mol. Microbiol.*, **28**, 1381–1389.

62 Krishnan, H.B., Lorio, J., Kim, W.S., Jiang, G., Kim, K.Y., DeBoer, M., and Pueppke, S.G. (2003) Extracellular proteins involved in soybean cultivar-specific nodulation are associated with pilus-like surface appendages and exported by a type III protein secretion system in *Sinorhizobium fredii* USDA257. *Mol. Plant-Microbe Interact.*, **16**, 617–625.

63 de Lyra, M.d.C., López-Baena, F.J., Madinabeitia, N., Vinardell, J.M., Espuny, M.R., Cubo, M.T., Bellogín, R.A., Ruiz-Sainz, J.E., and Ollero, F.J. (2006) Inactivation of the *Sinorhizobium fredii* HH103 *rhcJ* gene abolishes nodulation outer proteins (Nops) secretion and decreases the symbiotic capacity with soybean. *Int. Microbiol.*, **9**, 125–133.

64 López-Baena, F.J., Vinardell, J.M., Pérez-Montaño, F., Crespo-Rivas, J.C., Bellogín, R.A., Espuny, M.R., and Ollero, F.J. (2008) Regulation and symbiotic significance of nodulation outer proteins secretion in *Sinorhizobium fredii* HH103. *Microbiology*, **154**, 1825–1836.

65 Wassem, R., Kobayashi, H., Kambara, K., Le, Q.A., Walker, G.C., Broughton, W.J., and Deakin, W.J. (2008) TtsI regulates symbiotic genes in *Rhizobium* species NGR234 by binding to *tts* boxes. *Mol. Microbiol.*, **68**, 736–748.

66 Zehner, S., Schober, G., Wenzel, M., Lang, K., and Göttfert, M. (2008) Expression of the *Bradyrhizobium japonicum* type III secretion system in legume nodules and analysis of the associated *tts* box promoter. *Mol. Plant-Microbe Interact.*, **21**, 1087–1093.

67 Deakin, W.J., Marie, C., Saad, M.M., Krishnan, H.B., and Broughton, W.J. (2005) NopA is associated with cell surface appendages produced by the type III secretion system of *Rhizobium* sp. strain NGR234. *Mol. Plant-Microbe Interact.*, **18**, 499–507.

68 Saad, M.M., Kobayashi, H., Marie, C., Brown, I.R., Mansfield, J.W., Broughton, W.J., and Deakin, W.J. (2005) NopB, a type III secreted protein of *Rhizobium* sp. strain NGR234, is associated with pilus-like surface appendages. *J. Bacteriol.*, **187**, 1173–1181.

69 Saad, M.M., Staehelin, C., Broughton, W.J., and Deakin, W.J. (2008) Protein-protein interactions within type three secretion system-dependent pili of *Rhizobium* sp. strain NGR234. *J. Bacteriol.*, **190**, 750–754.

70 Süß, C., Hempel, J., Zehner, S., Krause, A., Patschkowski, T., and Göttfert, M. (2006) Identification of genistein-inducible and type III-secreted proteins of *Bradyrhizobium japonicum*. *J. Biotechnol.*, **126**, 69–77.

71 Caldelari Baumberger, I., Fraefel, N., Göttfert, M., and Hennecke, H. (2003) New NodW- or NifA-regulated *Bradyrhizobium japonicum* genes. *Mol. Plant-Microbe Interact.*, **16**, 342–351.

72 Dai, W.J., Zeng, Y., Xie, Z.P., and Staehelin, C. (2008) Symbiosis-promoting and deleterious effects of NopT, a novel type 3 effector of *Rhizobium*

sp. strain NGR234. *J. Bacteriol.*, **190**, 5101–5110.

73 Bartsev, A.V., Boukli, N.M., Deakin, W.J., Staehelin, C., and Broughton, W.J. (2003) Purification and phosphorylation of the effector protein NopL from *Rhizobium* sp. NGR234. *FEBS Lett.*, **554**, 271–274.

74 Bartsev, A.V., Deakin, W.J., Boukli, N.M., McAlvin, C.B., Stacey, G., Malnoe, P., Broughton, W.J., and Staehelin, C. (2004) NopL, an effector protein of *Rhizobium* sp. NGR234, thwarts activation of plant defense reactions. *Plant Physiol.*, **134**, 871–879.

75 Skorpil, P., Saad, M.M., Boukli, N.M., Kobayashi, H., Ares-Orpel, F., Broughton, W.J., and Deakin, W.J. (2005) NopP, a phosphorylated effector of *Rhizobium* sp. strain NGR234, is a major determinant of nodulation of the tropical legumes *Flemingia congesta* and *Tephrosia vogelii*. *Mol. Microbiol.*, **57**, 1304–1317.

76 Dakora, F.D. and Phillips, D.A. (1996) Diverse functions of isoflavonoids in legumes transcend anti-microbial definitions of phytoalexins. *Physiol. Mol. Plant Pathol.*, **49**, 1–20.

77 Pankhurst, C.E. and Biggs, D.R. (1980) Sensitivity of *Rhizobium* to selected isoflavonoids. *Can. J. Microbiol.*, **26**, 542–545.

78 Parniske, M., Ahlborn, B., and Werner, D. (1991) Isoflavonoid-inducible resistance to the phytoalexin glyceollin in soybean rhizobia. *J. Bacteriol.*, **173**, 3432–3439.

79 Barnett, M.J., Toman, C.J., Fisher, R.F., and Long, S.R. (2004) A dual-genome symbiosis chip for coordinate study of signal exchange and development in a prokaryote–host interaction. *Proc. Natl. Acad. Sci. USA*, **101**, 16636–16641.

80 Lang, K., Lindemann, A., Hauser, F., and Göttfert, M. (2008) The genistein stimulon of *Bradyrhizobium japonicum*. *Mol. Genet. Genomics*, **279**, 203–211.

81 Ma, D., Alberti, M., Lynch, C., Nikaido, H., and Hearst, J.E. (1996) The local repressor AcrR plays a modulating role in the regulation of *acrAB* genes of *Escherichia coli* by global stress signals. *Mol. Microbiol.*, **19**, 101–112.

82 Debellé, F., Moulin, L., Mangin, B., Dénarié, J., and Boivin, C. (2001) *nod* genes and Nod signals and the evolution of the *Rhizobium* legume symbiosis. *Acta Biochim. Pol.*, **48**, 359–365.

83 Sawada, H., Kuykendall, L.D., and Young, J.M. (2003) Changing concepts in the systematics of bacterial nitrogen-fixing legume symbionts. *J. Gen. Appl. Microbiol.*, **49**, 155–179.

84 Willems, A. (2006) The taxonomy of rhizobia: an overview. *Plant Soil*, **287**, 3–14.

85 Moulin, L., Munive, A., Dreyfus, B., and Boivin-Masson, C. (2001) Nodulation of legumes by members of the beta-subclass of Proteobacteria. *Nature*, **411**, 948–950.

86 Vandamme, P., Goris, J., Chen, W.M., De Vos, P., and Willems, A. (2002) *Burkholderia tuberum* sp. nov. and *Burkholderia phymatum* sp. nov., nodulate the roots of tropical legumes. *Syst. Appl. Microbiol.*, **25**, 507–512.

87 Chen, W.M., de Faria, S.M., Straliotto, R., Pitard, R.M., Simoes-Araujo, J.L., Chou, J.H., Chou, Y.J., Barrios, E., Prescott, A.R., Elliott, G.N. *et al.* (2005) Proof that *Burkholderia* strains form effective symbioses with legumes: a study of novel *Mimosa*-nodulating strains from South America. *Appl. Environ. Microbiol.*, **71**, 7461–7471.

88 Barcellos, F.G., Menna, P., da Silva Batista, J.S., and Hungria, M. (2007) Evidence of horizontal transfer of symbiotic genes from a *Bradyrhizobium japonicum* inoculant strain to indigenous *Sinorhizobium* (*Ensifer*) *fredii* and *Bradyrhizobium elkanii* in a Brazilian savannah soil. *Appl. Environ. Microbiol.*, **73**, 2635–2643.

89 Sullivan, J.T., Patrick, H.N., Lowther, W.L., Scott, D.B., and Ronson, C.W. (1995) Nodulating strains of *Rhizobium loti* arise through chromosomal symbiotic gene transfer in the environment. *Proc. Natl. Acad. Sci. USA*, **92**, 8985–8989.

90 Crossman, L.C., Castillo-Ramírez, S., McAnnula, C., Lozano, L., Vernikos, G.S., Acosta, J.L., Ghazoui, Z.F., Hernández-González, I., Meakin, G., Walker, A.W. *et al.* (2008) A common genomic framework for a diverse assembly of plasmids in the symbiotic

nitrogen fixing bacteria. *PLoS ONE*, **3**, e2567.

91 Giraud, E., Moulin, L., Vallenet, D., Barbe, V., Cytryn, E., Avarre, J.C., Jaubert, M., Simon, D., Cartieaux, F., Prin, Y. *et al.* (2007) Legumes symbioses: absence of *nod* genes in photosynthetic bradyrhizobia. *Science*, **316**, 1307–1312.

92 Lee, K.B., De, B.P., Aono, T., Liu, C.T., Suzuki, S., Suzuki, T., Kaneko, T., Yamada, M., Tabata, S., Kupfer, D.M. *et al.* (2008) The genome of the versatile nitrogen fixer *Azorhizobium caulinodans* ORS571. *BMC Genomics*, **9**, 271.

93 Messens, E., van Geelen, D., and Holsters, M. (1991) 7,4′-Dihydroxyflavanone is the major *Azorhizobium nod* gene-inducing factor present in *Sesbania rostrata* seedling exudate. *Mol. Plant-Microbe Interact.*, **4**, 262–267.

94 Kaneko, T., Nakamura, Y., Sato, S., Minamisawa, K., Uchiumi, T., Sasamoto, S., Watanabe, A., Idesawa, K., Iriguchi, M., Kawashima, K. *et al.* (2002) Complete genomic sequence of nitrogen-fixing symbiotic bacterium *Bradyrhizobium japonicum* USDA110. *DNA Res.*, **9**, 189–197.

95 Kosslak, R.M., Bookland, R., Barkei, J., Paaren, H.E., and Appelbaum, E.R. (1987) Induction of *Bradyrhizobium japonicum* common *nod* genes by isoflavones isolated from *Glycine max*. *Proc. Natl. Acad. Sci. USA*, **84**, 7428–7432.

96 Taurian, T., Morón, B., Soria-Díaz, M.E., Angelini, J.G., Tejero-Mateo, P., Gil-Serrano, A., Megías, M., and Fabra, A. (2008) Signal molecules in the peanut–bradyrhizobia interaction. *Arch. Microbiol.*, **189**, 345–356.

97 Dakora, F.D. (2000) Commonality of root nodulation signals and nitrogen assimilation in tropical grain legumes belonging to the tribe Phaseoleae. *Aust. J. Plant Physiol.*, **27**, 885–892.

98 González, V., Santamaría, R.I., Bustos, P., Hernández-González, I., Medrano-Soto, A., Moreno-Hagelsieb, G., Janga, S.C., Ramírez, M.A., Jiménez-Jacinto, V., Collado-Vides, J. *et al.* (2006) The partitioned *Rhizobium etli* genome: Genetic and metabolic redundancy in seven interacting replicons. *Proc. Natl. Acad. Sci. USA*, **103**, 3834–3839.

99 Hungria, M., Joseph, C.M., and Phillips, D.A. (1991) *Rhizobium nod* gene inducers exuded naturally from roots of common bean (*Phaseolus vulgaris* L.). *Plant Physiol.*, **97**, 759–764.

100 Young, J.P., Crossman, L.C., Johnston, A.W., Thomson, N.R., Ghazoui, Z.F., Hull, K.H., Wexler, M., Curson, A.R., Todd, J.D., Poole, P.S. *et al.* (2006) The genome of *Rhizobium leguminosarum* has recognizable core and accessory components. *Genome Biol.*, **7**, R34.

101 Zaat, S.A., Schripsema, J., Wijffelman, C.A., van Brussel, A.A., and Lugtenberg, B.J. (1989) Analysis of the major inducers of the *Rhizobium nodA* promoter from *Vicia sativa* root exudate and their activity with different *nodD* genes. *Plant Mol. Biol.*, **13**, 175–188.

102 Redmond, J.W., Batley, M., Djordjevic, M.A., Innes, R.W., Kuempel, P.L., and Rolfe, B.G. (1986) Flavones induce expression of nodulation genes in *Rhizobium*. *Nature*, **323**, 632–635.

103 Galibert, F., Finan, T.M., Long, S.R., Pühler, A., Abola, P., Ampe, F., Barloy-Hubler, F., Barnett, M.J., Becker, A., Boistard, P. *et al.* (2001) The composite genome of the legume symbiont *Sinorhizobium meliloti*. *Science*, **293**, 668–672.

104 Maxwell, C.A., Hartwig, U.A., Joseph, C.M., and Phillips, D.A. (1989) A chalcone and two related flavonoids released from alfalfa roots induce *nod* genes of *Rhizobium meliloti*. *Plant Physiol.*, **91**, 842–847.

105 Peters, N.K., Frost, J.W., and Long, S.R. (1986) A plant flavone, luteolin, induces expression of *Rhizobium meliloti* nodulation genes. *Science*, **233**, 977–980.

106 Pueppke, S.G. and Broughton, W.J. (1999) *Rhizobium* sp. strain NGR234 and *R. fredii* USDA257 share exceptionally broad, nested host ranges. *Mol. Plant-Microbe Interact.*, **12**, 293–318.

107 Mergaert, P., Van Montagu, M., Promé, J.-C., and Holsters, M. (1993) Three unusual modifications, a D-arabinosyl, an *N*-methyl, and a carbamoyl group,

are present on the Nod factors of *Azorhizobium caulinodans* strain ORS571. *Proc. Natl. Acad. Sci. USA*, **90**, 1551–1555.

108 Mergaert, P., Ferro, M., D'Haeze, W., Van Montagu, M., Holsters, M., and Prome, J.-C. (1997) Nod factors of *Azorhizobium caulinodans* strain ORS571 can be glycosylated with an arabinosyl group, a fucosyl group, or both. *Mol. Plant-Microbe Interact.*, **10**, 683–687.

109 Sanjuan, J., Carlson, R.W., Spaink, H.P., Bhat, U.R., Barbour, W.M., Glushka, J., and Stacey, G. (1992) A 2-O-methyl-fucose moiety is present in the lipo-oligosaccharide nodulation signal of *Bradyrhizobium japonicum*. *Proc. Natl. Acad. Sci. USA*, **89**, 8789–8793.

110 López-Lara, I.M., van den Berg, J.D., Thomas-Oates, J.E., Glushka, J., Lugtenberg, B.J., and Spaink, H.P. (1995) Structural identification of the lipo-chitin oligosaccharide nodulation signals of *Rhizobium loti*. *Mol. Microbiol.*, **15**, 627–638.

111 Cárdenas, L., Domínguez, J., Quinto, C., López-Lara, I.M., Lugtenberg, B.J., Spaink, H.P., Rademaker, G.J., Haverkamp, J., and Thomas-Oates, J.E. (1995) Isolation, chemical structures and biological activity of the lipo-chitin oligosaccharide nodulation signals from *Rhizobium etli*. *Plant Mol. Biol.*, **29**, 453–464.

112 Yang, G.P., Debellé, F., Savagnac, A., Ferro, M., Schiltz, O., Maillet, F., Promé, D., Treilhou, M., Vialas, C., Lindstrom, K. *et al.* (1999) Structure of the *Mesorhizobium huakuii* and *Rhizobium galegae* Nod factors: a cluster of phylogenetically related legumes are nodulated by rhizobia producing Nod factors with α,β-unsaturated *N*-acyl substitutions. *Mol. Microbiol.*, **34**, 227–237.

113 Spaink, H.P., Sheeley, D.M., van Brussel, A.A., Glushka, J., York, W.S., Tak, T., Geiger, O., Kennedy, E.P., Reinhold, V.N., and Lugtenberg, B.J. (1991) A novel highly unsaturated fatty acid moiety of lipo-oligosaccharide signals determines host specificity of *Rhizobium*. *Nature*, **354**, 125–130.

114 Lerouge, P., Roche, P., Faucher, C., Maillet, F., Truchet, G., Promé, J.C., and Dénarié, J. (1990) Symbiotic host-specificity of *Rhizobium meliloti* is determined by a sulphated and acylated glucosamine oligosaccharide signal. *Nature*, **344**, 781–784.

5
Communication Between Pathogens and Eukaryotic Cells
Jürgen Heesemann

5.1
Introduction

The human body lives in a permanent stage of communication and interaction with its indigenous microbiota (physiological microbial flora) that consists predominantly of bacteria. This microbiota colonizes body surfaces (e.g., the skin) and accessible cavities that are covered by epithelial cells and a mucus layer (called mucosa). In humans the mucosa of the respiratory, gastrointestinal, and urogenital tract comprise about 300 m^2. Together with the roughly 2 m^2 of the skin, these interfaces and cavities provide about 10^{14} microbial cells with nutrients and space. Thus, the entire human ecosystem comprises 10-fold more bacterial cells than host cells, justifying the term "homo bacteriens." This microbiota–host consortium has a long history of coevolution and coadaptation leading to a fine-tuned and stable coexistence of microbiota and host (also known as homeostasis). This is a result of communication and close interaction between the host innate immune system (including epithelial cells, forming physical barriers and immune surveillance by humoral and cellular components) and the microbiota [1, 2]. These two partners, the mammalian host and the microbiota, communicate by using different types of languages, of which only a few have been decoded during the last decade [3, 4]. The major aim of the host is (i) to keep away the microorganisms from the mucosal surfaces, and (ii) to control growth and composition of the beneficial microbiota. By this strategy, the host takes advantage of microbial metabolites (vitamins, short-chain fatty acids) and the capability of the microbiota to prevent the colonization of malevolent microorganisms [2]. This is achievable as long as the host defense is effective and fine-tuned to prevent autoreactive inflammation/impairment. It may happen that the innate host defense cannot control microbial invasion of the mucosal barrier or prevent dissemination of microbes into deeper tissue, leading to localized or systemic infection of the host. In this case the host defense may be compromised (loss of control of harmless or facultative pathogenic bacteria) or the invading microorganism is virulent (i.e. capable of evading the innate defense barrier). Major differences between nonpathogenic and pathogenic microorganisms are that pathogens are endowed

Bacterial Signaling. Edited by Reinhard Krämer and Kirsten Jung
Copyright © 2010 WILEY-VCH Verlag GmbH & Co. KGaA, Weinheim
ISBN: 978-3-527-32365-4

5 Communication Between Pathogens and Eukaryotic Cells

with an antihost defense armament (pathogenicity factors) and host-adapted metabolism (fitness factors) that enables the pathogen to translocate across the mucosal barrier and to multiply intracellularly or extracellularly in host tissue, leading to infectious disease with disturbances in the signaling system of the host (Figure 5.1). In general, the infectious process is complex and associated with sophisticated communication between pathogen and host. Pathogens have evolved individual strategies of conversation with host cells as well as with the microbial community using a wide variety of signals. Signaling systems consist of at least four major functional components: (i) signaling molecules/ligands, (ii) specific receptors at the cell surface or in the cytosol, (iii) signal transduction systems, and (iv) signal response modules. Interestingly, both bacteria and eukaryotic cells share many similarities in the signal transduction machinery (interkingdom signaling) [4]. It is not surprising that signaling molecules emanating from bacteria can be recognized by microorganisms as well as by eukaryotic cells. However, the final consequences of signal translation will be different. Microbial signaling molecules comprise soluble macromolecules such as polypeptides (modulins, toxins), nucleic acids, cell wall/envelope components, and small hormone-like molecules such as catecholates, acyl-homoserine lactones (AHLs), and iron chelators/siderophores. In addition to

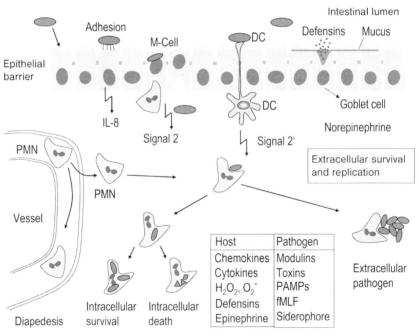

Figure 5.1 Schematic overview showing the initial steps of the infection process of bacterial pathogens invading the epithelial barrier of the intestinal mucosa consisting of brush border cells and microfold (M) cells. Protrusions of DCs of the submucosa can penetrate the mucosa and contact bacteria. Host–pathogen communica-tion leads to various signals that activate host defense as well as pathogen counter-response. PMNs are the first line of cellular defense. For further description, see text.

these diffusible ligands there are also cell membrane-associated ligands, fimbrial and nonfimbrial adhesins that act on surface-exposed receptors of the target cell through bacterium–host cell contact.

The language of pathogens is the language of war, which is directed towards the control on the innate immune defense of the host. In turn, the innate immune system of the host receives the microbial signals and responses rapidly with an antimicrobial defense program including components of the humoral response such as soluble bactericidal proteins (e.g., defensins, lysozyme), opsonins (complement system), cytokines, and chemokines for recruitment of professional phagocytes (cellular response: polymorphonuclear leukocyte/neutrophil PMNs, macrophages, dendritic cells (DCs)). In general, one can differentiate between long-distance communication (diffusible signaling molecules) and short-distance communication (intercellular contact, direct communication). Direct communication involves bacterial adhesins interacting with host cell receptors (such as integrins or carcinoembryonic antigen cell adhesion molecules (CEACAMs)) and contact-induced translocation/injection of pathogenicity factors from the cytosol of the bacterium to the cytosol of the target cell through type III or type IV secretion systems (type III secretion system (T3SS) or type IV secretion system (T4SS); for this latter issue, the reader is referred to [5]). For long-distance communication pathogens release diffusible signaling polypeptides or small nonpeptide molecules by different secretion systems. As shown in Figure 5.1, these different communication systems are the base of the inflammatory defense program of the host and the pathogenicity of the invading microorganism. Signal transduction processes result in different cellular alterations, for example: (i) cell metabolism [6, 7], (ii) transcription of selected genes [6, 7], (iii) cell membrane permeability, and (iv) rearrangement of the cytoskeleton. This is a dynamic process inducing a ping-pong of actions and reactions until the infectious disease has reached its final destination: (i) multiplication and transmission of the pathogen to new hosts and/or persistence within a hidden niche of the host, and (ii) survival and recovering of the host. Killing of the host would be a dead end for the pathogen, and can be considered as disruptive or unbalanced communication.

This chapter focuses on major principles of communication between bacterial pathogens and mammalian host cells by neglecting communication between pathogens and the microbiota. We start with long-distance communication followed by the diverse strategies of direct communication.

5.2
Long-Distance Communication

5.2.1
Language of Pathogen-Associated Molecular Patterns

The innate immune system is able to perceive the presence of microorganisms at interfaces and in deeper tissue by sensing highly conserved microbial molecules,

Table 5.1 PAMP signaling molecules, host receptors, and signaling effect.

PAMP signaling molecules	Host receptors/targets	Signaling effect
Lipoproteins, peptidoglycan	TLR2/1, TLR2/6	NF-κB ↑, MAPK ↑
LPSs	TLR4	NF-κB ↑, MAPK ↑, IRE3 ↑
Flagellin	TLR5	NF-κB ↑, MAPK ↑
DNA	TLR9	NF-κB ↑, IRF3 ↑
Peptidoglycan fragments		
MDP	NOD2	NF-κB ↑, MAPK ↑
iE-DAP	NOD1	NF-κB ↑, MAPK ↑
Cytosolic flagellin	NLRC4/ICE protease-activating factor	caspase 1 ↑, inflammasome ↑, IL-1β ↑, pyroptosis ↑
N-formyl peptides (prototype fMLF)	FPR (GPCR)	PI3Kγ, PLCβ, PKC, Rac
	$G_{\alpha i}$	chemoattraction ↑ oxidative burst ↑

called pathogen-associated molecular patterns (PAMPs), through pattern-recognition receptors (PRRs) [8]. However, PRRs cannot differentiate between PAMPs of nonpathogenic and pathogenic microorganisms (therefore, the term PAMP should be replaced by microbial-associated molecular pattern (MAMP)). Moreover, PRRs are also involved in sensing damage or death of host cells by recognition damage-associated molecular-pattern (DAMP) molecules/components [9]. The language of PAMPs/MAMPs comprise a large heterogeneous collection of microbial molecules such as lipopolysaccharide (LPS), lipoprotein, peptidoglycan, flagellin, nucleic acids (DNA, RNA), glycan, oligopeptides, and so on (Table 5.1). The PRRs can be broadly classified into two groups based on their location: (i) cell-associated PRRs (membrane or cytoplasmic located) or (ii) non-cell-associated PRRs (serum/interstitial space). Further subdivision is based on structure–function characteristics (e.g., Toll-like receptor (TLR) group versus non-TLR groups). Leukocytes and, to a certain degree, epithelial cells of the mucosa use PRRs to recognize the provoking microorganisms, and to induce a defense program for controlling the microbial density at interfaces and elimination of microbial invaders (see Figure 5.2).

5.2.1.1 TLRs

The Toll receptor was initially discovered as important protein in embryogenesis of *Drosophila melanogaster* in 1985 [10]. About 10 years later it was recognized that Toll is involved in defense against fungal and bacterial infection of flies [11]. This observation prompted the search for homologs in mammals and led to the term TLR. Presently, in *Homo sapiens* 10 functional TLRs (TLR1–TLR10) and in mice (*Mus musculus*) 12 functional TLRs (TLR1–TLR9, TLR11–TLR13) have been identified (reviewed in [8, 12]). The TLRs are type I transmembrane proteins comprising an N-terminal ectodomain, an α-helical membrane domain, and a C-terminal cytoplasmic domain. The ectodomain of TLRs consist of 18–25 tandem copies of a leucine-rich-repeat (LRR) motif which form horseshoe-like solenoid shapes that

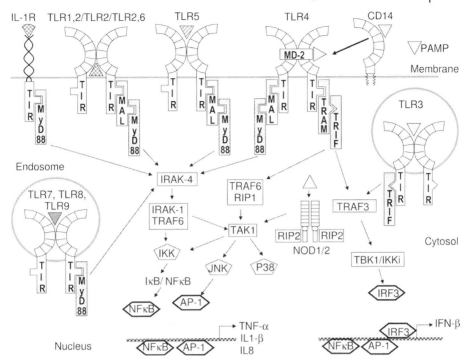

Figure 5.2 The large diversity of PAMP molecules is sensed by a set of membrane-integrated TLRs and cytosolic receptors (only NOD1 and NOD2 are shown). PAMP signaling leads to a proinflammatory response including release of cytokines, chemokines and activation of other defense programs. For detailed description of the signal transduction components, see text.

have been confirmed by crystal structures of TLR1/TLR2 heterodimers, and TLR3 and TLR4 homodimers [13, 14]. The cytoplasmic domain is homologous to that of the interleukin (IL)-1 receptor (R) and is therefore called the Toll/IL-1R homology (TIR) domain (Figure 5.2). According to this, it is not surprising that TLRs and IL-1R share several adapter proteins and signal transduction components leading to nuclear translocation of the transcriptional activator complexes of nuclear factor-κB (NF-κB) and activator protein-1 (AP-1), and of subsequent expression of a set of genes involved in innate immune defense (e.g., proinflammatory cytokines such as tumor necrosis factor (TNF)-α, IL-1β, and IL-6, chemokines such as IL-8, and interferon (IFN)-α, -β, and -γ, and bactericidal small cationic peptides such as defensins; see Figures 5.1 and 5.2).

In general, the TLRs form homodimers through the membrane-proximal portion of the ectodomain within the cytoplasmic or endosomal membrane of the cell (TLR2 is able to heterodimerize with TLR1 and TLR6, respectively) leading to different ligand specificities. Specific interaction between TLR ectodomains and cognate ligands causes conformational changes of the receptors and activation of the

TIR domains which initiates a cascade of interactions with TIR-containing adapter proteins (Figure 5.2): MyD88 (myeloid differentiation primary response gene 88), Mal (MyD88 adaptor-like, also known as TIRAP), TRIF (TIR domain-containing adapter inducing IFN-β, also known as TICAM1), and TRAM (TRIF-related adapter molecule, also known as TICAM2). Mal is utilized by TLRs that signal from the cytoplasmic membrane such as TLR4, TLRR1/2, and TLR2/6 after binding to phosphatidylinositol 4,5-bisphosphate (PIP2, lipid of the cytoplasmic leaflet of the membrane). The next components of downstream TLR signaling are IL-R-associated serine/threonine kinases (IRAK1/4), transforming growth factor-β (TGF-β)-activated kinase (TGF-β-activated kinase TAK1), TAK1-binding protein 1/2 (TAB1/2) and TNF-R-associated factors 3/6 (TRAF3/6), and receptor-interacting protein-1 (RIP1). The kinase complexes IRAK1/TRAF6 and TAK1/TAB1/2 activate the mitogen-activated protein kinases (MAPKs) p38 and c-Jun N-terminal kinase (JNK), and the inhibitor of NF-ϰB (IϰB) kinase complex (IKKα/β/γ). The IKK complex then phosphorylates IϰB, which leads to its ubiquitylation and subsequent degradation by the proteosome [15]. This allows the NF-ϰB to translocate to the nucleus and to activate genes involved in the proinflammatory response. Similarly JNK phosphorylates c-Jun and c-Fos, which are components of AP-1, and p38 activates activating transcription factor-2 (ATF2) and Elk-1. Together JNK and p38 are involved in IL-12 production and T helper (T_h) 1 cell differentiation. TLR4 and TLR3 are able to bind the adapter protein TRIF, which leads to activation of TRAF family member-associated NF-ϰB activator TANK-binding kinase-1 (TBK1) and IKKε by TRAF3, leading to activation of the IFN-regulatory factor-3 (IRF3) and the induction of IFN-β. As shown in Figure 5.2, all TLRs have in common the activation of NF-ϰB/AP-1 pathway, but only TLR4 and TLR3 activate additionally IFN-β production [8].

The TLRs location depends on the TLR type (Figure 5.2). TLR1, 2, 4, 5 and 6 are inserted in the cytoplasmic membrane and sense PAMPs of the exterior environment, whereas TLR3, 7, 8 and 9 are anchored in endosomal membranes and sense PAMPs in vesicular compartments (endosomes and phagosomes). It is of note that TLR4 initially activates the MyD88 pathway from the cell surface and after endocytosis the TRIF pathway, demonstrating spatial and temporal regulation of TLR signaling.

The main ligands/PAMPs recognized by the different TLRs are summarized in Table 5.1. There are specific and shared microbial patterns recognized by TLRs in some instances (it is also possible that PAMP preparations were contaminated by other PAMP types). However, in general, bacterial cell wall (peptidoglycan and lipoproteins) and outer membrane components (LPS and lipoproteins) and flagellin are recognized by TLRs on the host cell surface, whereas endosomal TLRs recognize nucleic acid-like structures (bacterial nonmethylated DNA carrying CpG motifs, viral double-stranded and single-stranded RNA).

5.2.1.2 Cytosolic PAMP Receptors

The cytosolic PAMP receptors comprise the nucleotide-binding and oligomerization domain (NOD)-like receptors (NLRs) and the retinoic acid-inducible gene I (RIG-I)-like receptors (RLRs) (reviewed in [16]). In contrast to TLR-mediated signaling pathways and ligands, much less is understood about the NLR-mediated signaling pathways. NLRs are defined by their tripartite domain structure: an N-terminal protein–protein interaction domain such as the caspase recruitment domain (CARD), pyrin domain (PYD), or the baculoviral inhibitory repeat (BIR), an intermediary NOD domain, and a C-terminal LRR domain. The NOD domain binds nucleotide (ATP), confers ATPase activity, and regulates oligomerization. The LRR domain is involved in PAMP sensing and autoregulation of protein–protein interaction and signaling. NLRs are highly conserved with orthologs in animals and plants (plant disease-resistant protein). The NLR proteins have been subclassified into five groups according to their N-terminal domains: NLRA (also known as CIITA), NLRB (NAIP), NLRC3-5 (including NOD1- and NOD2-containing CARD, see Figure 5.2), LNRP1-14 (NALP) and NLRX. The NLR family members detect cytosolic PAMPs and DAMPs, and can be also central components of inflammasomes (activation of caspase 1, IL-1 secretion, cell death) that play a key role in innate immune response against pathogens [17]. The first identified NLRs were NOD1 and NOD2 which sense the bacterial peptidoglycan fragments γ-D-glutamyl-meso-diaminopimelic acid (*meso*-DAP or iE-DAP, present in all Gram-negative and certain Gram-positive bacteria) and muranyl dipeptide (MDP, present in all Gram-positive and Gram-negative bacteria) (see Table 5.2). Ligand recognition induces conformational changes and self-oligomerization of NOD1 or NOD2 followed by recruitment of the receptor-interacting protein-2 (RIP2), and activation of NF-κB and the MAPK signaling pathways (p38, JNK, and extracellular signal-regulated kinase (ERK); see Figure 5.2). NOD1/2 activation alone does not lead to inflammasome formation as it will occur with activation of other NLR family members [17]. Recently, it has been shown that inflammasome activation even through MDP activation of NOD2 can occur by a two-step process which requires the NOD1/2 agonists and external ATP that activates the P2X7R purinergic receptor resulting in formation of the pannexin-1 pore which promotes K^+ efflux and Ca^{2+} influx [18]. These events stimulate the NLR inflammasome which leads to caspase-1 activation and release of activated proinflammatory cytokines such as IL-1 and IL-18, and eventually cell death by pyroptosis [19].

5.2.1.3 PAMPs as Chemoattractants

The migration of professional phagocytes (neutrophils and macrophages) from the blood circulation towards sites of bacterial infection is controlled by the gradient of chemoattractants released from the pathogen [20, 21]. On the one hand, the pathogen may activate the host complement system (major fluid-phase components called C1–C9) on the bacterial surface, and release C3a and C5a, which are cleaved fragments of complement factors C3 and C5, respectively [22]. C3a and C5a act as host-derived chemoattractants for phagocytes. Apart from the host intrinsic safeguard component, prokaryotes release *N*-formylated peptide chemoattractants derived from cleavage of the N-terminal oligopeptide intrinsic proteins [23, 24]. The prototype is *N*-formyl-methionyl-leucyl-phenylalanine (fMLF) which is recognized by formyl-peptide receptors (FPR, FPR-like-1 and -2 receptor) (see Table 5.2). The starter amino acid in ribosomal peptide biosynthesis of prokaryotes is

N-formyl-methionine. Therefore, it is not surprising that N-formylated peptides can also be released from disrupted mitochondria of damaged host cells. The FPR receptor family as well as the receptor for C5a (C5aR) and the receptors for host chemokines (e.g., CXCR1/2 for IL-8) transduce an activation signal through the subunit $G_{\alpha i}$ of heterotrimeric $G_{\alpha,\beta,\gamma}$-proteins, which are coupled to the seven transmembrane domain G-protein-coupled receptor (GPCR; see Figure 5.3). A characteristic feature of the $G_{\alpha i}$ family members is the sensitivity to pertussis toxin (PTX; inhibition of $G_{\alpha i}$ subunit by ADP ribosylation). Activation of GPCRs by chemoattractants leads to dissociation of the GTP-binding protein $G_{\alpha i}$ from the $G_{\beta\gamma}$ subunit and activation of signal transduction cascades involving MAPKs ERK1/2 and p38, phospholipase (PLC) β, phosphoinositide 3-kinase γ (PI3Kγ), phosphorylation of PIP2 to phoshpoinositol-3,4,5-triphosphate (PIP3), hydrolysis of PIP2 into diacylglycerol (DAG, activates protein kinase C (PKC)) and inositol triphosphate (IP3, induces Ca^{2+} release from the endoplasmic reticulum (ER); see Figure 5.3). Finally, these activated pathways lead to activation of the low-molecular-weight G-proteins of the Rho family (RhoA, Rac1/2, and Cdc42) that regulate leukocyte migration towards the chemoattractant gradient, activation of the NADPH oxidase by Rac2 (production of reactive oxygen species (ROS)), and phagocytosis of microbial pathogens or damaged cells [24].

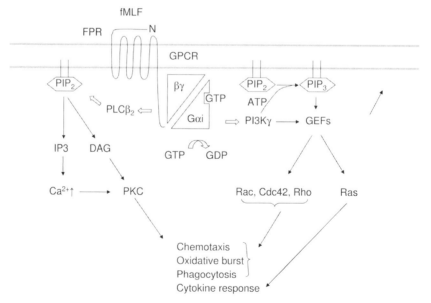

Figure 5.3 Scheme of GPCR for fMLF involving G-protein subunit $G_{\alpha i}$. Agonist binding to the seven transmembrane receptor FPR leads to exchange of GDP and binding of GTP to $G_{\alpha i}$ (activation). The consequence is activation of PLCβ2, PKC, PI3Kγ, G-protein exchange factor (GEF), and RhoGTPases, inducing cytoskeleton rearrangements. A similar signal transduction pathway is induced by receptors for C5a or chemokines.

5.2.2
Language of Hormones

The term hormone (Greek: *hormao* = to power/drive) is used for small fluid-phase signaling molecules. They are produced by specific tissue organs (glands) of animals, and are destined for coordination of the physiology of the organism and fine-tuning of the hormonal network (subject: endocrinology). Apart from the bacterial unicellular lifestyle, bacteria can form multicellular communities (e.g., microcolonies and biofilms) in a process of quorum sensing (QS). When approaching a high bacterial cell density (quorum) bacteria release hormone-like signaling molecules that will be sensed by themselves (autoinducers AI), but also by microorganisms in the environment and alter gene expression or even induce a cell differentiation program (see Chapters 1 and 2) [4, 25, 26]. Gram-negative bacteria utilize lipid-based molecules for QS such as acyl homoserine lactones (AHLs), alkyl hydroxyquinolones (AHQs), and tetrahydroxy methyl furan (THMF or autoinducer AI-2/LuxS). AI-2 class members are shared by Gram-negative and Gram-positive bacteria, whereas autoinducing cyclic peptides (autoinducing peptideAIP-I and -II) are exclusively produced and utilized by Gram-positive bacteria. AI-2 class members and AIP class members activate two-component signaling pathways, whereas lipid-based AI molecules translocate across the cytoplasmic membrane and are recognized in the cytoplasm by its cognate response regulator. The QS systems are involved not only in biofilm formation, but also control the expression of virulence genes of pathogenic bacteria. More recently, it became evident that AI molecules act also on host cells and mediate the cross-talk between pathogen and host. Vice versa, the host is also able to communicate with the pathogen by releasing catecholamines such as epinephrine or noradrenalin into the circulation and gut lumen where these hormones are recognized by bacteria, and control growth and expression of virulence genes [4]. It is suggested that these catecholamines are structurally related to the autoinducer AI-3 (presumably amidated aromatic compound) and, thus, this group of AIs could be the prototype in interkingdom signaling. The effect of noradrenaline (NA) on enterohemorrhagic *E. coli* (pathotype: EHEC) has been studied in more detail. As catecholamines carry two vicinal hydroxyl groups that are able to bind Fe^{3+}, they can probably provide *E. coli* with iron for growth (siderophore effect). For upregulation of the virulence program in the gut lumen EHEC senses three signaling molecules: the bacterial signal AI-3, generated by the gut microbiota, and the host-produced signaling molecules epinephrine and NA. All three signals can be blocked by adrenergic antagonists, suggesting signaling convergence. Indeed, two sensor kinases of the cytoplasmic membrane could be identified (see also Chapter 8). QseC recognizes AI-3, epinephrine, and NA, and transduces its signal as a kinase to the response regulator QseB. Phosphorylated QseB acts as transcriptional activator on a set of genes including *flhDC* (master regulator for motility), *stxAB* (encoding Shiga toxin), locus of enterocyte effacement (LEE, encoding type 3 secretion system, adhesin intimin and its receptor, effector proteins) and the operon of a second two-component signal transduction system QseE/QseF that recognizes probably adrenaline/NA and supports QseB in activation of LEE and induction of pedestral

Table 5.2 Secreted microbial signaling molecules.

Signaling molecules	Host receptor targets/molecules	Signaling effect	References
Small signaling molecules			
AHLs	?	NFκB ↓, p38 ↑, elF2α ↓ attenuation of TLR4-signaling, inhibition of mRNA translation	[4] [26]
AHQs	HIF-P4H FIH	Fe^{3+} chelation → HIF-1α ↑ inhibition of Fe^{2+}-dependent HIF proly-4-hydroxylases (HIF-P4Hs) and asparaginyl hydroxylase (FIH)	[7] [28]
Siderophores	HIF-P4H FIH	high-affinity Fe^{3+} chelation HIF-1α ↑, glucose transport ↑ host bactericidal response ↑	[9] [28–30]
Exotoxins/modulins			
Superantigen	TCR, MHC class II	bridging TCR and MHC class II molecule of APCs release of inflammatory cytokines → toxic shock syndrome	[31]
CTX	ganglioside GM1 ADP ribosylation of $G_{\alpha s}$	AC, c-AMP ↑ dampening bacterial response of phagocytes supporting $T_h 2$ response	[32]
PTX	glycoprotein ADP ribosylation of $G_{\alpha i}$	AC ↑, c-AMP ↑ dampening phagocytes supporting $T_h 2$ response	[33]
Pertussis CyaA	$\alpha_M\beta_2$ calmodulin	CyA, c-AMP ↑ anti-inflammatory response, $T_h 2$ ↑	[34]
Helicobacter VacA	$\alpha\beta_2$ calcineurin	NFAT ↓, IL-2 ↓ inhibition of calcineurin phosphatase in T cells → NFAT ↓, IL-2 ↓	[35]
C. difficile toxins A and B	RhoGTPases	RhoA ↓, Rac ↓, Cdc42 ↓ damage of mucosal barrier function, proinflammatory response ↑	[36]

formations by intestinal epithelial cells for EHEC binding on the surface of the brush border. As adrenaline/NA act as an α_2-agonists on G-protein subunit $G_{\alpha i}$ of GPCRs (similar to fMLF receptor, Figure 5.3), it is attractive to speculate whether the aromatic compound AI-3 produced by EHEC and gut microbiota is also able to act as an α_2-

agonist on host cells (unfortunately, AI-3 has not yet been be purified). Apart from EHEC, other enteric pathogens such as *Shigella*, *Salmonella*, and *Yersinia* as well as commensal *E. coli* also produce AI-3 and carry the QseC sensor system, which may explain their growth response to NA.

The lipid-like AHL molecules, which are produced by *Pseudomonas aeruginosa* (causing infections of wounds and respiratory tract as opportunistic pathogen) and enteropathogenic *Yersinia* species (but not by *E. coli* and *Salmonella*), are involved in bacterial QS and immunosuppression of the host (Table 5.2). The most active AHL appears to be N-(3-oxo-dodecanoyl) homoserine lactone (AHL-C12), which suppresses selectively NF-κB activation of PAMP-activated macrophages by sparing activation of MAPK p38, which leads to phosphorylation of the eukaryotic translation initiation factor-2α (elF2α) and subsequent inhibition of ribosomal protein biosynthesis [27]. Thus, AHL-C12 modulates the response of PAMP-activated macrophages by silencing the expression of proinflammatory cytokines and protein biosynthesis, and thus contributes obviously to pathogenicity as a messenger molecule. The AHL-C12 molecules can be degraded to 3-(1-hydroxydecylidene)-5-(2-hydroxyethyl) pyrrolidine-2,4-dione (tetramic acid), which binds iron with an affinity comparable to moderate-affinity microbial Fe^{3+} chelators (range of normalized Fe^{3+} chelating constants for siderophores 20–40 pM) [28]. The *Pseudomonas* quinolone signal (PQS) molecule 2-heptyl-3-hydroxy-4-quinolone, which is interconnected with the AHL-mediated QS systems RhlR/I and LasR/I, is directly involved in iron binding and entrapment in the membrane [29] (Table 5.2).

Due to the Fe^{3+}-binding function of tetramic acid and PQS, these iron chelators are suggested to act also on host cells. This assumption is based on the observation that released microbial siderophores, which are the major high-affinity Fe^{3+} chelators for iron uptake by bacteria, are able to induce a hypoxic response in host cells [6]. Host cells and, in particular, professional phagocytes respond to changes in oxygen tension by the oxygen-sensitive heterodimer transcriptional regulator hypoxia-inducible factor (HIF) [7]. Myeloid-targeted HIF-1α knockout mice (specific deletion of the HIF-1α gene in neutrophils and macrophages) display several defects that are responsible for decreased bactericidal activity: (i) reduction of survival time of neutrophils, (ii) reduced ATP pool because of ineffective glycolysis, (iii) reduced granule protease activity, nitric oxide (NO) release, and TNF-α production, and (iv) reduced expression of β₂ integrin and chemokine receptors of neutrophils and monocytes/macrophages, leading to impaired recruitment of phagocytes to sites of tissue infection and damage [30].

The constitutively expressed level of the subunit HIF-1α is regulated by oxygen-sensitive Fe^{2+}-containing dioxgenases that hydroxylate HIF-1α in two ways: (i) prolyl hydroxylation of HIF-1α by prolyl hydroxylase domain (PHD)-containing enzymes (PHD 1–3) leads to proteasomal degradation of HIF-1α and (ii) asparaginyl hydroxylation of factor inhibiting HIF (FIH, which inhibits binding of HIF to hypoxia response elements (HREs)). Microbial iron chelators such as siderophores reduce the activity of PHDs and FIH, which leads to accumulation of active HIF and upregulation of the HIF regulon in spite of normoxic growth conditions [6, 7]. As the transcription factor NF-κB is also involved in HIF-1α expression, recognition of

PAMPs lead also to upregulation of HIF-1α and, thus, to enhancement of the host microbial defense program.

5.2.3
Extracellular Bacterial Toxins in Pathogen–Host Cell Communication

Bacterial toxins, which are secreted into extracellular space (interstitium, mucus layer), can act on host cell surfaces or on intracellular signaling cascades. The former group comprises pore-forming toxins and so-called superantigens. The pore-forming toxins punch holes in the cytoplasmic membrane leading usually to cell death and interruption of pathogen–host cell communication. Therefore, these toxins are not discussed further. In contrast, extracellular toxins acting on intracellular signaling pathways aim at reprogramming host cells probably in favor of the pathogen. Here, several paradigmatic bacterial toxins (also called modulins) are discussed [3] (see also Table 5.3).

5.2.3.1 Superantigens
Superantigens are produced by Gram-positive and a few Gram-negative bacteria. As bivalent proteins they are able to cross-link the T cell receptor (TCR) of T lymphocytes with the major histocompatibility complex (MHC) class II receptor of antigen-presenting cells (APCs: DCs and macrophages) even in the absence of the specific peptide [31]. This bridging of T cells and APCs results in strong T cell proliferation and massive release of proinflammatory cytokines (e.g., IL-1, TNF-α, and IFN-γ). The superantigen-induced "cytokine storm" ultimately leads to toxic shock syndrome, including fever, hypotension, vomiting, diarrhea and so on. The prototype of superantigen is the toxic shock syndrome toxin-1 (TSST-1) produced by certain *Staphylococcus aureus* strains. It is not clear yet whether the pathogen benefits from this effect of generalized T cell activation by superantigens.

5.2.3.2 Cholera Toxin
Vibrio cholerae secretes the phage-encoded cholera toxin (CTX) sec-dependent into the periplasmic space where five B subunits (receptor-binding function) and one A subunit (enzymatic activity) are assembled and then translocated across the outer membrane by the type II secretion system (T2SS). The released CTX then binds to target cells, expressing the ganglioside receptor GM1. Subsequently, bound CTX is endocytosed, processed in the ER, and released as active toxin subunit A1 into the cytoplasm [32]. The subunit A1 has an ADP ribosyltransferase activity and ADP ribosylates preferentially the stimulating $G_{\alpha s}$ subunit of heterotrimeric G-proteins. This modification stabilizes GTP binding of $G_{\alpha s}$ by inhibiting intrinsic GTPase activity. In the intestinal mucosa the CTX-modified $G_{\alpha s}$ stimulates persistently the adenyl cyclase (AC) and, thus, increases the level of cAMP. This second messenger activates protein kinase A which subsequently activates the cystic fibrosis transmembrane conductance regulator (CFTR) chloride channel, leading to the known massive watery diarrhea of cholera (CFTR is impaired in cystic fibrosis patients). By a similar mechanism of AC stimulation, CTX silences

phagocytes. Thus, *V. cholerae* benefits from CTX induction of host exudates in the gut lumen as growth medium and additionally from washing out of the microbiota of the colon.

5.2.3.3 Bordetella Modulins

The whooping cough causative agent, *Bordetella pertussis*, secretes two toxins that upregulate cAMP in host cells [33]. PTX consists of five B subunits and one A subunit (similar to CTX). It binds to glycoproteins of the respiratory epithelium, carrying mannose and N-acetylglucosamine residues. The A subunit acts as ADP ribosyl-transferase on the inhibiting subunit $G_{\alpha i}$ of heterotrimeric G-proteins by preventing the GDP exchange for GTP that stabilizes the inactive state of $G_{\alpha i}$. Thus, upregulated AC in host cells cannot be inhibited by active $G_{\alpha i}$-GTP (physiologically destined for this function) and, thus, accumulation of c-AMP affects the cell similarly as in the case of CTX-treated cells.

B. pertussis is also able to secrete by the type I secretion system (T1SS) a calmodulin-dependent AC (CyaA). Secreted CyaA binds to adjacent $\alpha_M\beta_2$ integrin (expressed by myeloid cells like macrophages, DCs, and neutrophils) and the N-terminal catalytic domain translocates across the cytoplasmic membrane [34]. Ca^{2+} binding to the regulatory protein calmodulin leads to activation of the adenylate cyclase domain of CyaA. The increase of cAMP in myelotic cells suppresses production of ROS, chemotaxis, phagocytosis and induction of the anti-inflammatory cytokine IL-10 with the ultimate consequence of the expansion of regulatory T cells and T_h2-directed immune response (enhanced IgA response).

5.2.3.4 Helicobacter

5.2.3.5 VacA

The gastric ulcer-inducing pathogen *Helicobacter pylori* secretes the vacuolating toxin VacA, which also acts as an immunosuppressing modulin directly on T cells [35]. VacA is secreted to the surface of the bacterium by the type V secretion system (T5SS) and is released by proteolytic processing on the bacterial surface. VacA enters migrating primary human T lymphocytes expressing β_2 (CD18) integrin (VacA is not recognized by mouse β_2). In the cytosol VacA prevents the Ca^{2+}-dependent phosphatase calcineurin to dephosphorylate and thus, to inactivate the nuclear factor of activated T cells (NFAT). This inhibition leads to downregulation of IL-2 transcription in T cells and consequently to T cell suppression.

5.2.3.6 *Clostridium difficile* Toxins

The small GTPases of the Rho protein family (Rho, Rac, and Cdc42) are involved in integrin-dependent cell adhesion and migration, phagocytosis, production of ROS, exocytosis of granule protease, regulation of stress kinases, and T and B cell function (see Figure 5.3). This multifunctionality of Rho proteins make them to attractive targets of secreted bacterial modulins, such as *C. difficile* toxins [36]. *C. difficile* is a Gram-positive anaerobic pathogen which is involved in antibiotic-associated pseudomembranous colitis. The pathogen secrete two large toxins (toxin A and B with masses of 250 and 308 kDa), which are endocytosed by the intestinal epithelium, proteolytically processed, and released into the cytosol. The glucosyltransferase activity of the N-terminal region of the toxins glycosylate the Rho family members RhoA, Rac1, and Cdc42 at a specific threonine residue, resulting in irreversible inactivation. As the control of redistribution of the cytoskeleton of intoxicated cells is lost, the cells round up and cell barrier functions of enterocytes are impaired. Tissue damage of the intestinal epithelium by the toxins and recruitment of neutrophils might explain the manifestation as pseudomembranous colitis.

5.3
Short-Distance Communication

Bacterial pathogens are able to contact directly host cells through bacterial adhesin–host cell receptor interaction. This kind of "hand shaking" communication is more focused and cell-specific than that of long-distance communication through fluid mediator molecules. Moreover, short-distance communication initiates frequently the pathogenicity program of the microbe for invasion of host tissue and subsequent or intracellular multiplication. This strategy of the pathogen also includes reprograming of host cells in favor of the pathogen. In this chapter we summarize important host cell receptors and bacterial adhesins involved in host–pathogen communication (Table 5.3).

5.3.1
Bacterial Adhesins and Host Cell Receptors

Bacterial adhesins can be divided into two classes: fimbrial adhesins and nonfimbrial adhesins. Here, we present several typical examples of bacterial adhesins that are recognized by host cell receptors and are involved in activation of signal transduction pathways of host cells [37, 38].

Fimbrial adhesins are long surface appendages of about 2 µm length and 2–8 nm in diameter. They consist of more than 1000 major protein subunits arranged helically to form a pilus-like structure. The pilus ends in a thin and flexible rod, consisting of minor subunits and the tip adhesin, which has lectin-like binding function for carbohydrate residues. Thus, the tip adhesin determines the binding specificity for the cell type targeted by the pathogen (tissue tropism) and is, therefore, characteristic for the pathotype of a bacterial species such as *E. coli* (see Table 5.3). Enterotoxic *E. coli* (pathotype: ETEC) bind preferentially to asialogangliosides of brush border epithelium of the jejunum and ileum by fimbriae called colonization factor-1/2 (CF1/2). Uropathogenic *E. coli* (pathotype: UPEC) colonize the bladder epithelial layer by P-fimbriae-mediated adherence to galactosyl residues of glycosphingolipids [39]. This interaction appears to activate acid sphingomyelinase (ASM) and cleavage of the P-fimbriae receptor, glycosphingomyelin, resulting in release of ceramide and stimulation of TLR4 signal transduction (cytokine response). Recently, it has been demonstrated that also the tip protein FimH of type 1 fimbriae of UPECs may interact

Table 5.3 Bacterial adhesins and host receptors.

Structure	Pathogen	Host receptor	References
Fimbrial adhesions			
Type 1 fimbriae	E. coli (all)	mannosyl residues of glycolipids	[40]
P-fimbriae	E. coli (UPEC)	galactosyl α(1–4)β-galactosyl residues of glycosphingolipids	[39]
S-fimbriae	E. coli (MENEC)	sialyl α(1–4)β-galactosyl residues of glycolipids	[39]
CFA/1	E. coli (ETEC)	asialo-GM1 gangliosides, glycoproteins	[38]
Type IV fimbriae	P. aeruginosa	asialo-GM1 ganglioside	[41]
	Neisseria spp.	glycoprotein CD46	[42]
	E. coli (EPEC)	?	
	V. cholerae	?	
Nonfimbrial adhesions			
YadA	Y. enterocolitica	fibronectin/β_1 integrin	[48]
	Y. pseudotuberculosis		
UspA	M. catarrhalis	CEACAM1	[43]
OmpP5	H. influenzae	CEACAM1	[43]
NadA	N. meningitidis	β_1 (?)	[43]
Opa	Neisseria spp.	CEACAM1, 3 and 6	[43]
Invasin	Y. enterocolitica	β1 integrin	[45]
	Y. pseudotuberculosis		[46]
Intimin	E. coli (EHEC, EPEC)	Tir	[50]
Internalin A	L. monocytogenes	E-cadherin	[49]
Internalin B	L. monocytogenes	Met, C1qR	[49]
FnbP	S. aureus	fibronectin/β_1 integrin	[47]

directly with TLR4 and independently on the presence of the LPS-binding proteins CD14 and MD-2 [40]. Some UPEC strains are additionally endowed with S-fimbriae that recognize sialic acid (N-acetylneuraminic acid) residues of glycolipids, present on microvascular endothelial cells. This adhesin enables those E. coli that have reached the blood circulation to traverse the blood–brain barrier and cause meningitis (pathotype: meningitis-associated E. coli MENEC) in neonates.

Type IV pili have several distinctive features with respect to the biogenesis and structures of the fimbriae group discussed above [41]. They are thinner (about 6 nm in diameter) and longer (1–20 μm in length) and require a specific T2SS that is involved in pili elongation and retraction, and thus enables twitching motility (movement on cell surface at rates of 0.05–1 μm/s). Pathogenic Neisseria species (N. meningitidis and N. gonorhoeae) appear to exploit type IV pili extension for cellular attachment (in particular when the pathogen is encapsulated by polysaccharides) followed by retraction of the pili to decrease the distance between pathogen and host cell, and to bring nonfimbrial bacterial adhesins into contact with cellular receptors [42]. The major short-range adhesins of pathogenic neisseriae comprises a set of outer membrane proteins called opacity proteins (Opa, involved in the opacity phenotype of colonies) [43]. These eight-stranded β-barrel proteins of the outer membrane expose four loops to the surface that are highly variable and mediate adhesion to heparan sulfate proteoglycans and CEACAMs. CEACAMs belong to the immunoglobulin superfamily of cell adhesion molecules. The surface-exposed domain organization of CEACAM comprises one to six immunoglobulin-like domains that are highly glycosylated. Some CEACAM members are anchored to the cytoplasmic membrane by a helical transmembrane domain ending with a C-terminal cytoplasmic domain (e.g., CEACAM1 and 3) and others are attached to glycosylphosphatidylinositol (GPI) anchors (CEACAM5–8). CEACAM1 is found on epithelial and endothelial cells as well as on leukocytes, and it is present in humans and rodents. The cytoplasmic domain carries two tyrosine phosphorylatable residues and constitutes a functional immunoreceptor tyrosine-based inhibitory motif (ITIM).

In contrast to CEACAM1, CEACAM3 is restricted to human granulocytes and carries a cytoplasmic immunoreceptor tyrosine-based activation motif (ITAM), triggering nonopsonic uptake of attached bacterial pathogens [43]. As shown in Table 5.3, bacterial pathogens that colonize the human mucosa such as *Moraxella catarrhalis*, *Haemophilus influenzae*, and pathogenic *Neisseria* spp. utilize CEACAM1 as receptor for their nonfimbrial adhesins. Attachment can support surface colonization of the mucosa, but it can also trigger internalization and translocation of the pathogen across the mucosa. This occurs because CEACAM1 transmembrane domain directs the attached cargo to cholesterol-rich membrane microdomains (lipid rafts), the signaling platform for Rac-dependent endocytosis [44]. When *Neisseria* spp. interact with CEACAM1 of T lymphocytes they inhibit T cell proliferation due to the cytoplasmic ITAM which recruits SH2 domain-containing protein tyrosine phosphatase (SHP-1 or -2) which slow down activating signals.

Integrins are another group of cell receptors targeted by nonfimbrial adhesins. In the context of tissue integrity, they are the principle cell surface adhesion receptors mediating cell–extracellular matrix (ECM) adhesion [45]. They form heterodimeric receptors generated by selective pairing between eighteen α and eight β subunits expressed in a tissue- or stage-restricted manner. Integrin $β_1$ is ubiquitously expressed, whereas integrins $β_2$ and $β_3$ are preferentially expressed by leukocytes. Integrins can be engaged directly by bacterial adhesins or indirectly by bridging integrins with bacterial adhesins via recruitment of ECM proteins (e.g., fibronectin and collagen). The enteric pathogens *Yersinia enterocolitica* and *Y. pseudotuberculosis* express an outer membrane protein called invasin which interacts specifically with $αβ_1$ integrin and induces internalization [46]. As $αβ_1$ integrins are accessible on M cells covering the mucosa of Peyer's patches of the intestine, it is believed that *Yersinia* target these cells for translocation across the intestinal mucosa via invasin-$β_1$ integrin. *Staphylococcus* spp., but also *Yersinia* spp. and other pathogens are endowed with fibronectin-binding proteins (e.g., FnbP or YadA) that mediate indirect binding and signaling through $β_1$ and $β_3$ integrins [47, 48].

The intracellular pathogen *Listeria monocytogenes* uses several tissue specific receptors called internalins for invasion and internalization [49]. Internalin A interacts with E-cadherin of gut epithelial cells and, thus, initiates cell entry. Internalin B interacts with receptors Met (receptor for hepatocyte growth factor (HGF)) and gC1qR (receptor for complement component C1q) of fibroblasts, hepatocytes, and other stromal cells, resulting in establishing systemic infection of this intracellular pathogen.

A special case of pathogen–host cell communication is represented by EPEC or EHEC. These enteric pathogens express a *Yersinia* invasin-like adhesin called intimin, which does not preferentially interact with host cell receptors such as integrins, but with the translocated intimin receptor (Tir) of the pathogen that is integrated into the brush border membrane of intestinal cells by the bacterial T3SS. Translocated Tir attached to intimin subsequently activates the host cell actin assembly machinery resulting in actin-mediated pedestal formation underneath the bacterial-contacted cell [50]. This is an example how pathogens mimic signaling processes of the host by manipulation and imitation.

5.4 Conclusions

The study of host–pathogen communication highlights the sophisticated relationship between bacterial pathogens and host as a result of a long coevolution and reciprocal adaptation [51]. Unbalanced communication that leads to damage of tissue and symptomatic infectious disease can be due to the virulence activity of the pathogen or to excessive immune and inflammatory responses of the host. We have discussed major long-distance and short-distance communication pathways by excluding T3SS- and T4SS-mediated injection of antihost effector proteins. A better understanding of the languages of the two conflicting partners during the infection process would be crucial for the development of novel and more appropriate therapeutic strategies. A major handicap for studying bacterial–host communication *in vivo* has been the lack of appropriate mouse models simulating mucosal and systemic infection processes observed in humans. The generation of "humanized" transgenic mouse models might be a promising approach.

Acknowledgments

I am grateful to Mrs. Monika Krämer for the graphic work, thoroughness, and patience during preparing the manuscript.

References

1 Rakoff-Nahoum, S., Paglino, J., Eslami-Varzaneh, F., Edberg, S., and Medzhitov, R. (2004) Recognition of commensal microflora by Toll-like receptors is required for intestinal homeostasis. *Cell*, **118**, 229–241.

2 Stecher, B. and Hardt, W.D. (2008) The role of microbiota in infectious disease. *Trends Microbiol.*, **16**, 107–114.

3 Henderson, B., Poole, S., and Wilson, M. (1996) Bacterial modulins: a novel class of virulence factors which cause host tissue

pathology by inducing cytokine synthesis. *Microbiol. Rev.*, **60**, 316–341.
4 Hughes, D.T. and Sperandio, V. (2008) Inter-kingdom signalling: communication between bacteria and their hosts. *Nat. Rev. Microbiol.*, **6**, 111–120.
5 Bhavsar, A.P., Guttman, J.A., and Finlay, B.B. (2007) Manipulation of host-cell pathways by bacterial pathogens. *Nature*, **449**, 827–834.
6 Hartmann, H., Eltzschig, H.K., Wurz, H., Hantke, K., Rakin, A., Yazdi, A.S., Matteoli, G., Bohn, E., Autenrieth, I.B., Karhausen, J., Neumann, D., Colgan, S.P., and Kempf, V.A.J. (2008) Hypoxia-independent activation of HIF-1 by *Enterobacteriaceae* and their siderophores. *Gastroenterology*, **134**, 756–767.
7 Dehne, N. and Brüne, B. (2009) HIF-1 in the inflammatory microenvironment. *Exp. Cell Res.*, **315**, 1791–1797.
8 Ishii, K.J., Koyama, S., Nakagawa, A., Coban, C., and Akira, S. (2008) Host innate immune receptors and beyond: making sense of microbial infections. *Cell Host Microbe*, **3**, 352–363.
9 Bianchi, M.E. (2007) DAMPs, PAMPs and alarmins: all we need to know about danger. *J. Leukoc. Biol.*, **81**, 1–5.
10 Anderson, K.V., Jurgens, G., and Nusslein-Volhard, C. (1985) Establishment of dorsal–ventral polarity in the *Drosophila* embryo: genetic studies on the role of the Toll gene product. *Cell*, **42**, 779–789.
11 Lemaitre, B., Nicolas, E., Michaut, L., Reichhart, J., and Hoffmann, J. (1996) The dorsoventral regulatory gene cassette spätzle/Toll/cactus controls the potent antifungal response in *Drosophila* adults. *Cell*, **86**, 973–983.
12 Leulier, F. and Lemaitre, B. (2008) Toll-like receptors – taking an evolutionary approach. *Nat. Rev. Genet.*, **9**, 165–178.
13 Jin, M.S., Kim, S.E., Heo, J.Y., Lee, M.E., Kim, H.M., Paik, S.G., Lee, H., and Lee, J.O. (2007) Crystal structure of the TLR1–TLR2 heterodimer induced by binding of a tri-acylated lipopeptide. *Cell*, **130**, 1071–1082.
14 Park, B.S., Song, D.H., Kim, H.M., Choi, B.S., Lee, H., and Lee, J.O. (2009) The structural basis of lipopolysaccharide recognition by the TLR4–MD-2 complex. *Nature*, **458**, 1191–1196.
15 Ghosh, S. and Hayden, M.S. (2008) New regulators of NF-κB in inflammation. *Nat. Rev. Immunol.*, **8**, 837–848.
16 Fanchi, L., Park, J.H., Shaw, M.H., Marina-Garcia, N., Chen, G., Kim, Y.G., and Nunez, G. (2008) Intracellular NOD-like receptors in innate immunity, infection and disease. *Cell Microbiol.*, **10**, 1–8.
17 Chen, G., Shaw, M.H., Kim, Y.G., and Nunez, G. (2009) NOD-like receptors: role in innate immunity and inflammatory disease. *Annu. Rev. Pathol. Mech. Dis.*, **4**, 365–398.
18 Di Virgilio, F. (2007) Liaisons dangereuses: $P2X_7$ and the inflammasome. *Trends Pharmacol. Sci.*, **28**, 465–472.
19 Fink, S.L. and Cookson, B.T. (2005) Apoptosis, pyroptosis and necrosis: mechanistic description of dead and dying eukaryotic cells. *Infect. Immun.*, **73**, 1907–1916.
20 Kay, R.R., Landridge, P., Traynor, D., and Hoeller, O. (2008) Changing direction in the study of chemotaxis. *Nat. Rev. Mol. Cell Biol.*, **9**, 455–463.
21 Murdoch, C. and Finn, A. (2000) Chemokine receptors and their role in inflammation and infectious diseases. *Blood*, **95**, 3032–3043.
22 Lambris, J.D., Ricklin, D., and Geisbrecht, B.V. (2008) Complement evasion by human pathogens. *Nat. Rev. Microbiol.*, **6**, 132–142.
23 Le, Y., Murphy, P.M., and Wang, J.M. (2002) Formylpeptide receptors revisited. *Trends Immunol.*, **23**, 541–548.
24 Rabiet, M.J., Huet, E., and Boulay, F. (2007) The N-formyl peptide receptors and the anaphylatoxin C5a receptors: an overview. *Biochimie*, **89**, 1089–1106.
25 Aldridge, P.D., Gray, M.A., Hirst, B.H., and Khan, C.M.A. (2005) Who's talking to whom? Epithelial–bacterial pathogen interactions. *Mol. Microbiol.*, **55**, 655–663.
26 Shiner, E.K., Rumbaugh, K.P., and Williams, S.C. (2005) Interkingdom signaling: deciphering the language of acyl homoserine lactones. *FEMS Microbiol. Rev.*, **29**, 935–947.
27 Kravchenko, V.V., Kaufmann, G.F., Mathison, J.C., Scott, D.A., Katz, A.Z., Grauer, D.C., Lehmann, M., Meijler, M.M., Janda, K.D., and Ulevitch, R.J. (2008)

Modulation of gene expression via disruption of NF-κB signaling by a bacterial small molecule. *Science*, **321**, 259–263.

28 Schertzer, J.W., Boulette, M.L., and Whiteley, M. (2009) More than a signal: non-signaling properties of quorum sensing molecules. *Trends Microbiol.*, **17**, 189–195.

29 Diggle, S.P., Matthijs, S., Wright, V.J., Fletcher, M.P., Chhabra, S.R., Lamont, I.L., Kong, X., Hider, R.C., Cornelis, P., Camara, M., and Williams, P. (2007) The *Pseudomonas aeruginosa* 4-quinolone signal molecules HHQ and PQS play multifunctional roles in quorum sensing and iron entrapment. *Chem. Biol.*, **14**, 87–96.

30 Peyssonnaux, C., Datta, V., Cramer, T., Doedens, A., Theodorakis, E.A., Gallo, R.L., Hurtado-Ziola, N., Nizet, V., and Johnson, R.S. (2005) HIF-1α expression regulates the bactericidal capacity of phagocytes. *J. Clin. Invest.*, **115**, 1806–1815.

31 Kotb, M. (1998) Superantigens of Gram-positive bacteria: structure–function analyses and their implications for biological activity. *Curr. Opin. Microbiol.*, **1**, 56–65.

32 Vanden Broeck, D., Horvath, C., and De Wolf, M.J. (2007) *Vibrio cholerae* cholera toxin. *Int. J. Biochem. Cell Biol.*, **39**, 1771–1775.

33 Mattoo, S. and Cherry, J.D. (2005) Molecular pathogenesis, epidemiology and clinical manifestations of respiratory infections due to *Bordetella pertussis* and other *Bordetella* subspecies. *Clin. Microbiol. Rev.*, **18**, 326–382.

34 Vojtova, J., Kamanova, J., and Sebo, P. (2006) *Bordetella* adenylate cyclase toxin: a swift saboteur of host defense. *Curr. Opin. Microbiol.*, **9**, 69–75.

35 Sewald, X., Gebert-Vogl, B., Prassl, S., Barwig, I., Weiss, E., Fabbri, M., Osicka, R., Schiermann, M., Busch, D.H., Semmrich, M., Holzmann, B., Sebo, P., and Haas, R. (2008) Integrin subunit CD18 is the T-lymphocyte receptor for the *Helicobacter pylori* vacuolating cytotoxin. *Cell Host Microbe*, **3**, 20–29.

36 Aktories, K. and Barbieri, J.T. (2005) Bacterial cytotoxins: targeting eukaryotic switches. *Nat. Rev. Microbiol.*, **3**, 397–410.

37 Hauck, C.R., Agerer, F., Muenzner, P., and Schmitter, T. (2006) Cellular adhesion molecules as targets for bacterial infection. *Eur. J. Cell Biol.*, **85**, 235–242.

38 Viswanathan, V.K., Hodges, K., and Hecht, G. (2009) Enteric infection meets intestinal function: how bacterial pathogens cause diarrhoea. *Nat. Rev. Microbiol.*, **7**, 110–119.

39 Svanborg, C., Bergsten, G., Fischer, H., Goaly, G., Gustafsson, M., Karpman, D., Lundstedt, A.C., Ragnarsdottir, B., Svensson, M., and Wullt, B. (2006) Uropathogenic *Escherichia coli* as a model of host–parasite interaction. *Curr. Opin. Microbiol.*, **9**, 33–39.

40 Mossman, K.L., Mian, M.F., Lauzon, N.M., Gyles, C.L., Lichty, B., Mackenzie, R., Gill, N., and Ashkar, A.A. (2008) Cutting edge: FimH adhesin of type 1 fimbriae is a novel TLR4 ligand. *J. Immunol.*, **181**, 6702–6706.

41 Jarrell, K.F. and McBride, M.J. (2008) The surprising diverse ways that prokaryotes move. *Nat. Rev. Microbiol.*, **6**, 466–476.

42 Virji, M. (2009) Pathogenic neisseriae: surface modulation, pathogenesis and infection control. *Nat. Rev. Microbiol.*, **7**, 274–286.

43 Kuespert, K., Pils, S., and Hauch, C.R. (2006) CEACAMs: their role in physiology and pathophysiology. *Curr. Opin. Cell Biol.*, **18**, 565–571.

44 Muenzner, P., Bachmann, V., Kuespert, K., and Hauck, C.R. (2008) The CEACAM1 transmembrane domain, but not the cytoplasmic domain, directs internalization of human pathogens via membrane microdomains. *Cell Microbiol.*, **10**, 1074–1092.

45 DeMali, K.A., Wennerberg, K., and Burridge, K. (2003) Integrin signaling to the actin cytoskeleton. *Curr. Opin. Cell Biol.*, **15**, 572–582.

46 Isberg, R.R., Hamburger, Z., and Dersch, P. (2000) Signaling and invasin-promoted uptake via integrin receptors. *Microb. Infect.*, **2**, 793–801.

47 Hauck, C.R. and Ohlsen, K. (2006) Sticky connections: extracellular matrix protein recognition and integrin-mediated cellular invasion by *Staphylococcus aureus*. *Curr. Opin. Microbiol.*, **9**, 5–11.

48 Heesemann, J., Sing, A., and Trülzsch, K. (2006) Yersinia's stratagem: targeting innate and adaptive immune defense. *Curr. Opin. Microbiol.*, **9**, 55–61.

49 Hamon, M., Bierne, H., and Cossart, P. (2006) *Listeria monocytogenes*: a multi-faceted model. *Nat. Rev. Microbiol.*, **4**, 423–434.

50 Hayward, R.D., Leong, J.M., Koronakis, V., and Campellone, K.G. (2006) Exploiting pathogenic *Escherichia coli* to model transmembrane receptor signalling. *Nat. Rev. Microbiol.*, **4**, 358–370.

51 Roy, C.R. and Mocarski, E.S. (2007) Pathogen subversion of cell-intrinsic innate immunity. *Nat. Rev. Immunol.*, **8**, 1179–1187.

6
Identification of Bacterial Autoinducers –
Methods Chapter

Agnes Fekete, Michael Rothballer, Anton Hartmann, and Philippe Schmitt-Kopplin

6.1
Introduction

This is a specific chapter to describe various methods for the target analysis of known structures of quorum sensing (QS) molecules. Additionally, a general identification strategy for the characterization of unknown signal molecules and the basis of the commonly used analytical techniques are briefly discussed.

Gene regulation is often influenced by environmental concentrations of released diffusible substances – termed autoinducers (AIs) – to coordinate certain behaviors, such as the production of secondary metabolite(s), biofilm formation, virulence, or bioluminescence. These AIs are very common among bacterial species living in a variety of ecosystems [1, 2]. In many cases, several different AIs are released and perceived in bacteria [1]. This autoinduction is dependent on the concentration and often combination of AIs within a bacterial cell, which is a matter of cell density and distribution, as well as diffusion space depending on the surrounding microenvironment [3]. Since the distribution and concentration of small diffusible AIs reflects the fate of excreted, more costly biosynthesis products (like hydrolyzing enzymes or complex secondary metabolites), excretion and perception of AIs are the basis of bacterial "efficiency sensing" to optimize the adaptability to the given microhabitat, especially in colonized surfaces or biofilms [3]. The AI influences the gene activity by forming a complex with the receptor protein R, which, after a certain AI threshold concentration is reached, activates the expression of the AI biosynthesis gene I (positive feedback), but also of other AI-regulated genes. This phenomenon was termed QS or bacterial cell–cell communication [4]. The known QS signals are chemically diverse, which is shown in Figure 6.1. Cyclic oligopeptides are used mainly by Gram-positive bacteria, whereas *N*-acyl-homoserine lactones (AHLs) are produced by many Gram-negative bacteria. A furanon derivative–borate complex (AI-2) seems to act as a more general AI across phylogenetic borders and 4-hydroxy-2-alkyl-quinolones (HAQ) have been identified in a few Gram-negative species. However,

Bacterial Signaling. Edited by Reinhard Krämer and Kirsten Jung
Copyright © 2010 WILEY-VCH Verlag GmbH & Co. KGaA, Weinheim
ISBN: 978-3-527-32365-4

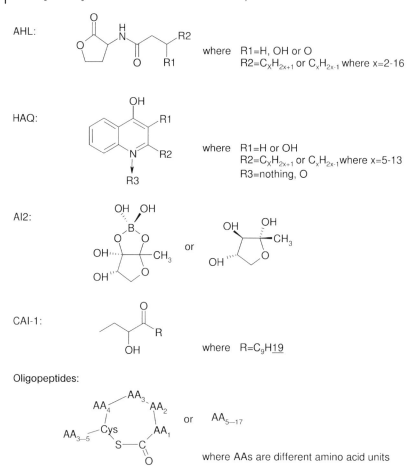

Figure 6.1 Structures of known AIs.

molecules with other structures might be involved in hitherto unknown QS systems.

Identifying a molecular structure that behaves as an AI is a difficult task and a combination of several analytical steps is required as summarized in Figure 6.2. The identification strategy is divided into two successive steps: (i) a screening step to recognize potential AIs followed by a preliminary classification with microbial biosensors and (ii) a structural characterization requiring an appropriate isolation of the signal molecule. The structure of the AI is concluded from the measurement of different physicochemical properties of the fraction containing the active components. In most cases a chemical synthesis of the identified molecules is needed for the final confirmation of the proposed chemical structure, and for investigating its stability during the analysis and in the biological system. Applying this identification strategy, several AI structures have been identified, like a novel derivative of AHLs [5] or CAI-1 [6].

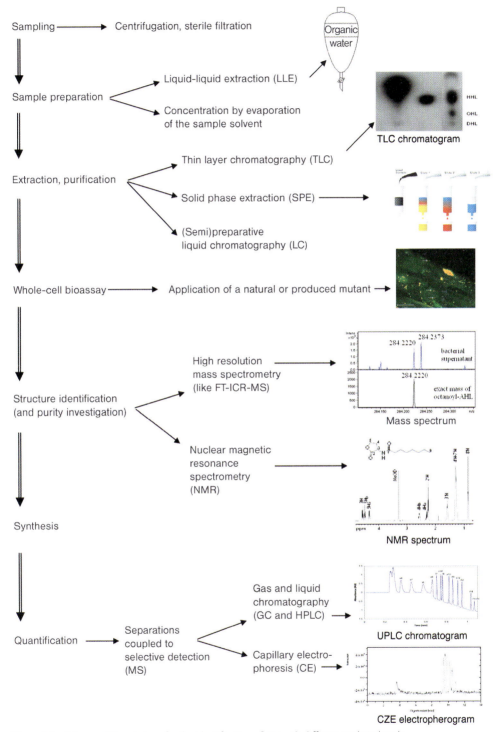

Figure 6.2 Scheme of the strategy for the identification of AIs with different analytical tools.

6.2
Biosensors

6.2.1
Biosensor Construction

The construction of an AI biosensor requires a bacterial strain as a host, which does not produce a specific AI itself. It can either be a mutant strain of an AI-producing wild-type or a bacterial species that is not able to synthesize AIs at all. If no natural AI-negative mutant can be selected, several methods exist to perform site-directed mutagenesis to knockout the AI synthesis gene (e.g., [7]), but a detailed description is beyond the scope of this chapter. If an AI-negative mutant exhibits AI-controlled fluorescence, bioluminescence, or production of other marker molecules when AIs are added externally, it can be used directly as a sensor strain. One example is the violacein production of *Chromobacterium violaceum* CV026 that is induced in the presence of AHLs with a short acyl side-chain [8, 9].

Another more elegant way to construct a biosensor for, for example, a certain range of different AHLs is to transfer the genes involved in one specific QS regulation system to a plasmid and replace the AI synthesis gene by a marker gene. The expression of this marker gene is then under the control of an AI-regulated promoter. Several suitable biological marker genes are available, such as genes expressing Green Fluorescent Protein, Red Fluorescent Protein, or other fluorescent proteins, the luciferase (*lux*) gene resulting in bioluminescence, or the β-galactosidase (*lac*), β-glucoronidase (*gus*), and violacein (*vio*) genes producing colored pigments. This kind of reporter construct is then introduced into a non-AI-producing bacterial host. In the case of AHL, this could be, for example, the gene for the AI-binding R protein and a marker gene (e.g., *lux* or *gfp*) fused to the corresponding AHL-R regulated promoter [10–15]. This sensor construct shows more or less the same AI specificities as the bacterial strain the QS promoter originated from, but which can be easily transferred to a suitable bacterial host. With the help of a transposon this sensor cassette may even be integrated into the genome of the host strain, making the construct more stable without antibiotic pressure. Biosensor constructs can be developed not only for AHLs, but also for the detection of other AIs, like quinolones, using the same principle. For example, a bioluminescent-based *lecA::luxCDABE* whole-cell biosensor has been developed that was sensitive for different derivatives of quinolones [16, 17].

6.2.2
AI Screening with Biosensors

Biosensor strains are usually applied for a first rapid screening of samples, which are expected to contain AIs. For this preliminary screening the samples may be extracts, culture supernatants, or whole bacterial cells and they must be incubated in close physical contact with the biosensor (e.g., on an agar plate). If the concentration of the AI in the sample is sufficient and the sensor construct shows the right specificity, the expression of the marker gene in the sensor strain will document the presence of AI molecules.

However, it must be emphasized that a selected biosensor strain will only detect a limited number of AIs with high sensitivity. Thus, it is advisable to apply several biosensor constructs with different specificities in order to ensure that nothing critical is missed. Also, the detection of false-positive substances generated from medium constituents, like diketopiperazines [18], must be taken into account. Therefore, a verification of this first biosensor screening is necessary by more elaborate methods described in the following sections.

6.3
Sample Preparation Prior to Analysis

A raw sample, such as bacterial culture supernatants or any environmental or clinical specimen, may contain tens of thousands of components originating, for example, from the growth medium, from the specific bacterial metabolism, or from the myriads of solutes in the complex natural sample. Therefore, careful sample pretreatments and conditionings, shown in Figure 6.2, are required to (i) remove unwanted nonactive components (also called matrix constituents or matrices) – the cleaning method must be highly selective toward the AI – and (ii) to optimize AI concentration ranging from picomolar to millimolar in the raw sample to a level that can be analyzed with the chosen analytical technology.

6.3.1
Liquid–Liquid Extraction

In liquid–liquid extraction (LLE), organic solvents like hexane or ethyl acetate are added to the aqueous raw sample. Compounds with hydrophobic or lipophilic character are concentrated in the organic solvent layer and the hydrophils remain in the aqueous phase. In this way not only the solubility of the target AI component can be characterized, but additionally matrix constituents can be partially removed (clean-up). Moreover, LLE is a classical preconcentration technique reaching factors between 10 and 5000, since the extraction solvent is evaporated and the sample is resolved in the required minimal volume of a compatible solvent before further analysis.

6.3.2
Principles of Liquid Chromatography

Fractionation by liquid chromatography (LC) is a further or alternative clean-up/isolation step generally applied after LLE. However, LC is not only used for purification but for analytical purposes as well (for further details, refer to textbooks such as [19, 20] or http://www.chromatography-online.org). The LC separation of molecules with similar physicochemical properties is based on their distribution between the stationary solid and the moving mobile phase (also termed the eluent). Thus, their velocity differs – the stronger the adsorption of the component to the solid phase, the slower it migrates with the eluent (and the higher its retention time).

One possibility to group the LC techniques is based on the type of adsorbent and the composition of the mobile phase that specify the type of separable molecules and the separation order (selectivity). These groups are termed normal-phase liquid chromatography (NPLC), reversed-phase liquid chromatography (RPLC), and ion-exchange chromatography (IEC). In the case of NPLC, the solid phase (e.g., silica gel or aluminum oxide) is more polar than the mobile phase (hexane, dichloromethane, ethyl acetate, etc.). The interaction is mainly based on dipole–dipole interaction and the formation of hydrogen bridges, and thus separation is restricted to the polar organic components that are soluble in water-free solvents. In the reversed set-up (RPLC) the stationary phase (such as silica gels functionalized with alkanes or polymer based materials) is more nonpolar than the eluent (mainly mixture of water, methanol, and acetonitrile). The separation is mostly based on dispersive interaction between the sample constituents and the solid material, and thus nonpolar components can be determined. IEC enables separation based on the ionic interaction between the target components and the resin with low ion-exchange capacity.

The LC set-up also plays an important role in determining the separation efficiency. The higher the efficiency is – defined by the value of theoretical number plate (N) – the more components can be separated into narrow peaks allowing higher purity of the fraction. The set-up of an LC system comprises the (i) injection of the sample, (ii) proper movement of the eluent, (iii) separation and elution of the sample compartments from the column, and (iv) detection of the separated fractions. Different classes of LC set-ups can be defined: solid-phase extraction (SPE), thin-layer chromatography (TLC), (semi)preparative liquid chromatography and high-performance liquid chromatography (HPLC), which all allow for the application of NPLC, RPLC and IEC. As only SPE, TLC and preparative LC are mostly used for sample purification, these three methods are discussed in this section, while HPLC, used for analytical (quantification) purposes, is explained in Section 6.4.2.

6.3.2.1 SPE

In SPE, 50–1000 mg of porous solid material (mostly 60 Å pore size) is packed into a short column between two frits. The transferred liquids used for sample injection and the elution of the components from the stationary phase are processed by gentle suction. The material for the stationary phase (NPLC, RPLC or IEC) has to be selected according to the characteristics of the active QS components. If the organic phase after LLE extraction shows activity in the bioassay, RPLC materials are recommended for further fractionation. When the AI does not interact with the reverse-phase material and flows through the packing, normal-phase SPE is an alternative method for purification, but the sample solvent has to be strictly water-free. If the active component does not dissolve in the organic phase of LLE or its solubility is strongly pH-dependent, the component may contain (de)protonizable groups and thus IEC materials might be the most effective isolation tool. If the solute behaves as a cation (better soluble in the organic solvent when the pH of the water is basic), cation-exchange IEC material is recommended. Likewise, when the active components contain a deprotonizable group, anion exchange could provide efficient clean up.

SPE provides quite poor efficiency ($N<50$), but it is often used in the sample pretreatment because no particular instrumentation is needed. It can be simultaneously used for concentration purposes with the advantage of low solvent consumption and easy fraction collection for further structure investigation of the active molecule. Recently, different formats like disks, microplates or Sep-Pack are commercially available that enable its easier use since they tackle several problems like slow sample processing rates, low tolerance of blockage by particles, inadequate and variable packing density of the cartridges, and the possibility of processing small sample volumes.

6.3.2.2 TLC

Although TLC can be classified as an analytical technique, we grouped it as a tool used for the isolation of AIs since, on the one hand, it has poor separation efficiency and thus high probability of the co-elution of sample constituents; on the other hand, the separation zones can be easily recovered for further purification and/or analysis. The separation of the sample takes place on a flat surface covered by the stationary phase. The sample is spotted (injected) onto the plate at the lowest possible volume and then it is placed vertically in a chamber containing the eluent, which should cover less than 10% of the plate at the lower end. The driving force of the solvent in the planar stationary phase is the capillary action and the separation of the sample components results in concentric spots. The stronger the interaction between the components and the stationary phase, the slower the sample migrates with the solvent on the plate. After the separation run different detection systems can be applied, like visualization with ultraviolet/visible (UV/Vis), fluorescence, and/or chemiluminescence spectrophotometry, but also biosensors are routinely used. For this, a culture of the reporter strain in the logarithmic growth phase is mixed with a semisolid agar medium and spread on top of the TLC plate. After several hours of incubation, the location of the different AIs on the plate is indicated by corresponding spots of activated biosensor. In spite of the drawbacks, mentioned above, this method is often used for identification of AIs like AHL and HAQ derivatives, since it provides high selectivity for AIs, the equipment required is simple and inexpensive, many separations can be carried out simultaneously by multiple spotting, and thus relatively short analysis time can be achieved.

6.3.2.3 (Semi)Preparative Liquid Chromatography

This approach has the highest separation efficiency for AI among the applied LC set-ups, and thus relatively pure fractions can be collected for further structure verification and for measurement of QS activity by bioassays. Moreover, due to the high purification level of the fraction it provides the opportunity for fast and easier molecule identification by structural analysis techniques. The high separation efficiency ($N > 1000$) is provided by the homogenous and tight packing of the stationary phase in the column, as well as the particle size (3–5 μm) and pore size (100–300 Å) increasing the surface:volume ratio. The similar type of stationary material (NPLC, RPLC, or IEC) is applicable as for SPE or TLC. However, complex instrumentation including the (i) injector system providing the plug-like delivery of the sample onto the stationary phase, (ii) the pump that delivers the eluent, (iii) a column, and (iv) an on-line detection and collection system is required and has relatively high solvent consumption.

6.4
Techniques for the Structural Analysis of AIs

Several techniques have been developed and applied for the structure identification of unknown components from different types of samples, but only those most frequently used for AI analysis are summarized (see Figure 6.2). The purity of the samples plays a very important role in the reliability of the molecule identification and thus purification possibilities were described in detail above. The higher the purity, the higher the reliability of the structure identification of the active component eliminating false-positive identification. The other key point that has to be taken into consideration is the stability of the active component. Until the stability of the AI is not characterized, its handling and storage has to be as mild as possible by, for example, storing the active fraction at low temperature in the dark or in lyophilized form.

6.4.1
Mass Spectrometry

Mass spectrometry (MS) is a generally used initial technique for structure recognition. In MS the mass:charge ratio (m/z) is determined utilizing the differences of the "movement" ability of the gaseous ion in the electric and/or magnetic field under vacuum. The intensities of the separated ions are continuously measured resulting in mass spectra (intensities in the function of m/z values) that also provide additional information on the purity of the fraction. Several complex set-ups of the MS instruments with different ionization, analysis and detection possibilities have been developed, and have to be selected according to the aim of the investigation (for more details refer to review books such as [21]). For identification purposes, high mass accuracy instruments with soft ionization like time-of-flight-MS or Fourier transform ion cyclotron resonance (FTICR)-MS is advisable. The softer the ionization is, the higher the intensity of the molecule peak, and the higher the mass accuracy of the instrument, the more accurately the elemental composition (type and number of the atoms constructing the molecule) of the molecule peak can be determined. However, more than one structure may exist with an identical elemental composition (termed isomers). Thus, further investigation is needed to be able to characterize the molecule structure. One possibility is the fragmentation of the molecule peak, which is termed MS/MS. Since the fragmentation process is structure dependent and thus follows the fragmentation routes and rules the molecule structure can be reconstructed. Chemical synthesis of the proposed molecular structure or reference material is frequently needed for reliable results. Another possibility is to use nuclear magnetic resonance (NMR) spectroscopy, UV/Vis spectrophotometry, and/or Fourier transform infrared (FT-IR) spectroscopy to confirm the results obtained from the MS studies.

6.4.2
NMR Spectroscopy

Another generally applied technique for structure identification of AIs is high-resolution, liquid-based NMR spectroscopy. It is based on the interaction of the atomic nuclei of the analyzed molecules with an electromagnetic radiation in the wavelength range of radio waves. The proton and carbon spectra, which are most important for structure analysis of organic components, give information about the number of hydrogen and carbon atoms in a molecule, and how those hydrogen and carbon atoms are connected. Additionally, information about functional groups is provided that can be registered in NMR spectra. By interpreting all bindings determined from these spectra, the structure of the molecule can be determined. However, complex and expensive instrumentation, as well as experience in measurement and in translation of the spectra into a possible molecule structure, is required.

6.5
Techniques for the Quantification of AIs

6.5.1
Principles of the Analysis Methods

After structure identification, the AI can be synthesized and used as a standard to develop a quantification method for the purified samples used for structure identification. A quantification method aiming at the analysis of a particular AI has to be specific and sensitive enough for a successful application in natural samples. Since most AIs are organic components, separation techniques including gas chromatography (GC), HPLC, and capillary electrophoresis (CE) are suitable tools to separate the target components form matrix constituents with similar physicochemical behavior as all of these methods provide high separation efficiency ($N > 100\,000$). Alternatively, molecule-specific detection – mostly MS – is used after the separation improving the specificity or selectivity of the quantification. The difference between the three separation techniques lies in the type of molecules to be analyzed and thus the physicochemical properties of the targeted AI molecules. Properties like volatility, hydrophilicity (octanol:water partition coefficient, K_{ow}), and acid–base association constant (pK_a) of the AIs have to be known to be able to select the proper analysis method.

In GC, the mobile phase is gas (nitrogen, argon, or helium) and the solid phase is packed or absorbed mostly on the surface of a long capillary. Highly hydrophobic and (semi)volatile components (boiling point higher than 250 °C) can be determined with GC. Most AIs are not volatile and have hydrophilic character. Thus, derivatization is often needed before the analysis to ensure the required property of the analyte. Several specific and sensitive detection methods like electron capture, flame ionization, nitrogen phosphorus detection or MS are compatible with GC, which increases the reliability of the target analysis. The most frequently applied technique for AI analysis is HPLC coupled to spectrophotometric and MS detection since most AIs are middle-polar components and thus easily separable without derivatization. If the AI contains (de)protonizable groups, CE is a selective and fast separation tool. Although concentration is required before the analysis due to its lower sensitivity compared to the other separation methods, this technique provides the highest selectivity since only the ions with similar electrophoretic mobility are separable.

The retention time in GC and HPLC, and migration time in CE, provides the identification parameter that can be expanded if selective detection is applied after the separation. For quantification, mostly the peak height or area is used that has to be linearly dependent on the analyte concentration (analytical range). Thus, standard addition or an internal standard has to be applied for calibration to determine the exact amount of the target molecule in a given sample. Prior to the analysis of unknown samples, the performance characteristics like repeatability, sensitivity, selectivity, accuracy, interference, or matrix effects have to be determined to achieve reliable results.

6.5.2
Quantification Methods of the Known AIs

Different analytical processes allowing direct identification and quantification of AHLs and HAQs. For the other AIs (AI-2, CAI-1, and oligopeptides) no valid quantification methods have been published so far. The structure of AI-2 was characterized as complex with its receptor protein [22] and not directly since AI-2 is a "cocktail of related molecules that exist in equilibrium and rapidly interconvert" [23, 24]. Therefore, its specific and reliable quantification has not been possible so far. The oligopeptide based AIs are cleaved from larger precursor peptides, which then are modified to contain lactone and thiolactone, lanthionines, and isoprenyl groups [25–28]. Their identification is based mainly on sequence analysis after their purification, but until now no quantification method has been developed for the peptide based AIs. The structure of CAI-1 has been recently identified as 3-hydroxytridecane-4-one with NMR and GC-MS analysis after fractionation by LC, although no performance characteristics of the GC-MS method for identification and quantification have been published [6].

6.5.2.1 Analysis of AHL-Based QS Signals

AHLs consist of a furen-2-on group with an acyl chain consisting of 4–14 (less frequently up to 18) carbon atoms (see Figure 6.1). The side-chain can be substituted at the β-carbon by a keto or hydroxyl function and the fatty acid side-chain may be unsaturated. Their polarity depends on the length of the acyl side-chain although they belong to the middle-polar class of organic compounds.

Since AHLs are not stable at basic conditions, the bacterial supernatants have to be acidified to prevent the hydrolysis of the AHL during the sample treatment [29]. Generally, the extraction of AHL from the supernatant starts with LLE followed by semipreparative HPLC [9, 14, 30–35], TLC or SPE [29, 36, 37] have been integrated in the method for purification purposes. RPLC was used for fractionation of different bacterial supernatants in order to identify AHLs in a facultative intracellular pathogen [31] or in *Sinorhizobium meliloti* [32]. Semipreparative LC with a reversed-phase column applying a mobile phase of a water:acetonitrile mixture is frequently used for fractionation [9, 14, 30, 33, 35]. The application of SPE instead of LLE is another possibility to separate matrix constituents from the AIs and the method can be used for additional preconcentration as well. Sensitivity improvements between 2- and 10-fold were reached by application of SPE instead of LLE when different solid phases were systematically tested [36, 37]. Normal-phase SPE was also used for the specification of AHL synthases (EsaI and LasI) through determination of three derivatives of AHLs by MS and HPLC-MS [29].

Methods have been developed using GC for the determination of AHL [34, 38–40]. However, in some cases poor sensitivity was achieved caused by the masking effects of matrix constituents like diketopiperazines [39], the polar nature of AHLs and/or their decomposition in the injector [41]. In order to increase the sensitivity of the method, several parameters were improved like splitless injection [34, 39, 40], fractionation prior to GC analysis [34], selected ion monitoring of the MS detection, or derivatization [40]. In the latter case only the substituted AHLs can be determined since the β-oxo group of the oxo-HSLs is derivatized to an oxime [40]. The GC separations were generally performed on a fused silica capillary at a length of 20 m under helium gas applying a temperature gradient from 100 to 300 °C. In previously published studies, AHL produced by different bacterial isolates were determined by GC-MS analysis [38, 39, 42].

The developed analytical procedure mostly based on reversed-phase HPLC coupled to MS for selective detection. The targets were generally separated on a C_{18} silica column with a mobile phase of water:methanol or water:acetonitrile applying isocratic and/or gradient elution [43–50]. Different quantification techniques like external and internal standards were also studied to improve the reliability of the quantification. The selected internal standard was *N*-heptanoyl-AHL that is rarely produced in nature. However, other standards like deuterium-labeled C6-AHL have also been applied previously [29].

Developments in the analysis technique have also been reported for LC, which is now achieving better performance characteristics in terms of analysis speed, resolution, method sensitivity, accuracy, and/or reliability. A simple microelectrospray interface for MS was developed following a nano-LC separation where the sample constituents were concentrated online in the beginning of the column and due to the capillary dimension the sample and solvent consumption was lower [51]. Another possibility to improve the characteristics of the HPLC separation is to speed-up the chromatography by using small particle diameters in the stationary phase which is termed ultra-performance liquid chromatography (UPLC). UPLC provided enough separation power to allow nonspecific detection resulting in a fast and easy analysis of AHLs [52], which was also applicable for a general screening [53]. Cataldi *et al.* [54] developed an HPLC method using high-resolution MS as detection of AHLs. This method allows an accurate mass measurement because it reduces the effects of interference with the matrix constituents having identical retention properties as the targets. Li *et al.* [55] combined the advantage of UPLC and high-resolution MS (FTICR-MS) developing and applying their at line coupling with the help of a nano-electrospray source.

However, not only chromatography-based methods have been developed for AHL analysis. Capillary zone electrophoresis coupled to MS was also applied for the analysis of the corresponding serines after hydrolysis of the AHLs, which provided a highly selective separation for the target components [56]. To improve the performance of the method, C7-AHL was used as internal standard for the analysis of AHLs produced by *Burkholderia* sp. Mmi 1537. A method applying micellar electrokinetic chromatography was also developed for the determination of AHLs [57] and the retention behavior of the analytes was studied as a function of the AHL structure and the concentration of the surfactant. Since the detection was based on MS, the surfactant was injected partial amounts into the capillary to avoid sensitivity loss of the electrospray by ion suppression.

These above-mentioned methods allow the identification and quantification of AHLs not only from cell-free supernatants (for further information, refer to the review by Fekete *et al.* [53]), but also from clinical samples, which is reviewed elsewhere [49].

6.5.2.2 Analysis of HAQ-Based QS Signals

HAQs contain an quinoline group that has an alkyl side chain on the second carbon and one or two hydroxyl group(s) on the third and fourth carbon. The best known example within this substance class is the so-called *Pseudomonas* quinolone signal (PQS, 3,4-dihydroxy-2-heptyl-quinolone) that has been shown to have QS activity [57]. Other derivatives of HAQs have also been identified from bacterial cultures that differ in length, saturation, and substitution of the alkyl side-chain, number of hydroxyl groups, and oxidation of the quinoline nitrogen [58]. They are highly hydrophobic (nonpolar) components, which has to be taken into consideration in the sample extraction since they adsorb to the cell walls [59]. The quinoline group is UV and fluorescence active, and thus a sensitive and relatively selective detection by spectrophotometry is possible.

A GC-MS-based method for the chromatographic separation of HAQs was developed although interference with the targets were observed and caused difficulties in the determination of HAQ derivatives with oxidized nitrogen of the ring [60].

Lepine et al. [58, 59] developed and applied a reversed-phase HPLC method coupled to MS detection for the quantification of PQS and for the identification of other HAQ derivatives from the supernatant of *P aeruginosa*. For quantification purposes deuterium-labeled internal standards were synthesized and applied, allowing an accurate determination of the concentration.

6.6
Conclusions and Future Perspectives

This chapter summarizes biological and chemical technologies and methods for the identification, structural analysis, and quantification of QS signaling molecules. Bioassays, LC-based purification technologies, MS, and spectroscopic methods play an important role for the separation and structural identification of AIs, and a high level of analytical accuracy has been reached in recent years. In the future, analytical technologies must provide even higher levels of selectivity and sensitivity to enable the detection of AIs in an even wider range of complex samples such as soil or sputum, in order to gather more information about the presence of bacterial AIs and quorum sensing in different environments.

References

1 Demuth, D. and Lamont, R.J. (2006) *Bacterial Cell-to-Cell Communication*, Cambridge University Press, Cambridge.

2 Williams, P., Winzer, K., Chan, W.C., and Camara, M. (2007) Look who's talking: communication and quorum sensing in the bacterial world. *Philos. Trans. R. Soc. Lond. B Biol. Sci.*, **362**, 1119–1134.

3 Hense, B.A., Kuttler, C., Muller, J., Rothballer, M., Hartmann, A., and Kreft, J.U. (2007) Does efficiency sensing unify diffusion and quorum sensing? *Nat. Rev. Microbiol.*, **5**, 230–239.

4 Fuqua, W., Winans, S.C., and Greenberg, E.P. (1994) Quorum sensing in bacteria: the LuxR–LuxI family of cell-density responsive transcriptional regulators. *J. Bacteriol.*, **176**, 269–275.

5 Schaefer, A.L., Greenberg, E.P., Oliver, C.M., Oda, Y., Huang, J.J., Bittan-Banin, G., Peres, C.M., Schmidt, S., Juhaszova, K., Sufrin, J.R. *et al.* (2008) A new class of homoserine lactone quorum-sensing signals. *Nature*, **454**, U595–U596.

6 Higgins, D.A., Pomianek, M.E., Kraml, C.M., Taylor, R.K., Semmelhack, M.F., and Bassler, B.L. (2007) The major *Vibrio cholerae* autoinducer and its role in virulence factor production. *Nature*, **450**, 883–886.

7 Hoang, T.T., Karkhoff-Schweizer, R.R., Kutchma, A.J., and Schweizer, H.P. (1998) A broad-host-range Flp-FRT recombination system for site-specific excision of chromosomally-located DNA sequences: application for isolation of unmarked *Pseudomonas aeruginosa* mutants. *Gene*, **212**, 77–86.

8 Latifi, A., Winson, M.K., Foglino, M., Bycroft, B.W., Stewart, G.S., Landzundski, A., and Williams, P. (1995) Multiple homologues of LuxR and LuxI control expression of virulence determinants and secondary metabolites through quorum sensing in *Pseudomonas aeruginosa* PAO1. *Mol. Microbiol.*, **17**, 333–343.

9 McClean, K.H., Winson, M.K., Fish, L., Taylor, A., Chhabra, S.R., Camara, M., Daykin, M., Lamb, J.H., Swift, S., Bycroft, B.W. *et al.* (1997) Quorum sensing and *Chromobacterium violaceum*: exploitation of violacein production and inhibition for the

detection of N-acylhomoserine lactones. *Microbiology*, **143**, 3703–3711.

10 Winson, M.K., Swift, S., Fish, L., Throup, J.P., Jorgensen, F., Chhabra, S.R., Bycroft, B.W., Williams, P., and Stewart, G.S.A.B. (1998) Construction and analysis of luxCDABE-based plasmid sensors for investigating N-acyl homoserine lactone-mediated quorum sensing. *FEMS Microbiol. Lett.*, **163**, 185–192.

11 Riedel, K., Hentzer, M., Geisenberger, O., Huber, B., Steidle, A., Wu, H., Hoiby, N., Givskov, M., Molin, S., and Eberl, L. (2001) N-Acylhomoserine-lactone-mediated communication between *Pseudomonas aeruginosa* and *Burkholderia cepacia* in mixed biofilms. *Microbiology*, **147**, 3249–3262.

12 Kuo, A., Callahan, S.M., and Dunlap, P.V. (1996) Modulation of luminescence operon expression by N-octanoyl-L-homoserine lactone in ainS mutants of *Vibrio fischeri*. *J. Bacteriol.*, **178**, 971–976.

13 Kumari, A., Pasini, P., Deo, S.K., Flomenhoft, D., Shashidhar, H., and Daunert, S. (2006) Biosensing systems for the detection of bacterial quorum signaling molecules. *Anal. Chem.*, **78**, 7603–7609.

14 Middleton, B., Rodgers, H.C., Camara, M., Knox, A.J., Williams, P., and Hardman, A. (2002) Direct detection of N-acylhomoserine lactones in cystic fibrosis sputum. *FEMS Microbiol. Lett.*, **207**, 1–7.

15 Steidle, A., Sigl, K., Schuhegger, R., Ihring, A., Schmid, M., Gantner, S., Stoffels, M., Riedel, K., Givskov, M., Hartmann, A. *et al.* (2001) Visualization of N-acylhomoserine lactone-mediated cell–cell communication between bacteria colonizing the tomato rhizosphere. *Appl. Environ. Microbiol.*, **67**, 5761–5770.

16 Fletcher, M.P., Diggle, S.P., Camara, M., and Williams, P. (2007) Biosensor-based assays for PQS, HHQ and related 2-alkyl-4-quinolone quorum sensing signal molecules. *Nat. Protoc.*, **2**, 1254–1262.

17 Fletcher, M.P., Diggle, S.P., Crusz, S.A., Chhabra, S.R., Camara, M., and Williams, P. (2007) A dual biosensor for 2-alkyl-4-quinolone quorum-sensing signal molecules. *Environ. Microbiol.*, **9**, 2683–2693.

18 Holden, M.T.G., Chhabra, S.R., de Nys, R., Stead, P., Bainton, N.J., Hill, P.J., Manefield, M., Kumar, N., Labatte, M., England, D. *et al.* (1999) Quorum-sensing cross talk: isolation and chemical characterization of cyclic dipeptides from *Pseudomonas aeruginosa* and other Gram-negative bacteria. *Mol. Microbiol.*, **33**, 1254–1266.

19 Scott, R.P.W. (1994) *Liquid Chromatography for the Analyst*, Dekker, New York.

20 Kazakevich, Y. and McNair, H.M. (2007) *HPLC for Pharmaceutical Scientists*, John Wiley & Sons, Inc., New York.

21 Gross, J.H. (2007) *Mass Spectrometry: A Textbook*, Springer, New York.

22 Chen, X., Schauder, S., Potier, N., Van Dorsselaer, A., Pelczer, I., Bassler, B.L., and Hughson, F.M. (2002) Structural identification of a bacterial quorum-sensing signal containing boron. *Nature*, **415**, 545–549.

23 Hardie, K.R. and Heurlier, K. (2008) Establishing bacterial communities by "word of mouth": LuxS and autoinducer 2 in biofilm development. *Nat. Rev. Microbiol.*, **6**, 635–643.

24 Waters, C.M. and Bassler, B.L. (2005) Quorum sensing: cell-to-cell communication in bacteria. *Annu. Rev. Cell Dev. Biol.*, **21**, 319–346.

25 Ansaldi, M., Marolt, D., Stebe, T., Mandic-Mulec, I., and Dubnau, D. (2002) Specific activation of the *Bacillus* quorum-sensing systems by isoprenylated pheromone variants. *Mol. Microbiol.*, **44**, 1561–1573.

26 Booth, M.C., Bogie, C.P., Sahl, H.G., Siezen, R.J., Hatter, K.L., and Gilmore, M.S. (1996) Structural analysis and proteolytic activation of *Enterococcus faecalis* cytolysin, a novel lantibiotic. *Mol. Microbiol.*, **21**, 1175–1184.

27 Mayville, P., Ji, G.Y., Beavis, R., Yang, H.M., Goger, M., Novick, R.P., and Muir, T.W. (1999) Structure–activity analysis of synthetic autoinducing thiolactone peptides from *Staphylococcus aureus* responsible for virulence. *Proc. Natl. Acad. Sci. USA*, **96**, 1218–1223.

28 Nakayama, J., Cao, Y., Horii, T., Sakuda, S., Akkermans, A.D.L., de Vos, W.M., and Nagasawa, H. (2001) Gelatinase

biosynthesis-activating pheromone: a peptide lactone that mediates a quorum sensing in *Enterococcus faecalis*. *Mol. Microbiol.*, **41**, 145–154.

29 Gould, T.A., Herman, J., Krank, J., Murphy, R.C., and Churchill, M.E.A. (2006) Specificity of acyl-homoserine lactone synthases examined by mass spectrometry. *J. Bacteriol.*, **188**, 773–783.

30 Chambers, C.E., Visser, M.B., Schwab, U., and Sokol, P.A. (2005) Identification of *N*-acylhomoserine lactones in mucopurulent respiratory secretions from cystic fibrosis patients. *FEMS Microbiol. Lett.*, **244**, 297–304.

31 Taminiau, B., Daykin, M., Swift, S., Boschiroli, M.L., Tibor, A., Lestrate, P., De Bolle, X., O'Callaghan, D., Williams, P., and Letesson, J.J. (2002) Identification of a quorum-sensing signal molecule in the facultative intracellular pathogen *Brucella melitensis*. *Infect. Immun.*, **70**, 3004–3011.

32 Teplitski, M., Eberhard, A., Gronquist, M.R., Gao, M., Robinson, J.B., and Bauer, W.D. (2003) Chemical identification of *N*-acyl homoserine lactone quorum-sensing signals produced by *Sinorhizobium meliloti* strains in defined medium. *Arch. Microbiol.*, **180**, 494–497.

33 Milton, D.L., Hardman, A., Camara, M., Chhabra, S.R., Bycroft, B.W., Stewart, G.S.A.B., and Williams, P. (1997) Quorum sensing in *Vibrio anguillarum*: characterization of the *vanI/vanR* locus and identification of the autoinducer *N*-(3-oxodecanoyl)-L-homoserine lactone. *J. Bacteriol.*, **179**, 3004–3012.

34 Pomini, A.M., Manfio, G.P., Araujo, W.L., and Marsaioli, A.J. (2005) Acyl-homoserine lactones from *Erwinia psidii* R. IBSBF 435(T), a guava phytopathogen (*Psidium guajava* L.). *J. Agric. Food Chem.*, **53**, 6262–6265.

35 Jiang, Y., Camara, M., Chhabra, S.R., Hardie, K.R., Bycroft, B.W., Lazdunski, A., Salmond, G.P.C., Stewart, G.S.A.B., and Williams, P. (1998) *In vitro* biosynthesis of the *Pseudomonas aeruginosa* quorum-sensing signal molecule *N*-butanoyl-L-homoserine lactone. *Mol. Microbiol.*, **28**, 193–203.

36 Schupp, P.J., Charlton, T.S., Taylor, M.W., Kjelleberg, S., and Steinberg, P.D. (2005) Use of solid-phase extraction to enable enhanced detection of acyl homoserine lactones (AHLs) in environmental samples. *Anal. Bioanal. Chem.*, **383**, 132–137.

37 Li, X., Fekete, A., Englmann, M., Götz, C., Rothballer, M., Frommberger, M., Buddrus, K., Fekete, J., Cai, C., Schröder, P., and Schmitt-Kopplin, P. (2006) Development and application of a method for the analysis of *N*-acylhomoserine lactones by solid-phase extraction and ultra high pressure liquid chromatography. *J. Chromatogr. A*, **1134**, 186–193.

38 Cataldi, T.R.I., Bianco, G., Frommberger, M., and Schmitt-Kopplin, P. (2004) Direct analysis of selected *N*-acyl-L-homoserine lactones by gas chromatography/mass spectrometry. *Rapid Commun. Mass Spectrom.*, **18**, 1341–1344.

39 Wagner-Dobler, I., Thiel, V., Eberl, L., Allgaier, M., Bodor, A., Meyer, S., Ebner, S., Hennig, A., Pukall, R., and Schulz, S. (2005) Discovery of complex mixtures of novel long-chain quorum sensing signals in free-living and host-associated marine alphaproteobacteria. *Chembiochem*, **6**, 2195–2206.

40 Charlton, T.S., de Nys, R., Netting, A., Kumar, N., Hentzer, M., Givskov, M., and Kjelleberg, S. (2000) A novel and sensitive method for the quantification of *N*-3-oxoacyl homoserine lactones using gas chromatography-mass spectrometry: application ito a model bacterial biofilm. *Environ. Microbiol.*, **2**, 530–541.

41 Llamas, I., Quesada, E., Martinez-Canovas, M., Gronquist, M., Eberhard, A., and Gonzalez, J.E. (2005) Quorum sensing in halophilic bacteria: detection of *N*-acyl-L-homoserine lactones in the exopolysaccharide-producing species of *Halomonas*. *Extremophiles*, **9**, 333–341.

42 Cataldi, T.R.I., Blanco, G., Palazzo, L., and Quaranta, V. (2007) Occurrence of *N*-acyl-L-homoserine lactones in extracts of some Gram-negative bacteria evaluated by gas chromatography-mass spectrometry. *Anal. Biochem.*, **361**, 226–235.

43 Morin, D., Grasland, B., Vallee-Rehel, K., Dufau, C., and Haras, D. (2003) On-line high-performance liquid chromatography-mass spectrometric detection and

quantification of *N*-acylhomoserine lactones, quorum sensing signal molecules, in the presence of biological matrices. *J. Chromatogr. A*, **1002**, 79–92.

44 Flodgaard, L.R., Dalgaard, P., Andersen, J.B., Nielsen, K.F., Givskov, M., and Gram, L. (2005) Nonbioluminescent strains of *Photobacterium phosphoreum* produce the cell-to-cell communication signal *N*-(3-hydroxyoctanoyl)homoserine lactone. *Appl. Environ. Microbiol.*, **71**, 2113–2120.

45 Bruhn, J.B., Christensen, A.B., Flodgaard, L.R., Nielsen, K.F., Larsen, T.O., Givskov, M., and Gram, L. (2004) Presence of acylated homoserine lactones (AHLs) and AHL-producing bacteria in meat and potential role of AHL in spoilage of meat. *Appl. Environ. Microbiol.*, **70**, 4293–4302.

46 Huang, J.J., Han, J.I., Zhang, L.H., and Leadbetter, J.R. (2003) Utilization of acyl-homoserine lactone quorum signals for growth by a soil pseudomonad and *Pseudomonas aeruginosa* PAO1. *Appl. Environ. Microbiol.*, **69**, 5941–5949.

47 Lin, Y.H., Xu, J.L., Hu, J.Y., Wang, L.H., Ong, S.L., Leadbetter, J.R., and Zhang, L.H. (2003) Acyl-homoserine lactone acylase from *Ralstonia* strain XJ12B represents a novel and potent class of quorum-quenching enzymes. *Mol. Microbiol.*, **47**, 849–860.

48 Michels, J.J., Allain, E.J., Borchardt, S.A., Hu, P.F., and McCoy, W.F. (2000) Degradation pathway of homoserine lactone bacterial signal molecules by halogen antimicrobials identified by liquid chromatography with photodiode array and mass spectrometric detection. *J. Chromatogr. A*, **898**, 153–165.

49 Kumari, A., Pasini, P., and Daunert, S. (2008) Detection of bacterial quorum sensing *N*-acyl homoserine lactones in clinical samples. *Anal. Bioanal. Chem.*, **391**, 1619–1627.

50 Englmann, M., Fekete, A., Gebefugi, I., and Schmitt-Kopplin, P. (2007) The dosage of small volumes for chromatographic quantifications using a drop-on-demand dispenser system. *Anal. Bioanal. Chem.*, **388**, 1109–1116.

51 Frommberger, M., Schmitt-Kopplin, P., Ping, G., Frisch, H., Schmid, M., Zhang, Y., Hartmann, A., and Kettrup, A. (2004) A simple and robust set-up for on-column sample preconcentration – nano-liquid chromatography–electrospray ionization mass spectrometry for the analysis of *N*-acylhomoserine lactones. *Anal. Bioanal. Chem.*, **378**, 1014–1020.

52 Li, X.J., Chen, G.N., Fekete, J., Yang, F., Fekete, A., Englmann, M., and Schmitt-Kopplin, P. (2007) Optimization of gradient elution in UPLC: a core study on the separation of homoserine lactones produced by *Bukholderia ubonensis* and structure confirmation with ultra high resolution mass spectrometry. *J. Liq. Chromatogr. Rel. Technol.*, **30**, 2515–2531.

53 Fekete, A., Frommberger, M., Rothballer, M., Li, X.J., Englmann, M., Fekete, J., Hartmann, A., Eberl, L., and Schmitt-Kopplin, P. (2007) Identification of bacterial *N*-acylhomoserine lactones (AHLs) with a combination of ultra-performance liquid chromatography (UPLC), ultra-high-resolution mass spectrometry, and *in-situ* biosensors. *Anal. Bioanal. Chem.*, **387**, 455–467.

54 Cataldi, T.R.I., Bianco, G., and Abate, S. (2008) Profiling of *N*-acyl-homoserine lactones by liquid chromatography coupled with electrospray ionization and a hybrid quadrupole linear ion-trap and Fourier-transform ion-cyclotron-resonance mass spectrometry (LC-ESI-LTQ-FTICR-MS). *J. Mass. Spectrom.*, **43**, 82–96.

55 Li, X., Fekete, A., Englmann, M., Frommberger, M., Lv, S., Chen, G., and Schmitt-Kopplin, P. (2007) At-line coupling of UPLC to chip-electrospray-FTICR-MS. *Anal. Bioanal. Chem.*, **389**, 1439–1446.

56 Frommberger, M., Hertkorn, N., Englmann, M., Jakoby, S., Hartmann, A., Kettrup, A., and Schmitt-Kopplin, P. (2005) Analysis of *N*-acylhomoserine lactones after alkaline hydrolysis and anion-exchange solid-phase extraction by capillary zone electrophoresis-mass spectrometry. *Electrophoresis*, **26**, 1523–1532.

57 Frommberger, M., Schmitt-Kopplin, P., Menzinger, F., Albrecht, V., Schmid, M., Eberl, L., Hartmann, A., and Kettrup, A.

(2003) Analysis of *N*-acyl-L-homoserine lactones produced by *Burkholderia cepacia* with partial filling micellar electrokinetic chromatography – electrospray ionization-ion trap mass spectrometry. *Electrophoresis*, **24**, 3067–3074.

58 Lepine, F., Milot, S., Deziel, E., He, J.X., and Rahme, L.G. (2004) Electrospray/ mass spectrometric identification and analysis of 4-hydroxy-2-alkylquinolines (HAQs) produced by *Pseudomonas aeruginosa*. *J. Am. Soc. Mass Spectrom.*, **15**, 862–869.

59 Lepine, F., Deziel, E., Milot, S., and Rahme, L.G. (2003) A stable isotope dilution assay for the quantification of the *Pseudomonas* quinolone signal in *Pseudomonas aeruginosa* cultures. *Biochem. Biophys. Acta*, **1622**, 36–41.

60 Taylor, G., Machan, Z.A., Mehmet, P.J., Cole, R., and Wilson, R. (1995) *J. Chromatogr. B*, **664**, 458–462.

Part II
Transmembrane Signaling

Introduction

Reinhard Krämer

The cytoplasmic membrane of bacterial cells separates the cytoplasm from the outer world and thus defines the internal from the external compartment. On the one hand, this allows maintaining the cellular integrity and individuality with respect to metabolic state, composition of solutes, and homeostatic functions. On the other hand, it creates a permeability barrier across which matter, energy, and information must somehow be transferred. The latter, namely transfer of information useful for the cell across the barrier of the cell membrane, is the main topic of Part II of the book.

In Gram-negative bacteria, in particular, the cell envelope actually consists of two permeability barriers, located at both the outer and the cytoplasmic membrane; consequently, this section starts with a chapter on solute and information transfer across the bacterial outer membrane, which necessarily is combined with consecutive steps at the cytoplasmic membrane (Chapter 7).

The most frequently used mechanism of transmembrane signaling in bacteria is represented by histidine (sensor) kinase/response regulator systems. The membrane-integrated receptor histidine kinase acts in general as a sensor for extracellular stimuli and is responsible for the information transfer across the membrane, whereas the response regulator represents a transmitter device coupling the phosphorylation signal originating from the kinase to the final target of signal transduction – transcriptional regulation of specific genes at the DNA level. Several chapters within this section focus on this type of transmembrane signaling (Chapters 8 to 11). Chapter 8 specifically deals with the first step of this process – stimulus perception and consecutive signal transduction by membrane-integrated histidine kinases. Information transfer across the membrane by conformational events within membrane-integrated receptors is discussed in detail in Chapter 11. Bacterial chemotaxis is a highly relevant and probably the most intensively studied mechanism of this kind of signaling processes, both with respect to the components involved and the underlying mechanistic aspects (Chapter 9). Perception of a physical stimulus (i.e., light) by bacterial photoreceptors is dealt with in Chapter 10.

The next two chapters describe two further aspects of transmembrane signaling. In some cases, membrane proteins whose major job is the transmembrane translocation of chemical compounds (i.e., transport systems) harbor a second function –

sensing. Prominent examples of such sensory transport proteins are discussed in Chapter 12. Membrane-related mechanisms of signaling sometimes do not occur in a transmembrane fashion, but may primarily happen within the hydrophobic core of the membrane itself. An important example of this kind of mechanisms is regulated intramembrane proteolysis, which is dealt with in Chapter 13.

When elucidating functional aspects of transmembrane signaling mechanisms, biochemical and biophysical aspects of membrane proteins are of key relevance for understanding these processes. Chapter 14 overviews modern methods to analyze dynamic aspects of membrane protein function, thus providing a core methodical access to the questions raised in Part II.

7
Outer Membrane Signaling in Gram-Negative Bacteria

Volkmar Braun

7.1
Introduction

Bacteria communicate with their environment by signaling. The signaling cascade involves signal generation, perception, transmission, and response. Signals can be generated by small chemicals or through protein–receptor interactions.

For some signaling events in Gram-negative bacteria, the outer membrane is instrumental. Signaling through outer membranes could conceivably occur when bacteria interact to transfer proteins and nucleoprotein complexes from donor cells into recipient cells during conjugation (type IV secretion), when bacteria transfer effector proteins into eukaryotic cells (type III secretion), during fruiting body formation, in contact-dependent growth inhibition, during avoidance of swarm merging by two different strains, and during biofilm formation (some of these topics are discussed in other chapters of this book). For most of these systems, a complete signaling cascade has not been shown and single steps or components of the cross-membrane signaling (e.g., the nature of signals, signal receptors, signal transmission, and final signal targets) are not known. In contrast, complete substrate signaling cascades for transcription regulation of substrate transport systems have been elucidated. The outer membrane proteins recognize the substrates and transmit the signals from the outer membrane across the periplasm and the cytoplasmic membrane into the cytoplasm. These systems are reviewed here.

7.2
A Sophisticated Mechanism: A Signaling Cascade Across the Outer Membrane in Transcriptional Regulation of the Ferric Citrate Transport Genes

The first understood system in which signaling occurs from the cell surface into the cytoplasm is the transcriptional regulation of the ferric citrate transport genes of *Escherichia coli* K-12 (Figure 7.1) [1–4]. The *fecABCDE* operon is required for ferric citrate transport. *fecA* encodes an outer membrane protein, *fecB* encodes a

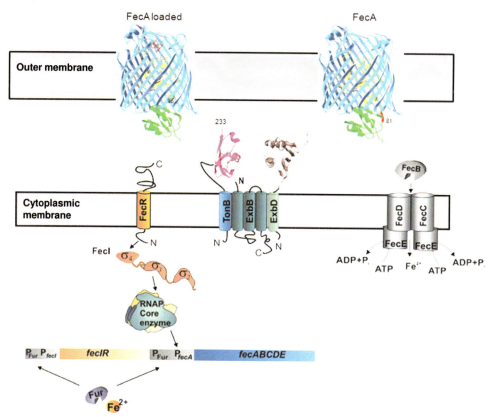

Figure 7.1 Model of the ferric citrate transport and regulatory system of *E. coli* K-12. The signaling pathway from FecA to FecI; the involvement of TonB, ExbB, and ExbD in signaling and transport; and transport of iron through the periplasmic FecB protein and the ABC transporter FecCDE proteins are shown. Available crystal structure are shown: FecA $(Fe^{3+}\text{-citrate})_2$ from residues 95 to 741 of the mature form [6, 7], unloaded FecA from residues 81 to 741, the NMR structure of the signaling domain (residues 1–74) [17, 18] modeled into the structure of FecA unloaded and loaded with $(Fe^{3+}\text{-citrate})_2$, the crystal structure of a C-terminal TonB fragment (residues 153–233) as derived from the FhuA–TonB [11] and BtuB–TonB cocrystals [12], and the NMR structure of the periplasmic portion of ExbD (residues 44–141) [59]. The FecA TonB box (red, residues 80–85) is seen in the crystal structure of unloaded FecA, but not in the crystal structure of loaded FecA. The structure of the connection between FecA and its signaling domain is unknown. For steric reasons, binding of TonB to the TonB box of FecA requires displacement of the signaling domain. Fe^{2+}-loaded Fur repressor binds to the promoter upstream of *fecI* and *fecA*. Under low-iron conditions, unloaded Fur dissociates from the promoter. Signaling induces transcription initiation of the *fecABCDE* transport genes. σ_2 and σ_4 indicate FecI domains involved in binding to DNA and FecR, respectively. The ABC transporter of iron across the cytoplasmic membrane consists of the FecBCDE proteins.

periplasmic binding protein, and *fecCDE* encode cytoplasmic membrane proteins, which together with FecB form an ATP-binding cassette (ABC) transporter. Two *fec*-specific regulatory genes, *fecIR*, are located upstream of *fecABCDE*. FecR is a cytoplasmic transmembrane protein, and FecI is an extracytoplasmic function (ECF) sigma factor of the σ^{70} family and binds to RNA polymerase core enzyme to transcribe the *fecABCDE* operon [5].

For the transcriptional regulation, the cells must sense the iron status. If the intracellular iron content is low, the *fecIR* genes are transcribed. However, iron deprivation is not sufficient to induce transcription of the *fecABCDE* transport genes. The substrate, ferric citrate, must also be present in the growth medium. When enough iron has entered the cells, the *fec* transport system is rapidly downregulated by the Fur repressor loaded with Fe^{2+}; Fe^{2+}-Fur represses both *fecIR* and *fecABCDE* transcription. In the absence of Fe^{2+}, Fur is inactive.

The mature form of FecA lacking the signal sequence consists of 741 amino acid residues that form a β-barrel completely closed by a globular domain, designated as the plug [6, 7]. The crystal structure of FecA reveals that among the various iron citrate forms in solution [8], $(Fe^{3+}\text{-citrate})_2$ is bound to FecA and thus serves as the inducer. Binding of ferric citrate to FecA induces transcription of the ferric citrate transport genes; however, even though FecA also transports ferric citrate into the periplasm, transport is not required for induction (Figure 7.1). The following data support this conclusion: (i) four *fecA* point mutants are constitutively (i.e., without ferric citrate) transcribed; two of these mutants do not grow on ferric citrate as sole iron source because they do not transport ferric citrate [9] (i.e., induction is uncoupled from transport); (ii) if the ferric citrate concentration in the growth medium of a *fecA* deletion mutant is high, enough ferric citrate diffuses through the porins into the periplasm to support growth; however, periplasmic ferric citrate does not induce *fec* transport gene transcription; and (iii) ABC transporter mutants are transport inactive, but fully inducible.

The structure of the FecA protein was studied in detail. Crystal structures were determined with bound $(Fe^{3+}\text{-citrate})_2$, without substrate, and with bound $(\text{citrate})_2$. $(Fe^{3+}\text{-citrate})_2$ binds to several residues of FecA located in a cavity that lies well above the outer boundary of the outer membrane lipid bilayer (Figure 7.1). Binding of $(Fe^{3+}\text{-citrate})_2$ induces strong long-range structural transitions in FecA. Surface loops 7 and 8 are translated up to 11 and 15 Å, respectively, and cover the entry of the surface cavity, preventing the escape of $(Fe^{3+}\text{-citrate})_2$ back to the external milieu. In the FecA region exposed to the periplasm, a short helix is unwound. $(Fe^{3+}\text{-citrate})_2$ and $(\text{citrate})_2$ bind at a similar site, but unlike $(Fe^{3+}\text{-citrate})_2$, $(\text{citrate})_2$ does not induce transcription and does not change the FecA conformation [6, 7].

The structural changes in FecA caused by ferric citrate binding are not sufficient to induce transcription of the *fec* transport genes. Induction requires energized cells and the Ton system, which is composed of the proteins TonB, ExbB, and ExbD. The region around residue 160 of TonB binds to the TonB box of FecA [10]. The TonB box consists of approximately seven residues usually close to the N-terminus of TonB-dependent transporters and of colicins (bacterial protein toxins that require TonB to enter cells). Since FecA contains an N-terminal extension for signaling (see below), the TonB box is located further from the N-terminus, between residues 80 and 85. Cocrystals of a C-proximal TonB fragment with the outer membrane transporters FhuA and BtuB demonstrate that the TonB box forms a β-strand when it is bound to a C-proximal β-sheet of

TonB composed of three antiparallel β-strands [11, 12]. These two regions interact, as shown by suppressor mutations and cysteine bridges formed between introduced cysteine residues [10]. The flexibility of the TonB box predicted from the crystal structure may facilitate binding of the TonB box to TonB. In response to the proton motive force across the cytoplasmic membrane, TonB changes its structure [13]. This change is thought to cause an interaction of TonB with FecA, leading to structural transitions in FecA that release ferric citrate from its binding site and open a pore in FecA to allow ferric citrate to pass through the β-barrel. For induction, the structural changes in the plug and fixation of the flexible TonB box to TonB are required, but ferric citrate does not have to diffuse through FecA.

In contrast to unloaded FecA, the TonB box of loaded FecA is not seen in the crystal structures because it is flexible. Electron paramagnetic resonance spectroscopy of FecA spin-labeled in the TonB box reveals a ferric citrate-induced disorder transition [14]. Molecular dynamic simulations of the complex between TonB and BtuB reveal that a mechanical force can be transmitted from TonB to BtuB without disruption of the small interface between the two proteins [15]. Unfolding of the plug requires only 10% of the force required to pull the plug out of the β-barrel. It has been proposed that interaction of energized TonB with BtuB mechanically moves the plug and thus allows diffusion of vitamin B_{12}, the substrate of BtuB. In FecA, the massive structural changes presumably generate a signal that is transmitted across the outer membrane to TonB and FecR.

The FecA sequence contains an N-terminal extension, residues 1–79; other homologous outer membrane transporters that have no regulatory activity, such as FhuA and BtuB, do not contain such an extension. Deletion of the FecA N-terminal extension abolishes induction, but retains transport [16]. Since the N-terminal extension is required for induction, it is called the signaling domain [16]. The signaling domain is flexible and is not observed in the FecA crystal structures. The structure has been determined by nuclear magnetic resonance (NMR) spectroscopy and shows a unique fold [17, 18]. Random mutagenesis of the signaling domain yields mutants with amino acid replacements and reduced induction [19]. The mutations are all located along one surface of the signaling domain and this surface might therefore form the interface to FecR. This conclusion is supported by two mutations in the predicted FecA interface, glycine replaced by arginine (G39R) and D43E, which partially restore induction of two induction-inactive mutants with mutations located in the C-proximal region of FecR (L269G, F284L) [20]. Site-directed mutations in which residues in the signaling domain are replaced by alanine are located in the same region, thereby supporting the conclusion that this region forms an interface between FecA and FecR [18]. Of the nine FecA mutants with reduced induction, five are strongly impaired in ferric citrate transport under conditions in which the transport proteins are synthesized [19]. This result is unexpected since deletion of the signaling domain does not affect the transport rate. However, since structural changes in FecA upon binding of ferric citrate and TonB cause induction, it is reasonable to conclude that structural changes in the signaling domain affect the structure of FecA such that transport is impaired.

7.3
Transfer of the Signal Across the Cytoplasmic Membrane

Since the signal initiated in FecA needs to reach the cytoplasm, where transcription of the *fec* genes takes place, the signal must cross the periplasmic space and the cytoplasmic membrane. The signal crosses the periplasm through interaction of the N-terminal end of FecA with the C-terminal end of FecR – both of which are localized in the periplasm. Residues 1–84 of FecR are localized in the cytoplasm, residues 85–100 span the cytoplasmic membrane, and residues 101–317 are localized in the periplasm [21]. The above-mentioned mutations in the FecA signaling domain that suppress allele-specific mutations in the C-proximal periplasmic domain of FecR support the interaction of the FecA N-terminus with the FecR C-terminus. This interaction is further demonstrated by a bacterial two-hybrid system, which *in vivo* reveals specific interaction of $FecA_{1-79}$ (residues 1–79 of mature FecA) with $FecR_{101-317}$ [22]. In addition, $(His)_{10}$-FecR (FecR with 10 histidine residues fused to the N-terminal end) binds to a Ni-agarose column and retains FecA on the column, and the two proteins coelute from the column. Furthermore, FecA with a deletion in the periplasmic N-terminus is unable to bind to $(His)_{10}$-FecR. These data clearly indicate that signal transmission from the outer membrane across the periplasm into the cytoplasm occurs through interaction of the FecA N-terminus with the FecR C-terminus. Interaction occurs without signaling.

FecR contains a motif composed of repeating heptapeptides (residues 247–268) flanked by one valine and three leucine residues. The motif resembles leucine zipper motifs and is highly conserved in FecR-like proteins. FecR mutants in which leucine and valine are replaced by proline exhibit a strongly decreased interaction with FecA and a low *fecA* transcription in response to ferric citrate. The region flanking the leucine zipper motif is also important for FecR activity, as the FecR(L269G, F284L) mutants demonstrate [20]. This region is close to the C-terminus of FecR, which might interact with FecA.

7.4
Signal Transfer into the Cytoplasm

Synthesis of the cytoplasmic $FecR_{1-85}$ fragment from a cloned *fecR* gene fragment results in constitutive transcription of the *fec* transport genes. Random polymerase chain reaction mutagenesis yielded three mutations in which W19, W39, or W50 is replaced by arginine [23]. These mutants display 4% of the wild-type *fec* transport gene transcription level. The tryptophan residues are highly conserved in proteins homologous to FecR. These results show that the cytoplasmic FecR portion is essential for signal transmission and that the tryptophan residues play an important role.

Interaction of $FecR_{1-85}$ with FecI has been demonstrated *in vivo* and *in vitro* [22]. Using a two-hybrid system, interaction of $FecR_{1-85}$, $FecR_{1-58}$, and $FecR_{9-85}$ with FecI was shown. *In vitro*, FecR-$(His)_6$ (six histidine residues fused to the C-terminal end) bound to a Ni-agarose column binds FecI; in contrast, $(His)_{10}$-FecR (10 histidine residues fused to the N-terminal end) does not retain FecI because the N-terminal His tag interferes with binding to FecI.

Signal transfer across the outer membrane by FecA, interaction of FecA and FecR in the periplasm, signal transfer across the cytoplasmic membrane by FecR, and interaction of FecR with FecI in the cytoplasm form a complete signaling cascade from the cell surface to the cytoplasm, where transcription initiation occurs. A number of models for the molecular mechanism of transmembrane signal transduction, as in the case of signal transmission by FecR across the cytoplasmic membrane, have been proposed: (i) association–dissociation by lateral translation of transmembrane helices, (ii) translation perpendicular to the membrane (piston motion), (iii) rotation along an axis parallel to the membrane (pivot motion), (iv) rotation along an axis perpendicular to the membrane, and (v) inactivation of a transmembrane anti-sigma factor by proteolysis. Mechanism (i) was excluded by cysteine scanning of the transmembrane region, in which no FecR dimers or multimers were revealed. FecR translation and rotation within the cytoplasmic membrane (mechanisms (ii) to (iv)) have not been studied. However, mechanism (v), proteolysis of a transmembrane anti-sigma factor, seems reasonable for FecR. In the σ^E stress response, accumulation of unfolded outer membrane proteins in the periplasm activates the DegS protease, which cleaves the RseA transmembrane regulatory protein. This in turn leads to further degradation of RseA within or at the inner side of the cytoplasmic membrane by the RseP protease. RseA/σ^E is released into the cytoplasm, where RseA is further degraded by ClpXP, and active σ^E is liberated and initiates transcription of stress response genes. Studies with mutants of $rseP$ reveal that the amount of FecA is strong reduced [24]. Complementation of $rseP$ mutants by plasmid-encoded wild-type $rseP$ restores the FecA wild-type level. Constitutive expression of the fec transport genes by $FecR_{1-85}$ suggests that this fragment or a similar fragment is formed during induction. The high expression of FecA by $FecR_{1-85}$ is independent of RseP. DegS is not required, but FecR is prone to proteolytic cleavage in the periplasm by an unidentified protease [25]. It is likely that signal transmission across the cytoplasmic membrane involves release of $FecR_{1-85}$ or a fragment of similar size by cleavage of FecR by RseP. Interaction of FecR with substrate-loaded FecA causes a structural change in FecR so that it is cleaved by RseP. However, it is unclear whether $FecR_{1-85}$ is further degraded to release active FecI.

7.5
FecI is an ECF Sigma Factor

The signal elicited by binding of ferric citrate to FecA activates FecI in the cytoplasm, which then recruits the core enzyme of the RNA polymerase and initiates transcription of the fec transport genes. FecI can be subdivided into structural and functional domains similar to those of σ^{70}, except that it lacks region 1.1 [26]. Amino acid replacements in region 4 abolish FecI activity [27]. A FecI deletion analysis using a two-hybrid system revealed that $FecR_{1-85}$ interacts with regions 4.1 and 4.2 of FecI. Overexpressed region 4 competes with chromosomally encoded FecI for FecR interaction, resulting in reduction of FecI activity. FecI derivatives with sequential deletions of regions 1–3 retain interaction with FecR$_{1-85}$, but lose activity. The inactive W19R, W39R, and W50R mutants of FecR fail to bind to FecI. Northern blotting identified a *fecA* DNA fragment whose synthesis is enhanced in the presence of citrate and under iron-limiting growth conditions [28]. The results demonstrate that region 4 of FecI specifically interacts with the N-proximal region of FecR and that this interaction is necessary for FecI to function as a sigma factor. Binding of FecI to the RNA polymerase core enzyme has been shown by a bacterial two-hybrid system and by retention of FecI by a fragment of the β′ subunit of the RNA polymerase bound via a His-tag to a Ni-agarose column [29].

7.6
Mechanism of Ferric Citrate Transcription Regulation

The activity of most ECF sigma factors seems to be controlled by anti-sigma factors. In the absence of the anti-sigma factors, the sigma factors initiate transcription without extracytoplasmic signals. There is no evidence that FecR acts as an anti-sigma factor. In the absence of FecR, there is virtually no *fecABCDE* transcription. In contrast, cells containing FecR$_{1-85}$ and even fragments of FecR$_{1-85}$ constitutively transcribe the *fecABCDE* genes at a high level. Cells containing longer FecR derivatives extending from residue 1 up to residue 273 (out of a total of 317 residues) that do not interact with FecA [22] also transcribe the *fec* transport genes constitutively but to a lower level. Although FecR is necessary for FecI activity, it cannot be ruled out that FecR acts as an anti-sigma factor. If FecI is unstable, spontaneously denatures, precipitates, or is degraded by proteases, binding to FecR could maintain FecI in a stable conformation. When the signal from FecA occupied by (Fe^{3+}-citrate)$_2$ arrives through FecR, FecR undergoes a conformational change, which may result in FecI dissociation from FecR and immediate binding of FecI to the RNA polymerase core enzyme. In this model, FecR acts as a chaperone since FecI is kept in an active conformation or assumes an active conformation. FecR also acts as an anti-sigma factor when FecI is inhibited while bound to FecR.

The Fec-type signaling cascade became the paradigm of later-characterized signaling cascades in bacteria. In all except one case, Fe^{3+} complexes serve as signaling molecules. In 2005, application of a hidden Markov model to 110 completely sequenced genomes revealed 84 genomes that contain TonB-dependent outer membrane transporters (receptors), 26 of which might function as signal transducers because they contain an N-terminal extension [30]. Only a few signaling transporters were studied experimentally.

7.7
Transcription Regulation of the Fec Type in *Pseudomonas putida*

In *P. putida* WCS358, the siderophore pseudobactin BN8 induces synthesis of the PupB outer membrane transporter [31]. The involvement of PupB in induction was shown using a hybrid protein in which the N-terminal end of PupB is fused to PupA, which transports pseudobactin 358. The hybrid protein induces *pupB* transcription in response to pseudobactin 358, although pseudobactin 358 does not induce transcription of *pupA*. Transcription induction of *pupB* by pseudobactin BN8 requires *pupI* and *pupR*, which map upstream of *pupB*. *pupI* shows 42.8% and *pupR* 36.6% sequence identity to *fecI* and *fecR*, respectively. Although regulation of *pupB* transcription resembles regulation of *fecABCDE* transcription, PupR might function only as an anti-sigma factor since PupI in the absence of PupR induces *pupB* transcription constitutively. In analogy to the Fec system, it has been concluded that the structural change in PupA upon binding of pseudobactin 358 is mediated through the N-terminal extension of PupB to PupR, which then dissociates from PupI and the released PupI then acts as a *pupB*-specific sigma factor.

7.8
Transcription Regulation of the Fec Type in *Pseudomonas aeruginosa*

P. aeruginosa expresses a high number (around 14) of ECF sigma factors that are regulated by surface signaling through outer membrane transporters [32–35]. *P. aeruginosa* synthesis of the siderophore pyoverdine and transport of ferric pyoverdine is controlled by a FecIR-like regulatory device [36]. Pyoverdine (presumably after loading with Fe^{3+}) functions as inducer of the pyoverdine synthesis and Fe^{3+}-pyoverdine transport genes and in addition induces formation of exotoxin A and an extracellular endoproteinase.

In this system, the *fecR* homolog *fpvR* and the *fpvA* gene (equivalent to *fecA*), which encodes the Fe^{3+}-pyoverdine outer membrane transporter, map close to the pyoverdine synthesis operon. In contrast, the *fecI* homolog *pvdS* is located some distance away. PvdS is an ECF sigma factor and is required for the synthesis of pyoverdine, exotoxin A, and endoproteinase, whereas transcription of *fpvA* is under the control of another sigma factor, FpvI, which is controlled by the same anti-sigma factor as PvdS, namely FpvR. Synthesis is high in *fpvR* mutants, whereas overexpressed FpvR inhibits synthesis. This indicates that FpvR functions as an anti-sigma factor. Similar to interaction of FecI with FecR, the periplasmic N-terminal 67-amino-acid domain of FpvR interacts with PvdS and region 4 of FpvI [36].

Furthermore, FpvA is required for the induction by pyoverdine. Of 36 peptide insertion mutants in FpvA, three mutants in the β-barrel are deficient in transport but retain signaling, which shows that transport is not required for signaling [37]. Deletion of the N-terminal domain abolishes induction of pyoverdine synthesis in response to pyoverdine in the growth medium but retains pyoverdine transport. In contrast to the crystal structure of FecA, which does not show the signaling domain, the signaling domain is seen in the crystal structure of FpvA (Figure 7.2) [38] and FpvA loaded with Fe^{3+}-pyoverdine [39]. Residues of the TonB box form a β-strand that interacts with a three-stranded β-sheet of the signaling domain. The signaling domain assumes a different location in the two FpvA structures. However,

Figure 7.2 Stereo view of the superposition of the FpvA crystal structure (gray) with signaling domain (blue) [38] and the Fe^{3+}-pyoverdine-loaded FpvA crystal structure with signaling domain (yellow) [39]. Note the different positions of the signaling domains. (Reproduced with permission by D. Cobessi).

different locations of the signaling domain are also observed in the two unloaded FpvA molecules in the asymmetric unit. It is therefore not clear which location, if any, reflects the position of the signaling domain in the induced and uninduced state. The different structures reflect the flexibility of the connection between the signaling domain and the plug of FpvA. Such a flexibility is required so that the TonB box of FpvA can either bind to the signaling domain or to TonB. Since both signaling and transport require TonB, a sequential binding of the TonB box to the signaling domain and to TonB is suggested. It is feasible that the signal elicited by loading FpvA with substrate displaces the signaling domain from the TonB box and induces preferential binding to TonB to open the pore in FpvA so that the substrate can pass through.

PvdS binds to the RNA polymerase core enzyme in a 1 : 1 ratio and this complex binds to a DNA promoter fragment of the *pvdA* pyoverdine biosynthesis gene, as shown by DNA retardation experiments. Under iron limitation, 578 [40] and 730 [41] PvdS molecules were determined per cell, in contrast to 930 molecules of the major sigma factor RpoD and 2221 molecules of RNA polymerase. The concentration of PvdS is much higher than that of FecI, of which only a few molecules are found [42]. Thirty percent of PvdS is bound to the cytoplasmic membrane and most of that is released upon signaling. Despite the high concentration of PvdS, only 1% of the RNA polymerase carries PvdS (compared to RpoD, 27%) [40]. The proposed transcription model is similar to the Fec transcription model. Fe^{3+}-pyoverdine binds to the FpvA outer membrane transport protein, which interacts with the FpvR protein. The signal is transmitted across the cytoplasmic membrane via the predicted FecR transmembrane segment, and then PvdS dissociates from FpvR and functions as an ECF sigma factor. A second *fecI* homolog, *fpvI*, maps adjacent to *fpvR* but is transcribed in the opposite direction to *fpvR*. FpvI also receives a transcription initiation signal from FpvR.

In a study demonstrating interactions of FecI and FecR in *E. coli*, two pairs of FecIR homologs of *P. aeruginosa* that map adjacent to each other were included [27]. As demonstrated by a bacterial two-hybrid system, the FecI homolog PA2468 and its truncated form $PA2468_{110-172}$ dimerize with the related $FecR_{1-85}$ homolog $PA2467_{1-90}$ but not with the unrelated $FecR_{1-85}$ homolog $PA3900_{1-85}$. $PA3900_{1-85}$ only dimerizes with the related FecI homolog PA3899 and the truncated $PA3899_{105-170}$. The truncated FecI-like fragments cover region 4, which demonstrates their involvement in the interaction with the FecR homologs of *P aeruginosa*. As with *fecIR*, the *fecIR*-like *Pseudomonas* genes are preceded by Fur boxes. Fourteen *fecI* type ECF sigma factors are adjacent to *fecR*-type regulatory genes. Ten of these *fecIR*-like genes are adjacent to *fecA*-like genes, which encode proteins with N-terminal extensions, as does FecA.

7.9
Transcriptional Regulation of the Fec Type in *Bordetella*

Regulatory systems analogous to FecAIR have been identified experimentally in *Bordetella pertussis* [43], *Bordetella bronchiseptica* [44], and *Bordetella avium* [45, 46]. They all regulate iron transport systems for which the iron ligand responsible for induction and regulation by iron has been shown. *B. pertussis* and *B. bronchiseptica* encode the two regulatory genes *hurI* and *hurR* upstream of the heme transport gene cluster *bhuRSTUV*. Synthesis of the BhuR outer membrane transport protein for heme is enhanced when cells are grown in a medium supplemented with hemin. Expression of an additional gene, *hurP*, is essential for heme-dependent induction of *bhuR* transcription [47]. HurP is a protease homologous to RseP of *E. coli*, which is involved in *fec* transport gene transcription. *hurP* also complements a mutant in YaeL of *Vibrio cholerae*; YaeL is homologous to RseP and HurP. YaeL cleaves TcP, a membrane-bound transcriptional activator of the virulence genes *toxT* and *ctxAB*. In addition, synthesis of a putative ferric siderophore outer membrane transport protein (BfrZ) of *B. bronchiseptica* is regulated by two proteins, designated *bupI* and *bupR*, which are homologous to *fecI* and *fecR*. *bupI* overexpression induces *bfrZ* transcription and the BfrZ protein is found in the outer membrane fraction. *B. avium* contains a heme-utilization system in which the synthesis of the related outer membrane transport protein BhuR is induced by heme and requires RhuI (homologous to FecI). BhuR synthesis is enhanced in cells that overexpress RhuI. Overexpression of RhuI reduces transcription of a σ^S-dependent gene, which suggests competition between RhuI and σ^S for the RNA polymerase core enzyme.

7.10
ECF Signaling in *Serratia marcescens*

In the high-affinity hemin transport system of *S. marcescens*, the hemophore HasA donates heme to the outer membrane transporter HasR. Binding of heme-loaded HasA to HasR induces transcription of the *has* operon, which encodes a heme transport system [48]. Unloaded HasA or heme are induction negative. A detailed study has revealed that for induction both heme and heme-free HasA must be bound to HasR. Heme must be transferred to HasR and the unloaded HasA together with heme on HasR elicit the inducing signal. Deletion of the N-proximal HasR signaling domain inactivates signaling but retains HasR-mediated heme transport. Signaling requires the TonB, ExbB, and ExbD proteins, and is regulated by iron via the Fur repressor. The signal elicited at the cell surface inactivates HasS, an anti-sigma factor structurally similar to FecR, resulting in activation of the HasI sigma factor. In contrast to *fecR*, transcription of *hasS* depends on HasI. The heme-induced signal inactivates anti-sigma HasS and HasI induces transcription of *hasS*. Accumulated HasS becomes active upon heme-induced signaling. *hasI* is preceded by a Fur box and its transcription is regulated by the iron status of the cells via Fur.

7.11
ECF Signaling in *Ralstonia solanacearum*

R. solanacearum elicits a hypersensitive response on nonhost plants. At the beginning of the regulatory cascade, which eventually induces the *hrp* hypersensitivity genes, stands a regulatory device of the Fec type [49]. Particularly interesting in the *R. solanacearum* system is the initial signal generated by physical contact between the bacteria and the plant cells, without involvement of a diffusible substance. A signaling system that starts at the bacterial cell surface is perfectly suited to respond to cell–cell contact. For the expression of the *prhJ* regulatory gene, the PrhA outer membrane protein, PrhI (homologous to FecI), and PrhR (homologous to FecR) are required. As with the *fecIR* genes, the *prhIR* genes form a separate regulatory unit that does not require PrhA for expression and it is not autoregulated. In addition, cells synthesizing a truncated PrhR protein are fully pathogenic on host plants. This is reminiscent of the constitutive expression of the *fec* transport genes in cells synthesizing C-terminally truncated FecR proteins.

7.12
Signaling in Outer Membrane Transport

Signaling across the outer membrane is not only required for regulation of gene transcription, but also for transport. Structural changes upon substrate binding are observed also in outer membrane transporters that do not regulate gene transcription, such as FhuA, FepA, and BtuB [50–53]. Loops at the cell surface and regions exposed to the periplasm assume positions that differ from the positions in the unloaded states. Movements can range over large distances as far as 19 Å. These structural changes are thought to enhance interaction with TonB. Interaction of TonB with FecA does not need ferric citrate and is only slightly enhanced in the presence of ferric citrate [10]. It seems that it is not the interaction as such but the mode of interaction that elicits signaling.

Signaling does not occur only from the cell surface into the periplasm, but also from the periplasm to the cell surface. The first case of outward signaling was shown with FhuA of *E. coli* in which binding of the phages T1 and φ80 at the cell surface requires energized TonB wild-type cells [54]. In *P. aeruginosa*, the exchange of pyoverdine (in this experiment presumably Al^{3+}-pyoverdine [60]) by Fe^{3+}-pyoverdine at the FpvA outer membrane transporter is slow in TonB mutants (approximately 25 h) and fast (1 min) in energized TonB wild-type cells [35]. The *S. marcescens* hemophore HasA dissociates from the HasR transporter only in energized TonB wild-type cells. Binding of the hemophore loaded with heme or unloaded and transfer of heme from HasA to HasR do not require TonB [48].

7.13
Assumed Outer Membrane Signaling

Outer membrane signaling is certainly involved in biofilm formation and cell–cell communication between bacteria (conjugation, fruiting body formation, contact-dependent growth inhibition), and between bacteria and eukaryotic cells. However, in these systems, very little is known about outer membrane signaling. For example, transfer of bacterial proteins into eukaryotic cells via the type III secretion system is initiated by bacterial contact with the target cells [55]. This regulatory mechanism prevents futile secretion of proteins into the medium and ensures that the secretion machine is activated only on bacterial contact with the cells in which the proteins are transferred. Host cell contact triggers (i) secretion of the pore-forming proteins in the target cell membrane, (ii) translocation of the effector proteins into the host cells, and (iii) increased synthesis of secretion-related genes. It is usually assumed that the stimulus is a soluble factor, but as the above-discussed hypersensitivity reaction to *R. solanacearum* shows [49], it could as well be specific receptors at the bacterial cell surface that recognize distinct structures on the surface of eukaryotic cells. Cell–cell contact would transduce a signal into the cytoplasm and the periplasm to activate the secretion system. In fact, mutations have been found in the needle protein of the type III secretion apparatus that result in constitutive type III secretion. The mutated needle could assume a conformation that resembles the stimulated structure [56, 57].

7.14
Conclusions

Apart from being a protective skin, outer membranes of Gram-negative bacteria serve complex functions in communication with the environment. Since its identification [5], the ECF sigma factor family has rapidly grown [58]. In the FecIR subgroup, all studied systems except one regulate the synthesis of Fe^{3+} transport systems. The reason may be found in the very low availability of Fe^{3+}, which requires intricate transport systems across the outer membrane, through the periplasm, and across the cytoplasmic membrane. Control starting from the cell surface enables transcription initiation without formation of the transport system in the cytoplasmic membrane.

It is economical to synthesize the iron import systems only when they are needed. In cells with several iron transport systems, regulation by iron starvation alone would turn all systems on. It is more sophisticated to induce only the iron transport system whose iron substrate is available in the growth medium. This is the case for Fec-type regulation. Iron starvation induces synthesis of the regulatory proteins. When the cognate substrate is in the medium, synthesis of the outer membrane inducers and transporters is strongly increased. For example, in the uninduced state, FecA forms a basal level of less than 100 molecules per cell; in the induced state, the number of molecules increases to 80 000, making FecA one of the most highly synthesized proteins in E. coli. Despite the high synthesis of FecA, induction is not self-promoted since the same concentration of ferric citrate, 0.1 µM, that is required to induce the system is required to maintain it. Uninduced low synthesis of TonB, ExbB, and ExbD, and the very low synthesis of FecIR might be the limiting factors in induction. The high levels of the outer membrane proteins facilitate capture of the scarce substrates in the medium. Binding of the substrates initiates a signal that eventually ends in the cytoplasm, where transcription of the transport genes is initiated. Only two proteins are required to transfer the signal across three compartments – the outer membrane, the periplasm, and the cytoplasmic membrane. Both proteins have functions in addition to signal transfer. The outer membrane protein is also a transporter and the inner membrane protein regulates the activity of the sigma factor that functions specifically in transport gene transcription.

The iron transport systems discussed in this chapter are virulence factors of pathogenic or potentially pathogenic bacteria. Regulation of the genes that control the hypersensitivity response in plants by R. solanacearum, and exoproteinase and exotoxin synthesis in P. aeruginosa extend the functions that are controlled by the FecIRA-type regulation beyond regulation of iron transport systems. It will be of interest to see whether initiation of a signaling cascade by cell–cell contact, as observed in the R. solanacearum/plant system, occurs also during bacterial infections of animals and human.

Acknowledgments

Generous support by Andrei Lupas and critical reading of the manuscript by Karen A. Brune are gratefully acknowledged. The author's work was supported by the Max Planck Gesellschaft, the Deutsche Forschungsgemeinschaft, and the Fonds der Chemischen Industrie.

References

1 Braun, V. (1997) Surface signaling: novel transcription initiation mechanism starting from the cell surface. *Arch. Microbiol.*, **167**, 325–331.

2 Braun, V., Mahren, S., and Ogierman, M. (2003) Regulation of the FecI-type ECF sigma factor by transmembrane signaling. *Curr. Opin. Microbiol.*, **6**, 173–180.

3 Braun, V. and Mahren, S. (2005) Transmembrane transcription control (surface signaling) of the *Escherichia coli* Fec type. *FEMS Microbiol. Lett.*, **29**, 673–684.

4 Brooks, B.E. and Buchanan, S.K. (2007) Signaling mechanisms for activation of extracytoplasmic function (ECF) sigma factors. *Biochim. Biophys. Acta*, **1778**, 1930–1945.

5 Lonetto, M., Gribskov, M., and Gross, C. (1992) The σ^{70} family: sequence conservation and evolutionary relationships. *J. Bacteriol.*, **174**, 3843–3849.

6 Ferguson, A.D., Chakraborty, R., Smith, B.S., Esser, L., van der Helm, D., and Deisenhofer, J. (2002) Structural basis of gating by the outer membrane transporter FecA. *Science*, **295**, 1715–1719.

7 Yue, W.W., Grizot, S., and Buchanan, S.K. (2003) Structural evidence for iron-free citrate and ferric citrate binding to the TonB-dependent outer membrane transporter FecA. *J. Mol. Biol.*, **332**, 353–368.

8 Pierre, J.L. and Gautier-Luneau, I. (2000) Iron and citric acid: a fuzzy chemistry of ubiquitous biological relevance. *Biometals*, **13**, 91–96.

9 Härle, C., Insook, K., Angerer, A., and Braun, V. (1995) Signal transfer through three compartments: transcription initiation of the *Escherichia coli* ferric citrate transport system from the cell surface. *EMBO J.*, **14**, 1430–1438.

10 Ogierman, M. and Braun, V. (2003) *In vivo* cross-linking of the outer membrane ferric citrate transporter FecA and TonB: studies of the FecA TonB box. *J. Bacteriol.*, **185**, 1870–1885.

11 Pawelek, P.D., Croteau, N., Ng-Thow-Hing, C., Khursigara, C.M., Moiseeva, N., Allaire, M., and Coulton, J.W. (2006) Structure of the TonB in complex with FhuA, *E. coli* outer membrane receptor. *Science*, **321**, 1399–1402.

12 Shultis, D.D., Purdy, M.D., Banchs, C.N., and Wiener, M.C. (2006) Outer membrane active transport: structure of the BtuB: TonB complex. *Science*, **312**, 1396–1399.

13 Postle, K. and Larsen, R.A. (2007) TonB-dependent energy transduction between outer and cytoplasmic membranes. *Biometals*, **20**, 453–465.

14 Kim, M., Fanucci, G.E., and Cafiso, D.S. (2007) Substrate-dependent transmembrane signaling in TonB-dependent transporters is not conserved. *Proc. Natl. Acad. Sci. USA*, **104**, 11975–11980.

15 Gumbart, J., Wiener, M.C., and Tajkhorshid, E. (2007) Mechanics of force propagation in TonB dependent outer membrane transport. *Biophys. J.*, **93**, 496–504.

16 Kim, I., Stiefel, A., Plantör, S., Angerer, A., and Braun, V. (1997) Transcription induction of the ferric citrate transport genes via the N terminus of the FecA outer membrane protein, the Ton system and the electrochemical potential of the cytoplasmic membrane. *Mol. Microbiol.*, **23**, 333–344.

17 Garcia-Herrero, A. and Vogel, H. (2005) Nuclear magnetic resonance solution structure of the periplasmic signalling domain of the TonB-dependent outer membrane transporter FecA from *Escherichia coli*. *Mol. Microbiol.*, **58**, 1226–1237.

18 Ferguson, A.D., Amezcua, C.A., Halabi, N.M., Chellia, Y., Rosen, M.K., Ranganathan, R., and Deisenhofer, J. (2007) Signal transduction pathway of TonB-dependent transporters. *Proc. Natl. Acad. Sci. USA*, **104**, 513–518.

19 Breidenstein, E., Mahren, S., and Braun, V. (2006) Residues involved in FecR binding are localized on one side of the FecA signaling domain in *Escherichia coli*. *J. Bacteriol.*, **188**, 6440–6442.

20 Enz, S., Brand, H., Orellana, C., Mahren, S., and Braun, V. (2003) Sites of interaction between the FecA and FecR signal transduction proteins of ferric citrate

transport in *Escherichia coli* K-12. *J. Bacteriol.*, **185**, 3745–3752.

21 Welz, D. and Braun, V. (1998) Ferric citrate transport of *Escherichia coli*: functional regions of the FecR transmembrane regulatory protein. *J. Bacteriol.*, **180**, 2387–2394.

22 Enz, S., Mahren, S., Stroeher, U.H., and Braun, V. (2000) Surface signaling in ferric citrate transport gene induction: interaction of the FecA, FecR and FecI regulatory proteins. *J. Bacteriol.*, **182**, 637–646.

23 Stiefel, A., Mahren, S., Ochs, M., Schindler, P.T., Enz, S., and Braun, V. (2001) Control of the ferric citrate transport system of *Escherichia coli*: mutations in region 2.1 of the FecI extracytoplasmic-function sigma factor suppress mutations in the FecR transmembrane regulatory protein. *J. Bacteriol.*, **183**, 162–170.

24 Braun, V., Mahren, S., and Sauter, A. (2006) Gene regulation by transmembrane signaling. *Biometals*, **19**, 103–106.

25 Wriedt, K., Angerer, A., and Braun, V. (1995) Transcriptional regulation from the cell surface: conformational changes in the transmembrane protein FecR lead to altered transcription of the ferric citrate transport genes in *Escherichia coli*. *J. Bacteriol.*, **177**, 3320–3322.

26 Angerer, A., Enz, S., Ochs, M., and Braun, V. (1995) Transcriptional regulation of ferric citrate transport in *Escherichia coli* K-12. FecI belongs to a new subfamily of σ^{70}-type factors that respond to extracytoplasmic stimuli. *Mol. Microbiol.*, **18**, 163–174.

27 Mahren, S., Enz, S., and Braun, V. (2002) Functional interaction of region 4 of the extracytoplasmic function sigma factor FecI with the cytoplasmic portion of the FecR transmembrane protein of the *Escherichia coli* ferric citrate transport system. *J. Bacteriol.*, **184**, 3704–3711.

28 Enz, S., Braun, V., and Crosa, J.H. (1995) Transcription of the region encoding the ferric dicitrate-transport system in *Escherichia coli*: similarity between promoters for *fecA* and for extracytoplasmic function sigma factors. *Gene*, **163**, 13–18.

29 Mahren, S. and Braun, V. (2003) The FecI extracytoplasmic-function sigma factor of *Escherichia coli* interacts with the β' subunit of RNA polymerase. *J. Bacteriol.*, **185**, 1796–1802.

30 Koebnik, R. (2005) TonB-dependent *trans*-envelope signalling: the exception or the rule? *Trends in Microbiol.*, **13**, 343–347.

31 Koster, M., van Klompenburg, W., Bitter, W., Leong, J., and Weisbeek, P. (1994) Role for the outer membrane ferric siderophore receptor PupB in signal transduction across the bacterial cell envelope. *EMBO J.*, **13**, 2805–2813.

32 Visca, P., Leoni, L., Wilson, M.J., and Lamont, I.L. (2002) Iron transport and regulation, cell signalling and genomics: lessons from *Escherichia coli* and *Pseudomonas*. *Mol. Microbiol.*, **45**, 1177–1190.

33 Llamas, M.A., Mooij, M.J., Sparrius, M., Vandenbroucke-Grauls, C.M., Ratledge, C., and Bitter, W. (2008) Characterization of five novel *Pseudomonas aeruginosa* cell-surface signalling systems. *Mol. Microbiol.*, **67**, 458–472.

34 Vasil, M.L. (2007) How we learned about iron acquisition in *Pseudomonas aeruginosa*: a series of very fortunate events. *Biometals*, **20**, 587–601.

35 Schalk, I. (2008) Metal trafficking via siderophores in Gram-negative bacteria: specificities and characteristics of the pyoverdine pathway. *J. Inorg. Biochem.*, **102**, 1159–1169.

36 Redly, G.A. and Poole, K. (2005) FpvIR control of ferric pyoverdine receptor gene expression in *Pseudomonas aeruginosa*: demonstration of an interaction between FpvI and FpvR and identification of mutations in each compromising interaction. *J. Bacteriol.*, **187**, 5648–5657.

37 James, H.E., Beare, P.A., Martin, L., and Lamont, I. (2005) Mutational analysis of a bifunctional ferrisiderophore receptor and signal-transducing protein from *Pseudomonas aeruginosa*. *J. Bacteriol.*, **187**, 4514–4520.

38 Brillet, K., Journet, L., Celia, H., Paulus, L., Stahl, A., Pattus, F., and Cobessi, D. (2007) A β-strand lock exchange for signal transduction in TonB-dependent

transducers on the basis of a common structural motif. *Structure*, **15**, 1383–1391.

39 Wirth, C., Meyer-Klaucke, W., Pattus, F., and Cobessi, D. (2007) From the periplasmic signaling domain to the extracellular face of an outer membrane signal transducer of *Pseudomonas aeruginosa*: crystal structure of the ferric pyoverdine outer membrane receptor. *J. Mol. Biol.*, **368**, 398–406.

40 Tiburzi, F., Imperi, F., and Visca, P. (2008) Intracellular levels and activity of PvdS, the major iron starvation sigma factor of *Pseudomonas aeruginosa*. *Mol. Microbiol.*, **67**, 213–227.

41 Spencer, M.R., Beare, P.A., and Lamont, I.L. (2008) Role of cell surface signaling in proteolysis of an alternative sigma factor in *Pseudomonas aeruginosa*. *J. Bacteriol.*, **190**, 4865–4869.

42 Maeda, H., Jishage, M., Nomura, T., Fujita, N., and Ishihama, A. (2000) Two extracytoplasmic function sigma subunits, σ^E and σ^{FecI}, of *Escherichia coli*: promoter selectivity and intracellular levels. *J. Bacteriol.*, **182**, 1181–1184.

43 Vanderpool, C.K. and Armstrong, S.K. (2003) Heme-responsive transcriptional activation of *Bordetella bhu* genes. *J. Bacteriol.*, **185**, 909–917.

44 Pradel, E. and Locht, C. (2001) Expression of the putative siderophore receptor gene *bfrZ* is controlled by the extracytoplasmic-function sigma factor BupI in *Bordetella bronchiseptica*. *J. Bacteriol.*, **183**, 2910–2917.

45 Kirby, A.E., King, N.D., and Connell, T.D. (2004) RhuR, an extracytoplasmic function sigma factor activator, is essential for heme-dependent expression of the outer membrane heme and hemoprotein receptor of *Bordetella avium*. *Infect. Immun.*, **72**, 896–907.

46 Kirby, A.E., Metzger, D.J., Murphy, E.R., and Connell, T.D. (2001) Heme utilization in *Bordetella avium* is regulated by RhuI, a heme-responsive extracytoplasmic function sigma factor. *Infect. Immun.*, **69**, 6951–6961.

47 King-Lyons, N.D., Smith, k.F., and Connell, T.D. (2007) Expression of *hurP*, a gene encoding a prospective site 2 protease, is essential for heme-dependent induction of *bhuR* in *Bordetella bronchiseptica*. *J. Bacteriol.*, **189**, 6266–6275.

48 Cescau, S., Cwreman, H., Letoffe, S., Delepelaire, P., Wandersman, C., and Biville, F. (2007) Heme acquisition by hemophores. *Biometals*, **20**, 603–610.

49 Brito, B., Aldon, D., Barberis, P., Boucher, C., and Genin, S. (2002) A signal transfer system through three compartments transduces the plant cell contact-dependent signal controlling *Ralstonia solanacearum hrp* genes. *Mol. Plant. Microbiol.*, **15**, 109–119.

50 Ferguson, A.D., Hofmann, E., Coulton, J.W., Diederichs, K., and Welte, W. (1989) Siderophore-mediated iron transport: crystal structure of FhuA with bound lipopolysaccharide. *Science*, **282**, 2215–2220.

51 Locher, K., Rees, B., Koebnik, R., Mitschler, L., Moulinier, A., Rosenbusch, J., and Moras, D. (1998) Transmembrane signaling across the ligand-induced FhuA receptor: crystal structures of the free and ferrichrome-bound states reveal allosteric changes. *Cell*, **95**, 771–778.

52 Ma, L., Kaserer, W., Annamalai, R., Scott, D.C., Jin, B., Jiang, X., Xiao, Q., Maymani, H., Massis, L.M., Ferreira, L.C. *et al.* (2007) Evidence for a ball-and-chain transport of ferric enterobactin through FepA. *J. Biol. Chem.*, **282**, 397–406.

53 Wiener, M.C. (2005) TonB-dependent outer membrane transport: going for Baroque. *Curr. Opin. Struct. Biol.*, **15**, 394–400.

54 Hancock, R.G.E. and Braun, V. (1976) Nature of the energy requirement for the irreversible adsorption of bacteriophages T1 and φ80 to *Escherichia coli*. *J. Bacteriol.*, **125**, 409–415.

55 Galan, J.E. and Wolf-Watz, H. (2006) Protein delivery into eukaryotic cells by type III secretion machines. *Nat. Rev.*, **444**, 567–573.

56 Torruelas, J., Jackson, M., Pennock, J., and Plano, G. (2005) The *Yersinia pestis* type III secretion needle plays a role in the regulation of Yop secretion. *Mol. Microbiol.*, **57**, 1719–1733.

57 Kenjale, R., Wilson, J., Zenk, S.F., Saurya, S., Picking, W.L., Picking, W. D., and

Blocker, A. (2005) The needle component of the type III secretion of *Shigella* regulates the activity of the secretion apparatus. *J. Biol. Chem.*, **280**, 42929–42937.

58 Helman, J.D. (2002) The extracytoplasmic function (ECF) sigma factors. *Adv. Microb. Physiol.*, **46**, 47–110.

59 Garcia-Herrero, A., Peacock, R.S., Howard, S.P., and Vogel, H.J. (2007) The solution structure of the periplasmic domain of the TonB system ExbD protein reveals an unexpected structural homology with siderophore-binding proteins. *Mol. Microbiol.*, **66**, 872–889.

60 Greenwald, J., Zeder-Lutz, G., Hagege, A., Celia, H., and Pattus, F. (2008) The metal dependence of pyoverdine interactions with its outer membrane receptor FpvA. *J. Bacteriol.*, **190**, 6548–6558.

8
Stimulus Perception and Signaling in Histidine Kinases
Ralf Heermann and Kirsten Jung

8.1
Introduction

Bacteria are exposed to constantly varying environmental conditions, like variations in nutrient availability, osmolarity, pH, temperature, or cell density. To survive, cells have to monitor and to adapt adequately to these changing conditions. Hence, bacteria are equipped with numerous signal transduction systems that mediate the response to varying environmental stimuli. The most widely distributed type of signal transduction systems are the histidine kinase/response regulator systems, also known as sensor kinase/response regulator systems. In their prototype, these systems consist of a membrane-integrated histidine kinase that senses the stimulus and a cytoplasmic response regulator, which is often a transcriptional regulator, that mediates the output response. Signaling is initiated when the histidine kinase autophosphorylates at a histidine upon stimulus perception. Subsequently, the phosphoryl group is transferred to an aspartate within the response regulator, resulting in modulation of its activity and ultimately in the adaptation of the cell to the changed environment (see [1] for review).

8.2
Histidine Kinase Family

The exploding availability of more and more complete genomes allows definitive assessment of the prevalence of histidine kinase/response regulator systems. In 365 bacterial and archaeal genomes analyzed so far, approximately 23 000 genes were found encoding histidine kinases and response regulators [2]. The number of histidine kinase/response regulator systems differs enormously from species to species. It ranges from zero in *Mycoplasma genitalium* [3], over 30/32 (histidine kinases/response regulators) in *Escherichia coli* [4], 36/34 in *Bacillus subtilis* [5] to 131/80 in *Anabaena* sp. strain PCC7120 [6] and 132/119 in *Myxococcus xanthus* [7]. The varying number of these systems between species seems to be related to the

Bacterial Signaling. Edited by Reinhard Krämer and Kirsten Jung
Copyright © 2010 WILEY-VCH Verlag GmbH & Co. KGaA, Weinheim
ISBN: 978-3-527-32365-4

number of different environmental cues and the frequency of environmental changes the corresponding bacteria are exposed to.

Histidine kinases are the key proteins in bacterial signaling. These proteins sense specific stimuli and transduce them into an intracellular signal by initiating a phosphorylation cascade. Although hundreds of different histidine kinases are well characterized, our understanding how the stimuli are perceived and transduced is still incomplete. Moreover, for many histidine kinases not even the primary stimulus, which activates the protein, is known. In general, there are three mechanisms how histidine kinases perceive stimuli (Figure 8.1A) [8]. The majority of histidine kinases sense extracellular stimuli. These histidine kinases have a periplasmic/extracellular sensing/input domain that is framed by at least two transmembrane helices and the corresponding transmitter domain is found in the cytoplasm. In these histidine kinases the input and transmitter domains are located in two different cellular compartments separated by a membrane necessitating transmembrane signal transduction. A second group of histidine kinases perceives stimuli by the membrane-spanning helices alone or by a combination of the membrane helices and a N-terminal cytoplasmic domain. The input domains of these histidine kinases contain up to 13 transmembrane helices that are interconnected via short loops. These proteins perceive stimuli that come from the membrane interface or the cytoplasm. A third mechanism of stimulus perception is realized by cytoplasmic sensor proteins that are either membrane associated or soluble. The stimulus sensed by this group is exclusively of intracellular nature. Independent of how a certain stimulus is perceived, all histidine kinases share a similar mechanism of signal transduction. Once the stimulus is sensed, the histidine kinase is activated, resulting in autophosphorylation at a highly conserved histidine with the γ-phosphoryl group originating from ATP. Histidine kinases are homodimers, and occurs in *trans*, implying that one monomer binds ATP and phosphorylates the histidine residue of the other monomer [9, 10]. Rapidly, the phosphoryl group is transferred to a highly conserved aspartate in the receiver domain of the response regulator, resulting in activation of the output domain. In many cases, histidine kinases function also as phosphatases, resulting in the dephosphorylation of the response regulator and, concurrently, in the termination of the signaling cascade [1].

8.2.1
Basic Structure of Histidine Kinases

All histidine kinases share unique sequence motifs in their transmitter domains, designated as H, N, G1, F, and G2 boxes (Figure 8.1B) [11, 12]. The transmitter domain can be further dissected into two parts: the H-box containing the *d*imerization and *h*istidine *p*hosphotransfer domain (DHp), and the *h*istidine kinase-type *ATPase c*atalytic (HATPase_c) domain, better known as the *c*atalytic and *A*TP-binding (CA) domain [13]. On the basis of their domain organization, histidine kinases have been grouped into two classes: class I and class II histidine kinases [14]. In class I histidine kinases, the DHp domain is connected via a more or less flexible linker with the CA domain. In class II histidine kinases, the H box is located in a

8.2 Histidine Kinase Family

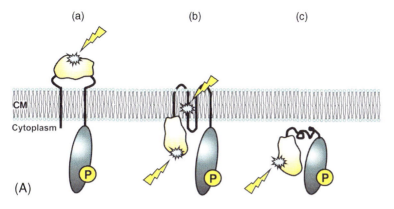

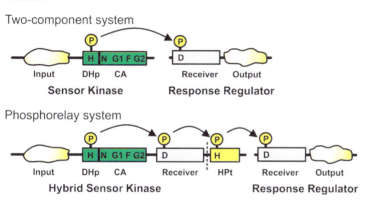

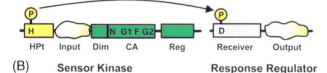

Figure 8.1 Domain organization of histidine kinases. (A) Different mechanisms of stimulus perception by histidine kinases: extracellular stimuli are sensed by histidine kinases that have a periplasmic/extracellular input domain (a); histidine kinases with several transmembrane helices and/or a N-terminal cytoplasmic domain sense stimuli at the membrane interface and/or intracellular stimuli (b); soluble histidine kinases sense intracellular stimuli (c). Upon stimulus perception by the input domain (white with star), the signal is transduced to the transmitter domain (gray) resulting in its autophosphorylation (marked with "P"). Generally, histidine kinases are homodimers, and autophosphorylation occurs in *trans*. To simplify matters, monomers are shown in this figure. (B) Domains and sequence motifs of histidine kinase/response regulator systems: the proteins are symbolized by black lines. Characteristic domains and sequence motifs are indicated (see text for details). In class I histidine kinases, the DHp domain is connected via a linker with the CA domain. In class II histidine kinases, the H-box containing HPt domain is separated from the CA domain by both an input and a Dim domain. An orthodox histidine kinase/response regulator system is characterized by one phosphoryl transfer (His → Asp). Phosphorelays catalyze multiple phosphotransfer reactions (His → Asp → His → Asp). Reg, regulatory domain; CM, cytoplasmic membrane.

HPt (Histidine Phosphotransfer) domain, which is separated from the CA domain by both an input and a *dimerization* (Dim) domain (Figure 8.1B) [13]. Most of the histidine kinases known thus far belong to class I. In *Escherichia coli*, for example, only one example for a class II histidine kinase exists and this is the soluble histidine kinase CheA that is involved in chemotaxis [4].

Apart from the orthodox histidine kinase/response regulator systems, which are characterized by one phosphoryl transfer between the two proteins (His → Asp), phosphorelays are known that catalyze multiple phosphotransfer reactions (His → Asp → His → Asp) (Figure 8.1B) [3]. Hybrid histidine kinases are representatives for phosphorelays. In these proteins the transmitter domain is fused to a receiver domain and the phosphoryl group is transferred from histidine to aspartate within one protein. From the receiver domain, the phosphoryl group goes to a HPt domain (histidine), which can be a separate protein or a covalently bound domain of the hybrid kinase. The HPt domain transfers the phosphoryl group to the receiver domain (aspartate) of the response regulator. The more complex phosphorelays bear multiple points of regulation and might facilitate a better fine-tuning than orthodox phosphorylation step systems. In *E. coli*, five histidine kinases (ArcB, BarA, EvgS, RcsC, and TorS) out of 30 are hybrid kinases [4]; in *Bacillus* there are no hybrid kinases [5]. *Anabaena* contains 55 hybrid sensor kinases, some catalyze more than four phosphorylation steps [6]. In *M. xanthus*, five hybrid kinases are found that contain at least two HPt domains and three receiver domains [15]. Phosphorelays are suspected to reflect the need to integrate various signals into the output of a pathway [16].

In *Bacteroides thetaiotaomicron*, 32 members of a novel class of membrane-integrated hybrid sensor kinases were identified that, in addition to the above described domains, contain one or two DNA-binding domains of the AraC type [17].

8.2.2
Specifics of Histidine Kinases in Comparison to Serine/Threonine/Tyrosine Kinases

Histidine kinase/response regulator systems are predominantly used by bacteria for gene regulation, but they are also present in Archaea and eukaryotes, like fungi and plants [18, 19]. Nevertheless, in eukaryotes, signaling is predominantly mediated by serine/threonine/tyrosine kinases, which are also present in prokaryotes, but with less abundance (see Chapter 24). The basic principle of both systems is utilization of the high-energy phosphoryl groups as transferable means for signal transduction cascades. However, there are substantial differences between histidine/aspartate and serine/threonine/tyrosine phosphorylation-dependent signaling pathways. First, the high-energy phosphoryl group at histidine and, particularly, at aspartate is intrinsically instable, and therefore has a half-life ranging from a few seconds in some systems to several minutes in some others. Owing to the high free energy stored in mixed acid anhydride bonds, phosphorylation of an aspartyl side-chain is optimally suited to elicit conformational changes [20, 21]. Moreover, it has to be emphasized that for thermodynamic reasons phosphorylation of histidine by ATP and phosphotransfer to aspartate are reversible reactions. This explains why most of the histidine kinases are bifunctional enzymes that in addition to their (auto)kinase activity also function as phosphatase for the cognate phosphorylated response regulator. Therefore, the ratio between kinase and phosphatase activity is crucial for the extent of response to the received stimulus. In contrast, phosphorylated serine, threonine, or tyrosine residues are extremely stable. This stability requires a separate phosphatase to remove the phosphoryl group from the kinase to switch off the signaling cascade. Furthermore, serine/threonine/tyrosine kinases are arranged in cascades. Thus, an upstream kinase can phosphorylate a large number of downstream kinases to amplify the signal exponentially. In contrast, histidine/aspartate signaling pathways are more specific and allow a precise coordination between input and output.

8.3
Stimulus Perception and Signaling by Histidine Kinases

8.3.1
Chemical Stimuli

The group of transmembrane histidine kinases includes numerous examples for which the chemical stimuli are known (Table 8.1). Most of them have a periplasmic/extracellular sensing domain that might contain specific folds, such as Per/Arnt/Sim (PAS), cyclase/histidine kinase-associated sensing extracellular (CHASE), four-helix bundle (4HB), and nitrate and nitrite sensing (NIT) (see [22] for review). The best-characterized chemosensory histidine kinases are now described in more detail.

CitA and DcuS respond to external available tri- and dicarboxylates under anaerobic conditions. Structural and biochemical analyses revealed that the periplasmic domain of CitA from *E. coli* or *Klebsiella pneumoniae* is a high-affinity citrate receptor [23], while the periplasmic domain of DcuS has a high affinity for C_4-dicarboxylates and citrate [24]. Both domains have similar structures, composed of a PAS-like fold with a core structure consisting of four or five β-strands [25–27]. After ligand binding, the signal is transduced across the membrane resulting in autophosphorylation of CitA or DcuS, respectively. There is recent evidence that structural disorder within the cytoplasmic PAS domain of DcuS is important to transduce the signal to the transmitter domain and regulating kinase activity [28] (see also Chapter 11). The structure of the periplasmic sensing domain of *E. coli* PhoQ is similar to those of DcuS and CitA. Importantly, the PhoQ, DcuS, and CitA sensor folds are found to be distinct from the superficially similar PAS domain [29]. The PhoQ/PhoP systems of enteric bacteria are signaling systems that regulate virulence and stress response including intracellular survival, invasion, lipid A structure, resistance to antimicrobial peptides, and phagosome alteration (see also Figure 8.2D) [30, 31]. PhoQ binds divalent cations, particularly Mg^{2+}, that form a metal bridge between the acidic surface of the sensing domain and the surface of the membrane, and consequently hold PhoQ in an inactive state [32]. Low Mg^{2+} concentrations, antimicrobial peptides, and acidic pH activate PhoQ, and these stimuli, found within phagosomes and intestinal tissues, may also serve as specific signatures of the host environment for these Gram-negative pathogens.

Table 8.1 Histidine kinase/response regulator systems with known molecular signals.

System	Stimulus	Physiological link	Organism	Reference
AgrC/AgrA	AIP	pathogenesis	S. aureus	[56]
AlgR2/AlgR1	Cu^{2+}	alginate synthesis	P. syringae	[142]
ApsS/ApsR	cationic antimicrobial peptides	antimicrobial peptide resistance	S. epidermidis	[143]
ArcB/ArcA	quinones, lactate	(an)aerobiosis	E. coli	[34]
ArsS/ArsR	low pH	urease and other acid-resistance genes	Helicobacter pylori	[144]
AtoS/AtoC	acetoacetate	acetoacetate metabolism	E. coli	[10]
BarA/UvrY	pH	pilus adherence	E. coli	[43]
BasS/BasR	$Fe^{2+/3+}$	iron response	E. coli	[145]
BceR/BceS	bacitracin	bacitracin resistance	B. subtilis	[146]
BvgA/BvgS	temperature, redox state of quinones, SO_4^{2-}, nicotinic acid	toxin and adhesin expression, biofilm formation	B. pertussis	[147]
CbrA/CbrB	amino acids	carbon and nitrogen source utilization	P. aeruginosa	[148]
CdsS/CdsR		chitin degradation	Pseudoalteromonas piscicida	[149]
CheA/CheB, CheY	methylation of methyl-accepting chemotaxis proteins	chemotaxis	E. coli	[150]
ChiS/ChiR	chitin	chitinase production	Streptomyces thermoviolaceus	[151]
CitA/CitB	citrate	citrate uptake	E. coli	[23]
ComP/ComA	peptide ComX	competence, sporulation	B. subtilis	[152]
CopS/CopR	Cu^{2+}	copper resistance	P. syringae	[153]
CorS/CorR, CorP	temperature	phytotoxin production	P. syringae	[109]
CqsS-LuxU/LuxO	CAI-1	QS	V. harveyi	[68]
CpxA/CpxR	periplasmic protein CpxP, misfolded envelope proteins (pH, osmotic stress)	regulation of envelope stress	E. coli	[53]
CreC/CreB	glycolytic carbon compounds	regulation of intermediary metabolism	E. coli	[154]
CsrS/CsrR (CovS/CovR)	Mg^{2+}	capsule synthesis, virulence genes ska, sagA	Streptococcus pyogenes	[155]
CusS/CusR	Cu^{2+}	copper resistance	E. coli	[156]
DctB/DctD	C_4-dicarboxylates	C_4-metabolism	Sinorhizobium meliloti	[157]

Table 8.1 (Continued)

System	Stimulus	Physiological link	Organism	Reference
DcuS/DcuR	C_4-dicarboxylates	C_4-metabolism	E. coli	[158]
DevS/DevR	O_2, NO, CO	adaptation to NO and hypoxic conditions	Mycobacterium tuberculosis	[159]
DosS, DosT/DosR	hypoxia, nitric acid, ethanol, H_2O_2	dormancy	M. tuberculosis	[160]
EnvZ/OmpR	osmolarity, K^+	outer membrane porin production	E. coli	[88]
EvgS/EvgA	ubiquinone	multiple drug resistance	E. coli	[161]
FeuQ/FeuP	osmolarity	cyclic glucan production; cell adhesion	S. meliloti	[162]
FixL/FixJ	gaseous ligands (O_2, NO, CO)	N_2 fixation	B. japonicum, S. meliloti	[38]
FsrC/FsrA	GBAP	QS	E. faecalis	[62]
HK17/RR17	ethanolamine	ethanolamine degradation	E. faecalis	[163]
HrrS/HrrR ChrS/ChrA	heme	regulation of heme homeostasis	C. diphthtriae	[164]
HssS/HssR	heme	heme toxicity, virulence	S. aureus	[165]
KdpD/KdpE	K^+, ATP, ionic strength	high-affinity K^+ transport	E. coli	[166]
LiaS/LiaR	cell envelope damage	bacitracin resistance	B. subtilis	[115]
LovK/LovR	flavin, light	surface attachment	C. crescentus	[106]
LuxN-LuxU/LuxO	β-hydroxy-butyryl-homoserine lactone	QS	V. harveyi	[70]
LuxQ-LuxU/LuxO	LuxP/AI-2	QS	V. harveyi	[67]
MtrA/MtrB	sugars, amino acids, polyethylene glycols	osmoregulation	C. glutamicum	[89]
NarQ/NarP NarX/NarL	nitrate/nitrite	anaerobic respiration	E. coli	[20]
NifA/NifL	O_2, N_2	nitrogen fixation	K. pneumoniae	[167]
NisK/NisR	nisin	nisin production and immunity	Lactococcus lactis	[168]
NodV/NodW	flavonoid	symbiosis	B. japonicum	[169]
NtrB/NtrC	P(II) protein	nitrogen metabolism	E. coli	[170]
PfeS/PfeR	enterobactin	ferric enterobactin receptor	P. aeruginosa	[171]
PhoP/PhoQ	divalent cations, antimicrobial peptides	bacterial virulence	E. coli, S. enterica	[32]

Table 8.1 (Continued)

System	Stimulus	Physiological link	Organism	Reference
PmrB/PmrA	Fe^{3+}	lipid A modification	S. enterica	[47]
QseC/QseB	epinephrine, norepinephrine, AI-3	interkingdom signaling	E. coli	[78]
RcsC, RcsD/RcsB	undecaprenyl-(pyro)phosphate, outer membrane protein RcfF	capsule synthesis, motility, chemotaxis	E. coli, S. enterica	[172]
RegB/RegA	ubiquinone	photosynthesis, carbon fixation, nitrogen fixation, hydrogen utilization, aerobic and anaerobic respiration, denitrification, electron transport, aerotaxis	Rhodobacter capsulatus, R. sphaeroides	[173]
RprX/RprY	tetracycline	tetracycline resistance	Bacteroides fragilis	[174]
TodS/TodR	toluene, o-xylene	toluene dioxygenase pathway	P. putida	[175]
TorS/TorR	trimethylamine N-oxide	thrimethylamine metabolism	E. coli	[176]
UhpB/UhpA	glucose-6-phosphate- receptor UhpC	glucose-6-phosphate transport	E. coli	[111]
VanS/VanR	cell wall compounds	vancomycin resistance	S. coelicolor, E. faecalis	[177]
VirA/VirG	phenolic compounds (plant wound factors), pH, periplasmic sugar-binding protein ChvE	virulence, plant infection	A. tumefaciens	[42]
ZraS/ZraR	Zn^{2+}, Pb^{2+}	Zn^{2+} tolerance	E. coli	[178]

Histidine kinases are at the first position followed by the cognate response regulator separated by a slash. Two proteins of the same family are separated by a comma. Histidine kinases and HPt proteins are separated by a hyphen.

The ArcB/ArcA system is important for regulation of numerous genes encoding enzymes of respiratory and fermentative metabolism in response to oxygen availability. Under anaerobic conditions, ArcB is in its active state and phosphorylates the response regulator ArcA. ArcB is a hybrid histidine kinase and anchored with two transmembrane domains connected by a small periplasmic loop in the cytoplasmic membrane. This sensor kinase has a cytoplasmic PAS domain within its input domain [33]. It is suggested that ArcB senses oxygen availability indirectly since the soluble membrane quinone analogs ubiquinone-0 (Q_0) and menadione (MK_3) inhibit the autophosphorylation activity [34]. Moreover, the quinone-dependent inhibition requires the formation of one or two intermolecular disulfide bonds between the PAS domains of two ArcB monomers. The completely reduced histidine kinase has full kinase activity, whereas the completely oxidized protein lacks kinase activity. The disulfide bonds are believed to cause domain rearrangements disturbing the interaction of the transmitter domains within the dimer [35]. Furthermore, the fermentative metabolite D-lactate seems to be an effector that is additionally sensed by ArcB under anaerobiosis resulting in enhanced kinase activity [36].

In the root-nodulating bacteria *Bradyrhizobium japonicum* or *Rhizobium meliloti*, the FixL/FixJ histidine kinase/response regulator system regulates the transcription of N_2 fixation genes, a process which is strictly anaerobic [37]. FixL also has a cytoplasmic PAS domain within its input domain that binds a ferrous heme molecule. At high oxygen concentrations, oxygen is bound to the ligand resulting in inhibition of the kinase activity. In its deoxy state, FixL exhibits kinase activity switching on the signaling cascade [37, 38]. FixL activity is also modulated by binding of other gaseous ligands such as CO and NO, but to a far lesser extent [39]. These examples show that the PAS domains in histidine kinases fulfill different functions in stimulus perception.

Numerous histidine kinases bind ligands, but lack a PAS domain. The NarX/NarL and NarQ/NarP systems, which control anaerobic respiratory gene expression in *E. coli* or *Salmonella enterica*, are phosphorylated in response to nitrate or nitrite, respectively. During anaerobic growth these dual histidine kinase/response regulator systems control hierarchal use of alternative electron acceptors in the order of their electron potential. NarX and NarQ share a characteristic periplasmic input domain (NIT), for which two P-box sequences (18 residues each) are characteristic that flank the two transmembrane helices. The P-box region is involved in the discrimination between nitrate and nitrite [20]. In addition to the periplasmic P-boxes, both NarQ and NarX contain an extended cytoplasmic linker region that is important to transmit the signal [40].

The VirA/VirG system of the Gram-negative plant pathogen *Agrobacterium tumefaciens* controls expression of virulence genes in response to acidic pH, monosaccharides and phenolic compounds like acetosyringone (for review, see [41]). Upon stimulus perception, VirA/VirG regulates the production of both the transferred oncogenic DNA and the DNA transfer machinery from the tumor-inducing Ti plasmid. The input domain of VirA consists of two transmembrane domains, a periplasmic domain, and a large cytoplasmic linker domain. VirA senses phenolic compounds, which are produced by wounded plants, in the cytoplasmic linker domain resulting in kinase activity. Sugars, also released by wounded plants, act as strong enhancers of kinase activity. Sugars are not directly sensed by VirA, but bind to the periplasmic sugar-binding protein, ChvE, which then interacts with the periplasmic input domain of VirA. The periplasmic domain is also involved in pH sensing. This process is somehow coupled to sugar sensing, because deletion of *chvE* abolishes pH sensing [42].

Another pH-responsive system is the BarA/UvrY system, which regulates central carbon metabolism, motility, and biofilm formation by controlling the expression of noncoding RNAs, and therefore virulence of avian pathogenic *E. coli* (APEC) or insect pathogenic *Photorhabdus luminescens* [43, 44]. The sensor kinase BarA consists of two transmembrane domains, which flank a periplasmic sensor domain. BarA activity is inhibited at an external pH of 5.0 [45]. It is discussed whether BarA kinase activity is additionally affected by other yet unknown chemical molecules under conditions of low pH and/or whether BarA senses the pH directly by detecting the pH difference across the cytoplasmic membrane [45].

The PmrB/PmrA system controls lipopolysaccharide modifications mediating resistance to the antibiotic polymyxin B in *S. enterica* depending on the extracellular iron concentration [46]. In addition, wild-type *Salmonella* grown at pH 5.8 was more resistant to polymyxin B than organisms grown at pH 7.7. Substitution of a conserved histidine or either one of four conserved glutamic acid residues in the periplasmic domain of PmrB severely affected the mild acid-promoted transcription of PmrA-activated genes. Consequently, it is proposed that protonation of the periplasmic histidine and/or of the glutamic acid residues activates PmrB [47].

The Cpx system of *E. coli* responds to external pH, too. The system consists of the membrane-bound histidine kinase CpxA, the cognate response regulator CpxR, and the periplasmic protein CpxP, which is somehow a signal carrier for CpxA (see Figure 8.2C and Section 8.4). CpxA generally senses different kinds of envelope stress, and triggers the expression of periplasmic folding factors and proteases [48]. Apart from external pH, the system responds to altered membrane composition, overproduction of membrane lipoproteins, accumulation of misfolded proteins, such as the maltose-binding protein or pilus subunits, indole, and increased osmolarity [49–53].

Bacteria are able to communicate via diffusible chemical signals referred to as quorum sensing (QS) (see previous chapters). Some of these chemical signals are recognized by histidine kinases. Gram-positive bacteria secrete peptides via dedicated ATP-binding cassette (ABC) export proteins. These peptides are recognized by histidine kinase/response regulator systems, such as ComP/ComA and ComD/ComE regulating competence for DNA uptake in *B. subtilis* [54] and *Streptococcus pneumoniae* [55], or AgrC/AgrA in *Staphylococcus aureus* [56], and FsrC/FscA in *Enterococcus faecalis* [57] regulating virulence. ComP of *B. subtilis* is a histidine kinase with six to eight membrane-spanning segments and two extracytoplasmic loops [58] that senses the extracellular ComX pheromone. ComX is released at a constant rate per cell during growth [59]. ComX is produced in inactive pre-form, activated after proteolytic cleavage by the protease ComQ [58], and released as an isoprenylated peptide [60].

In *E. faecalis*, biosynthesis of the virulence factor gelatinase, is regulated by cell density. *E. faecalis* secretes a gelatinase biosynthesis-activating pheromone (GBAP), an 11-residue cyclic peptide containing a lactone structure, in which the α-carboxyl group of the C-terminal amino acid is linked to the hydroxyl group of the serine of the third residue. The precursor FsrD is processed by the cysteine protease FsrB to GBAP [61]. GBAP binds to the periplasmic domain of the sensor kinase FsrC and activates signaling by FsrC/FsrA [62].

In *S. aureus* the synthesis of an untranslated RNA molecule called RNAIII is under the control of peptide QS mediated by the Agr system. This QS system consists of four proteins. AgrB is a transmembrane protein that appears to be involved in processing of the propeptide AgrD into the autoinducer peptide (AIP) pheromone, its secretion, and modification to form a thiolactone ring. The histidine kinase AgrC, which is anchored with six transmembrane segments in the membrane, binds the extracellular AIP and in turn modulates the activity of the response regulator AgrA. Phosphorylated AgrA activates transcription of the RNAIII molecule, which is the regulatory effector of the system and responsible for both positive and negative regulation of a variety of downstream targets including genes encoding protein A, coagulase, enterotoxins, and hemolysins [63, 64].

A more complex QS regulation is found in the free-living marine bioluminescent bacterium *Vibrio harveyi*. The three hybrid sensor kinases LuxN, LuxQ, and CqsS sense the autoinducers 3-hydroxybutanoyl homoserine lactone (HAI-1), (2S,4S)-2-methyl-2,3,3,4-tetrahydroxytetrahydrofuran-borate (AI-2), and (S)-3-hydroxytridecan-4-one (CAI-1) [65–68]. These chemical signals are directly detected by LuxN and CqsS, while AI-2 is sensed by LuxQ after binding to the periplasmic protein LuxP [69, 70]. LuxN, the sensor kinase for HAI-1, has nine transmembrane domains and belongs to the rare examples of membrane proteins with a N-terminus located in the periplasm [71]. Biochemical characterization of the enzymatic activities of LuxN demonstrated that the presence of HAI-1 downregulates the kinase activity, while the phosphatase activity remains unaffected [72]. LuxQ, the sensor kinase for AI-2, is anchored with two transmembrane domains in the cytoplasmic membrane, which are connected by a large periplasmic loop containing two PAS domains. Cocrystallization of the periplasmic domain of LuxQ and LuxP implied binding of LuxP to LuxQ that was independent of the presence of AI-2 [73]. Instead, binding of AI-2 causes conformational changes within LuxQ [69] (see Chapter 11). The phosphorylation cascade is active when the extracellular autoinducer concentration is low [74]. The phosphoryl group is transferred via phosphorelay including the HPt protein LuxU to the σ^{54}-dependent transcriptional activator LuxO. Phosphorylated LuxO activates transcription of four regulatory small RNAs (*qrr1–4*). Together with the RNA chaperone Hfq these small RNAs destabilize the transcript of the master regulator LuxR [75, 76]. LuxR controls genes of the QS regulon, including those required for bioluminescence, type III secretion, and production of siderophores and an exoprotease [77].

Enterohemorrhagic *E. coli* uses epinephrine/norepinephrine and QS AI-3 (autoinducer-3) as chemical signals to regulate expression of virulence and flagella/motility genes. These signals are sensed by the hybrid sensor kinase/response regulator system QseC/QseB. QseC is the only known hybrid sensor kinase that responds to eukaryotic hormones, and thus mediates interkingdom signaling by functioning as a bacterial receptor [78, 79].

8.3.2
Physical Stimuli

For bacteria sensing of physical stimuli is much more difficult compared to the sensing of chemical stimuli. Nevertheless, some well-characterized examples of histidine kinase exist which detect alterations in osmolarity, turgor, light, or temperature.

An increase in external osmolarity triggers a water flux along the osmotic gradient, causing dehydration and plasmolysis of the cells. Osmolarity is a physicochemical parameter that cannot be deduced to a single molecule. Thus, under this condition, cells might sense alterations in turgor, membrane strain or shrinkage, which affect the curvature of the membrane, external or internal osmolarity, external or internal ion concentration or ionic strength, changes of the concentration of specific cytoplasmic or periplasmic solutes, the transmembrane osmotic gradient, or macromolecular crowding [80–83]. One paradigm for an osmosensitive histidine kinase response/regulator system is EnvZ/OmpR of *E. coli*, which has been analyzed exhaustively (see also Chapter 11), but the nature of the primary stimuli perceived remains elusive. EnvZ/OmpR regulates the expression of *ompC* and *ompF* encoding two outer membrane porins depending on external osmolarity. The histidine kinase EnvZ consists of a periplasmic input domain flanked by two transmembrane domains, similarly to histidine kinases detecting chemical molecules at the extracellular side. However, truncation of the periplasmic domain was without effect on osmosensing [84]. A protein or ligand interacting with the periplasmic domain was not found [85]. In addition, EnvZ of the insect pathogen *Xenorhabdus nematophila*, which has a much shorter periplasmic domain compared to the *E. coli* protein, responds very well to alterations in the external osmolarity [86]. Thus, the question remains how alterations of external osmolarity might be sensed? One of the first responses of bacteria to counteract an increased external osmolarity is the rapid accumulation of K^+ [87]. An increase of the K^+ concentration stimulated the kinase activity of EnvZ. Thus, hyperosmotic stress might be sensed by EnvZ indirectly by measuring alterations of the intracellular K^+ concentration [88]. A rise in the cytoplasmic K^+ concentration is also discussed as a stimulus for the osmosensing transport protein BetP (see Chapter 12).

The recently discovered osmoresponsive MtrB/MtrA system of *Corynebacterium glutamicum* also responds to chemical rather than to physicochemical stimuli. The histidine kinase MtrB was found to be activated *in vitro* by an increase of the osmolarity of buffer imposed by various osmolytes, like sugars, amino acids, or polyethylene glycols. Owing to the different chemical nature of the solutes, it seemed unlikely that they bind directly to MtrB. Moreover, deletion of the periplasmic domain of MtrB had no influence on the osmosensory activity of the protein indicating that the primary stimuli sensed by MtrB are related to cytoplasmic alterations. It is proposed that MtrB is activated by changes of the hydration state of the protein [89].

In response to K^+ limitation or salt stress, the histidine kinase KdpD together with the response regulator KdpE regulates the expression of *kdpFABC* encoding a high-affinity K^+ uptake system in *E. coli*. The sensing domain of KdpD consists of a large (about 400-amino-acid) N-terminal cytoplasmic domain, four transmembrane domains, and about 100 C-terminally amino acids at the cytoplasmic side [90]. Since K^+ plays a major role to maintain turgor, it was proposed that a decrease in turgor or some effect thereof activates the Kdp system [91]. This model has been challenged by the finding that *kdpFABC* expression is only significantly induced when the osmolarity of the medium is increased by a salt and not by a sugar [92–94]. In addition, recent studies indicate that KdpD activation does not correlate with changes in the cytoplasmic volume [95]. The level of *kdpFABC* transcription is at least 10-fold higher under K^+ limitation than in response to salt stress [95–97]. Furthermore, the four transmembrane domains are not essential for sensing of K^+ limitation, but are rather important for the correct positioning of the large cytoplasmic domains to each other [98]. Instead of being a turgor sensor, there are several indications that KdpD senses alterations of chemical, intracellular parameters that are changed under hyperosmotic stress or K^+ limitation: (i) KdpD is an exceptional histidine kinase as it contains a large cytoplasmic input domain [90]. (ii) K^+ ions have an inhibitory effect on the KdpD kinase activity [97]. A putative K^+-binding site is suspected within the cytoplasmic extension of transmembrane helix IV [99, 100]. The effect of K^+ on KdpD activation was corroborated by the observation that transcription of *kdpFABC* is also induced when cells are grown in the presence of Cs^+, a known inhibitor of K^+ uptake systems [101]. (iii) Studies with KdpD inserted in right-side-out membrane vesicles, in which the cytoplasmic domains are unidirectionally exposed to the lumen of the vesicles, indicated that an increase of the intraluminal ionic strength stimulates the kinase activity. (iv) A regulatory ATP-binding site was identified within the input domain of KdpD [102, 103]. An osmotic upshift is accompanied by a transient increase of intracellular ATP [104], which might be sensed by KdpD [95, 102, 103].

For the osmoresponsive hybrid sensor kinase Sln1p of *Saccharomyces cerevisiae*, a mechanical stimulus has been suggested to activate the protein [105]. A decrease of turgor, caused by hyperosmotic stress, addition of the membrane-permeabilizing antifungal drug nystatin (which induces cell volume shrinkage) or removal of the cell wall, activate the HOG pathway directly via the Sln branch, consistent with a concept of a turgor-dependent regulation of Sln1p. Thus, Sln1p can be regarded as a direct osmosensor, whereas EnvZ and KdpD sense alterations of intracellular parameters in consequence of specific and unspecific first responses of the cells to counteract hyperosmotic stress. It should be noted that osmosensing is also mediated by transport proteins (see Chapter 12).

The differentiating bacterium *Caulobacter crescentus* contains a light response LovK/LovR system that is required for light-dependent intercellular attachment [106]. LovK has a so-called LOV (light, oxygen, and *v*oltage) domain in its input domain, which is a flavin-binding blue-light photosensory domain. In response to visible light LovK binds a flavin cofactor, undergoes a reversible photocycle, and displays increased autophosphorylation activity. A conserved cysteine residue in the LOV domain of LovK, which forms a covalent adduct with the flavin cofactor upon absorption of visible light, is necessary for the light-dependent regulation of LovK activity [106]. Genes predicted to encode LOV histidine kinases are conserved across a broad range of bacterial taxa, from aquatic oligotrophs to plant and mammalian pathogens. However, the function of these putative photoreceptors is still unclear.

Many bacterial gene regulatory circuits are controlled by temperature. Bacterial virulence is often temperature regulated. Thus, many examples for thermoregulation come from pathogenic organisms. Temperature-mediated regulation occurs at the level of transcription and translation. Supercoiling, changes in mRNA conformation and protein conformation are all implicated in thermosensing [107] (see Chapter 23). The best example for a temperature-sensing histidine kinase/response regulator system is CorS/CorR, regulating coronatine biosynthesis in *Pseudomonas syringae* [108]. This plant pathogenic bacterium synthesizes high levels of the phytotoxin coronatine at the virulence-promoting temperature of 18 °C, but not at the optimal growth temperature 28 °C. The Cor system comprises a second response regulator CorP, which lacks a DNA-binding helix–turn–helix motif and which is not involved in temperature-dependent regulation of coronatine biosynthesis. The N-terminal input domain of CorS comprises six transmembrane helices, and CorS is believed to sense the fluidity and/or the fatty acid composition of the membrane via the transmembrane anchor. A temperature-dependent conformational change of CorS is proposed during which the sixth transmembrane helix leaves the membrane at 28 °C by flipping into the periplasm, so that the DHp domain is pulled into the membrane avoiding phosphorylation [109].

8.4
Accessory Proteins of Histidine Kinases

Histidine kinase/response regulator systems are often referred to as "two-component systems" based on the assumption that they consist of only two components. Meanwhile, many systems are known that include further components responsible for stimulus perception (cosensing proteins), fine-tuning (scaffolding proteins), cross-talk (connector proteins), or signal integration (allosteric effects) (Figure 8.2). In the following we will explain these principles with examples.

UhpABC represents a perfect example for cosensing. In this system UhpC perceives the external stimulus and transfers the signal to the histidine kinase UhpB (Figure 8.2A). The UhpABC regulatory system controls the expression of *uhpT* encoding the hexose phosphate transporter UhpT in enterobacteria. The signaling pathway is triggered by external glucose-6-phosphate, which is recognized by the membrane-integrated receptor UhpC. Upon sensing of glucose-6-phosphate, UhpC interacts with the membrane-integrated histidine kinase UhpB [110, 111] and stimulates phosphorylation of the UhpB/UhpA system that in turn induces expression of *uhpT* encoding the transport protein UhpT [112]. Glucose-6-phosphate is the only ligand that activates the signaling through UhpABC, which results in *uhpT* expression and in turn in the uptake of a broad range of phosphorylated sugars [113]. UhpC and UhpT are homologous to each other, and belong to the Major Facilitator Superfamily [114]. It has been suggested that the primordial unregulated gene, that encoded the transport protein, was duplicated and then modified to become a sensor while losing transport function. The input domain of the histidine kinase UhpB consists of eight putative transmembrane domains. It is supposed that UhpB interacts with the receptor UhpC via the transmembrane domains resulting in UhpB/UhpC complex formation upon stimulus perception [110, 111].

In *B. subtilis*, sensing of the cell wall antibiotic bacitracin depends on the bacitracin ABC transporter BceAB that acts as a bacitracin detoxification pump and regulator

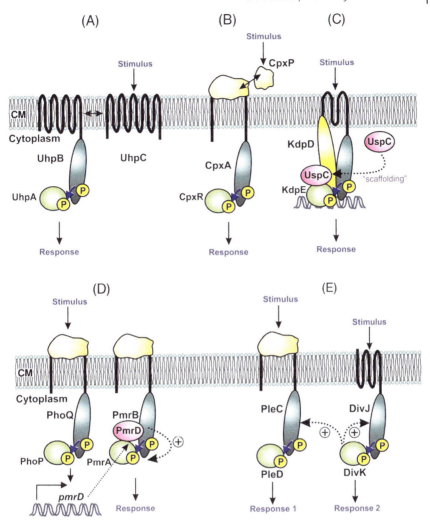

Figure 8.2 Signal integration and accessory proteins of histidine kinases. (A) Cosensing by membrane-integrated receptor proteins. (B) Cosensing by periplasmic receptor proteins. (C) Signal integration by scaffolding proteins. (D) Signal integration by connector proteins. (E) Signal integration via allosteric regulation by cognate response regulators. For each principle, one example is illustrated; see text for details. CM, cytoplasmic membrane.

of the two-component system BceS/BceR [115]. To date it is still unclear in which direction bacitracin is transported. Originally, BceAB was described as an efflux pump because of the lack of a substrate-binding protein [116], but its architecture and the existence of a large extracellular loop important for stimulus perception allude to active bacitracin uptake. The two-component system BceS/BceR, activated by BceAB upon stimulus perception, induces expression of *bceAB* in the presence of bacitracin. In the interplay between BceAB and BceS, ATP binding and hydrolysis by BceAB, and thus active transport are crucial for stimulus perception. A connection between ABC transporters and two-component systems seems to be quite frequent in the Firmicutes like *Bacillus*, because to date about 70 examples are described for genes encoding histidine kinase/response regulator that are adjacent to those encoding ABC transporters [8].

The receptor protein can also be a periplasmic protein instead of a transporter, which transduces the stimulus to the periplasmic input domain of the histidine kinase. This regulatory scheme was acquired by the sensor kinase VirA and the corresponding periplasmic protein ChvE or the Cpx envelope stress system (see Section 8.3.1 for details). Biochemical studies provided direct evidence for the protein–protein interaction between the histidine kinase CpxA and the periplasmic protein CpxP resulting in a downregulation of the autokinase activity of CpxA (Figure 8.2B) [53].

Scaffolding proteins are known from eukaryotes. These proteins bring together proteins in a signaling pathway and thus influence signal transduction [117–119]. Recently, the first scaffolding protein influencing a histidine kinase/response regulator system was characterized in detail. The universal stress protein UspC acts as a scaffolding protein of the KdpD/KdpE signaling cascade by interacting with a Usp domain in KdpD under salt stress (Figure 8.2C) [120]. Usp proteins are small soluble proteins that accumulate under diverse stress conditions. They are widespread in living organisms, but their physiological role is rather unclear [121]. KdpD/KdpE controls expression of the *kdpFABC* operon in response to K^+ limitation or salt stress (see Section 8.3.2). Under K^+-limiting conditions the Kdp system restores the intracellular K^+ concentration, while in response to salt stress K^+ is accumulated far above the normal content. The kinase activity of KdpD is inhibited at high concentrations of K^+, so it has been puzzling how the sensor can be activated in response to salt stress. It was demonstrated that UspC, which is significantly upregulated under salt stress, interacts with KdpD and circumvents the inhibitory effect of K^+ by stabilizing the KdpD/KdpE/DNA complex. UspC had no effect under K^+-limiting growth conditions indicating that UspC fine-tunes the KdpD/KdpE system under specific stress conditions [120]. Another example for signal integration by a putative scaffolding protein is the ApsS/ApsR/ApsX system of *S. aureus* and *S. epidermidis* that regulates response upon detection of antimicrobial peptides. ApsX is a small supposable cytoplasmic protein that is essential for the activation of the histidine kinase/response regulator system. However, the function and molecular mechanism of ApsX are still unclear [122].

Other accessory proteins of histidine kinase/response regulator systems function as connectors between two signaling pathways (Figure 8.2D). The best characterized connector protein is PmrD that links the PhoP/PhoQ and the PmrA/PmrB systems in *S. enterica* [123]. PmrD accomplishes this task by binding to the phosphorylated form of the response regulator PmrA and protects it from dephosphorylation by the histidine kinase PmrB [124]. In this way, PmrD indirectly influences the activation of PmrA-regulated genes. Expression of *pmrD* is under control of the PhoP/PhoQ system that responds to low extracellular Mg^{2+} (see Section 8.3.1) [125, 126]. Three of the PmrA-activated loci mediate resistance to the antibiotic polymyxin B. The PmrD-mediated connection enables *S. enterica* to be resistant to polymyxin B not only in response to the Fe^{3+} signal that is sensed by the PmrB protein [127], but also in a low-Mg^{2+} environment, which is sensed by PhoP/PhoQ [128]. Another example of a connector protein is B1500 in *E. coli*. B1500 is a small membrane protein and there is evidence that B1500 activates the PhoQ/PhoP system in *E. coli* via direct interaction with the sensor kinase PhoQ [129]. Expression of *b1500* is under control of the EvgS/EvgA system, which is similar to the virulence-associated BvgS/BvgA system of *Bordetella pertussis* and confers acid resistance as well as multidrug resistance to *E. coli* [130, 131]. By using B1500 as connector, PhoP-regulated genes are expressed under conditions of EvgS/EvgA activating stimuli [129]. Mathematical simulations revealed that the connector-mediated regulatory pathway allows both signal amplification and persistence of expression that can hardly be realized by two separately transcriptionally controlled pathways [132].

Recently, an allosteric regulation of two histidine kinases by their cognate response regulator was shown [133]. In *Caulobacter crescentus* the single domain response regulator DivK stimulates the autokinase activities of PleC and DivJ (Figure 8.2E). Phosphorylated DivK switches PleC from a phosphatase into an autokinase state. Subsequently, the response regulator PleD is phosphorylated and initiates a c-di-GMP-dependent morphogenetic program. Moreover, phosphorylated DivK also acts as an allosteric activator of the autokinase of DivJ, and thereby stimulates its own phosphorylation and polar localization. These results demonstrate that in addition to the one-to-one relationship between histidine kinases and their cognate response regulators, there exists also interconnectivity between apparently isolated signaling systems through retrograde information transfer from a response regulator to two histidine kinases.

8.5
Conclusions and Outlook

Usually bacteria are studied in the laboratory as monoculture exposed to a specific stress situation. In their natural habitat bacteria have to respond to a wide range of simultaneous and fluctuating stimuli, and are forced to make an "informed decision" [134]. Therefore, it is important to analyze how all histidine kinase/response regulator systems work within a bacterium under natural multistress conditions. It is often discussed that the entity of these systems forms a neuronal network [135, 136]. These systems fulfill the requirements for such a network for the following reasons: (i) The majority of the histidine kinases and response regulators are constitutively expressed [137]. Most of these systems function in parallel; a cascade-wise organization is relatively rare. (ii) Many histidine kinase/response regulator systems show the phenomenon of autoamplification, which means that a stimulus perceived by the signaling components leads to a boost of their own expression [138, 139]. (iii) Histidine kinase/response regulator systems carry out "logical operations." Response regulators can be considered as very simple two-state (i.e., digital) signaling devices. (iv) There is cross-talk between the pathways [140]. However, the efficiency of the interaction between a response regulator and a noncognate histidine kinase and/or low-molecular-weight phosphodonors appears to be too low for cross-talk events under physiological conditions [134]. Nonetheless, it becomes more and more evident that cross-talk is mediated by connector proteins, scaffolding proteins, or allosteric effects as exemplified above. It can be envisaged that *in silico* analysis [141] will complement our experimental efforts to understand the functionality of such a neuronal network.

Despite recent achievements in the structural analysis of domains of histidine kinases, our understanding of the question how signals are transmitted within the sensors and across the membrane remains unsatisfactory as long as the structure of a full-length kinase is unsolved. Structural information might also be the prerequisite to develop new antimicrobial strategies using histidine kinase/response regulator systems as targets.

Acknowledgments

This work was financially supported by the Deutsche Forschungsgemeinschaft (Exc114-1) and the Bundesministerium für Bildung und Forschung (SysMO, project KOSMOBAC). We thank Tobias Kraxenberger and Luitpold Fried for their excellent support to generate Table 8.1, as well as Dr. Sabine Hunke for critical reading of the manuscript.

References

1 Stock, A.M., Robinson, V.L., and Goudreau, P.N. (2000) Two-component signal transduction. *Annu. Rev. Biochem.*, **69**, 183–215.

2 Ulrich, L.E. and Zhulin, I.B. (2007) MiST: a microbial signal transduction database. *Nucleic Acids Res.*, **35**, D386–D390.

3 Mizuno, T. (1998) His–Asp phosphotransfer signal transduction. *J. Biochem.*, **123**, 555–563.

4 Mizuno, T. (1997) Compilation of all genes encoding two-component phosphotransfer signal transducers in the genome of *Escherichia coli*. *DNA Res.*, **4**, 161–168.

5 Fabret, C., Feher, V.A., and Hoch, J.A. (1999) Two-component signal transduction in *Bacillus subtilis*: how one organism sees its world. *J. Bacteriol.*, **181**, 1975–1983.

6 Wang, L., Sun, Y.P., Chen, W.L., Li, J.H., and Zhang, C.C. (2002) Genomic analysis of protein kinases, protein phosphatases and two-component regulatory systems of the cyanobacterium *Anabaena* sp. strain PCC 7120. *FEMS Microbiol. Lett.*, **217**, 155–165.

7 Shi, X., Wegener-Feldbrugge, S., Huntley, S., Hamann, N., Hedderich, R., and Sogaard-Andersen, L. (2008) Bioinformatics and experimental analysis of proteins of two-component systems in *Myxococcus xanthus*. *J. Bacteriol.*, **190**, 613–624.

8 Mascher, T., Helmann, J.D., and Unden, G. (2006) Stimulus perception in bacterial signal-transducing histidine kinases. *Microbiol. Mol. Biol. Rev.*, **70**, 910–938.

9 Heermann, R., Altendorf, K., and Jung, K. (1998) The turgor sensor KdpD of

Escherichia coli is a homodimer. *Biochim. Biophys. Acta*, **1415**, 114–124.

10 Filippou, P.S., Kasemian, L.D., Panagiotidis, C.A., and Kyriakidis, D.A. (2008) Functional characterization of the histidine kinase of the *E. coli* two-component signal transduction system AtoS–AtoC. *Biochim. Biophys. Acta*, **1780**, 1023–1031.

11 Parkinson, J.S. and Kofoid, E.C. (1992) Communication modules in bacterial signaling proteins. *Annu. Rev. Genet.*, **26**, 71–112.

12 Swanson, R.V., Alex, L.A., and Simon, M.I. (1994) Histidine and aspartate phosphorylation: two-component systems and the limits of homology. *Trends Biochem. Sci.*, **19**, 485–490.

13 Dutta, R., Qin, L., and Inouye, M. (1999) Histidine kinases: diversity of domain organization. *Mol. Microbiol.*, **34**, 633–640.

14 Bilwes, A.M., Alex, L.A., Crane, B.R., and Simon, M.I. (1999) Structure of CheA, a signal-transducing histidine kinase. *Cell*, **96**, 131–141.

15 Whitworth, D.E. and Cock, P.J. (2008) Two-component systems of the myxobacteria: structure, diversity and evolutionary relationships. *Microbiology*, **154**, 360–372.

16 Hoch, J.A. (2000) Two-component and phosphorelay signal transduction. *Curr. Opin. Microbiol.*, **3**, 165–170.

17 Xu, J., Chiang, H.C., Bjursell, M.K., and Gordon, J.I. (2004) Message from a human gut symbiont: sensitivity is a prerequisite for sharing. *Trends Microbiol.*, **12**, 21–28.

18 Deutscher, J. and Saier, M.H. Jr. (2005) Ser/Thr/Tyr protein phosphorylation in bacteria – for long time neglected, now well established. *J. Mol. Microbiol. Biotechnol.*, **9**, 125–131.

19 Thomason, P. and Kay, R. (2000) Eukaryotic signal transduction via histidine–aspartate phosphorelay. *J. Cell Sci.*, **113**, 3141–3150.

20 Williams, S.B. and Stewart, V. (1997) Discrimination between structurally related ligands nitrate and nitrite controls autokinase activity of the NarX transmembrane signal transducer of *Escherichia coli* K-12. *Mol. Microbiol.*, **26**, 911–925.

21 Stock, J.B., Ninfa, A.J., and Stock, A.M. (1989) Protein phosphorylation and regulation of adaptive responses in bacteria. *Microbiol. Rev.*, **53**, 450–490.

22 Galperin, M.Y. (2004) Bacterial signal transduction network in a genomic perspective. *Environ. Microbiol.*, **6**, 552–567.

23 Kaspar, S. and Bott, M. (2002) The sensor kinase CitA (DpiB) of *Escherichia coli* functions as a high-affinity citrate receptor. *Arch. Microbiol.*, **177**, 313–321.

24 Krämer, J., Fischer, J.D., Zientz, E., Vijayan, V., Griesinger, C., Lupas, A., and Unden, G. (2007) Citrate sensing by the C4-dicarboxylate/citrate sensor kinase DcuS of *Escherichia coli*: binding site and conversion of DcuS to a C4-dicarboxylate- or citrate-specific sensor. *J. Bacteriol.*, **189**, 4290–4298.

25 Pappalardo, L., Janausch, I.G., Vijayan, V., Zientz, E., Junker, J., Peti, W., Zweckstetter, M., Unden, G., and Griesinger, C. (2003) The NMR structure of the sensory domain of the membranous two-component fumarate sensor (histidine protein kinase) DcuS of *Escherichia coli*. *J. Biol. Chem.*, **278**, 39185–39188.

26 Reinelt, S., Hofmann, E., Gerharz, T., Bott, M., and Madden, D.R. (2003) The structure of the periplasmic ligand-binding domain of the sensor kinase CitA reveals the first extracellular PAS domain. *J. Biol. Chem.*, **278**, 39189–39196.

27 Cheung, J. and Hendrickson, W.A. (2008) Crystal structures of C4-dicarboxylate ligand complexes with sensor domains of histidine kinases DcuS and DctB. *J. Biol. Chem.*, **283**, 30256–30265.

28 Etzkorn, M., Kneuper, H., Dunnwald, P., Vijayan, V., Krämer, J., Griesinger, C., Becker, S., Unden, G., and Baldus, M. (2008) Plasticity of the PAS domain and a potential role for signal transduction in the histidine kinase DcuS. *Nat. Struct. Mol. Biol.*, **15**, 1031–1039.

29 Cheung, J., Bingman, C.A., Reyngold, M., Hendrickson, W.A., and Waldburger, C.D. (2008) Crystal structure of a

29 functional dimer of the PhoQ sensor domain. *J. Biol. Chem.*, **283**, 13762–13770.
30 Prost, L.R. and Miller, S.I. (2008) The Salmonellae PhoQ sensor: mechanisms of detection of phagosome signals. *Cell Microbiol.*, **10**, 576–582.
31 Gunn, J.S. and Richards, S.M. (2007) Recognition and integration of multiple environmental signals by the bacterial sensor kinase PhoQ. *Cell Host Microbe*, **1**, 163–165.
32 Cho, U.S., Bader, M.W., Amaya, M.F., Daley, M.E., Klevit, R.E., Miller, S.I., and Xu, W. (2006) Metal bridges between the PhoQ sensor domain and the membrane regulate transmembrane signaling. *J. Mol. Biol.*, **356**, 1193–1206.
33 Georgellis, D., Kwon, O., De Wulf, P., and Lin, E.C. (1998) Signal decay through a reverse phosphorelay in the Arc two-component signal transduction system. *J. Biol. Chem.*, **273**, 32864–32869.
34 Georgellis, D., Kwon, O., and Lin, E.C. (2001) Quinones as the redox signal for the Arc two-component system of bacteria. *Science*, **292**, 2314–2316.
35 Malpica, R., Franco, B., Rodriguez, C., Kwon, O., and Georgellis, D. (2004) Identification of a quinone-sensitive redox switch in the ArcB sensor kinase. *Proc. Natl. Acad. Sci. USA*, **101**, 13318–13323.
36 Rodriguez, C., Kwon, O., and Georgellis, D. (2004) Effect of D-lactate on the physiological activity of the ArcB sensor kinase in *Escherichia coli*. *J. Bacteriol.*, **186**, 2085–2090.
37 Gilles-Gonzalez, M.A., Ditta, G.S., and Helinski, D.R. (1991) A haemoprotein with kinase activity encoded by the oxygen sensor of *Rhizobium meliloti*. *Nature*, **350**, 170–172.
38 Rodgers, K.R. and Lukat-Rodgers, G.S. (2005) Insights into heme-based O_2 sensing from structure–function relationships in the FixL proteins. *J. Inorg. Biochem.*, **99**, 963–977.
39 Tuckerman, J.R., Gonzalez, G., Dioum, E.M., and Gilles-Gonzalez, M.A. (2002) Ligand and oxidation-state specific regulation of the heme-based oxygen sensor FixL from *Sinorhizobium meliloti*. *Biochemistry*, **41**, 6170–6177.

40 Appleman, J.A., Chen, L.L., and Stewart, V. (2003) Probing conservation of HAMP linker structure and signal transduction mechanism through analysis of hybrid sensor kinases. *J. Bacteriol.*, **185**, 4872–4882.
41 Gelvin, S.B. (2006) *Agrobacterium* virulence gene induction. *Methods Mol. Biol.*, **343**, 77–84.
42 Gao, R. and Lynn, D.G. (2005) Environmental pH sensing: resolving the VirA/VirG two-component system inputs for *Agrobacterium* pathogenesis. *J. Bacteriol.*, **187**, 2182–2189.
43 Herren, C.D., Mitra, A., Palaniyandi, S.K., Coleman, A., Elankumaran, S., and Mukhopadhyay, S. (2006) The BarA–UvrY two-component system regulates virulence in avian pathogenic *Escherichia coli* O78: K80: H9. *Infect. Immun.*, **74**, 4900–4909.
44 Krin, E., Derzelle, S., Bedard, K., Adib-Conquy, M., Turlin, E., Lenormand, P., Hullo, M.F., Bonne, I., Chakroun, N., Lacroix, C., and Danchin, A. (2008) Regulatory role of UvrY in adaptation of *Photorhabdus luminescens* growth inside the insect. *Environ. Microbiol.*, **10**, 1118–1134.
45 Mondragon, V., Franco, B., Jonas, K., Suzuki, K., Romeo, T., Melefors, O., and Georgellis, D. (2006) pH-dependent activation of the BarA–UvrY two-component system in *Escherichia coli*. *J. Bacteriol.*, **188**, 8303–8306.
46 Gunn, J.S. (2008) The *Salmonella* PmrAB regulon: lipopolysaccharide modifications, antimicrobial peptide resistance and more. *Trends Microbiol.*, **16**, 284–290.
47 Perez, J.C. and Groisman, E.A. (2007) Acid pH activation of the PmrA/PmrB two-component regulatory system of *Salmonella enterica*. *Mol. Microbiol.*, **63**, 283–293.
48 Raivio, T.L. and Silhavy, T.J. (2001) Periplasmic stress and ECF sigma factors. *Annu. Rev. Microbiol.*, **55**, 591–624.
49 Danese, P.N. and Silhavy, T.J. (1998) CpxP, a stress-combative member of the Cpx regulon. *J. Bacteriol.*, **180**, 831–839.
50 Mileykovskaya, E. and Dowhan, W. (1997) The Cpx two-component signal

transduction pathway is activated in *Escherichia coli* mutant strains lacking phosphatidylethanolamine. *J. Bacteriol.*, **179**, 1029–1034.

51 Snyder, W.B., Davis, L.J., Danese, P.N., Cosma, C.L., and Silhavy, T.J. (1995) Overproduction of NlpE, a new outer membrane lipoprotein, suppresses the toxicity of periplasmic LacZ by activation of the Cpx signal transduction pathway. *J. Bacteriol.*, **177**, 4216–4223.

52 Hunke, S. and Betton, J.M. (2003) Temperature effect on inclusion body formation and stress response in the periplasm of *Escherichia coli*. *Mol. Microbiol.*, **50**, 1579–1589.

53 Fleischer, R., Heermann, R., Jung, K., and Hunke, S. (2007) Purification, reconstitution, and characterization of the CpxRAP envelope stress system of *Escherichia coli*. *J. Biol. Chem.*, **282**, 8583–8593.

54 Turgay, K., Hahn, J., Burghoorn, J., and Dubnau, D. (1998) Competence in *Bacillus subtilis* is controlled by regulated proteolysis of a transcription factor. *EMBO J.*, **17**, 6730–6738.

55 Suntharalingam, P. and Cvitkovitch, D.G. (2005) Quorum sensing in streptococcal biofilm formation. *Trends Microbiol.*, **13**, 3–6.

56 Boles, B.R. and Horswill, A.R. (2008) Agr-mediated dispersal of *Staphylococcus aureus* biofilms. *PLoS Pathog.*, **4**, e1000052.

57 Nakayama, J., Cao, Y., Horii, T., Sakuda, S., Akkermans, A.D., de Vos, W.M., and Nagasawa, H. (2001) Gelatinase biosynthesis-activating pheromone: a peptide lactone that mediates a quorum sensing in *Enterococcus faecalis*. *Mol. Microbiol.*, **41**, 145–154.

58 Piazza, F., Tortosa, P., and Dubnau, D. (1999) Mutational analysis and membrane topology of ComP, a quorum-sensing histidine kinase of *Bacillus subtilis* controlling competence development. *J. Bacteriol.*, **181**, 4540–4548.

59 Bacon, S.K., Palmer, T.M., and Grossman, A.D. (2002) Characterization of *comQ* and *comX*, two genes required for production of ComX pheromone in *Bacillus subtilis*. *J. Bacteriol.*, **184**, 410–419.

60 Okada, M., Sato, I., Cho, S.J., Iwata, H., Nishio, T., Dubnau, D., and Sakagami, Y. (2005) Structure of the *Bacillus subtilis* quorum-sensing peptide pheromone ComX. *Nat. Chem. Biol.*, **1**, 23–24.

61 Nakayama, J., Chen, S., Oyama, N., Nishiguchi, K., Azab, E.A., Tanaka, E., Kariyama, R., and Sonomoto, K. (2006) Revised model for *Enterococcus faecalis fsr* quorum-sensing system: the small open reading frame *fsrD* encodes the gelatinase biosynthesis-activating pheromone propeptide corresponding to staphylococcal *agrD*. *J. Bacteriol.*, **188**, 8321–8326.

62 Nishiguchi, K., Nagata, K., Tanokura, M., Sonomoto, K., and Nakayama, J. (2008) Structure–activity relationship of gelatinase biosynthesis-activating pheromone of *Enterococcus faecalis*. *J. Bacteriol.*, **191**, 641–650.

63 Novick, R.P. and Geisinger, E. (2008) Quorum sensing in staphylococci. *Annu. Rev. Genet.*, **42**, 541–564.

64 Yarwood, J.M. and Schlievert, P.M. (2003) Quorum sensing in *Staphylococcus* infections. *J. Clin. Invest*, **112**, 1620–1625.

65 Henke, J.M. and Bassler, B.L. (2004) Three parallel quorum-sensing systems regulate gene expression in *Vibrio harveyi*. *J. Bacteriol.*, **186**, 6902–6914.

66 Surette, M.G., Miller, M.B., and Bassler, B.L. (1999) Quorum sensing in *Escherichia coli*, *Salmonella typhimurium*, and *Vibrio harveyi*: a new family of genes responsible for autoinducer production. *Proc. Natl. Acad. Sci. USA*, **96**, 1639–1644.

67 Chen, X., Schauder, S., Potier, N., Van Dorsselaer, A., Pelczer, I., Bassler, B.L., and Hughson, F.M. (2002) Structural identification of a bacterial quorum-sensing signal containing boron. *Nature*, **415**, 545–549.

68 Higgins, D.A., Pomianek, M.E., Kraml, C.M., Taylor, R.K., Semmelhack, M.F., and Bassler, B.L. (2007) The major *Vibrio cholerae* autoinducer and its role in virulence factor production. *Nature*, **450**, 883–886.

69 Neiditch, M.B., Federle, M.J., Pompeani, A.J., Kelly, R.C., Swem, D.L., Jeffrey, P.D., Bassler, B.L., and Hughson, F.M. (2006) Ligand-induced asymmetry in histidine

sensor kinase complex regulates quorum sensing. *Cell*, **126**, 1095–1108.

70 Swem, L.R., Swem, D.L., Wingreen, N.S., and Bassler, B.L. (2008) Deducing receptor signaling parameters from *in vivo* analysis: LuxN/AI-1 quorum sensing in *Vibrio harveyi*. *Cell*, **134**, 461–473.

71 Jung, K., Odenbach, T., and Timmen, M. (2007) The quorum-sensing hybrid histidine kinase LuxN of *Vibrio harveyi* contains a periplasmically located N terminus. *J. Bacteriol.*, **189**, 2945–2948.

72 Timmen, M., Bassler, B.L., and Jung, K. (2006) AI-1 influences the kinase activity but not the phosphatase activity of LuxN of *Vibrio harveyi*. *J. Biol. Chem.*, **281**, 24398–24404.

73 Neiditch, M.B., Federle, M.J., Miller, S.T., Bassler, B.L., and Hughson, F.M. (2005) Regulation of LuxPQ receptor activity by the quorum-sensing signal autoinducer-2. *Mol. Cell*, **18**, 507–518.

74 Freeman, J.A. Bassler, B.L. Sequence and function of LuxU: a two-component phosphorelay protein that regulates quorum sensing in *Vibrio harveyi*. *J. Bacteriol*, **181**, (1999) 899–906.

75 Pompeani, A.J., Irgon, J.J., Berger, M.F., Bulyk, M.L., Wingreen, N.S., and Bassler, B.L. (2008) The *Vibrio harveyi* master quorum-sensing regulator, LuxR, a TetR-type protein is both an activator and a repressor: DNA recognition and binding specificity at target promoters. *Mol. Microbiol.*, **70**, 76–88.

76 Tu, K.C. and Bassler, B.L. (2007) Multiple small RNAs act additively to integrate sensory information and control quorum sensing in *Vibrio harveyi*. *Genes Dev.*, **21**, 221–233.

77 Waters, C.M. and Bassler, B.L. (2005) Quorum sensing: cell-to-cell communication in bacteria. *Annu. Rev. Cell Dev. Biol.*, **21**, 319–346.

78 Clarke, M.B., Hughes, D.T., Zhu, C., Boedeker, E.C., and Sperandio, V. (2006) The QseC sensor kinase: a bacterial adrenergic receptor. *Proc. Natl. Acad. Sci. USA*, **103**, 10420–10425.

79 Hughes, D.T. and Sperandio, V. (2008) Inter-kingdom signalling: communication between bacteria and their hosts. *Nat. Rev. Microbiol.*, **6**, 111–120.

80 Sleator, R.D. and Hill, C. (2002) Bacterial osmoadaptation: the role of osmolytes in bacterial stress and virulence. *FEMS Microbiol. Rev.*, **26**, 49–71.

81 Morbach, S. and Krämer, R. (2002) Body shaping under water stress: osmosensing and osmoregulation of solute transport in bacteria. *ChemBioChem*, **3**, 384–397.

82 Wood, J.M. (1999) Osmosensing by bacteria: signals and membrane-based sensors. *Microbiol. Mol. Biol. Rev.*, **63**, 230–262.

83 Wood, J.M., Bremer, E., Csonka, L.N., Kraemer, R., Poolman, B., van der, H.T., and Smith, L.T. (2001) Osmosensing and osmoregulatory compatible solute accumulation by bacteria. *Comp. Biochem. Physiol. A Mol. Integr. Physiol.*, **130**, 437–460.

84 Leonardo, M.R. and Forst, S. (1996) Re-examination of the role of the periplasmic domain of EnvZ in sensing of osmolarity signals in *Escherichia coli*. *Mol. Microbiol.*, **22**, 405–413.

85 Egger, L.A. and Inouye, M. (1997) Purification and characterization of the periplasmic domain of EnvZ osmosensor in *Escherichia coli*. *Biochem. Biophys. Res. Commun.*, **231**, 68–72.

86 Tabatabai, N. and Forst, S. (1995) Molecular analysis of the two-component genes, *ompR* and *envZ*, in the symbiotic bacterium *Xenorhabdus nematophilus*. *Mol. Microbiol.*, **17**, 643–652.

87 Epstein, W. (2003) The roles and regulation of potassium in bacteria. *Prog. Nucleic Acid Res. Mol. Biol.*, **75**, 293–320.

88 Jung, K., Hamann, K., and Revermann, A. (2001) K^+ stimulates specifically the autokinase activity of purified and reconstituted EnvZ of *Escherichia coli*. *J. Biol. Chem.*, **276**, 40896–40902.

89 Möker, N., Reihlen, P., Krämer, R., and Morbach, S. (2007) Osmosensing properties of the histidine protein kinase MtrB from *Corynebacterium glutamicum*. *J. Biol. Chem.*, **282**, 27666–27677.

90 Zimmann, P., Puppe, W., and Altendorf, K. (1995) Membrane topology analysis of the sensor kinase KdpD of *Escherichia coli*. *J. Biol. Chem.*, **270**, 28282–28288.

91 Laimins, L.A., Rhoads, D.B., and Epstein, W. (1981) Osmotic control of *kdp* operon

expression in *Escherichia coli*. *Proc. Natl. Acad. Sci. USA*, **78**, 464–468.

92 Asha, H. and Gowrishankar, J. (1993) Regulation of *kdp* operon expression in *Escherichia coli*: evidence against turgor as signal for transcriptional control. *J. Bacteriol.*, **175**, 4528–4537.

93 Gowrishankar, J. (1985) Identification of osmoresponsive genes in *Escherichia coli*: evidence for participation of potassium and proline transport systems in osmoregulation. *J. Bacteriol.*, **164**, 434–445.

94 Sutherland, L., Cairney, J., Elmore, M.J., Booth, I.R., and Higgins, C.F. (1986) Osmotic regulation of transcription: induction of the *proU* betaine transport gene is dependent on accumulation of intracellular potassium. *J. Bacteriol.*, **168**, 805–814.

95 Hamann, K., Zimmann, P., and Altendorf, K. (2008) Reduction of turgor is not the stimulus for the sensor kinase KdpD of *Escherichia coli*. *J. Bacteriol.*, **190**, 2360–2367.

96 Sugiura, A., Hirokawa, K., Nakashima, K., and Mizuno, T. (1994) Signal-sensing mechanisms of the putative osmosensor KdpD in *Escherichia coli*. *Mol. Microbiol.*, **14**, 929–938.

97 Jung, K., Veen, M., and Altendorf, K. (2000) K^+ and ionic strength directly influence the autophosphorylation activity of the putative turgor sensor KdpD of *Escherichia coli*. *J. Biol. Chem.*, **275**, 40142–40147.

98 Heermann, R., Fohrmann, A., Altendorf, K., and Jung, K. (2003) The transmembrane domains of the sensor kinase KdpD of *Escherichia coli* are not essential for sensing K^+ limitation. *Mol. Microbiol.*, **47**, 839–848.

99 Zimmann, P., Steinbrügge, A., Schniederberend, M., Jung, K., and Altendorf, K. (2007) The extension of the fourth transmembrane helix of the sensor kinase KdpD of *Escherichia coli* is involved in sensing. *J. Bacteriol.*, **189**, 7326–7334.

100 Rothenbücher, M.C., Facey, S.J., Kiefer, D., Kossmann, M., and Kuhn, A. (2006) The cytoplasmic C-terminal domain of the *Escherichia coli* KdpD protein functions as a K^+ sensor. *J. Bacteriol.*, **188**, 1950–1958.

101 Jung, K., Krabusch, M., and Altendorf, K. (2001) Cs^+ induces the *kdp* operon of *Escherichia coli* by lowering the intracellular K^+ concentration. *J. Bacteriol.*, **183**, 3800–3803.

102 Heermann, R., Altendorf, K., and Jung, K. (2000) The hydrophilic N-terminal domain complements the membrane-anchored C-terminal domain of the sensor kinase KdpD of *Escherichia coli*. *J. Biol. Chem.*, **275**, 17080–17085.

103 Jung, K. and Altendorf, K. (1998) Truncation of amino acids 12–128 causes deregulation of the phosphatase activity of the sensor kinase KdpD of *Escherichia coli*. *J. Biol. Chem.*, **273**, 17406–17410.

104 Ohwada, T. and Sagisaka, S. (1987) An immediate and steep increase in ATP concentration in response to reduced turgor pressure in *Escherichia coli* B. *Arch. Biochem. Biophys.*, **259**, 157–163.

105 Reiser, V., Raitt, D.C., and Saito, H. (2003) Yeast osmosensor Sln1 and plant cytokinin receptor Cre1 respond to changes in turgor pressure. *J. Cell Biol.*, **161**, 1035–1040.

106 Purcell, E.B., Siegal-Gaskins, D., Rawling, D.C., Fiebig, A., and Crosson, S. (2007) A photosensory two-component system regulates bacterial cell attachment. *Proc. Natl. Acad. Sci. USA*, **104**, 18241–18246.

107 Schumann, W. (2007) Thermosensors in eubacteria: role and evolution. *J. Biosci.*, **32**, 549–557.

108 Ullrich, M., Penaloza-Vazquez, A., Bailey, A.M., and Bender, C.L. (1995) A modified two-component regulatory system is involved in temperature-dependent biosynthesis of the *Pseudomonas syringae* phytotoxin coronatine. *J. Bacteriol.*, **177**, 6160–6169.

109 Braun, Y., Smirnova, A.V., Weingart, H., Schenk, A., and Ullrich, M.S. (2007) A temperature-sensing histidine kinase: function, genetics, and membrane topology. *Methods Enzymol.*, **423**, 222–249.

110 Island, M.D., Wei, B.Y., and Kadner, R.J. (1992) Structure and function of the *uhp* genes for the sugar phosphate transport system in *Escherichia coli* and *Salmonella typhimurium*. *J. Bacteriol.*, **174**, 2754–2762.

111 Island, M.D. and Kadner, R.J. (1993) Interplay between the membrane-

associated UhpB and UhpC regulatory proteins. *J. Bacteriol.*, **175**, 5028–5034.

112 Wright, J.S. and Kadner, R.J. (2001) The phosphoryl transfer domain of UhpB interacts with the response regulator UhpA. *J. Bacteriol.*, **183**, 3149–3159.

113 Verhamme, D.T., Postma, P.W., Crielaard, W., and Hellingwerf, K.J. (2002) Cooperativity in signal transfer through the Uhp system of *Escherichia coli*. *J. Bacteriol.*, **184**, 4205–4210.

114 Lalonde, S., Boles, E., Hellmann, H., Barker, L., Patrick, J.W., Frommer, W.B., and Ward, J.M. (1999) The dual function of sugar carriers. Transport and sugar sensing. *Plant Cell*, **11**, 707–726.

115 Rietkötter, E., Hoyer, D., and Mascher, T. (2008) Bacitracin sensing in *Bacillus subtilis*. *Mol. Microbiol.*, **68**, 768–785.

116 Bernard, R., Guiseppi, A., Chippaux, M., Foglino, M., and Denizot, F. (2007) Resistance to bacitracin in *Bacillus subtilis*: unexpected requirement of the BceAB ABC transporter in the control of expression of its own structural genes. *J. Bacteriol.*, **189**, 8636–8642.

117 Pawson, T. and Scott, J.D. (1997) Signaling through scaffold, anchoring, and adaptor proteins. *Science*, **278**, 2075–2080.

118 Garrington, T.P. and Johnson, G.L. (1999) Organization and regulation of mitogen-activated protein kinase signaling pathways. *Curr. Opin. Cell Biol.*, **11**, 211–218.

119 Burack, W.R. and Shaw, A.S. (2000) Signal transduction: hanging on a scaffold. *Curr. Opin. Cell Biol.*, **12**, 211–216.

120 Heermann, R., Weber, A., Mayer, B., Ott, M., Hauser, E., Gabriel, G., Pirch, T., and Jung, K. (2009) The universal stress protein UspC scaffolds the KdpD/KdpE signaling cascade of *Escherichia coli* under salt stress. *J. Mol. Biol.*, **386**, 134–148.

121 Kvint, K., Nachin, L., Diez, A., and Nystrom, T. (2003) The bacterial universal stress protein: function and regulation. *Curr. Opin. Microbiol.*, **6**, 140–145.

122 Li, M., Cha, D.J., Lai, Y., Villaruz, A.E., Sturdevant, D.E., and Otto, M. (2007) The antimicrobial peptide-sensing system *aps* of *Staphylococcus aureus*. *Mol. Microbiol.*, **66**, 1136–1147.

123 Kox, L.F., Wosten, M.M., and Groisman, E.A. (2000) A small protein that mediates the activation of a two-component system by another two-component system. *EMBO J.*, **19**, 1861–1872.

124 Kato, A. and Groisman, E.A. (2004) Connecting two-component regulatory systems by a protein that protects a response regulator from dephosphorylation by its cognate sensor. *Genes Dev.*, **18**, 2302–2313.

125 Shin, D., Lee, E.J., Huang, H., and Groisman, E.A. (2006) A positive feedback loop promotes transcription surge that jump-starts *Salmonella virulence* circuit. *Science*, **314**, 1607–1609.

126 Shin, D. and Groisman, E.A. (2005) Signal-dependent binding of the response regulators PhoP and PmrA to their target promoters *in vivo*. *J. Biol. Chem.*, **280**, 4089–4094.

127 Wosten, M.M., Kox, L.F., Chamnongpol, S., Soncini, F.C., and Groisman, E.A. (2000) A signal transduction system that responds to extracellular iron. *Cell*, **103**, 113–125.

128 Garcia, V.E., Soncini, F.C., and Groisman, E.A. (1996) Mg^{2+} as an extracellular signal: environmental regulation of *Salmonella* virulence. *Cell*, **84**, 165–174.

129 Eguchi, Y., Itou, J., Yamane, M., Demizu, R., Yamato, F., Okada, A., Mori, H., Kato, A., and Utsumi, R. (2007) B1500, a small membrane protein, connects the two-component systems EvgS/EvgA and PhoQ/PhoP in *Escherichia coli*. *Proc. Natl. Acad. Sci. USA*, **104**, 18712–18717.

130 Eguchi, Y., Oshima, T., Mori, H., Aono, R., Yamamoto, K., Ishihama, A., and Utsumi, R. (2003) Transcriptional regulation of drug efflux genes by EvgAS, a two-component system in *Escherichia coli*. *Microbiology*, **149**, 2819–2828.

131 Utsumi, R., Katayama, S., Taniguchi, M., Horie, T., Ikeda, M., Igaki, S., Nakagawa, H., Miwa, A., Tanabe, H., and Noda, M. (1994) Newly identified genes involved in the signal transduction of *Escherichia coli* K-12. *Gene*, **140**, 73–77.

132 Kato, A., Mitrophanov, A.Y., and Groisman, E.A. (2007) A connector of two-component regulatory systems promotes

signal amplification and persistence of expression. *Proc. Natl. Acad. Sci. USA*, **104**, 12063–12068.

133 Paul, R., Jaeger, T., Abel, S., Wiederkehr, I., Folcher, M., Biondi, E.G., Laub, M.T., and Jenal, U. (2008) Allosteric regulation of histidine kinases by their cognate response regulator determines cell fate. *Cell*, **133**, 452–461.

134 Bijlsma, J.J. and Groisman, E.A. (2005) The PhoP/PhoQ system controls the intramacrophage type three secretion system of *Salmonella enterica*. *Mol. Microbiol.*, **57**, 85–96.

135 Hellingwerf, K.J., Postma, P.W., Tommassen, J., and Westerhoff, H.V. (1995) Signal transduction in bacteria: phospho-neural network(s) in *Escherichia coli*? *FEMS Microbiol. Rev.*, **16**, 309–321.

136 Hellingwerf, K.J. (2005) Bacterial observations: a rudimentary form of intelligence? *Trends Microbiol.*, **13**, 152–158.

137 Aiso, T. and Ohki, R. (2003) Instability of sensory histidine kinase mRNAs in *Escherichia coli*. *Genes Cells*, **8**, 179–187.

138 Hoffer, S.M., Westerhoff, H.V., Hellingwerf, K.J., Postma, P.W., and Tommassen, J. (2001) Autoamplification of a two-component regulatory system results in "learning" behavior. *J. Bacteriol.*, **183**, 4914–4917.

139 Polarek, J.W., Williams, G., and Epstein, W. (1992) The products of the *kdpDE* operon are required for expression of the Kdp ATPase of *Escherichia coli*. *J. Bacteriol.*, **174**, 2145–2151.

140 Laub, M.T. and Goulian, M. (2007) Specificity in two-component signal transduction pathways. *Annu. Rev. Genet.*, **41**, 121–145.

141 Igoshin, O.A., Alves, R., and Savageau, M.A. (2008) Hysteretic and graded responses in bacterial two-component signal transduction. *Mol. Microbiol.*, **68**, 1196–1215.

142 Fakhr, M.K., Penaloza-Vazquez, A., Chakrabarty, A.M., and Bender, C.L. (1999) Regulation of alginate biosynthesis in *Pseudomonas syringae* pv. *syringae*. *J. Bacteriol.*, **181**, 3478–3485.

143 Li, M., Lai, Y., Villaruz, A.E., Cha, D.J., Sturdevant, D.E., and Otto, M. (2007) Gram-positive three-component antimicrobial peptide-sensing system. *Proc. Natl. Acad. Sci. USA*, **104**, 9469–9474.

144 Wen, Y., Feng, J., Scott, D.R., Marcus, E.A., and Sachs, G. (2007) The HP0165–HP0166 two-component system (ArsRS) regulates acid-induced expression of HP1186 alpha-carbonic anhydrase in *Helicobacter pylori* by activating the pH-dependent promoter. *J. Bacteriol.*, **189**, 2426–2434.

145 Hagiwara, D., Yamashino, T., and Mizuno, T. (2004) A Genome-wide view of the *Escherichia coli* BasS–BasR two-component system implicated in iron-responses. *Biosci. Biotechnol. Biochem.*, **68**, 1758–1767.

146 Ohki, R., Giyanto, Tateno, K., Masuyama, W., Moriya, S., Kobayashi, K., and Ogasawara, N. (2003) The BceRS two-component regulatory system induces expression of the bacitracin transporter, BceAB, in *Bacillus subtilis*. *Mol. Microbiol*, **49**, 1135–1144.

147 Beier, D. and Gross, R. (2008) The BvgS/BvgA phosphorelay system of pathogenic Bordetellae: structure, function and evolution. *Adv. Exp. Med. Biol.*, **631**, 149–160.

148 Nishijyo, T., Haas, D., and Itoh, Y. (2001) The CbrA–CbrB two-component regulatory system controls the utilization of multiple carbon and nitrogen sources in *Pseudomonas aeruginosa*. *Mol. Microbiol.*, **40**, 917–931.

149 Miyamoto, K., Okunishi, M., Nukui, E., Tsuchiya, T., Kobayashi, T., Imada, C., and Tsujibo, H. (2007) The regulator CdsS/CdsR two-component system modulates expression of genes involved in chitin degradation of *Pseudoalteromonas piscicida* strain O-7. *Arch. Microbiol.*, **188**, 619–628.

150 Baker, M.D., Wolanin, P.M., and Stock, J.B. (2006) Signal transduction in bacterial chemotaxis. *Bioessays*, **28**, 9–22.

151 Tsujibo, H., Hatano, N., Okamoto, T., Endo, H., Miyamoto, K., and Inamori, Y. (1999) Synthesis of chitinase in *Streptomyces thermoviolaceus* is regulated by a two-component sensor-regulator system. *FEMS Microbiol. Lett.*, **181**, 83–90.

152 Tran, L.S., Nagai, T., and Itoh, Y. (2000) Divergent structure of the ComQXPA quorum-sensing components: molecular basis of strain-specific communication mechanism in *Bacillus subtilis*. *Mol. Microbiol.*, **37**, 1159–1171.

153 Mills, S.D., Jasalavich, C.A., and Cooksey, D.A. (1993) A two-component regulatory system required for copper-inducible expression of the copper resistance operon of *Pseudomonas syringae*. *J. Bacteriol.*, **175**, 1656–1664.

154 Cariss, S.J., Tayler, A.E., and Avison, M.B. (2008) Defining the growth conditions and promoter-proximal DNA sequences required for activation of gene expression by CreBC in *Escherichia coli*. *J. Bacteriol.*, **190**, 3930–3939.

155 Gryllos, I., Grifantini, R., Colaprico, A., Jiang, S., Deforce, E., Hakansson, A., Telford, J.L., Grandi, G., and Wessels, M.R. (2007) Mg^{2+} signalling defines the group A streptococcal CsrRS (CovRS) regulon. *Mol. Microbiol.*, **65**, 671–683.

156 Munson, G.P., Lam, D.L., Outten, F.W., and O'Halloran, T.V. (2000) Identification of a copper-responsive two-component system on the chromosome of *Escherichia coli* K-12. *J. Bacteriol.*, **182**, 5864–5871.

157 Zhou, Y.F., Nan, B., Nan, J., Ma, Q., Panjikar, S., Liang, Y.H., Wang, Y., and Su, X.D. (2008) C4-dicarboxylates sensing mechanism revealed by the crystal structures of DctB sensor domain. *J. Mol. Biol.*, **383**, 49–61.

158 Krämer, J., Fischer, J.D., Zientz, E., Vijayan, V., Griesinger, C., Lupas, A., and Unden, G. (2007) Citrate sensing by the C4-dicarboxylate/citrate sensor kinase DcuS of *Escherichia coli*: binding site and conversion of DcuS to a C4-dicarboxylate- or citrate-specific sensor. *J. Bacteriol.*, **189**, 4290–4298.

159 Ioanoviciu, A., Yukl, E.T., Moenne-Loccoz, P., and de Montellano, P.R. (2007) DevS, a heme-containing two-component oxygen sensor of *Mycobacterium tuberculosis*. *Biochemistry*, **46**, 4250–4260.

160 Kumar, A., Toledo, J.C., Patel, R.P., Lancaster, J.R. Jr., and Steyn, A.J. (2007) *Mycobacterium tuberculosis* DosS is a redox sensor and DosT is a hypoxia sensor. *Proc. Natl. Acad. Sci. USA*, **104**, 11568–11573.

161 Bock, A. and Gross, R. (2002) The unorthodox histidine kinases BvgS and EvgS are responsive to the oxidation status of a quinone electron carrier. *Eur. J. Biochem.*, **269**, 3479–3484.

162 Griffitts, J.S., Carlyon, R.E., Erickson, J.H., Moulton, J.L., Barnett, M.J., Toman, C.J., and Long, S.R. (2008) A *Sinorhizobium meliloti* osmosensory two-component system required for cyclic glucan export and symbiosis. *Mol. Microbiol.*, **69**, 479–490.

163 Del Papa, M.F. and Perego, M. (2008) Ethanolamine activates a sensor histidine kinase regulating its utilization in *Enterococcus faecalis*. *J. Bacteriol.*, **190**, 7147–7156.

164 Bibb, L.A., Kunkle, C.A., and Schmitt, M.P. (2007) The ChrA–ChrS and HrrA–HrrS signal transduction systems are required for activation of the *hmuO* promoter and repression of the *hemA* promoter in *Corynebacterium diphtheriae*. *Infect. Immun.*, **75**, 2421–2431.

165 Stauff, D.L., Torres, V.J., and Skaar, E.P. (2007) Signaling and DNA-binding activities of the *Staphylococcus aureus* HssR–HssS two-component system required for heme sensing. *J. Biol. Chem.*, **282**, 26111–26121.

166 Jung, K. and Altendorf, K. (2002) Towards an understanding of the molecular mechanisms of stimulus perception and signal transduction by the KdpD/KdpE system of *Escherichia coli*. *J. Mol. Microbiol. Biotechnol.*, **4**, 223–228.

167 Gloer, J., Thummer, R., Ullrich, H., and Schmitz, R.A. (2008) Towards understanding the nitrogen signal transduction for *nif* gene expression in *Klebsiella pneumoniae*. *FEBS J.*, **275**, 6281–6294.

168 Zhou, X.X., Li, W.F., Ma, G.X., and Pan, Y.J. (2006) The nisin-controlled gene expression system: construction, application and improvements. *Biotechnol. Adv.*, **24**, 285–295.

169 Loh, J., Garcia, M., and Stacey, G. (1997) NodV and NodW, a second flavonoid recognition system regulating nod gene expression in *Bradyrhizobium japonicum*. *J. Bacteriol.*, **179**, 3013–3020.

170 Jiang, P. and Ninfa, A.J. (1999) Regulation of autophosphorylation of *Escherichia coli* nitrogen regulator II by the PII signal transduction protein. *J. Bacteriol.*, **181**, 1906–1911.

171 Dean, C.R., Neshat, S., and Poole, K. (1996) PfeR, an enterobactin-responsive activator of ferric enterobactin receptor gene expression in *Pseudomonas aeruginosa. J. Bacteriol.*, **178**, 5361–5369.

172 Huang, Y.H., Ferrieres, L., and Clarke, D.J. (2006) The role of the Rcs phosphorelay in *Enterobacteriaceae. Res. Microbiol.*, **157**, 206–212.

173 Wu, J. and Bauer, C.E. (2008) RegB/RegA, a global redox-responding two-component system. *Adv. Exp. Med. Biol.*, **631**, 131–148.

174 Rasmussen, B.A. and Kovacs, E. (1993) Cloning and identification of a two-component signal-transducing regulatory system from *Bacteroides fragilis. Mol. Microbiol.*, **7**, 765–776.

175 Busch, A., Lacal, J., Martos, A., Ramos, J.L., and Krell, T. (2007) Bacterial sensor kinase TodS interacts with agonistic and antagonistic signals. *Proc. Natl. Acad. Sci. USA*, **104**, 13774–13779.

176 Bordi, C., Theraulaz, L., Mejean, V., and Jourlin-Castelli, C. (2003) Anticipating an alkaline stress through the Tor phosphorelay system in *Escherichia coli. Mol. Microbiol.*, **48**, 211–223.

177 Hong, H.J., Hutchings, M.I., and Buttner, M.J. (2008) Vancomycin resistance VanS/VanR two-component systems. *Adv. Exp. Med. Biol.*, **631**, 200–213.

178 Leonhartsberger, S., Huber, A., Lottspeich, F., and Bock, A. (2001) The *hydH/G* Genes from *Escherichia coli* code for a zinc and lead responsive two-component regulatory system. *J. Mol. Biol.*, **307**, 93–105.

9
Chemotaxis and Receptor Localization
Victor Sourjik

9.1
Introduction

Most motile bacteria are able to migrate towards higher concentrations of certain chemicals (attractants), which are usually nutrients or signaling molecules. At the same time, bacteria can avoid higher concentrations of repellents, potentially harmful chemicals. As the small size of a bacterial cell makes direct spatial detection of chemoeffector gradients inefficient, bacteria have evolved a chemotaxis strategy that differs substantially from that commonly employed by larger eukaryotic cells. Bacterial chemotaxis relies on temporal – rather than spatial – comparisons of chemoeffector concentrations, which are performed while moving in a gradient. Bacteria therefore have to swim first and only then can decide whether the chosen direction is favorable or not. *Escherichia coli* – the best-studied model for bacterial chemotaxis – has two types of swimming. A smooth swimming "run," which propels the cell forward, results from the counterclockwise rotation of the flagellar motors and bundling of the flagellar filaments. Reorienting "tumbles" are produced by the clockwise motor rotation and dissociation of the flagellar bundle. In the adapted state with no gradients present, runs (around 2 s duration) are constantly interrupted by short tumbles (around 0.1 s) and as a result the cell performs a random walk that allows it to efficiently explore its environment [1]. In the presence of a gradient cells bias their random walk (Figure 9.1A). Although the swimming direction after each tumble is chosen nearly randomly, swimming in a favorable direction in a gradient suppresses tumbles and thus results in longer runs.

The chemotactic response is mediated by a signaling system that relies on protein phosphorylation and is a member of a large class of bacterial two-component sensors. The chemotaxis pathway is well conserved across bacteria and Archaea, with the *E. coli* pathway (Figure 9.1B) being the best-studied example [2, 3]. The two central signaling proteins are the histidine kinase CheA and the response regulator CheY. In contrast to a typical two-component sensory kinase, CheA does not have a sensory domain, but instead – together with the "adaptor" protein CheW – associates with the dedicated chemosensory receptors at the membrane. Ternary complex

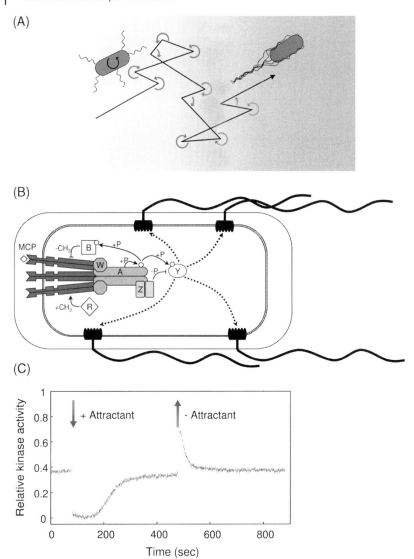

Figure 9.1 Chemotaxis of E. coli.
(A) Chemotaxis strategy of E. coli. Bacterial movement consists of short runs punctuated by tumbles. Runs in favorable directions (up the gradient of attractant, gray shading) are extended. (B) Chemotaxis pathway. Changes in attractant or repellent concentration are detected by sensory complexes consisting of transmembrane receptors (methyl-accepting chemotaxis proteins (MCPs)), an adapter protein CheW, and a histidine kinase CheA. A tumble signal is transmitted to the flagellar motors by phosphoryl group (open circle) transfer to CheY. Adaptation is mediated by addition or removal of methyl groups at four specific glutamyl residues on each receptor monomer. Grayscale levels indicate stability of protein association with the sensory complexes, with darker shading corre-sponding to more stable association. See text for details. (C) Changes in relative kinase activity (measured using fluorescence resonance energy transfer [57]) upon addition and removal of saturating a concentration of attractant (30 µM α-methyl-DL-aspartate).

formation stimulates CheA autophosphorylation and allows the receptors to modulate CheA activity upon chemoeffector binding. The CheA phosphoryl group is subsequently transferred to CheY, which can diffuse through the cytoplasm and transmit the signal to the flagellar motors. Binding of phosphorylated CheY to the motors enhances the probability of clockwise rotation and causes the cell to tumble. When the cell swims up an attractant gradient, the concentration increase results in CheA inactivation (Figure 9.1C) and suppression of tumbles, which allows the bacterium to continue swimming up-gradient.

Adaptation to continuous stimulation is mediated by the methyltransferase CheR and the methylesterase CheB, which tune the ability of receptors to activate CheA by adjusting the level of methylation at specific glutamyl residues, whereby receptor modification increases CheA activity and decreases sensitivity to attractants. The feedback from ternary complex activity to the adaptation system is provided both by the substrate specificity of the adaptation enzymes, whereby CheR preferentially methylates inactive receptors and CheB preferentially demethylates active receptors, and by CheB phosphorylation through CheA that increases CheB activity. Changes in receptor methylation levels result in precise sensory adaptation, returning kinase activity over time to the same intermediate level irrespective of the strength of the initial stimulus (Figure 9.1C). The adapted kinase activity brings the level of phosphorylated CheY into the narrow dynamic range of the flagellar motor so that an adapted cell stochastically alternates between running and tumbling. Adaptation thus enables the cell to detect further concentration changes as it swims up a chemical gradient and to follow gradients over several order of magnitude of ligand concentration. Since the process of receptor modification is slower than the initial response, receptor methylation at the same time provides a short-term memory about past conditions.

Together, the receptors CheA, CheW, CheR, and CheB represent an evolutionarily conserved core of the pathway. The chemotaxis system of *E. coli* includes only one additional protein, the phosphatase CheZ, which ensures a rapid turnover of CheY~P to quickly readjust swimming behavior. Chemotaxis systems in other bacteria are generally more complex, frequently having multiple sets of chemotaxis genes, cytoplasmic receptors, varying mechanisms of CheY dephosphorylation, and/or additional enzymes involved in adaptation [2]. Nevertheless, the basics of chemotactic signal processing, including spatial organization of the pathway, are conserved in all prokaryotes.

9.2
Architecture of the Sensory Complex

9.2.1
Structure and Function of Chemoreceptors

Receptor–CheW–CheA complexes are at the core of sensing and processing of stimuli in bacterial chemotaxis. Most bacteria have multiple types of receptors, which define the specificity of chemotaxis in a particular bacterium, with a single receptor typically mediating response to several chemicals. *E. coli* has five types of attractant-specific receptors, which fall into two classes according to their copy number: major receptors that sense aspartate (Tar) and serine (Tsr), and several other attractants and repellents, as well as pH and temperature, are highly abundant and number several thousand molecules per cell, whereas minor receptors specific for dipeptides and pyrimidines (Tap), ribose, glucose and galactose (Trg), and redox potential (Aer) are less abundant [4]. Another difference between major and minor receptors is that the former bind most of their ligands directly, whereas dipeptide and sugar binding to minor chemoreceptors Tap and Trg is mediated by periplasmic binding proteins.

Receptors are stable homodimers that typically span the cytoplasmic membrane. The cytoplasmic portion of the receptor dimer forms a four-helix bundle, producing an extended needle-like structure normal to the membrane [5–7]. Receptors contain three operational modules [8]: transmembrane sensing, signal conversion, and kinase control. Transmembrane sensing is performed by the ligand-specific N-terminal sensory domain. Binding of the ligand to the periplasmic pocket formed at the interface of two monomers is believed to induce a piston-like sliding of one transmembrane helix to transmit the signal across the membrane [9]. This signal, in turn, initiates a conformational change in the signal conversion module, a HAMP domain (found in *h*istidine kinases, *a*denylyl cyclases, *m*ethyl-accepting chemotaxis proteins, and *p*hosphatases [10]), which might involve helix rotation [11]. The HAMP conformational change in turn regulates the kinase control module, possibly through twisting and/or bending motions of individual receptor molecules [12] and through inter-receptor movements in higher-order clusters [13–15] (see Section 9.3.1 and Chapter 11). Receptor methylation, involved in adaptation, appears to reverse these changes, both in the cytoplasmic and in the transmembrane parts of the receptor [16]. An additional feature present in the major, but not minor, receptors of *E. coli* is a C-terminal binding sequence for CheR and CheB, which tethers them near the methylation sites. Such tethering is important to ensure efficient adaptation in *E. coli*, but is not evolutionarily conserved [17] and can be substituted by overexpression of the adaptation enzymes [18].

9.2.2
Protein Interactions in the Ternary Complex

The exact molecular nature of interactions that are involved in the assembly of the receptor–CheW–CheA complex, and protein orientation and stoichiometry in this complex *in vivo* remain controversial. Complexes with varying stoichiometry and functionality can form both *in vitro* [19, 20] and *in vivo* [21] depending on the relative concentrations of the components. Recent elegant experiments involving receptors embedded in nanodisks (soluble particles of lipid bilayer surrounded by an annulus of amphiphilic membrane scaffold protein) suggest that optimal kinase activation is achieved at a ratio of two to three receptor dimers to one CheA dimer [19]. This argues for a trimer of receptor dimers, which was observed in the crystal structure of the cytoplasmic domain of the serine receptor Tsr [5], as the basic element of the active ternary complex. *In vivo* evidence for trimer-of-dimers contacts between the signaling tips of receptor molecules has been provided by cross-linking and conformational suppression studies [22–25]. Receptor complexes are further stabilized by the binding of CheW and CheA to the signaling tip of receptors [24]. CheW is necessary to assist the association of full-length CheA with receptors, probably both by activating CheA to bind receptors directly and by providing additional binding interface [26, 27]. Interestingly, CheA and CheW might even compete for the same binding site on the receptor [20], in agreement with homology between CheW and the C-terminal domain of CheA [28, 29].

9.3
Clustering of Sensory Complexes

9.3.1
Chemoreceptor Clusters

Although transmembrane signal transduction and kinase regulation in chemotaxis could be performed by individual receptor–kinase complexes, these complexes, in fact, form one or a few discrete clusters in *E. coli* and all other bacteria and Archaea so far examined [30–32]. Receptor clusters of variable size, around 250 ± 120 nm, can be observed at cell poles and along the cell body with immunoelectron, cryoelectron, and fluorescence microscopy [30, 33, 34]. They are formed by large (partially) ordered arrays of thousands of receptors and associated chemotaxis proteins, with recent analysis of clusters in *Caulobacter crescentus* suggesting a hexagonal arrangement of trimers of dimers in the lattice [35, 36]. An observed degree of lattice disorder is consistent with the view that the assembly is relatively dynamic, and can rearrange upon stimulation and adaptation [14, 37]. It also confirms functional analysis (see Section 9.4.3), which suggests that the lattice is not perfectly regular, but rather consists of regions of tightly coupled receptors bridged by looser connections.

Receptors alone are able to form both trimers of dimers [23] and larger complexes that are visible as discrete foci in fluorescence microscopy images [26]. Binding of CheA and/or CheW is required to stabilize and order these larger structures [24, 26, 30]. The stabilizing effect of CheA and CheW could either result from bridging connections between receptor dimers or higher-order signaling teams, or from stabilizing receptors in a conformation that favors complex formation. As with sensory complexes, the stoichiometry of the cluster can apparently vary, and depends on the expression levels of receptors, CheW and CheA in the cell [21].

All remaining proteins localize to the clusters by interaction with receptors or CheA. An NWETF pentapeptide sequence at the C-terminus of the major receptors, Tsr and Tar, tethers CheR and CheB [38, 39]. CheB additionally binds to clusters through interaction with the P2 domain of CheA, which also serves to localize CheY [33]. CheZ binds specifically to the N-terminus of the short form of CheA, CheA$_S$, that lacks the first 97 amino acids, including the phosphorylation site [40].

9.3.2
Cluster Assembly and Positioning

Chemosensory clusters in *E. coli* are found both at the cell poles and along the cell body [30, 33]. Individual receptors are newly inserted along the entire cell membrane [41] and can then either join existing clusters or nucleate new ones, with the former process being primarily limited by receptor diffusion in the membrane and the latter by receptor concentration [42]. If a cluster already exists within a critical distance, the receptors are much more likely to collide with and bind to that cluster than to nucleate a new one, but above that distance nucleation dominates. The critical distance is determined by the relative rates of receptor diffusion and binding to existing clusters on one hand and nucleation of new clusters on the other hand. This mode of assembly produces an observed linear dependence of the cluster number per cell on cell length, with a constant cluster density, around one cluster per micrometer at wild-type expression levels of chemotaxis proteins [42, 43]. A typical cell of around 2–4 μm length thus has several clusters. As expected for a stochastic nucleation mechanism, the number of nucleated clusters depends on the expression level of chemotaxis proteins, with the Hill coefficient of around 3 possibly pointing to formation of trimer of receptor dimers as the critical nucleation step [42]. Clusters grow gradually after nucleation, but this growth slows down as they reach a certain size and eventually reverses, thus resulting in a steady-state cluster size distribution in a cell population.

9.3.3
Cluster Positioning

The stochastic mode of cluster assembly couples length-dependent nucleation to cell growth and division, so that additional clusters are nucleated before a cell divides. This alone, however, is not sufficient to ensure that clusters are equally distributed to both daughter cells. Individual receptors and small receptor complexes are highly mobile in the membrane [44] and even large polar clusters can move at rates above 0.5 μm/min [43]. If unrestricted, this movement of a small number of clusters could result in all clusters accumulating in one daughter cell, thus leading to asymmetric cluster distribution. Apparently to prevent this, lateral clusters anchor to a structure that labels future cell division sites [43] and are consequently found at periodic positions along the cell body. As a consequence, clusters become polar after one of the next cell divisions and are free to move, but remain restricted to the pole by some other mechanism, possibly the membrane curvature or lipid composition of the polar membrane.

In addition to providing a simple cluster segregation mechanism, the uniform cluster distribution along the cell body benefits signaling by ensuring that, even in longer cells, sites of CheY phosphorylation are not too remote from the laterally distributed flagellar motors and the kinetics of chemotactic signaling are not limited by the diffusion of phosphorylated CheY.

9.3.4
Cluster Stability

As mentioned in Section 9.3.1, all of the chemotaxis proteins localize to receptor clusters, but the kinetics of protein equilibration at clusters, as observed in "fluorescence recovery after photobleaching" experiments, are protein-specific and reflect characteristic timescales that correspond to cell division and adaptation and excitation in chemotaxis, respectively [44]. Receptors do not exchange between clusters even in 30 min, and CheW and CheA exchange very slowly, with characteristic times of 12 min, consistent with the exchange kinetics observed *in vitro* [45]. These proteins thus form a sensory core of the cluster that is stable on the timescale of chemotactic signaling, demonstrating that signaling is mediated by conformational changes in the cluster and not by its assembly or disassembly. Nevertheless, slow equilibration of the core components on the timescale of cell division could be beneficial, ensuring that the stoichiometry of sensory complexes is uniform throughout the cluster and among clusters, and that receptors of different specificities are well mixed. Yet another component of the stable cluster core is the CheA-associated phosphatase CheZ, with the exchange time of around 8 min. It is notable that in other two-component sensors the kinase and phosphatase activities are parts of the same protein.

Adaptation enzymes represent the next level of stability, equilibrating on the timescale of 15 s. This is much longer than the response (excitation) time of the chemotaxis system (around 0.1–1 s) or the typical run time of an adapted swimming cell (around 2 s), but is comparable to the time required for adaptation to saturating stimuli. Equilibration on this timescale ensures a uniform distribution of adaptation enzymes in a cluster, as well as between multiple clusters in the same cell. It might be particularly important because of the low copy numbers of CheR and CheB in the cell (200–400), compared to around 15 000 receptors and 3000–8000 copies of all other cytoplasmic proteins [4]. The low ratio of adaptation enzymes to receptors implies that they have to move around the cluster during the adaptation process to sample all available methylation sites [46], but their dwell time has to be long enough to allow slow adaptation kinetics. Notably, the stochastic binding and unbinding of CheR and CheB at the cluster on this timescale result in slow fluctuations in the adapted level of kinase activity, and consequently in the bias of the flagellar motors [47], which is believed to be important for optimizing bacterial search behavior [48].

CheY is thus the only protein that shows rapid exchange kinetics on the signaling timescale, in concert with its role as messenger between the spatially localized sensory clusters and flagellar motors.

9.4
Role of Clustering in Signal Processing

9.4.1
Role of Protein Localization

An obvious function of clustering in chemotaxis – as in many other pathways – might be to increase specificity and reaction rates by localizing all pathway components to a few large sensory complexes. Consistent with that, chemotaxis proteins tend to bind close to their functional interaction sites and mutations that affect protein localization but retain enzymatic activity can usually be compensated by overexpression of the affected proteins [18, 49]. The phosphatase CheZ localization to the receptor cluster, the site of CheY phosphorylation, seems counterintuitive. Its role might be more subtle, for example, in flattening the concentration gradient of phosphorylated CheY through the cell, so that flagellar motors at different distances from receptor cluster "see" similar levels of phosphorylated CheY [50].

9.4.2
Signal Amplification

The chemotactic response in *E. coli* is extremely sensitive. Cells respond to as little as 10 nM steps in the concentration of aspartate [51] – corresponding to less than 10 molecules of aspartate in the 1.4-fl volume of an *E. coli* cell – even though the affinity of Tar receptor for aspartate is relatively low, around 1 μM. An increase in receptor occupancy by 0.2% (i.e., 15 out of approximately 7500 receptor dimers in the cell) has been estimated to result in a 23% change in the bias of motor rotation [52], indicating signal amplification (or gain) by a factor of around 100. Similarly, minor receptors that constitute only a small percentage of the total receptor pool can mediate responses of the same amplitude as major receptors.

The nature of this amplification remained a long-standing puzzle in chemotaxis research, but allosteric signal amplification in clustered sensory complexes was proposed as a likely explanation [53]. Subsequent studies of signal processing by receptor clusters confirmed the existence of cooperative interactions between receptors *in vitro* [54, 55] and *in vivo* [15, 21, 56], and demonstrated an around 35-fold amplification of chemotactic signals at the clusters *in vivo*, meaning that a 1% increase in attractant binding to receptor results in a 35% inhibition of kinase activity [57]. The remaining signal amplification happens at the flagellar motor, which shows a highly cooperative dependence of switching on CheY$\sim$P concentration [58].

9.4.3
Allosteric Models and the Role of the Methylation System in High Sensitivity

Classically, regulation of CheA activity has been described in terms of a two-state model, where all receptor–kinase complexes are considered independent and stable.

The equilibrium between active (kinase-activating) and inactive (kinase-inactivating) states in these complexes can be shifted either by stimulation or by receptor methylation. The two-state model can well explain the basics of chemotaxis signaling, but fails to account for signal amplification, because the fractional change in the kinase activity in the model is the same as the fractional change in the receptor occupancy.

Allosteric receptor behavior can be described using two types of mathematical models: Monod–Wyman–Changeux (MWC)-type models and Ising-type models [21, 59–61], in which the two-state receptor model is extended by assuming that the activity states of individual receptors are coupled. The inactive state of a particular receptor–kinase complex can thus be stabilized not only by attractant binding to this complex, but also by inactivation of the neighboring receptors. Another important assumption is that the inactive state of a receptor has a higher affinity for attractant than does the active state. Additionally to these assumptions that are common to both types of allosteric models, the MWC models further assume that a cluster consists of many independent tightly coupled cooperative units, or signaling teams, with all receptors in one unit switching synchronously between inactive and active states. In contrast, Ising models describe the cluster as one lattice in which the activities of adjacent receptors are coupled with a finite strength, leading to an effective distance of the conformational spread. The number of effectively interacting receptors corresponds to the size of the cooperative unit in the former case and to the extent of the conformational spread in the latter case. Both models can account for most of the available data and their predictions differ only in subtle details, so distinguishing between them experimentally is difficult. It is likely, however, that real clusters may be better described by a hybrid model where groups of tightly coupled receptors are more loosely coupled to one another (Figure 9.2A).

Cooperative interactions between receptors in clusters can amplify changes in the kinase activity relative to ligand binding, with the strength of amplification depending on the degree of receptor coupling. If the prestimulus activity of an allosteric unit is low or moderate (i.e., if the complex has a similar probability of being active or inactive), then additional inactivation of only a few receptors will stabilize the entire complex in the inactive state (Figure 9.2B). This enables the cluster to respond to very low threshold concentration of ligands and yields dose–response curves with low apparent cooperativity, or Hill coefficient. Low-activity clusters thus behave like molecular antennas, with the detection threshold decreasing linearly with an increasing number of interacting subunits [21]. This regime is best suited for detection of weak gradients and indeed reflects the experimentally measured response of clusters in adapted cells [21, 57], with an estimated size for the cooperative units of 10–30 receptor dimers. In contrast, if the activity bias of receptors in the absence of ligand is high, the complex does not shift to the inactive state until most subunits are occupied, producing a steep response but only at higher ligand concentrations. Consistent with that, Hill coefficients of up to 10 have been observed in clusters of highly active receptors *in vitro* [55] and *in vivo* [21, 62].

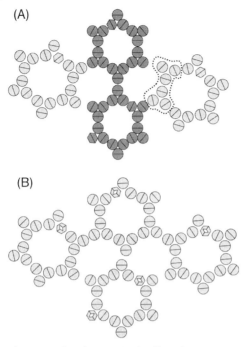

Figure 9.2 Signal processing by allosteric receptor units. (A) Receptor homodimers are arranged in strongly coupled allosteric units, which are in turn loosely coupled to each other. Units switch between active (dark gray) and inactive (light gray) states as a whole. Inactive complexes are less compact. Different shapes indicate different receptor types; dashed line shows the size of the adaptational assistance neighborhood. (B) Due to allosteric signal amplification, ligand (open diamonds) binding to a few receptors can inactivate entire signaling units.

9.4.4
Signal Integration

Most bacteria, including *E. coli*, use their chemotaxis system to respond to a wide variety of stimuli, including not only multiple chemicals, but also pH, temperature, and osmolarity, which ensures finding an optimal combination of nutrients and environmental conditions for growth. Allosteric interactions between receptors can account for integration of all these different stimuli. Receptor dimers of different types interact and appear to be randomly mixed in receptor complexes [23]. Receptor dimers of all types in mixed complexes appear to be "equal," that is, the output of the system depends simply upon the total number of receptors occupied by ligand, providing the simplest possible mechanism of signal integration. Moreover, due to allosteric interactions of receptors in mixed allosteric complexes, partial kinase inhibition by one attractant increases the response sensitivity to other attractants [21]. This means that a cell swimming up a weak gradient of one attractant is sensitized to gradients of other attractants in the same direction.

9.4.5
Adaptational Assistance Neighborhoods

Another type of functional units in receptor clusters are adaptational assistance neighborhoods [63]. As mentioned in Section 9.2.1, major receptors have a C-terminal tethering site for CheR and CheB, which is connected to the rest of the receptor by a flexible linker [64]. This linker allows the adaptation enzyme to perform covalent modification of the tethering receptor as well as multiple of neighboring, "assisted" receptors, while remaining bound at the same tethering site. This creates an assistance neighborhood with an estimated size of five to seven receptor dimers [63], which means that an allosteric unit comprises several such neighborhoods (Figure 9.2A). Such neighborhoods are important to ensure high precision of adaptation to a wide range of stimulus strengths, particularly in the presence of strong allosteric interactions between receptors [65]. This improvement comes from providing each enzyme with a much larger number of available methylation sites and thereby allowing it to gradually adjust the level of activity of the allosteric receptor unit without encountering saturation.

9.5
Conclusions and Outlook

Despite considerable recent progress in understanding receptor clustering and its role in bacterial chemotaxis, much remains to be learned. Interesting open questions are related to the structure and dynamics of clusters. Current structural studies do not yet provide sufficient resolution to unambiguously determine the protein arrangement in clusters and to determine the degree of order in the array lattice. Even more intriguing are the cluster dynamics during signaling, with mounting evidence that receptor arrangement in clusters and the degree of cluster compactness could be modulated by ligand stimulation and by sensory adaptation [14, 37, 62]. The effect of such changes on the signaling properties of sensory complexes and their role in the overall regulation of chemotaxis signaling awaits further analysis.

Acknowledgments

I thank John S. Parkinson for critical reading of the manuscript and David Kentner for help with the artwork.

References

1. Berg, H.C. and Brown, D.A. (1972) Chemotaxis in *Escherichia coli* analysed by three-dimensional tracking. *Nature*, **239**, 500–504.

2. Szurmant, H. and Ordal, G.W. (2004) Diversity in chemotaxis mechanisms among the bacteria and archaea. *Microbiol. Mol. Biol. Rev.*, **68**, 301–319.

3 Wadhams, G.H. and Armitage, J.P. (2004) Making sense of it all: bacterial chemotaxis. *Nat. Rev. Mol. Cell Biol.*, **5**, 1024–1037.

4 Li, M. and Hazelbauer, G.L. (2004) Cellular stoichiometry of the components of the chemotaxis signaling complex. *J. Bacteriol.*, **186**, 3687–3694.

5 Kim, K.K., Yokota, H., and Kim, S.H. (1999) Four-helical-bundle structure of the cytoplasmic domain of a serine chemotaxis receptor. *Nature*, **400**, 787–792.

6 Weis, R.M., Hirai, T., Chalah, A., Kessel, M., Peters, P.J., and Subramaniam, S. (2003) Electron microscopic analysis of membrane assemblies formed by the bacterial chemotaxis receptor Tsr. *J. Bacteriol.*, **185**, 3636–3643.

7 Park, S.Y., Borbat, P.P., Gonzalez-Bonet, G., Bhatnagar, J., Pollard, A.M., Freed, J.H., Bilwes, A.M., and Crane, B.R. (2006) Reconstruction of the chemotaxis receptor–kinase assembly. *Nat. Struct. Mol. Biol.*, **13**, 400–407.

8 Hazelbauer, G.L., Falke, J.J., and Parkinson, J.S. (2008) Bacterial chemoreceptors: high-performance signaling in networked arrays. *Trends Biochem. Sci.*, **33**, 9–19.

9 Falke, J.J. and Hazelbauer, G.L. (2001) Transmembrane signaling in bacterial chemoreceptors. *Trends Biochem. Sci.*, **26**, 257–265.

10 Aravind, L. and Ponting, C.P. (1999) The cytoplasmic helical linker domain of receptor histidine kinase and methyl-accepting proteins is common to many prokaryotic signalling proteins. *FEMS Microbiol. Lett.*, **176**, 111–116.

11 Hulko, M., Berndt, F., Gruber, M., Linder, J.U., Truffault, V., Schultz, A., Martin, J., Schultz, J.E., Lupas, A.N., and Coles, M. (2006) The HAMP domain structure implies helix rotation in transmembrane signaling. *Cell*, **126**, 929–940.

12 Winston, S.E., Mehan, R., and Falke, J.J. (2005) Evidence that the adaptation region of the aspartate receptor is a dynamic four-helix bundle: cysteine and disulfide scanning studies. *Biochemistry*, **44**, 12655–12666.

13 Vaknin, A. and Berg, H.C. (2006) Osmotic stress mechanically perturbs chemoreceptors in *Escherichia coli*. *Proc. Natl. Acad. Sci. USA*, **103**, 592–596.

14 Vaknin, A. and Berg, H.C. (2007) Physical responses of bacterial chemoreceptors. *J. Mol. Biol.*, **366**, 1416–1423.

15 Vaknin, A. and Berg, H.C. (2008) Direct evidence for coupling between bacterial chemoreceptors. *J. Mol. Biol.*, **82**, 573–577.

16 Lai, W.C., Beel, B.D., and Hazelbauer, G.L. (2006) Adaptational modification and ligand occupancy have opposite effects on positioning of the transmembrane signalling helix of a chemoreceptor. *Mol. Microbiol.*, **61**, 1081–1090.

17 Alexander, R.P. and Zhulin, I.B. (2007) Evolutionary genomics reveals conserved structural determinants of signaling and adaptation in microbial chemoreceptors. *Proc. Natl. Acad. Sci. USA*, **104**, 2885–2890.

18 Okumura, H., Nishiyama, S., Sasaki, A., Homma, M., and Kawagishi, I. (1998) Chemotactic adaptation is altered by changes in the carboxy-terminal sequence conserved among the major methyl-accepting chemoreceptors. *J. Bacteriol.*, **180**, 1862–1868.

19 Boldog, T., Grimme, S., Li, M., Sligar, S.G., and Hazelbauer, G.L. (2006) Nanodiscs separate chemoreceptor oligomeric states and reveal their signaling properties. *Proc. Natl. Acad. Sci. USA*, **103**, 11509–11514.

20 Levit, M.N., Grebe, T.W., and Stock, J.B. (2002) Organization of the receptor–kinase signaling array that regulates *Escherichia coli* chemotaxis. *J. Biol. Chem.*, **277**, 36748–36754.

21 Sourjik, V. and Berg, H.C. (2004) Functional interactions between receptors in bacterial chemotaxis. *Nature*, **428**, 437–441.

22 Ames, P., Studdert, C.A., Reiser, R.H., and Parkinson, J.S. (2002) Collaborative signaling by mixed chemoreceptor teams in *Escherichia coli*. *Proc. Natl. Acad. Sci. USA*, **99**, 7060–7065.

23 Studdert, C.A. and Parkinson, J.S. (2004) Crosslinking snapshots of bacterial chemoreceptor squads. *Proc. Natl. Acad. Sci. USA*, **101**, 2117–2122.

24 Studdert, C.A. and Parkinson, J.S. (2005) Insights into the organization and dynamics of bacterial chemoreceptor clusters through *in vivo* crosslinking

studies. *Proc. Natl. Acad. Sci. USA*, **102**, 15623–15628.

25 Ames, P. and Parkinson, J.S. (2006) Conformational suppression of inter-receptor signaling defects. *Proc. Natl. Acad. Sci. USA*, **103**, 9292–9297.

26 Kentner, D., Thiem, S., Hildenbeutel, M., and Sourjik, V. (2006) Determinants of chemoreceptor cluster formation in *Escherichia coli*. *Mol. Microbiol.*, **61**, 407–417.

27 Miller, A.S., Kohout, S.C., Gilman, K.A., and Falke, J.J. (2006) CheA Kinase of bacterial chemotaxis: chemical mapping of four essential docking sites. *Biochemistry*, **45**, 8699–8711.

28 Bilwes, A.M., Alex, L.A., Crane, B.R., and Simon, M.I. (1999) Structure of CheA, a signal-transducing histidine kinase. *Cell*, **96**, 131–141.

29 Griswold, I.J., Zhou, H., Matison, M., Swanson, R.V., McIntosh, L.P., Simon, M.I., and Dahlquist, F.W. (2002) The solution structure and interactions of CheW from *Thermotoga maritima*. *Nat. Struct. Biol.*, **9**, 121–125.

30 Maddock, J.R. and Shapiro, L. (1993) Polar location of the chemoreceptor complex in the *Escherichia coli* cell. *Science*, **259**, 1717–1723.

31 Alley, M.R., Maddock, J.R., and Shapiro, L. (1992) Polar localization of a bacterial chemoreceptor. *Genes Dev.*, **6**, 825–836.

32 Gestwicki, J.E., Lamanna, A.C., Harshey, R.M., McCarter, L.L., Kiessling, L.L., and Adler, J. (2000) Evolutionary conservation of methyl-accepting chemotaxis protein location in Bacteria and Archaea. *J. Bacteriol.*, **182**, 6499–6502.

33 Sourjik, V. and Berg, H.C. (2000) Localization of components of the chemotaxis machinery of *Escherichia coli* using fluorescent protein fusions. *Mol. Microbiol.*, **37**, 740–751.

34 Zhang, P., Khursigara, C.M., Hartnell, L.M., and Subramaniam, S. (2007) Direct visualization of *Escherichia coli* chemotaxis receptor arrays using cryo-electron microscopy. *Proc. Natl. Acad. Sci. USA*, **104**, 3777–3781.

35 Briegel, A., Ding, H.J., Li, Z., Werner, J., Gitai, Z., Dias, D.P., Jensen, R.B., and Jensen, G.J. (2008) Location and architecture of the *Caulobacter crescentus* chemoreceptor array. *Mol. Microbiol.*, **69**, 30–41.

36 Khursigara, C.M., Wu, X., and Subramaniam, S. (2008) Chemoreceptors in *Caulobacter crescentus*: trimers of receptor dimers in a partially ordered hexagonally packed array. *J. Bacteriol.*, **190**, 6805–6810.

37 Besschetnova, T.Y., Montefusco, D.J., Asinas, A.E., Shrout, A.L., Antommattei, F.M., and Weis, R.M. (2008) Receptor density balances signal stimulation and attenuation in membrane-assembled complexes of bacterial chemotaxis signaling proteins. *Proc. Natl. Acad. Sci. USA*, **105**, 12289–12294.

38 Banno, S., Shiomi, D., Homma, M., and Kawagishi, I. (2004) Targeting of the chemotaxis methylesterase/deamidase CheB to the polar receptor–kinase cluster in an *Escherichia coli* cell. *Mol. Microbiol.*, **53**, 1051–1063.

39 Shiomi, D., Zhulin, I.B., Homma, M., and Kawagishi, I. (2002) Dual recognition of the bacterial chemoreceptor by chemotaxis-specific domains of the CheR methyltransferase. *J. Biol. Chem.*, **277**, 42325–42333.

40 Cantwell, B.J., Draheim, R.R., Weart, R.B., Nguyen, C., Stewart, R.C., and Manson, M.D. (2003) CheZ phosphatase localizes to chemoreceptor patches via CheA-short. *J. Bacteriol.*, **185**, 2354–2361.

41 Shiomi, D., Yoshimoto, M., Homma, M., and Kawagishi, I. (2006) Helical distribution of the bacterial chemoreceptor via colocalization with the Sec protein translocation machinery. *Mol. Microbiol.*, **60**, 894–906.

42 Thiem, S. and Sourjik, V. (2008) Stochastic assembly of chemoreceptor clusters in *Escherichia coli*. *Mol. Microbiol.*, **68**, 1228–1236.

43 Thiem, S., Kentner, D., and Sourjik, V. (2007) Positioning of chemosensory clusters in *E. coli* and its relation to cell division. *EMBO J.*, **26**, 1615–1623.

44 Schulmeister, S., Ruttorf, M., Thiem, S., Kentner, D., Lebiedz, D., and Sourjik, V. (2008) Protein exchange dynamics at chemoreceptor clusters in *Escherichia coli*. *Proc. Natl. Acad. Sci. USA*, **105**, 6403–6408.

45 Gegner, J.A., Graham, D.R., Roth, A.F., and Dahlquist, F.W. (1992) Assembly of an

MCP receptor, CheW, and kinase CheA complex in the bacterial chemotaxis signal transduction pathway. *Cell*, **70**, 975–982.

46 Levin, M.D., Shimizu, T.S., and Bray, D. (2002) Binding and diffusion of CheR molecules within a cluster of membrane receptors. *Biophys. J.*, **82**, 1809–1817.

47 Korobkova, E., Emonet, T., Vilar, J.M., Shimizu, T.S., and Cluzel, P. (2004) From molecular noise to behavioural variability in a single bacterium. *Nature*, **428**, 574–578.

48 Emonet, T. and Cluzel, P. (2008) Relationship between cellular response and behavioral variability in bacterial chemotaxis. *Proc. Natl. Acad. Sci. USA*, **105**, 3304–3309.

49 Jahreis, K., Morrison, T.B., Garzon, A., and Parkinson, J.S. (2004) Chemotactic signaling by an *Escherichia coli* CheA mutant that lacks the binding domain for phosphoacceptor partners. *J. Bacteriol.*, **186**, 2664–2672.

50 Vaknin, A. and Berg, H.C. (2004) Single-cell FRET imaging of phosphatase activity in the *Escherichia coli* chemotaxis system. *Proc. Natl. Acad. Sci. USA*, **101**, 17072–17077.

51 Mao, H., Cremer, P.S., and Manson, M.D. (2003) A sensitive, versatile microfluidic assay for bacterial chemotaxis. *Proc. Natl. Acad. Sci. USA*, **100**, 5449–5454.

52 Segall, J.E., Block, S.M., and Berg, H.C. (1986) Temporal comparisons in bacterial chemotaxis. *Proc. Natl. Acad. Sci. USA*, **83**, 9486–9493.

53 Bray, D., Levin, M.D., and Morton-Firth, C.J. (1998) Receptor clustering as a cellular mechanism to control sensitivity. *Nature*, **393**, 85–88.

54 Lai, R.Z., Manson, J.M., Bormans, A.F., Draheim, R.R., Nguyen, N.T., and Manson, M.D. (2005) Cooperative signaling among bacterial chemoreceptors. *Biochemistry*, **44**, 14298–14307.

55 Li, G. and Weis, R.M. (2000) Covalent modification regulates ligand binding to receptor complexes in the chemosensory system of *Escherichia coli*. *Cell*, **100**, 357–365.

56 Gestwicki, J.E. and Kiessling, L.L. (2002) Inter-receptor communication through arrays of bacterial chemoreceptors. *Nature*, **415**, 81–84.

57 Sourjik, V. and Berg, H.C. (2002) Receptor sensitivity in bacterial chemotaxis. *Proc. Natl. Acad. Sci. USA*, **99**, 123–127.

58 Cluzel, P., Surette, M., and Leibler, S. (2000) An ultrasensitive bacterial motor revealed by monitoring signaling proteins in single cells. *Science*, **287**, 1652–1655.

59 Duke, T.A., Le Novere, N., and Bray, D. (2001) Conformational spread in a ring of proteins: a stochastic approach to allostery. *J. Mol. Biol.*, **308**, 541–553.

60 Keymer, J.E., Endres, R.G., Skoge, M., Meir, Y., and Wingreen, N.S. (2006) Chemosensing in *Escherichia coli*: two regimes of two-state receptors. *Proc. Natl. Acad. Sci. USA*, **103**, 1786–1791.

61 Mello, B.A. and Tu, Y. (2005) An allosteric model for heterogeneous receptor complexes: understanding bacterial chemotaxis responses to multiple stimuli. *Proc. Natl. Acad. Sci. USA*, **102**, 17354–17359.

62 Endres, R.G., Oleksiuk, O., Hansen, C.H., Meir, Y., Sourjik, V., and Wingreen, N.S. (2008) Variable sizes of *Escherichia coli* chemoreceptor signaling teams. *Mol. Syst. Biol.*, **4**, 211.

63 Li, M. and Hazelbauer, G.L. (2005) Adaptational assistance in clusters of bacterial chemoreceptors. *Mol. Microbiol.*, **56**, 1617–1626.

64 Li, M. and Hazelbauer, G.L. (2006) The carboxyl-terminal linker is important for chemoreceptor function. *Mol. Microbiol.*, **60**, 469–479.

65 Endres, R.G. and Wingreen, N.S. (2006) Precise adaptation in bacterial chemotaxis through "assistance neighborhoods". *Proc. Natl. Acad. Sci. USA*, **103**, 13040–13044.

10
Photoreception and Signal Transduction
Sonja Brandt and Nicole Frankenberg-Dinkel

10.1
Introduction

Light is a form of energy that influences life on Earth in a wide-ranging manner. Thus, light perception is one of the most essential features of nearly all organisms. During evolution sophisticated photoreceptors have evolved ranging from bacteria to fungi up to plants and animals required to obtain information about their ambient light environment. In particular, photoautotrophic organisms, dependent upon light as an energy source, have developed a complicated network of photoreceptors to respond to light of different quality, fluence rate, and duration over a broad spectral range. Light sensing also plays a crucial role in the bacterial kingdom to adapt to changing environmental conditions. Accordingly, cyanobacteria are able to react to changing light conditions via phototactic responses [1] or regulate differentiation processes by special light qualities [2]. Due to the capacity of ultraviolet (UV) light to generate reactive oxygen species [3, 4] and thus to expose the bacterial cell to oxidative stress, negative phototaxis is a convenient mechanism to prevent the cell from photo-oxidative damage [5, 6].

In general, photoreceptors are chromoproteins consisting of an apoprotein moiety that covalently or noncovalently assembles a chromophore as prosthetic group for photoperception. The chromophore enables the holoprotein to undergo light-induced conformational changes that typically stimulate downstream signaling cascades. Consequently, the spectral features are determined by the nature of the chromophoric group. In most cases chromophores arise in conjugated π-electron systems, whose atoms are covalently bonded with alternating single and double bonds, resulting in a delocalization of the electrons. By absorption of a photon of the appropriate wavelength and accordingly the right energy, the absorbed electron can be transferred to a higher energy level. When the electron relaxes back to the ground state the difference in energy is either emitted as fluorescence or transformed into chemical energy dependent on the electronic transition. Photoreceptors are designated as molecular light-switches owing to their ability to mainly generate two parent states. In the dark, the inactive ground state is present that converts to a light-induced signaling state upon irradiation, involving the first committed step of a photochemical reaction. Conditional on the chromophoric group, most often the primary photoreactions are cis/trans isomerizations or the chromophore serves as an electron donor for a chemical reaction. The photocycle is completed when the photoreceptor returns back to its ground state in the dark, which occurs non-photochemically.

To optimally adjust to their light environment, photosynthetic and nonphotosynthetic bacteria have developed various photoreceptors to cover the whole spectral range from UV-C to far-red light. These photosensory receptors can be mainly subdivided into UV- and blue-light-sensing chromoproteins and red-light-sensing photoreceptors, depending on the chromophore they use. For bacteria these are rhodopsins, cryptochromes, xanthopsins, the blue-light sensing using flavin (BLUF) domain-containing proteins and phototropin-like proteins as members of the UV- and blue-green-light absorbing photoreceptors, and the superfamily of the red/far-red-light-inducible phytochromes. Thereby rhodopsins, xanthopsins and phytochromes catalyze cis/trans isomerization of their chromophore, whereas different mechanisms of flavin-based photochemistry are used by cryptochromes, phototropin-like photoreceptors, and BLUF domain proteins to integrate the light signal.

10.2
Bacterial Blue-Light Photoreceptors

10.2.1
Microbial Rhodopsins

The best classified and characterized group of photosensory receptors is the family of rhodopsins. Rhodopsin, generally also known as visual purple, is a pigment of the retina of mammalian rod cells, where it is responsible for visual phototransduction. Rhodopsins are widely distributed not only in mammals, but also in the microbial world and serve as membrane embedded receptors for light. The first microbial rhodopsins were discovered in the haloarchaeon *Halobacterium salinarum* in the search for receptors that mediate phototaxis [7]. Accordingly, the best-studied proteins are the four haloarchaeal types of microbial rhodopsins: the light-driven ion pumps bacteriorhodopsin [8] and halorhodopsin [9, 10], and sensory rhodopsin (SR) I [7] and SRII [11], which are involved in phototactic responses. Rhodopsins consist of a protein moiety, the opsin, and an all-*trans*-retinal chromophore as cofactor for light perception (Figure 10.NaNA). Although the microbial rhodopsins do not show a significant sequence similarity with the human rhodopsin, rhodopsins in general share common protein architecture with seven-transmembrane α-helices forming an interior pocket for the retinal chromophore. Since the signaling mechanism of rhodopsin photoreceptors is highly conserved, the mechanism of phototransduction of the intensively studied *H. salinarum* bacteriorhodopsin is presented in detail as an example, including a short illustration of the function as an ion pump.

Following oxygen depletion, haloarchaeal growth is reduced due to low respiratory electron transport activity and, consequently, low energy production. Accordingly, the synthesis of the purple membrane is induced, which consists of up to around 75% of bacteriorhodopsin. Bacteriorhodopsin serves as a light-driven proton pump, generating a positive outside membrane potential and thus creating an inwardly directed proton motive force utilized for ATP synthesis. The absorption maximum of bacteriorhodopsin lies at 568 nm, which causes the purple appearance of the membrane [12]. Upon photoexcitation the all-*trans*-retinal chromophore, which is bound by a protonated Schiff-base linkage to the ε-amino group of a conserved lysine in the middle of the seventh helix, undergoes a cis/trans isomerization to the 13-*cis*-retinal [13]. This photoisomerization induces conformational changes within the protein moiety, which move the protonated Schiff-base to a new protein environment. This leads to a proton transfer from the Schiff-base to a nearby acceptor, a conserved aspartate, resulting in a release of a proton from the extracellular side of the protein. The Schiff-base becomes reprotonated via another conserved aspartate that assimilates a proton from the cytoplasm and the protein returns back to its initial state [14]. At least 50 protons/s are pumped to the other side of the membrane to generate an electrochemical gradient [15]. In contrast to bacteriorhodopsin, photoexcitation of halorhodopsin leads to the transport of chloride ions into the cell, generating a positive outside membrane potential for the purpose of providing an electrical potential for net proton uptake [16, 17].

SRI and SRII serve as phototaxis receptors exerting influences on the swimming behavior of the cells in response to changes in light intensity and color [11]; thereby, SRI controls the antagonistic behavior of SRII induction in dependence of different wavelengths [5]. In greater detail, SRI and SRII form 2 : 2 membrane-embedded complexes with their cognate transducers, HtrI and HtrII, respectively, which resemble the bacterial chemotaxis receptor-transducer proteins (or methyl-accepting chemotaxis proteins (MCPs)) [5, 11, 18, 19]. Activation of SRI by orange light results in an attractant response by transiently inhibiting the kinase activity of the sensor kinase CheA. This is mediated by the transducer protein HtrI, which is able to interact with CheA following conformational changes upon signal transduction. A decrease of the CheA phosphorylation state involves less phosphorylation of the response regulator CheY – the cytoplasmic flagellar motor switch regulator of *H. salinarum* (Figure 10.2A). The cells continue to swim. However, light can also exert a harmful effect on living cells due to high-energy radiation caused by UV and blue light. Therefore, blue-light activation of SRII initiates a repellent response conveying the signal via HtrII to CheA. Phosphorylated CheA transfers its phosphoryl group to CheY, which interacts in its phosphorylated state with the flagellar motor to induce clockwise flagellar rotation and tumbling. SRI also mediates a strong repellent response when activated by two sequential photons (first from orange-light and then from blue-light radiation) [5]. Genome sequencing analyses during recent decades have enabled the identification of several archaeal rhodopsin-like photoreceptors in diverse prokaryotic and eukaryotic organisms, revealing strikingly different modes of signaling [20–23].

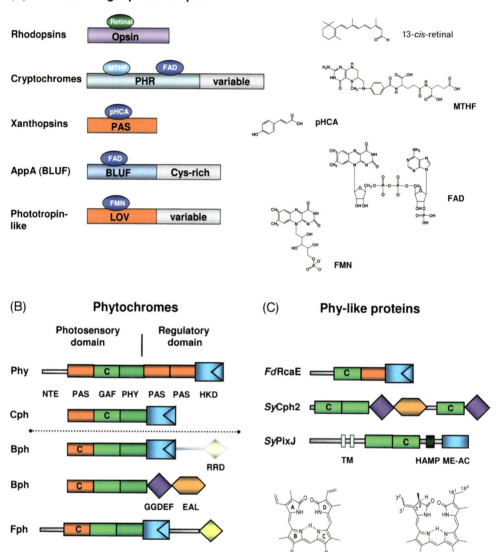

Figure 10.NaN Domain organization of representative members of the bacterial UV/blue-light photoreceptors, the red-light-sensing phytochromes, and the phytochrome-like (Phy-like) proteins. (A) UV/blue-light photoreceptors. The protein moiety of rhodopsins, the opsin, covalently attaches a retinal chromophore. The highest conservation among cryptochrome sequences is found in the PHR, which noncovalently binds FAD as cofactor. Known from plant cryptochromes, a second chromophore can be noncovalently bound which involves either MTHF or 8-hydroxy-5-deazariboflavin. However, in most bacterial cryptochromes identified thus far the second cofactor is absent. Additionally, cryptochromes carry a variable C-terminal domain. Xanthopsins (PYP) comprise a PAS fold and contain a thiol ester-linked p-coumaric acid (pHCA) chromophore at a cysteine residue. AppA, a BLUF

10.2.2
Cryptochromes

Cryptochromes are photoreceptors responding to UV-A and blue light, sharing sequence similarities to DNA photolyases that exhibit DNA repair activity by blue light. In spite of the strong similarities, cryptochromes do not display any double-stranded DNA photolyase activity [24]. For a long time cryptochromes were thought to be restricted to higher organisms, evolving from photolyases after the origin of eukaryotes [25]. Recently, bacterial cryptochromes were identified in the genomes of some eubacteria such as *Synechocystis* sp. PCC6803, *Cytophaga hutchinsonii*, or *Vibrio cholerae* [26]. Accordingly, these blue-light photoreceptors are widely distributed among bacteria and eukaryotes, and can be classified into three distinct groups – the plant cryptochromes, animal cryptochromes and CRY-DASH proteins, related to cryptochromes from *Drosophila*, *Arabidopsis*, *Synechocystis* and *Homo*, although CRY-DASH proteins themselves are not found in *Drosophila* or humans [27]. Most knowledge on bacterial cryptochromes has been gained from studies of *Synechocystis* sp. PCC6803 cryptochrome and *Vc*Cry1 from *V. cholerae*, both belonging to the class of CRY-DASH proteins [28, 29]. To date, investigations on these proteins have not

domain protein from *R. sphaeroides*, consists of an N-terminal BLUF domain harboring the noncovalently attached FAD chromophore and a C-terminal catalytic cysteine-rich domain. The bacterial phototropin-like proteins possess an N-terminal LOV domain binding a FMN chromophore and a variable C-terminal domain. The majority of bacterial phototropin-like proteins contain histidine kinase domains (HKD) as the output module. Some proteins also reveal transcriptional regulator domains, putative phosphodiesterases, and regulators of stress factors such as YtvA from *B. subtilis*. (B) Domain organization of plant (Phy), cyanobacterial (Cph), bacterial (Bph), and fungal (Fph) phytochromes. Phytochromes consist of two major domains – the photosensory and regulatory domain, respectively. Each of these domains can be subdivided into smaller domains designated NTE, variable N-terminal extension; PAS, PAS domain; GAF, GAF domain; PHY, phytochrome domain (GAF-related domain) as input domains and HKD; RRD, response regulator domain; GGDEF, domain with diguanylate cyclase activity; EAL, domain comprising phosphodiesterase activity, as output domains. C, conserved cysteine residue for chromophore attachment. The linear tetrapyrrole of cyanobacterial and plant phytochromes is covalently attached to a conserved cysteine residue in the GAF domain. By contrast, the bacterial and fungal phytochromes bind their chromophore to a conserved cysteine residue in the PAS domain. The dotted line indicates this differentiation. (C) Domain structures of representative members of the phytochrome-like proteins. *Fd*RcaE, *F. diplosiphon* sensor kinase of the Rca system comprising two sequential GAF domains with the first carrying a conserved cysteine for chromophore attachment, followed by a HKD. *Sy*Cph2, *Synechocystis* sp. PCC 6803 extraordinary phytochrome Cph2 containing three GAF domains and instead of a HKD GGDEF and EAL output domains. A red/far-red-light reversibility is only established via chromophore binding to the cysteine in the N-terminal protein moiety. *Sy*PixJ, *Synechocystis* sp. PCC 6803 cyanobacteriochrome PixJ composed of two transmembrane (TM) spanning helices, two GAF domains, the latter of which binds a bilin, a HAMP (found in *h*istidine kinases, *a*denylyl cyclases, *m*ethyl-accepting chemotaxis proteins, and *p*hosphatases) domain and a MCP signaling domain (ME-AC). *Sy*PixJ was shown to function as blue/green reversible photoreceptor. Structures of the employed chromophores are displayed.

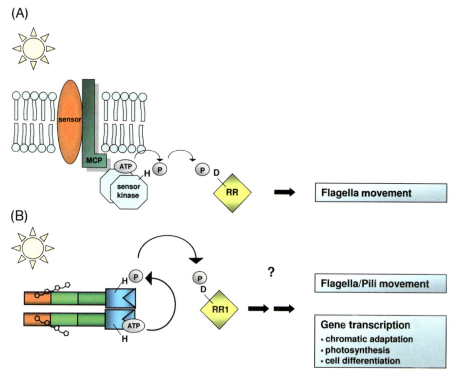

Figure 10.2 Schematic drawing for light-induced signal transduction of bacterial photoreceptors. (A) Some photosensory proteins use MCPs or MCP protein domains to transduce the signal light. Here, sensory rhodospin I (orange) serves as light sensor and initiates its transducer HtrI (green), a MCP, to interact with the sensor kinase CheA. This sensor kinase autophosphorylates and further transfers the phosphate to the associated response regulator (in this case CheY), which in turn triggers the flagellar motor switch. It should be noted that sensory rhodopsins form 2:2 membrane-embedded complexes with their cognate transducers Htr. (B) Light-activated Bph cross-phosphorylates the conserved histidine (H) of the second monomer, which is transferred to the conserved aspartate (D) of the cognate response regulator-1 (RR1). Since lots of histidine phosphotransferases and response regulators with effector domains appear in bacterial genomes, the final signal transduction pathways are most often still unknown. Diverse response regulators can either regulate the flagella motor switch or induce transcription of target genes affecting photosynthesis or chromatic adaptation.

exposed any functionality towards the organisms; however, it appears that they exhibit features of both photolyases (i.e., light-induced redox activity) and cryptochromes. Structural analyses as well as gel-shift assays revealed that the CRY-DASH protein from *Synechocystis* sp. PCC6803 can bind double-stranded DNA and microarray analyzes presented hints for its involvement in transcriptional regulation [27]. Furthermore, VcCry1 even exhibited photolyase activity with specificity for cyclobutane pyrimidine dimers in single-stranded DNA [29]. Most of the cryptochromes are composed of two domains – an N-terminal photolyase-related domain (PHR) and a C-terminal region variable in size (Figure 10.NaNA). The PHR region is the chromophore-binding domain, assembling a flavin adenine dinucleotide (FAD) and one molecule of 5,10-methenyltetrahydrofolate (MTHF or pterin) responsible for light absorption [30]. FAD binds noncovalently in a U-shaped conformation and is buried deeply in a cavity formed by the α/β domain of the PHR region of the protein [27]. The catalytic mechanism has not been clarified in detail, but most probably involves transfer of the excitation energy absorbed by the other chromophore to FAD. This could now serve as an electron donor as it is also the case in cyclobutane pyrimidine dimer (CPD) photolyases, repairing CPDs in double-stranded DNA [31]. Blue-light absorption via plant cryptochromes promotes phosphorylation in the serine-rich C-terminal part of the protein, indicative of another regulatory mechanism of these photoreceptors [32].

10.2.3
Photoactive Yellow Proteins (Xanthopsins)

The photoactive yellow protein (PYP), the prototype xanthopsin, constitutes a rather new type of blue-light photoreceptor and was discovered in halophilic phototrophic purple eubacteria such as *Ectothiorhodospira halophila* [33], and thus far has mostly been identified in proteobacteria. Very recently, six new PYPs were detected in the genomes of other proteobacteria and, surprisingly, in the nonphotosynthetic bacterium *Salinibacter ruber* from the phylum bacteriodetes [34]. PYPs comprise a Per/Arnt/Sim (PAS) fold and contain a thiol ester-linked *p*-coumaric acid chromophore at a cysteine residue (Figure 10.NaNA) [35]. A 1.4-Å crystallographic structure was solved in 1995 that demonstrated the PYP adopting an α/β-fold with a six-stranded antiparallel β-sheet as a scaffold flanked by α-helices [36]. Thus, PYP appears to prototypically exhibit the major structural and functional features characteristic of the PAS domain superfamily [37]. PYPs undergo a photocycle upon irradiation including a cis/trans isomerization that starts with a ground- or dark-adapted state with a deprotonated *p*-coumaryl chromophore in the trans configuration. After excitation by blue light PYP undergoes several transformations including a cis isomerization followed by a protonation of the chromophore, most probably the signaling state that returns back to the deprotonated ground state [38]. Since these photoreceptors are mostly present in halophilic phototrophic bacteria, they could serve in the phototactic negative response to blue light, which would fit to the PYP absorption spectra [39]. Thus far, not much is known about the downstream signaling cascade triggered by PYP activation in the different organisms. However, if PYP is involved in phototactic events, further signal transduction components could likely be part of the flagellar motor.

10.2.4
BLUF Domain Proteins

The BLUF domain represents a new class of blue-light-sensing minimal modules, present in several bacterial and algal proteins [40]. BLUF domains are α/β-folds burying the flavin cofactor noncovalently in the core of the domain. One of the best studied BLUF domain-containing proteins is the photoreceptor AppA from *Rhodobacter sphaeroides* [40]. AppA contains an N-terminal BLUF domain harboring the FAD chromophore prosthetic group and a C-terminal catalytic cysteine-rich domain (Figure 10.NaNA). Photoexcitation of AppA leads to formation of a red-shifted intermediate state, which is slowly reversed in the dark [41]. In *R. sphaeroides* AppA serves as blue-light-regulated antirepressor of the photosynthesis repressor PpsR. In its dark state, AppA inactivates PpsR by converting it from an active tetramer to an inactive dimer through reduction of a disulfide bond. Blue-light absorption by AppA has been assumed to induce conformational changes preventing it from interacting with PpsR [41]. Apart from its function as light receptor, AppA also exhibits PpsR antirepressor activity in response to oxygen mediated by a heme cofactor [42]. Both activities enable the organism to fine-tune transcriptional regulation of the photosynthetic genes.

10.2.5
Phototropin-Like Microbial Photoreceptors

Similar to plant phototropins, some bacterial species such as *Bacillus subtilis* or *Caulobacter crescentus* use phototropin-like chromoproteins for the detection of UV-A and blue light via LOV domains [43, 44]. LOV domains are members of the PAS superfamily and are named after the external signals they are regulated by such as *l*ight, *o*xygen or *v*oltage. Due to the ability of the LOV domain to bind a flavin mononucleotide (FMN) chromophore they function as the blue-light sensory part of the phototropins (Figure 10.NaNA). Upon excitation by UV-A and blue light, LOV domains are able to undergo a photocycle [45]. FMN is noncovalently bound to both domains in the dark, generating a form that maximally absorbs around 447 nm [45]. Illumination induces the formation of a transient covalent FMN-cysteinyl adduct creating a spectral species that strongly absorbs around 390 nm [45, 46]. In a nonphotochemical reaction this form returns to the ground state in the dark. Up to date, 29 sequences encoding LOV proteins have been detected in the genomes of 24 bacterial species [47], but very few of them have yet been analyzed in greater detail. It is noteworthy that bacterial LOV proteins can be distinguished by their great variety of effector domains that are associated to the conserved light-sensing module of the protein. The majority of bacterial phototropin-like proteins contain a histidine kinase domains (HKD) as an output module. The HKD is part of a two-component signal transduction system and an environmental stimulus usually triggers autophosphorylation at a conserved histidine residue. The phosphoryl group is subsequently transferred to an aspartate residue in the response regulator protein which in case of *Pseudomonas syingae* or *Xanthomonas campestris* phototropin-like proteins is directly fused to the HKD [47]. Phosphotransfer to the response regulator activates a downstream effector domain or initiates a phosphorelay eliciting a specific response. The formation of a transient flavoprotein adduct presumably induces a stimulation of the HKD that is located at the C-terminus of the protein. Apart from histidine kinase fusion proteins, domain analyses of bacterial phototropin-like proteins also revealed transcriptional regulator domains, putative phosphodiesterases, and regulators of stress factors. The latter being a component of the first and best studied bacterial LOV protein termed YtvA from B. subtilis shown to bind FMN and to undergo the same photochemistry as plant phototropins [43]. YtvA consists of an N-terminal LOV domain and a C-terminal NTP-binding sulfate transporter and anti-sigma antagonist (STAS) domain (Figure 10.NaNA) [47, 48]. When cells encounter environmental stress conditions, the alternative sigma factor σ^B is activated to induce the expression of genes responsible for the general stress response [49–51]. Due to its STAS domain YtvA shows extensive similarity to a group of five proteins that are regulators of σ^B activity, most likely functioning collectively in the environmental stress signaling pathway [52]. Recent studies indeed indicated YtvA as a photoreceptor, leading to photoenhancement of a σ^B-dependent transcription also by other stress signals than light [48]. Furthermore, YtvA was shown to be part of the σ^B regulator complex [53]. But how is the signal integrated? One possibility might be the NTP-binding STAS domain that could be responsible for signal transmission. It is noteworthy that YtvA without the STAS domain is not capable of enhancing the activation of σ^B [48]. Once the signal is perceived, the STAS domain could transfer the signal to other members of the complex of σ^B regulators by direct protein–protein interactions [48]. This group of positive and negative regulators of σ^B activity does not directly act on the alternative sigma factor, but work upstream in the signaling cascade.

10.3
Red-Light Sensing – Phytochromes

Until 1996 the existence of red/far-red-light photoreceptors was assumed to be restricted to plants. Only in 1997 was identification and characterization of bacterial phytochrome proteins started with the discovery of the first phytochrome-like sequence in the cyanobacterium *Synechocystis* sp. PCC6803. This cyanobacterial phytochrome strongly resembles that of plants and was the first phytochromes characterized *in vitro* beyond the plant kingdom [54–56]. Extensive genomic studies confirmed the widespread existence of phytochromes among cyanobacteria and, in addition, several phytochrome coding genes were found in the genomes of α-proteobacteria such as the phototrophic purple bacteria *Rhodopseudomonas palustris* and *R. sphaeroides* or in the closely related photosynthetic bacterium *Bradyrhizobium* sp. strain ORS278 [57–59], which strengthened the hypothesis that phytochrome proteins generally occur in phototrophic bacteria. More surprisingly, at the same time phytochrome-like sequences were detected in heterotrophic bacteria like *Agrobacterium tumefaciens*, *Deinococcus radiodurans*, or *Pseudomonas aeruginosa* [60, 61].

10.3.1
Principle of Phytochrome Action

Phytochromes belong to the family of red/far-red-light-sensing photoreceptors originally discovered in plants approximately 50 years ago [62]. Based on phylogenetic examinations, phytochrome proteins can be organized into five different clades, in plant phytochromes (plant phytochrome Phy), cyanobacterial (cyanobacterial phytochrome Cph), bacterial (bacterial phytochrome Bph) and fungal phytochrome photoreceptors (fungal phytochrome Fph), and into a cluster of phytochrome-like sequences (Figure 10.NaNB and C) [63, 64].

Phytochromes are able to reversibly photoisomerize between two stable conformations – a red-light absorbing Pr form and a far-red-light absorbing Pfr form. Both forms are spectrally distinct. The photointerconversion is mediated by a linear tetrapyrrole prosthetic group (bilin), which is autocatalytically assembled to the apoprotein via a covalent thioether linkage to a conserved cysteine residue [65].

Upon excitation by red or far-red light the chromophore undergoes a cis/trans isomerization around the C15–C16 double bond between the pyrrole rings C and D, followed by conformational changes within the protein. Since Pr and Pfr exhibit overlapping absorbance spectra, a photoequilibrium is formed under saturating light conditions consisting of a mixture of both forms. The Pr/Pfr ratio particularly varies within the versatile phytochromes. However, the Pr/Pfr ratio determines the magnitude of the biological response, whereas the Pfr form often is the active state. A still unanswered phenomenon, which has been extensively studied photobiochemically, but remains to be elucidated on the molecular level, is the process of nonphotochemical Pfr-to-Pr dark reversion. Known from several bacterial phytochromes, dark reversion occurs in time ranges from minutes to hours, indicative for controlling the signal output [66–68]. Owing to the capacity of switching between two spectrally distinct forms, phytochromes are often designated as "molecular light-switches."

10.3.2
Domain Organization of Phytochromes

Phytochromes in general share a common domain organization (Figure 10.NaNB). They typically consist of an N-terminal photosensory domain, harboring the bilin chromophore and a C-terminal output module. The photosensory domain is comprised of PAS, GAF, and PHY domains. In particular, PAS domains are involved in many signaling proteins where they are used as a signal sensor domain. Several PAS domain proteins are known to detect their signal by way of an associated cofactor such as heme, flavin, and a 4-hydroxycinnamyl chromophore. Furthermore PAS domains are involved in protein–protein interactions. GAF domains are renowned for attaching small planar aromatic ligands including tetrapyrroles, flavins, and nucleotide cofactors, and are present in vertebrate cGMP-specific phosphodiesterases, cyanobacterial *a*denylate cyclases and in the *f*ormate hydrogen lyase of *E. coli* [69]. PAS and GAF domains display structural similarities [69, 70]. PHY (*phy*trochrome) domains are evolutionary related to PAS and GAF domains, since all three domains evolved from a common ancestor [63]. Concerning the tetrapyrrole binding site in the N-terminal photosensory domain, a significant difference among the phytochrome clades becomes obvious. Interestingly, in bacteria and fungi the chromophore attachment site as well as the chromophore itself differ from cyanobacterial and plant phytochromes. Plant and cyanobacterial phytochromes bind their chromophore to a conserved cysteine residue in the GAF domain, whereas bacterial and fungal phytochromes covalently attach their linear tetrapyrrole prosthetic group to a cysteine residue in the PAS domain (Figure 10.NaNB).

The C-terminal output modules are usually more diverged but most bacterial and fungal phytochromes contain a HKD, which is sometimes light-regulated. Examples are *Synechocystis* Cph1 [56] and *A. tumefaciens* Agp2 [71]. The corresponding response regulator is often encoded separately within the *bph* operon [61, 71, 72]. Investigations on rather exceptional phytochromes such as BphG1 from *R. sphaeroides* or Cph2 from *Synechocystis* sp. PCC6803 containing GGDEF and EAL domains (Figure 10. NaNB and C), which are involved in the turnover of the second messenger c-di-GMP, revealed the variability of the C-terminal output modules for achieving specific light-dependent responses [59, 73].

10.3.3
Cyanobacterial Phytochromes and Phytochrome-Like Proteins

Cyanobacterial phytochromes most commonly use phycocyanobilin (PCB) as chromophore which, due to the high abundance in phycobilisomes, is easily available for the assembly with apophytochrome. The first and best characterized cyanobacterial phytochrome is Cph1 from *Synechocystis* sp. PCC6803. Cph1 and Cph1-related phytochromes like CphA from *Calothrix* sp. PCC7601 feature the common PAS–GAF–PHY architecture of the N-terminal photosensory domain and possess a HKD as output module. Cph1 acts as a typical red/far-red-light-inducible photoreceptor after assembly with PCB [56, 74]. The absorbance maxima of the Pr and Pfr form of Cph1-PCB are $Pr_{max} = 654$ nm and $Pfr_{max} = 706$ nm. Cph1 and CphA display light-dependent histidine kinase activities that are more activated under far-red-light conditions. It has also be shown that the phosphoryl groups are transferred to their cognate response regulators Rcp1 and RcpA, respectively [56, 75], constituting a functional bacterial two-component system (Figure 10.2). However, which modes of signaling cascades are stimulated after irradiation with red and far-red light are still unexplained.

More information about the functionality of cyanobacterial phytochromes is provided by the extraordinary phytochrome Cph2 from *Synechocystis* sp. PCC6803 (Figure 10.NaNC). Unlike Cph1, Cph2 contains three GAF domains and instead of a HKD GGDEF and EAL output domains [63]. It is able to attach a bilin at the N-terminal GAF domain as well as one at the C-terminal, but a red/far-red light reversibility is only established via chromophore binding to the cysteine in the N-terminal protein moiety [74, 76]. Early studies suggested Cph2 as being part of a light-stimulated signal transduction pathway for inhibiting the movement towards blue light [77]. Recent data even strengthened the hypothesis of Cph2 being involved in a blue-light-mediated signaling cascade [73].

Many other phytochrome-like proteins have evolved in cyanobacteria. This is in part due to their high dependence of light as an energy source and the necessity to be able to adapt to different light environments. Many species are able to adapt their phycobilisome antenna to maximize their light harvesting efficiency. This process is commonly known as complementary chromatic adaptation (CCA) and is best studied in the filamentous cyanobacterium *Fremyella diplosiphon* (also called *Calothrix* or *Tolypothrix* sp. PCC 7601). The central element of the CCA regulatory pathway(s) is the *r*egulator for *c*omplementary *c*hromatic *a*daptation (Rca) system [78], which is a photoreceptor-based two-component regulatory system [79]. The corresponding photoreceptor RcaE is a sensor kinase that N-terminally contains a chromophore-binding related domain similar to phytochromes and C-terminally a histidine kinase module [80]. RcaE was the founding member of the group of phytochrome-like sequences that diverges from the actual red/far-red light photosensors thus far they do not comprise a PHY domain and some even no PAS domain [64]. RcaE has been shown to be a chromophorylated biliprotein in *F. diplosiphon*, functioning as a red- and green-light-responsive protein during CCA [79, 80]. Together with the downstream response regulators RcaF and RcaC, RcaE is suggested to form a signal transduction phosphorelay that controls the phycobilisome composition in response to changing ambient light conditions [81]. Thus, RcaE is proposed to be active in both red and green light, probably acting as kinase in red light and as phosphatase in green light [80]. Though promoter-binding proteins of red- and green-light-activated genes have been identified [78, 82, 83], no molecular link between the RcaE response regulators and these DNA-binding proteins could be demonstrated up to now. A recent report has grouped RcaE within the cyanobacteriochrome family, a group of phytochrome-like proteins thus far restricted to cyanobacteria [84]. These new type of phytochrome-like photoreceptors require only a GAF domain for chromophore binding and photoconversion. Furthermore, they were shown to bind PCB or the unusual phycoviolobilin as a chromophore and thereby also exhibiting unusual spectroscopic features. One of them, CcaS, is a green-light receptor regulating the expression of phycobilisome linker genes in *Synechocystis* sp. PCC 6803. Like the prototypical cyanobacterial phytochrome Cph1, this photoreceptor uses PCB as a chromophore, but displays photoconversion between a green-light absorbing Pg form ($\lambda_{max} = 535$ nm) and a red-absorbing Pr form ($\lambda_{max} = 672$ nm). Furthermore, this cyanobacteriochrome displays light-dependent autophorphorylation and phosphoryl transfer to its cognate response regulator [85]. Other members of the cyanobacteriochrome family like *Sy*PixJ from *Synechocystis* sp. PCC6803 and *Te*PixJ from *Thermosynechococcus elongatus* regulate positive phototaxis on solid surfaces via type IV pili, and the orthologs were shown to function as blue/green reversible photoreceptors [86–88].

10.3.4
Phytochromes in Other Phototrophic Bacteria

The appearance of photoreceptors in phototrophic bacteria besides cyanobacteria is quite understandable in that they function in light-mediated energy generation. Among the phytochromes of photosynthetic bacteria some of them exhibit rather unusual spectroscopic features (near-red-light absorbing Pnr $= \lambda_{max} = 800$ nm) or possess a different output module like S-box domains as the regulatory domain. Many of these bacteriophytochromes trigger the expression of the entire photosynthetic apparatus including the synthesis of the required bacteriochlorophyll and carotenoid molecules during illumination with far-red light [57, 58]. Additionally, it has been shown recently that the bacteriophytochrome *Rp*BphP4 from *R. palustris* strain CGA009 regulating the synthesis of LH2 complexes lacks the chromophore-binding cysteine and displays a redox-sensing ability coupled to kinase activity instead [72].

10.3.5
Phytochromes in Heterotrophic Bacteria

The class of Bph photoreceptors (BphP) is widespread among photoautotrophic bacteria as described in the previous sections and surprisingly also among heterotrophic, nonphotosynthetic bacteria, raising the question to the physiological function of these phytochromes. The best photobiochemically characterized phytochromes of heterotrophic bacteria are those from *A. tumefaciens*, *D. radiodurans*, *P. aeruginosa*, *P. syringae*, and *Rhizobium leguminosarium*, whose *bphP* operons often include a coding region for heme oxygenase which produces the linear tetrapyrrole chromophore biliverdin IXα (BV). The common architecture of an N-terminal chromophore-binding domain, including the PAS, GAF, and PHY domains, and a C-terminal regulatory HKD is kept among the bacterial red/far-red-light photoreceptors [60, 89]. An essential feature that distinguishes the cyanobacterial and plant phytochromes from the bacterial and also fungal phytochromes is the covalent binding of BV. The spectra of phytochrome-BV adducts are considerably red-shifted as a result of the oxidized state of BV compared to PCB (Figure 10.NaNC). Additionally, bacterial phytochromes lack the canonical chromophore binding cysteine in the GAF domain known from cyanobacterial phytochromes and contain an unpolar amino acid (isoleucine, leucine or valine) at this position. As an alternative, they covalently bind BV to a conserved cysteine in the PAS domain (Figure 10.NaNB) [66, 70, 90]. The crystal structure of *D. radiodurans* photosensory core (PAS–GAF) in the Pr state, solved in 2005, was a major breakthrough and provided a wealth of information for understanding photochemistry and structure of phytochromes [70]. Spectroscopic analyses of bacteriophytochromes significantly show the classic behavior of phytochromes undergoing red/far-red-light reversible photochemistry which is explained in detail for Agp1 from *A. tumefaciens*. Agp1 quickly assembles with BV, generating a stable Pr form with an absorption maximum of 702 nm as the ground state in the dark. During irradiation with red light a photoequilibrium with high portion of Pfr comprising a maximum at 749 nm is established, thermally relaxing back to the Pr ground state in the dark within one hour. The Pr form displays a stronger histidine kinase activity than the Pfr form. To complete the two-component signaling cascade the phosphoryl group is subsequently transferred to its cognate response regulator RR1, also encoded within the *agp1* operon [71, 91].

Astonishingly, a rather new subfamily of bacteriophytochromes was identified to function in reverse – the so-called bathy BphPs. To date *P. aeruginosa Pa*BphP, Agp2 from *A. tumefaciens*, *Rp*BphP1 from *R. palustris* as well as *Bradyrhizobium* ORS278 *Br*BphP1 have been spectroscopically and biochemically characterized; *Pa*BphP is picked here for a detailed explanation [57, 66, 71, 91]. Assembly of apo-*Pa*BphP with BV in the dark also proceeds very quickly and generates a transient Pr-like state with an absorption maximum at 700 nm, which immediately converts nonphotochemically to a Pfr-enriched stable ground state. The absorbance maximum of the Pfr-enriched ground state forms a peak at 754 nm. In contrast to Agp2 presenting a higher histidine autophosphorylation activity under red-light conditions, no strong light- or bilin-dependent kinase activity could be observed with *Pa*BphP [66, 71]. More recent studies on the functionality of *Pa*BphP highlight a putative role in the stationary phase [92].

It is noteworthy to mention that all bacterial phytochromes thus far characterized display a remarkable chromophore-independent histidine kinase activity that can be intensely weakened after BV assembly and irradiation with red or far-red light depending on the phytochrome protein. These observations even strengthen the demand for answers concerning the physiological impact of bacteriophytochromes on the organisms.

By now phytochromes have been described to play various roles in bacteria like control of the light-harvesting complexes in *F. diplosiphon* [80], regulation of carotenoid synthesis in *D. radiodurans* [60], phototaxis in *Synechocystis* sp. PCC6803 [93], and control of photosystem synthesis in *Bradyrhizobium* ORS278 and *R. palustris* [57].

10.4
Conclusions

As seen from the data presented in this chapter, the biochemistry of microbial photoperception is already quite well understood. However, the downstream signal transduction and the function of many photoreceptors is still enigmatic, and remains a matter of current research in many laboratories.

Acknowledgments

Research in the authors laboratory was funded by the Volkswagen Stiftung, the Deutsche Forschungsgemeinschaft, and the Fonds der Chemischen Industrie, whose support is greatly acknowledged.

References

1 Choi, J.S., Chung, Y.H., Moon, Y.J., Kim, C., Watanabe, M., Song, P.S., Joe, C.O., Bogorad, L., and Park, Y.M. (1999) Photomovement of the gliding cyanobacterium *Synechocystis* sp. PCC 6803. *Photochem. Photobiol.*, **70**, 95–102.

2 Campbell, D., Houmard, J., and De Marsac, N.T. (1993) Electron transport regulates cellular differentiation in the filamentous cyanobactcrium *Calothrix*. *Plant. Cell*, **5**, 451–463.

3 Petersen, A.B., Gniadecki, R., Vicanova, J., Thorn, T., and Wulf, H.C. (2000) Hydrogen

peroxide is responsible for UVA-induced DNA damage measured by alkaline comet assay in HaCaT keratinocytes. *J. Photochem. Photobiol. B*, **59**, 123–131.

4 Tyrrell, R.M. (1995) Ultraviolet radiation and free radical damage to skin. *Biochem. Soc. Symp.*, **61**, 47–53.

5 Spudich, J.L. and Bogomolni, R.A. (1984) Mechanism of colour discrimination by a bacterial sensory rhodopsin. *Nature*, **312**, 509–513.

6 Takahashi, T., Yan, B., Mazur, P., Derguini, F., Nakanishi, K., and Spudich, J.L. (1990) Color regulation in the archaebacterial phototaxis receptor phoborhodopsin (sensory rhodopsin II). *Biochemistry*, **29**, 8467–8474.

7 Bogomolni, R.A. and Spudich, J.L. (1982) Identification of a third rhodopsin-like pigment in phototactic *Halobacterium halobium*. *Proc. Natl. Acad. Sci. USA*, **79**, 6250–6254.

8 Oesterhelt, D. and Stoeckenius, W. (1973) Functions of a new photoreceptor membrane. *Proc. Natl. Acad. Sci. USA*, **70**, 2853–2857.

9 Matsuno-Yagi, A. and Mukohata, Y. (1977) Two possible roles of bacteriorhodopsin; a comparative study of strains of *Halobacterium halobium* differing in pigmentation. *Biochem. Biophys. Res. Commun.*, **78**, 237–243.

10 Schobert, B. and Lanyi, J.K. (1982) Halorhodopsin is a light-driven chloride pump. *J. Biol. Chem.*, **257**, 10306–10313.

11 Hoff, W.D., Jung, K.H., and Spudich, J.L. (1997) Molecular mechanism of photo-signaling by archaeal sensory rhodopsins. *Annu. Rev. Biophys. Biomol. Struct.*, **26**, 223–258.

12 Oesterhelt, D. (1976) Bacteriorhodopsin as an example of a light-driven proton pump. *Angew. Chem. Int. Ed. Engl.*, **15**, 17–24.

13 Oesterhelt, D. (1998) The structure and mechanism of the family of retinal proteins from halophilic archaea. *Curr. Opin. Struct. Biol.*, **8**, 489–500.

14 Krebs, M.P. and Khorana, H.G. (1993) Mechanism of light-dependent proton translocation by bacteriorhodopsin. *J. Bacteriol.*, **175**, 1555–1560.

15 Fritsche, W. (2002) *Mikrobiologie*, Spektrum Akademischer Verlag, Heidelberg.

16 Varo, G. (2000) Analogies between halorhodopsin and bacteriorhodopsin. *Biochim. Biophys. Acta*, **1460**, 220–229.

17 Kulcsar, A., Groma, G.I., Lanyi, J.K., and Varo, G. (2000) Characterization of the proton-transporting photocycle of pharaonis halorhodopsin. *Biophys. J.*, **79**, 2705–2713.

18 Kamo, N., Shimono, K., Iwamoto, M., and Sudo, Y. (2001) Photochemistry and photoinduced proton-transfer by pharaonis phoborhodopsin. *Biochemistry (Mosc.)*, **66**, 1277–1282.

19 Klare, J.P., Gordeliy, V.I., Labahn, J., Buldt, G., Steinhoff, H.J., and Engelhard, M. (2004) The archaeal sensory rhodopsin II/transducer complex: a model for transmembrane signal transfer. *FEBS Lett.*, **564**, 219–224.

20 Spudich, J.L., Yang, C.S., Jung, K.H., and Spudich, E.N. (2000) Retinylidene proteins: structures and functions from archaea to humans. *Annu. Rev. Cell Dev. Biol.*, **16**, 365–392.

21 Brown, L.S., Dioumaev, A.K., Lanyi, J.K., Spudich, E.N., and Spudich, J.L. (2001) Photochemical reaction cycle and proton transfers in *Neurospora* rhodopsin. *J. Biol. Chem.*, **276**, 32495–32505.

22 Jung, K.H., Trivedi, V.D., and Spudich, J.L. (2003) Demonstration of a sensory rhodopsin in eubacteria. *Mol. Microbiol.*, **47**, 1513–1522.

23 Sineshchekov, O.A., Jung, K.H., and Spudich, J.L. (2002) Two rhodopsins mediate phototaxis to low- and high-intensity light in *Chlamydomonas reinhardtii*. *Proc. Natl. Acad. Sci. USA*, **99**, 8689–8694.

24 Sancar, A. (2003) Structure and function of DNA photolyase and cryptochrome blue-light photoreceptors. *Chem. Rev.*, **103**, 2203–2237.

25 Cashmore, A.R., Jarillo, J.A., Wu, Y.J., and Liu, D. (1999) Cryptochromes: blue light receptors for plants and animals. *Science*, **284**, 760–765.

26 Lin, C. and Todo, T. (2005) The cryptochromes. *Genome Biol.*, **6**, 220.

27 Brudler, R., Hitomi, K., Daiyasu, H., Toh, H., Kucho, K., Ishiura, M., Kanehisa, M., Roberts, V.A., Todo, T., Tainer, J.A. et al. (2003) Identification of a new cryptochrome class. Structure, function, and evolution. *Mol. Cell*, **11**, 59–67.

28 Hitomi, K., Okamoto, K., Daiyasu, H., Miyashita, H., Iwai, S., Toh, H., Ishiura, M., and Todo, T. (2000) Bacterial cryptochrome and photolyase: characterization of two photolyase-like genes of *Synechocystis* sp. PCC6803. *Nucleic Acids Res.*, **28**, 2353–2362.

29 Selby, C.P. and Sancar, A. (2006) A cryptochrome/photolyase class of enzymes with single-stranded DNA-specific photolyase activity. *Proc. Natl. Acad. Sci. USA*, **103**, 17696–17700.

30 Malhotra, K., Kim, S.T., Batschauer, A., Dawut, L., and Sancar, A. (1995) Putative blue-light photoreceptors from *Arabidopsis thaliana* and *Sinapis alba* with a high degree of sequence homology to DNA photolyase contain the two photolyase cofactors but lack DNA repair activity. *Biochemistry*, **34**, 6892–6899.

31 Park, H.W., Kim, S.T., Sancar, A., and Deisenhofer, J. (1995) Crystal structure of DNA photolyase from *Escherichia coli*. *Science*, **268**, 1866–1872.

32 Shalitin, D., Yang, H., Mockler, T.C., Maymon, M., Guo, H., Whitelam, G.C., and Lin, C. (2002) Regulation of *Arabidopsis* cryptochrome 2 by blue-light-dependent phosphorylation. *Nature*, **417**, 763–767.

33 Meyer, T.E. (1985) Isolation and characterization of soluble cytochromes, ferredoxins and other chromophoric proteins from the halophilic phototrophic bacterium *Ectothiorhodospira halophila*. *Biochim. Biophys. Acta*, **806**, 175–183.

34 Kumauchi, M., Hara, M.T., Stalcup, P., Xie, A., and Hoff, W.D. (2008) Identification of six new photoactive yellow proteins – diversity and structure–function relationships in a bacterial blue light photoreceptor. *Photochem. Photobiol. Sci.*, **84**, 956–969.

35 Hoff, W.D., Dux, P., Hard, K., Devreese, B., Nugteren-Roodzant, I.M., Crielaard, W., Boelens, R., Kaptein, R., van Beeumen, J., and Hellingwerf, K.J. (1994) Thiol ester-linked p-coumaric acid as a new photoactive prosthetic group in a protein with rhodopsin-like photochemistry. *Biochemistry*, **33**, 13959–13962.

36 Borgstahl, G.E., Williams, D.R., and Getzoff, E.D. (1995) 1.4 A structure of photoactive yellow protein, a cytosolic photoreceptor: unusual fold, active site, and chromophore. *Biochemistry*, **34**, 6278–6287.

37 Pellequer, J.L., Wager-Smith, K.A., Kay, S.A., and Getzoff, E.D. (1998) Photoactive yellow protein: a structural prototype for the three-dimensional fold of the PAS domain superfamily. *Proc. Natl. Acad. Sci. USA*, **95**, 5884–5890.

38 Thompson, M.J., Bashford, D., Noodleman, L., and Getzoff, E.D. (2003) Photoisomerization and proton transfer in photoactive yellow protein. *J. Am. Chem. Soc.*, **125**, 8186–8194.

39 Sprenger, W.W., Hoff, W.D., Armitage, J.P., and Hellingwerf, K.J. (1993) The eubacterium *Ectothiorhodospira halophila* is negatively phototactic, with a wavelength dependence that fits the absorption spectrum of the photoactive yellow protein. *J. Bacteriol.*, **175**, 3096–3104.

40 Anderson, S., Dragnea, V., Masuda, S., Ybe, J., Moffat, K., and Bauer, C. (2005) Structure of a novel photoreceptor, the BLUF domain of AppA from *Rhodobacter sphaeroides*. *Biochemistry*, **44**, 7998–8005.

41 Kraft, B.J., Masuda, S., Kikuchi, J., Dragnea, V., Tollin, G., Zaleski, J.M., and Bauer, C.E. (2003) Spectroscopic and mutational analysis of the blue-light photoreceptor AppA: a novel photocycle involving flavin stacking with an aromatic amino acid. *Biochemistry*, **42**, 6726–6734.

42 Moskvin, O.V., Kaplan, S., Gilles-Gonzalez, M.A., and Gomelsky, M. (2007) Novel heme-based oxygen sensor with a revealing evolutionary history. *J. Biol. Chem.*, **282**, 28740–28748.

43 Losi, A., Polverini, E., Quest, B., and Gartner, W. (2002) First evidence for phototropin-related blue-light receptors in prokaryotes. *Biophys. J.*, **82**, 2627–2634.

44 Crosson, S., Rajagopal, S., and Moffat, K. (2003) The LOV domain family: photoresponsive signaling modules coupled to

diverse output domains. *Biochemistry*, **42**, 2–10.
45 Salomon, M., Christie, J.M., Knieb, E., Lempert, U., and Briggs, W.R. (2000) Photochemical and mutational analysis of the FMN-binding domains of the plant blue light receptor, phototropin. *Biochemistry*, **39**, 9401–9410.
46 Kasahara, M., Swartz, T.E., Olney, M.A., Onodera, A., Mochizuki, N., Fukuzawa, H., Asamizu, E., Tabata, S., Kanegae, H., Takano, M. *et al.* (2002) Photochemical properties of the flavin mononucleotide-binding domains of the phototropins from *Arabidopsis*, rice, and *Chlamydomonas reinhardtii*. *Plant Physiol.*, **129**, 762–773.
47 Losi, A. (2004) The bacterial counterparts of plant phototropins. *Photochem. Photobiol. Sci.*, **3**, 566–574.
48 Suzuki, N., Takaya, N., Hoshino, T., and Nakamura, A. (2007) Enhancement of a sigma^B-dependent stress response in *Bacillus subtilis* by light via YtvA photoreceptor. *J. Gen. Appl. Microbiol.*, **53**, 81–88.
49 Hecker, M. and Volker, U. (1998) Non-specific, general and multiple stress resistance of growth-restricted *Bacillus subtilis* cells by the expression of the sigma^B regulon. *Mol. Microbiol.*, **29**, 1129–1136.
50 Kang, C.M., Brody, M.S., Akbar, S., Yang, X., and Price, C.W. (1996) Homologous pairs of regulatory proteins control activity of *Bacillus subtilis* transcription factor sigma^B in response to environmental stress. *J. Bacteriol.*, **178**, 3846–3853.
51 Vijay, K., Brody, M.S., Fredlund, E., and Price, C.W. (2000) A PP2C phosphatase containing a PAS domain is required to convey signals of energy stress to the sigmaB transcription factor of *Bacillus subtilis*. *Mol. Microbiol.*, **35**, 180–188.
52 Akbar, S., Gaidenko, T.A., Kang, C.M., O'Reilly, M., Devine, K.M., and Price, C.W. (2001) New family of regulators in the environmental signaling pathway which activates the general stress transcription factor sigma^B of *Bacillus subtilis*. *J. Bacteriol.*, **183**, 1329–1338.
53 Gaidenko, T.A., Kim, T.J., Weigel, A.L., Brody, M.S., and Price, C.W. (2006) The blue-light receptor YtvA acts in the environmental stress signaling pathway of *Bacillus subtilis*. *J. Bacteriol.*, **188**, 6387–6395.
54 Hughes, J., Lamparter, T., Mittmann, F., Hartmann, E., Gartner, W., Wilde, A., and Borner, T. (1997) A prokaryotic phytochrome. *Nature*, **386**, 663.
55 Wilde, A., Churin, Y., Schubert, H., and Borner, T. (1997) Disruption of a *Synechocystis* sp. PCC 6803 gene with partial similarity to phytochrome genes alters growth under changing light qualities. *FEBS Lett.*, **406**, 89–92.
56 Yeh, K.C., Wu, S.H., Murphy, J.T., and Lagarias, J.C. (1997) A cyanobacterial phytochrome two-component light sensory system. *Science*, **277**, 1505–1508.
57 Giraud, E., Fardoux, J., Fourrier, N., Hannibal, L., Genty, B., Bouyer, P., Dreyfus, B., and Vermeglio, A. (2002) Bacteriophytochrome controls photosystem synthesis in anoxygenic bacteria. *Nature*, **417**, 202–205.
58 Giraud, E., Zappa, S., Jaubert, M., Hannibal, L., Fardoux, J., Adriano, J.M., Bouyer, P., Genty, B., Pignol, D., and Vermeglio, A. (2004) Bacteriophytochrome and regulation of the synthesis of the photosynthetic apparatus in *Rhodopseudomonas palustris*: pitfalls of using laboratory strains. *Photochem. Photobiol. Sci.*, **3**, 587–591.
59 Tarutina, M., Ryjenkov, D.A., and Gomelsky, M. (2006) An unorthodox bacteriophytochrome from *Rhodobacter sphaeroides* involved in turnover of the second messenger c-di-GMP. *J. Biol. Chem.*, **281**, 34751–34758.
60 Davis, S.J., Vener, A.V., and Vierstra, R.D. (1999) Bacteriophytochromes: phytochrome-like photoreceptors from nonphotosynthetic eubacteria. *Science*, **286**, 2517–2520.
61 Bhoo, S.H., Davis, S.J., Walker, J., Karniol, B., and Vierstra, R.D. (2001) Bacteriophytochromes are photochromic histidine kinases using a biliverdin chromophore. *Nature*, **414**, 776–779.
62 Butler, W.L., Norris, K.H., Siegelman, H.W., and Hendricks, S.B. (1959) Detection, assay, and preliminary purification of the pigment controlling

photoresponsive development of plants. *Proc. Natl. Acad. Sci. USA*, **45**, 1703–1708.

63 Montgomery, B.L. and Lagarias, J.C. (2002) Phytochrome ancestry: sensors of bilins and light. *Trends Plant. Sci.*, **7**, 357–366.

64 Karniol, B., Wagner, J.R., Walker, J.M., and Vierstra, R.D. (2005) Phylogenetic analysis of the phytochrome superfamily reveals distinct microbial subfamilies of photoreceptors. *Biochem. J.*, **392**, 103–116.

65 Lagarias, J.C., Klotz, A.V., Dallas, J.L., Glazer, A.N., Bishop, J.E., O'Connell, J.F., and Rapoport, H. (1988) Exclusive A-ring linkage for singly attached phycocyanobilins and phycoerythrobilins in phycobiliproteins. Absence of singly D-ring-linked bilins. *J. Biol. Chem.*, **263**, 12977–12985.

66 Tasler, R., Moises, T., and Frankenberg-Dinkel, N. (2005) Biochemical and spectroscopic characterization of the bacterial phytochrome of *Pseudomonas aeruginosa*. *FEBS J.*, **272**, 1927–1936.

67 Lamparter, T., Michael, N., Mittmann, F., and Esteban, B. (2002) Phytochrome from *Agrobacterium tumefaciens* has unusual spectral properties and reveals an N-terminal chromophore attachment site. *Proc. Natl. Acad. Sci. USA*, **99**, 11628–11633.

68 Giraud, E., Zappa, S., Vuillet, L., Adriano, J.M., Hannibal, L., Fardoux, J., Berthomieu, C., Bouyer, P., Pignol, D., and Vermeglio, A. (2005) A new type of bacteriophytochrome acts in tandem with a classical bacteriophytochrome to control the antennae synthesis in *Rhodopseudomonas palustris*. *J. Biol. Chem.*, **280**, 32389–32397.

69 Aravind, L. and Ponting, C.P. (1997) The GAF domain: an evolutionary link between diverse phototransducing proteins. *Trends Biochem. Sci.*, **22**, 458–459.

70 Wagner, J.R., Brunzelle, J.S., Forest, K.T., and Vierstra, R.D. (2005) A light-sensing knot revealed by the structure of the chromophore-binding domain of phytochrome. *Nature*, **438**, 325–331.

71 Karniol, B. and Vierstra, R.D. (2003) The pair of bacteriophytochromes from *Agrobacterium tumefaciens* are histidine kinases with opposing photobiological properties. *Proc. Natl. Acad. Sci. USA*, **100**, 2807–2812.

72 Vuillet, L., Kojadinovic, M., Zappa, S., Jaubert, M., Adriano, J.M., Fardoux, J., Hannibal, L., Pignol, D., Vermeglio, A., and Giraud, E. (2007) Evolution of a bacteriophytochrome from light to redox sensor. *EMBO J.*, **26**, 3322–3331.

73 Fiedler, B., Borner, T., and Wilde, A. (2005) Phototaxis in the cyanobacterium *Synechocystis* sp. PCC 6803: role of different photoreceptors. *Photochem. Photobiol.*, **81**, 1481–1488.

74 Park, C.M., Kim, J.I., Yang, S.S., Kang, J.G., Kang, J.H., Shim, J.Y., Chung, Y.H., Park, Y.M., and Song, P.S. (2000) A second photochromic bacteriophytochrome from *Synechocystis* sp. PCC 6803: spectral analysis and down-regulation by light. *Biochemistry*, **39**, 10840–10847.

75 Hubschmann, T., Borner, T., Hartmann, E., and Lamparter, T. (2001) Characterization of the Cph1 holo-phytochrome from *Synechocystis* sp. PCC 6803. *Eur. J. Biochem.*, **268**, 2055–2063.

76 Wu, S.H. and Lagarias, J.C. (2000) Defining the bilin lyase domain: lessons from the extended phytochrome superfamily. *Biochemistry*, **39**, 13487–13495.

77 Wilde, A., Fiedler, B., and Borner, T. (2002) The cyanobacterial phytochrome Cph2 inhibits phototaxis towards blue light. *Mol. Microbiol.*, **44**, 981–988.

78 Sobczyk, A., Schyns, G., Tandeau de Marsac, N., and Houmard, J. (1993) Transduction of the light signal during complementary chromatic adaptation in the cyanobacterium *Calothrix* sp. PCC 7601: DNA-binding proteins and modulation by phosphorylation. *EMBO J.*, **12**, 997–1004.

79 Terauchi, K., Montgomery, B.L., Grossman, A.R., Lagarias, J.C., and Kehoe, D.M. (2004) RcaE is a complementary chromatic adaptation photoreceptor required for green and red light responsiveness. *Mol. Microbiol.*, **51**, 567–577.

80 Kehoe, D.M. and Grossman, A.R. (1996) Similarity of a chromatic adaptation sensor to phytochrome and ethylene receptors. *Science*, **273**, 1409–1412.

81 Kehoe, D.M. and Gutu, A. (2006) Responding to color: the regulation of

complementary chromatic adaptation. *Annu. Rev. Plant Biol.*, **57**, 127–150.

82 Noubir, S., Luque, I., Ochoa de Alda, J.A., Perewoska, I., Tandeau de Marsac, N., Cobley, J.G., and Houmard, J. (2002) Co-ordinated expression of phycobiliprotein operons in the chromatically adapting cyanobacterium *Calothrix* PCC 7601: a role for RcaD and RcaG. *Mol. Microbiol.*, **43**, 749–762.

83 Schmidt-Goff, C.M. and Federspiel, N.A. (1993) *In vivo* and *in vitro* footprinting of a light-regulated promoter in the cyanobacterium *Fremyella diplosiphon*. *J. Bacteriol.*, **175**, 1806–1813.

84 Ikeuchi, M. and Ishizuka, T. (2008) Cyanobacteriochromes: a new superfamily of tetrapyrrole-binding photoreceptors in cyanobacteria. *Photochem. Photobiol. Sci.*, **7**, 1159–1167.

85 Hirose, Y., Shimada, T., Narikawa, R., Katayama, M., and Ikeuchi, M. (2008) Cyanobacteriochrome CcaS is the green light receptor that induces the expression of phycobilisome linker protein. *Proc. Natl. Acad. Sci. USA*, **105**, 9528–9533.

86 Yoshihara, S., Katayama, M., Geng, X., and Ikeuchi, M. (2004) Cyanobacterial phytochrome-like PixJ1 holoprotein shows novel reversible photoconversion between blue- and green-absorbing forms. *Plant Cell Physiol.*, **45**, 1729–1737.

87 Yoshihara, S., Shimada, T., Matsuoka, D., Zikihara, K., Kohchi, T., and Tokutomi, S. (2006) Reconstitution of blue–green reversible photoconversion of a cyanobacterial photoreceptor, PixJ1, in phycocyanobilin-producing *Escherichia coli*. *Biochemistry*, **45**, 3775–3784.

88 Ishizuka, T., Shimada, T., Okajima, K., Yoshihara, S., Ochiai, Y., Katayama, M., and Ikeuchi, M. (2006) Characterization of cyanobacteriochrome TePixJ from a thermophilic cyanobacterium *Thermosynechococcus elongatus* strain BP-1. *Plant Cell Physiol.*, **47**, 1251–1261.

89 Vierstra, R.D. and Davis, S.J. (2000) Bacteriophytochromes: new tools for understanding phytochrome signal transduction. *Semin. Cell Dev. Biol.*, **11**, 511–521.

90 Lamparter, T., Carrascal, M., Michael, N., Martinez, E., Rottwinkel, G., and Abian, J. (2004) The biliverdin chromophore binds covalently to a conserved cysteine residue in the N-terminus of *Agrobacterium* phytochrome Agp1. *Biochemistry*, **43**, 3659–3669.

91 Lamparter, T., Michael, N., Caspani, O., Miyata, T., Shirai, K., and Inomata, K. (2003) Biliverdin binds covalently to agrobacterium phytochrome Agp1 via its ring A vinyl side chain. *J. Biol. Chem.*, **278**, 33786–33792.

92 Barkovits, K., Harms, A., Benkartek, C., Smart, J.L., and Frankenberg-Dinkel, N. (2008) Expression of the phytochrome operon in *Pseudomonas aeruginosa* is dependent on the alternative sigma factor RpoS. *FEMS Microbiol. Lett.*, **280**, 160–168.

93 Yoshihara, S., Suzuki, F., Fujita, H., Geng, X.X., and Ikeuchi, M. (2000) Novel putative photoreceptor and regulatory genes required for the positive phototactic movement of the unicellular motile cyanobacterium *Synechocystis* sp. PCC 6803. *Plant Cell Physiol.*, **41**, 1299–1304.

11
Transmembrane Signaling
Melinda D. Baker and Matthew B. Neiditch

11.1
Introduction

Two-component histidine–aspartate signaling systems are the principal mechanism of signal transduction in bacteria. These signaling systems regulate diverse cellular activities including nutrient uptake, quorum sensing (QS), sporulation, biofilm formation, motility, genetic competence, and virulence. The prototypical two-component signaling system consists of a transmembrane sensor histidine kinase and a cytoplasmic response regulator. Histidine kinases use ATP to autophosphorylate a conserved histidine residue, and this autophosphorylation is activated or repressed by conformational changes induced by environmental signals. The high-energy phosphoryl group is then transferred from the sensor kinase to a conserved aspartate residue in a protein called a response regulator. Response regulator activity that includes, for example, transcriptional regulation, is modulated by phosphorylation and returns to its basal state upon dephosphorylation. Response regulator dephosphorylation is catalyzed by phosphatase activity intrinsic to the response regulator protein, the extrinsic activity of aspartyl phosphate phosphatases, or the phosphatase activity of histidine kinases [1–5].

In contrast to the histidine kinases described above, the histidine kinases that regulate bacterial chemotaxis are soluble cytosolic proteins. Environmental signals indirectly regulate the activity of these kinases through transmembrane proteins called chemoreceptors that have no intrinsic catalytic activity. Like the membrane-bound histidine kinases, the histidine kinases that regulate chemotaxis autophosphorylate a conserved histidine and transfer phosphoryl groups to an aspartate residue in a response regulator protein. The chemotaxis response regulators do not, however, typically regulate gene transcription. Rather, the phosphorylated chemotaxis response regulator proteins interact with flagellar motor proteins, causing a change in bacterial swimming behavior.

Here, we discuss models of transmembrane signaling that follow from decades of biochemical, genetic, and structural analysis of histidine sensor kinases and chemoreceptors. Despite tremendous progress, our understanding of bacterial transmembrane signaling is incomplete due to the unique challenges of working with transmembrane receptors [6]. As a result, models of histidine kinase or chemoreceptor transmembrane signaling built from existing data remain somewhat speculative.

11.2
Transmembrane Receptor Domain Architecture

11.2.1
Transmembrane Histidine Kinase Domain Architecture

Transmembrane histidine sensor kinases form homodimeric receptor complexes (Figure 11.1). Each protomer contains an N-terminal extracellular sensory domain and a C-terminal cytoplasmic transmitter domain. The largest group of transmembrane histidine kinases encodes an extracellular sensory domain, flanked most commonly by two transmembrane helices (Figure 11.2A) [7]. Transmembrane helix 1 (TM1) is an N-terminal anchor, while transmembrane helix 2 (TM2) connects the extracellular sensory domain to the cytoplasmic transmitter domain. Histidine kinase extracellular sensory domains are predicted to be structurally diverse and can be composed of the following folds: cyclase/histidine kinase-associated sensing extracellular (CHASE), Ca^{2+} channels and chemotaxis receptor (CACHE), four-helix bundle (4HB), nitrate and nitrite sensing (NIT), periplasmic solute-binding proteins (PBP), and Per/Arnt/Sim (PAS) [8–13]. To date, only the structures of extracellular domains comprising PAS folds have been determined experimentally. Although discussed only briefly in this chapter, additional transmembrane histidine kinases lack a *bona fide* extracellular domain and detect signals using membrane or cytoplasmic domains (reviewed in [7]).

Common to all histidine kinases is a cytoplasmic dimerization domain [14]. Each monomer within a receptor dimer contributes two helices from its dimerization domain to the formation of a 4HB [15, 16]. In the vast majority of histidine kinases, the dimerization domain contains the autophosphorylated histidine residue; therefore, this domain is commonly referred to as the dimerization and histidine phosphotransfer (DHp) domain. An apparently flexible linker at the DHp domain C-terminus connects it to a histidine kinase-type ATPase catalytic (HATPase_c) domain, also known simply as the catalytic activity (CA) domain. The accepted dogma is that the CA domain of one monomer binds ATP in preparation for *trans*-phosphorylation of the conserved histidine located within the other monomer; however, the possibility that this phosphorylation can occur in *cis* has not been altogether ruled out [17, 18]. Together, the cytosolic DHp and CA domains form the histidine kinase core domain.

In stark contrast to the periplasmic and cytoplasmic domains, very little is known about the structure of histidine kinase transmembrane domains. The common belief is that histidine kinase transmembrane domains form 4HBs composed of transmembrane helices, TM1 and TM2. Support for this hypothesis stems largely from disulfide cross-linking studies of histidine kinases and chemoreceptors, as well as the

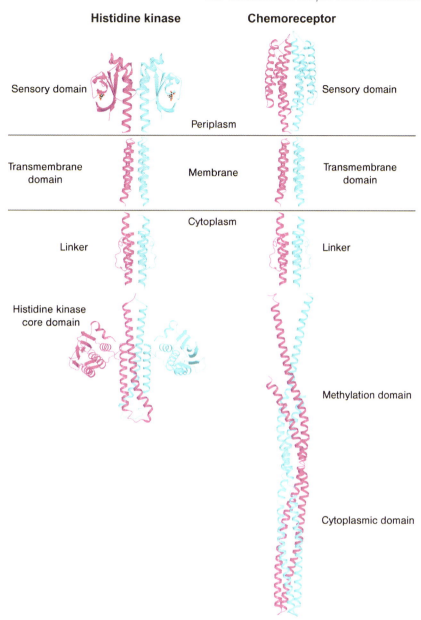

Figure 11.1 Histidine kinase and chemoreceptor domain architecture. Structures of the CitA$_p$-citrate PAS sensory domain (2J80), HtrII transmembrane domain (2F93), HAMP linker domain (2ASW), and HK853 histidine kinase core domain (2C2A) depict the domain architecture of the prototypical transmembrane histidine kinase. Structures of the Tar 4HB sensory domain (1LIH), HtrII transmembrane domain (2F93), HAMP linker domain (2ASW), and Tsr methylation and cytoplasmic domain (1QU7) depict the domain architecture of the prototypical transmembrane chemoreceptor. Figure produced using PyMOL [67].

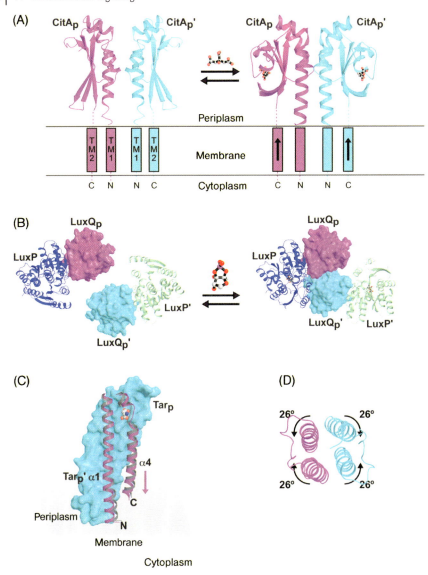

Figure 11.2 Histidine kinase and chemoreceptor domain motions. (A) Citrate binding induces contraction of the $CitA_p$ β-sheet. This conformational change is proposed to cause the piston-like movement of TM2 toward the periplasm. The ligand-free $CitA_p$ structures are depicted in a symmetric orientation (left panel), but the relative disposition of ligand-free $CitA_p$ domains in a signaling dimer is unknown. (B) In the absence of AI-2, no interaction between the periplasmic domains of the LuxPQ receptors (viewed from above looking down toward the membrane) was detected. AI-2 binding causes the LuxPQ periplasmic domains to interact asymmetrically. This ligand-induced asymmetry downregulates autophosphorylation of the LuxQ cytoplasmic domains (not pictured). (C) Each Tar periplasmic domain (Tar_p and Tar_p') in the chemoreceptor dimer is composed of a 4HB (α1–α4). For clarity, α2 and α3 are omitted from subunit Tar_p' (depicted in cartoon format). Tar_p subunits of the ligand-bound (2LIG) and

recently determined X-ray crystal structure of the transmembrane domain of the sensory rhodopsin transducer protein, HtrII [19–24].

11.2.2
Chemoreceptor Domain Architecture

The prototypical chemoreceptor (Figure 11.1) contains an N-terminal extracellular sensory domain flanked by a pair of transmembrane helices, TM1 and TM2. Chemoreceptors form homodimeric complexes in which TM1 and TM2 from each receptor interact to form a four-helix transmembrane bundle [19]. Sequence analysis shows that chemoreceptor sensory domains are structurally diverse and can consist of 4HB, CHASE3, CACHE, and NIT folds [8–11]. Extensive biochemical, genetic, biophysical, and computational modeling studies have primarily focused on the chemoreceptors containing 4HB sensory domains.

Chemoreceptor periplasmic sensory domains are connected by TM2 to a cytoplasmic linker domain (described in Section 11.3.4). The linker domain is in turn connected to a C-terminal domain, routinely referred to as the cytoplasmic domain. The cytoplasmic domain is a long, α-helix that folds back on itself to form a coiled-coil [25]. The coiled-coil cytoplasmic domains of each chemoreceptor dimer interact to form a supercoiled bundle. The cytoplasmic domain is further divided into methylation and signaling regions. The methylation region contains conserved glutamate residues that are methylated and demethylated by chemotaxis accessory proteins [26, 27]. Methylation decreases chemoreceptor sensitivity to attractant, while demethylation has the opposite effect [28, 29]. Methylation/demethylation enables chemoreceptor adaptation to changing concentrations of attractants and repellents. The signaling domain encompasses the region near the distal end of the cytoplasmic domain and is the site of histidine kinase interaction [30–32].

Chemoreceptors oligomerize into trimers of dimers and these trimers cluster into higher-order signaling arrays [33, 34]. The roles that chemoreceptor arrays play in signaling cooperativity and amplification are detailed in Chapter 9. Here, we limit our discussion of chemoreceptor transmembrane signaling to the conformational changes that occur within receptor dimers.

11.3
Structural Analysis of Transmembrane Signaling

In the following sections we primarily describe structural studies of truncated histidine kinases and chemoreceptors. The studies of periplasmic domains that are

---

ligand-free (1LIH) structures were aligned. For clarity, only the surface of ligand-free Tar_p is depicted. Alignment of the Tar_p subunits shows that aspartate (depicted as spheres) binding causes $Tar_p{'}$ helix α4 to rotate slightly and move in a piston-like manner toward the cytoplasm (in the direction indicated by the magenta arrow). Helix α4 is connected to TM2 (not pictured). (D) The HAMP domain (2ASW) can interconvert between a canonical knobs-into-holes configuration and an atypical knobs-to-knobs configuration by the rotation of each helix 26° in the indicated direction. Figure produced using PyMOL [67].

the subjects of discussion below have had their three-dimensional structures determined in both their ligand-bound and ligand-free conformations. Comparison of the ligand-bound and ligand-free periplasmic domain structures reveals the conformational changes that occur in the extracellular sensory domains upon ligand binding. These studies, in combination with the results of existing structural and functional studies of isolated cytoplasmic and transmembrane domains, as well as the results of *in vivo* functional analysis of full-length receptors, has enabled bacterial transmembrane signaling to be modeled in detail. However, a rigorous explanation of precisely how conformational changes in the extracellular domains of histidine kinases and chemoreceptors are transduced across the membrane and regulate cytoplasmic signaling activity in intact receptors will require additional studies.

11.3.1
CitA Transmembrane Signaling

The *Klebsiella pneumoniae* histidine kinase, CitA, modulates the activity of the response regulator CitB [35]. Under anaerobic conditions, CitA phosphorylates CitB, which in turn induces the transcription of genes involved in the uptake and fermentation of citrate. X-ray crystal structures of the periplasmic domain of CitA (CitA$_p$) revealed that it comprises a PAS fold (Figure 11.2A) [36, 37]. The PAS fold is a protein–ligand and protein–protein interaction domain and is widespread in all kingdoms of life. Prior to the first CitA$_p$ structure, PAS domains had not been shown to exist in bacteria extracytoplasmically.

Reinelt *et al.* observed MoO_3, derived from Na_2MoO_4 in the crystallization buffer, bound to both CitA$_p$ and citrate in the citrate-bound CitA$_p$ (CitA$_p$-citrate) X-ray crystal structure. An additional isopolymolybdate cluster (Mo_7O_{24}) was identified near the CitA$_p$ N- and C-termini. In part, because no role for molybdenum in citrate metabolism or transport had been described, Sevanna *et al.* hypothesized that the molybdenum bound to CitA$_p$ represented a nonphysiological interaction; therefore, they crystallized CitA$_p$ in complex with citrate in the absence of molybdenum [37].

The N- and C-termini as well as loops near the citrate-binding site in the molybdenum-free CitA$_p$-citrate structure differed somewhat from those in the molybdenum-bound CitA$_p$-citrate structure, but overall the CitA$_p$-citrate monomer structures were found to be quite similar (reported root-mean-squared deviation for all residues including side chains $= 1.78$ Å) [37]. The molybdenum-bound and molybdenum-free structures of CitA$_p$-citrate did, however, differ substantially in their proposed physiological CitA$_p$ dimer interfaces. For example, the dimer interface of the molybdenum-bound structure buries 1415 Å^2 surface area while the molybdenum-free structure is more tightly packed and buries 2123 Å^2 surface area.

Sevvana *et al.* also determined the crystal structure of citrate-free CitA$_p$ and compared it to the molybdate-free CitA$_p$-citrate crystal structure [37]. This comparison showed that citrate binding causes the ends of the CitA sensory domain β-sheet to move toward the citrate binding site (Figure 11.2A). The authors propose that the citrate-induced contraction of the CitA$_p$ β-sheet is likely to cause TM2 to move in a piston-like manner; however, the intermolecular conformational changes that occur within the CitA$_p$ dimer could not be discerned because a physiologically relevant structure of the ligand-free CitA$_p$ dimer has not been determined.

The histidine kinase VirA is also proposed to transmit a piston-like transmembrane signal, which is then converted into a rotational movement by its cytoplasmic linker domain [38]. It remains to be determined whether the CitA linker domain likewise converts transmembrane piston-like movements into rotational changes, how linker domain-mediated rotational movements regulate histidine kinase activity, and whether such rotational regulatory mechanisms are widespread among histidine kinases.

11.3.2
LuxPQ Transmembrane Signaling

Secreted bacterial cell–cell signaling molecules called autoinducers (AIs) accumulate in proportion to cell density (see also Chapter 1). Coordinated gene expression of the population is triggered when a threshold concentration of AIs is reached. This bacterial cell density dependent signaling process, called QS, regulates diverse phenotypes including virulence factor expression, antibiotic production, biofilm development, and genetic competence [39–42].

The *Vibrio harveyi* receptor for the interspecies QS signal AI-2 is an oligomeric complex consisting of the polypeptides LuxP and LuxQ (Figure 11.2B) (see also Chapters 6 and 8) [43]. LuxP is a periplasmic-binding protein that clamps AI-2 between two large domains connected by a flexible hinge [44]. LuxQ is an inner membrane histidine kinase that associates with LuxP both in the absence and presence of AI-2 [45]. In the absence of AI-2, the LuxQ cytoplasmic histidine kinase core domains autophosphorylate [46]. Phosphoryl groups are transferred from LuxQ along a phosphorelay pathway to the response regulator LuxO. At high cell density, AI-2 binding to LuxP turns off LuxQ kinase activity resulting in LuxO dephosphorylation and downregulation of its transcriptional activity. Downregulation of LuxO transcriptional activity results in the derepression of another transcription factor, LuxR, which induces the expression of *V. harveyi* high cell density phenotypes.

The X-ray crystal structure of ligand-free LuxP in complex with the periplasmic domain of LuxQ (LuxQ$_p$) revealed that LuxQ$_p$ is composed of tandem PAS folds [45]. Rather than binding ligand directly, like many other PAS folds, the LuxQ PAS folds interact with and monitor the ligand-bound state of LuxP. The ligand-free LuxPQ$_p$ crystal structure and biochemical experiments showed that the periplasmic domain of one LuxQ$_p$ monomer binds to one molecule of LuxP in the absence of AI-2 (Figure 11.2B). No interaction between the periplasmic domains of one LuxPQ monomer and the other was detected in the absence of AI-2.

Genetic analysis showed that the LuxQ cytoplasmic domains phosphorylate at low cell density in the absence of AI-2 and that the LuxQ periplasmic domains are entirely dispensable for this LuxQ kinase activity [47]. These data suggest that the LuxPQ periplasmic domains are not contributing to the stability of the kinase "on" state and that their role is instead to turn off LuxQ cytoplasmic kinase activity in response to

AI-2. It therefore seemed likely that the LuxPQ periplasmic domains undergo an AI-2-induced conformational change that turns off the cytoplasmic kinase activity of LuxQ. Consistent with this model, the crystal structure of LuxPQ$_p$ bound to AI-2 showed that the LuxPQ periplasmic domains interact upon AI-2 binding, forming strikingly asymmetric (LuxPQ$_p$)$_2$ dimers (Figure 11.2B) [47]. Comparison of the ligand-free LuxPQ$_p$ and the AI-2-bound (LuxPQ$_p$)$_2$ X-ray crystal structures show that LuxP traps AI-2 by closing like a venus flytrap. Only one molecule of LuxP, however, makes new contacts upon AI-2 binding. This LuxP contacts the periplasmic domains of both LuxQ molecules which, as a result, interact with one another asymmetrically (Figure 11.2B).

Comparison of the AI-2-bound (LuxPQ$_p$)$_2$ X-ray crystal structure to the structures of ligand-free LuxPQ$_p$ and LuxQ$_p$ alone show that the individual LuxQ periplasmic domains remain unchanged. In fact, AI-2 binding to LuxP causes no significant conformational rearrangements in the periplasmic domains of the LuxQ monomers (root mean square deviation for Cα carbons $= 0.7$–1.0 Å for all pair-wise comparisons). Instead, rather than introducing significant intramolecular changes in LuxQ$_p$, AI-2 binding to LuxP alters the relative disposition of the LuxQ monomers in a signaling dimer (Figure 11.2B). The LuxPQ$_p$ receptors are related to one another in the AI-2 bound structure by a 140° rotation, and complementary biochemical and genetic data suggest that it is this ligand-induced rotational asymmetry in the LuxPQ dimer that turns off LuxQ cytoplasmic phosphorylation.

The structure of the LuxQ cytoplasmic domain has not been determined; however, the X-ray crystal structure of the *Thermotoga maritima* histidine kinase HK853 cytoplasmic core domain showed that it forms a symmetric dimer (Figure 11.1) [15]. If the LuxQ core domains autophosphorylate within a similar symmetric receptor dimer, then the large 40° asymmetry in (LuxPQ$_p$)$_2$ induced by AI-2 binding could turn off LuxQ kinase activity. How the LuxPQ rotational asymmetry is transmitted across the membrane is unknown, but presumably the LuxQ transmembrane helices form a 4HB that undergoes an AI-2-induced rotational reorganization. X-ray crystal structures of ligand-bound and ligand-free full-length LuxPQ will be needed to reveal exactly how LuxPQ periplasmic conformational changes are transduced across the membrane and to show whether the 40° periplasmic asymmetry induced by AI-2 binding translates into a similarly large rotational symmetry in the cytoplasmic linker and core domains.

11.3.3
Chemotaxis Receptor Transmembrane Signaling

The *Salmonella* and *Escherichia coli* aspartate receptor, Tar, has been studied extensively and serves as the chemoreceptor prototype (Figures 11.1 and 11.2C) [33] (see also Chapter 9). The Tar 4HB periplasmic sensory domains dimerize to form two rotationally symmetric aspartate-binding sites [48–50]. Each receptor contributes residues to the aspartate-binding pockets; however, aspartate binding to one site induces asymmetry in the receptor dimer that precludes the simultaneous binding of ligand to both sites [51]. Comparison of the ligand-free and ligand-bound structures reveals that ligand binding causes periplasmic helix α4 in one receptor to rotate 5° and to move 1.6 Å in a piston-like fashion toward the membrane (Figure 11.2C) [48, 52].

The chemoreceptor extracellular domain piston-like motion appears to extend into the transmembrane domain. Extensive disulfide cross-linking and electron paramagnetic resonance analysis suggests that the ligand-induced conformational change causes a 1–2 Å piston-like movement in TM2 [52–54]. Consistent with the disulfide cross-linking studies, mutations that simulated TM2 piston-like sliding toward the periplasm or cytoplasm *in vivo* resulted in receptors that were locked in the kinase-on and kinase-off signaling states, respectively [55–57]. *In lieu* of ligand-bound and ligand-free crystal structures of an intact chemoreceptor, the disulfide cross-linking, electron paramagnetic resonance, and genetic analysis remain the most direct evidence of chemoreceptor transmembrane conformational changes.

Finally, the structure of the transmembrane domains of the *Natronomonas pharaonis* bacterial phototaxis receptor complex sensory rhodopsin II (SRII)–HtrII has been determined in its active and inactive signaling states (Figure 11.1) [21, 22]. HtrII, the transducer protein for sensory rhodopsin, is highly homologous to the chemotaxis receptors and regulates a homolog of the chemotaxis signaling pathway histidine kinase [58]. The SRII–HtrII structure revealed that the membrane domain is, as predicted, a 4HB [22]. The HtrII transducer transmembrane domain undergoes a signal-induced conformational change in TM2 consisting principally of a 15° rotation with a concomitant displacement of 0.9 Å near the cytoplasmic surface of the membrane [21]. It will be very interesting to see whether crystal structures of intact prototypical chemoreceptors reveal transmembrane conformational changes in line with those observed for the SRII–HtrII complex or more consistent with the primarily piston-like movements predicted from structural and functional analysis of the prototypical chemoreceptors.

Precisely how the transmembrane conformational changes discussed above regulate chemoreceptor cytoplasmic domain activity is not known, but studies of HAMP linker domains (see Section 11.3.4) suggest that it converts translational movements into rotational changes [59]. How HAMP domain rotation controls the proposed twisting, bending, and sliding regulatory conformational changes in the chemoreceptor cytoplasmic signaling domain, and how these conformational changes regulate the activity of the associated histidine kinase, remains to be determined [31, 60–62].

11.3.4
HAMP Linker Domain Structure

Jin and Inouye first described sequence conservation in the linker region connecting TM2 to the cytoplasmic core domains of chemoreceptors and histidine kinases [63]. The linker domain that they identified, and which they proposed to comprise a "structural joint" capable of transducing transmembrane conformational changes to changes in signaling output, was later named the HAMP domain due to its wide distribution among histidine kinases, adenylyl cyclases, methyl-accepting chemotaxis proteins, and phosphatases [64]. The NMR structure of the *Archaeoglobus fulgidus* protein Af1503 showed that the helix–loop–helix motifs of individual HAMP domains dimerize to form parallel, four-helical coiled-coils with atypical knobs-to-knobs rather than knobs-into-holes interhelical packing (Figure 11.2D) [59]. Analysis of the Af1503 HAMP domain sequence and structure led to the prediction that it could adopt the canonical knobs-into-holes configuration by rotation of each helix by 26° (Figure 11.2D). Indeed, nuclear magnetic resonance studies showed that the Af1503 HAMP domain, containing an engineered mutation predicted to destabilize knobs-to-knobs packing and favor knobs-into-holes packing, displayed conformational changes consistent with helical rotation.

HAMP linker domains may convert signal-induced transmembrane motions of different types into rotational movements that regulate the activity of the receptor signaling output domains. It is important to note, however, that according to the SMART database, only 28% of histidine kinases contain HAMP domains [65]. It will be interesting to determine whether other linker domains, such as the recently identified coiled-coil S-helix domain, also convert upstream transmembrane motions into rotational changes [66].

11.4
Conclusions

Two-component signal transduction has been extensively studied, yet what is arguably the most fundamental question still remains unanswered: how is the cytoplasmic signaling activity of transmembrane histidine kinases and chemoreceptors regulated by extracellular sensory domain signal perception? It seems likely that crystal structures of full-length histidine kinases and chemoreceptors in their active and inactive conformations will be required to answer this question. However, we suspect that while there may be few inter-subunit configurations that support an active receptor signaling state, a much larger number of conformations may have evolved to turn receptor activity off. Therefore, structural analysis of a particular full-length receptor may only reveal the transmembrane signaling mechanism of that receptor, together perhaps with those of highly homologous receptors. Despite this possibility, and the challenges of working with full-length membrane proteins, we believe that the insight into transmembrane receptor signaling that would be gained from X-ray crystal structures of full-length histidine kinases and chemoreceptors justifies the pursuit.

Acknowledgments

We are grateful to Frederick Hughson for critical review of the manuscript and to Carol Neiditch for editorial assistance. Research in the laboratory of M.B.N. is supported by National Institutes of Health grant AI-081736. M.D.B. is supported by a postdoctoral fellowship provided by the United Negro College Fund/Merck Science Initiative.

References

1 Lukat, G.S., Lee, B.H., Mottonen, J.M., Stock, A.M. and Stock, J.B. (1991) Roles of the highly conserved aspartate and lysine residues in the response regulator of bacterial chemotaxis. *J. Biol. Chem.*, **266**, 8348–8354.

2 Ohlsen, K.L., Grimsley, J.K. and Hoch, J.A. (1994) Deactivation of the sporulation transcription factor Spo0A by the Spo0E protein phosphatase. *Proc. Natl. Acad. Sci. USA*, **91**, 1756–1760.

3 Perego, M., Hanstein, C., Welsh, K.M., Djavakhishvili, T., Glaser, P. and Hoch, J.A. (1994) Multiple protein-aspartate phosphatases provide a mechanism for the integration of diverse signals in the control of development in *B. subtilis*. *Cell*, **79**, 1047–1055.

4 Zhao, R., Collins, E.J., Bourret, R.B. and Silversmith, R.E. (2002) Structure and catalytic mechanism of the *E. coli* chemotaxis phosphatase CheZ. *Nat. Struct. Biol.*, **9**, 570–575.

5 Yoshida, T., Cai, S. and Inouye, M. (2002) Interaction of EnvZ, a sensory histidine kinase, with phosphorylated OmpR, the cognate response regulator. *Mol. Microbiol.*, **46**, 1283–1294.

6 Matthews, E.E., Zoonens, M., and Engelman, D.M. (2006) Dynamic helix interactions in transmembrane signaling. *Cell*, **127**, 447–450.

7 Mascher, T., Helmann, J.D., and Unden, G. (2006) Stimulus perception in bacterial signal-transducing histidine kinases. *Microbiol. Mol. Biol. Rev.*, **70**, 910–938.

8 Anantharaman, V. and Aravind, L. (2001) The CHASE domain: a predicted ligand-binding module in plant cytokinin receptors and other eukaryotic and bacterial receptors. *Trends Biochem. Sci.*, **26**, 579–582.

9 Anantharaman, V. and Aravind, L. (2000) Cache – a signaling domain common to animal Ca^{2+}-channel subunits and a class of prokaryotic chemotaxis receptors. *Trends Biochem. Sci.* **25**, 535–537.

10 Ulrich, L.E. and Zhulin, I.B. (2005) Four-helix bundle: a ubiquitous sensory module in prokaryotic signal transduction. *Bioinformatics*, **21** (Suppl 3), iii45–iii48.

11 Shu, C.J., Ulrich, L.E. and Zhulin, I.B. (2003) The NIT domain: a predicted nitrate-responsive module in bacterial sensory receptors. *Trends Biochem. Sci.*, **28**, 121–124.

12 Tam, R. and Saier, M.H. Jr. (1993) Structural, functional, and evolutionary relationships among extracellular solute-binding receptors of bacteria. *Microbiol. Rev.*, **57**, 320–346.

13 Zhulin, I.B., Taylor, B.L. and Dixon, R. (1997) PAS domain S-boxes in Archaea, Bacteria and sensors for oxygen and redox. *Trends Biochem. Sci.*, **22**, 331–333.

14 Grebe, T.W. and Stock, J.B. (1999) The histidine protein kinase superfamily. *Adv. Microb. Physiol.*, **41**, 139–227.

15 Marina, A., Waldburger, C.D. and Hendrickson, W.A. (2005) Structure of the entire cytoplasmic portion of a sensor histidine-kinase protein. *EMBO J.*, **24**, 4247–4259.

16 Bilwes, A.M., Alex, L.A., Crane, B.R. and Simon, M.I. (1999) Structure of CheA, a signal-transducing histidine kinase. *Cell*, **96**, 131–141.

17 Heermann, R., Altendorf, K. and Jung, K. (1998) The turgor sensor KdpD of *Escherichia coli* is a homodimer. *Biochim. Biophys. Acta*, **1415**, 114–124.

18 Filippou, P.S., Kasemian, L.D., Panagiotidis, C.A. and Kyriakidis, D.A. (2008) Functional characterization of the histidine kinase of the *E. coli* two-component signal transduction system AtoS–AtoC. *Biochim. Biophys. Acta*, **1780**, 1023–1031.

19 Lynch, B.A. and Koshland, D.E. Jr. (1991) Disulfide cross-linking studies of the transmembrane regions of the aspartate sensory receptor of *Escherichia coli*. *Proc. Natl. Acad. Sci. USA*, **88**, 10402–10406.

20 Pakula, A.A. and Simon, M.I. (1992) Determination of transmembrane protein structure by disulfide cross-linking: the *Escherichia coli* Tar receptor. *Proc. Natl. Acad. Sci. USA*, **89**, 4144–4148.

21 Moukhametzianov, R., Klare, J.P., Efremov, R., Baeken, C., Goppner, A., Labahn, J., Engelhard, M., Buldt, G., and Gordeliy, V.I. (2006) Development of the signal in

sensory rhodopsin and its transfer to the cognate transducer. *Nature*, **440**, 115–119.

22 Gordeliy, V.I., Labahn, J., Moukhametzianov, R., Efremov, R., Granzin, J., Schlesinger, R., Buldt, G. *et al.* (2002) Molecular basis of transmembrane signalling by sensory rhodopsin II-transducer complex. *Nature*, **419**, 484–487.

23 Forst, S., Comeau, D., Norioka, S. and Inouye, M. (1987) Localization and membrane topology of EnvZ, a protein involved in osmoregulation of OmpF and OmpC in *Escherichia coli*. *J. Biol. Chem.*, **262**, 16433–16438.

24 Collins, L.A., Egan, S.M. and Stewart, V. (1992) Mutational analysis reveals functional similarity between NARX, a nitrate sensor in *Escherichia coli* K-12, and the methyl-accepting chemotaxis proteins. *J. Bacteriol.*, **174**, 3667–3675.

25 Kim, K.K., Yokota, H., and Kim, S.H. (1999) Four-helical-bundle structure of the cytoplasmic domain of a serine chemotaxis receptor. *Nature*, **400**, 787–792.

26 Terwilliger, T.C., Bogonez, E., Wang, E.A. and Koshland, D.E. Jr. (1983) Sites of methyl esterification on the aspartate receptor involved in bacterial chemotaxis. *J. Biol. Chem.*, **258**, 9608–9611.

27 Terwilliger, T.C. and Koshland, D.E. Jr. (1984) Sites of methyl esterification and deamination on the aspartate receptor involved in chemotaxis. *J. Biol. Chem.*, **259**, 7719–7725.

28 Sourjik, V. and Berg, H.C. (2002) Receptor sensitivity in bacterial chemotaxis. *Proc. Natl. Acad. Sci. USA*, **99**, 123–127.

29 Sourjik, V. and Berg, H.C. (2004) Functional interactions between receptors in bacterial chemotaxis. *Nature*, **428**, 437–441.

30 Gegner, J.A., Graham, D.R., Roth, A.F. and Dahlquist, F.W. (1992) Assembly of an MCP receptor, CheW, and kinase CheA complex in the bacterial chemotaxis signal transduction pathway. *Cell*, **70**, 975–982.

31 Surette, M.G. and Stock, J.B. (1996) Role of alpha-helical coiled-coil interactions in receptor dimerization, signaling, and adaptation during bacterial chemotaxis. *J. Biol. Chem.*, **271**, 17966–17973.

32 Francis, N.R., Wolanin, P.M., Stock, J.B., Derosier, D.J. and Thomas, D.R. (2004) Three-dimensional structure and organization of a receptor/signaling complex. *Proc. Natl. Acad. Sci. USA*, **101**, 17480–17485.

33 Hazelbauer, G.L., Falke, J.J., and Parkinson, J.S. (2008) Bacterial chemoreceptors: high-performance signaling in networked arrays. *Trends Biochem. Sci.*, **33**, 9–19.

34 Kentner, D. and Sourjik, V. (2006) Spatial organization of the bacterial chemotaxis system. *Curr. Opin. Microbiol.*, **9**, 619–624.

35 Bott, M., Meyer, M. and Dimroth, P. (1995) Regulation of anaerobic citrate metabolism in *Klebsiella pneumoniae*. *Mol. Microbiol.*, **18**, 533–546.

36 Reinelt, S., Hofmann, E., Gerharz, T., Bott, M. and Madden, D.R. (2003) The structure of the periplasmic ligand-binding domain of the sensor kinase CitA reveals the first extracellular PAS domain. *J. Biol. Chem.*, **278**, 39189–39196.

37 Sevvana, M., Vijayan, V., Zweckstetter, M., Reinelt, S., Madden, D.R., Herbst-Irmer, R., Sheldrick, G.M. *et al.* (2008) A ligand-induced switch in the periplasmic domain of sensor histidine kinase CitA. *J. Mol. Biol.*, **377**, 512–523.

38 Gao, R. and Lynn, D.G. (2007) Integration of rotation and piston motions in coiled-coil signal transduction. *J. Bacteriol.*, **189**, 6048–6056.

39 Zhu, J., Miller, M.B., Vance, R.E., Dziejman, M., Bassler, B.L. and Mekalanos, J.J. (2002) Quorum-sensing regulators control virulence gene expression in *Vibrio cholerae*. *Proc. Natl. Acad. Sci. USA*, **99**, 3129–3134.

40 Derzelle, S., Duchaud, E., Kunst, F., Danchin, A. and Bertin, P. (2002) Identification, characterization, and regulation of a cluster of genes involved in carbapenem biosynthesis in *Photorhabdus luminescens*. *Appl. Environ. Microbiol.*, **68**, 3780–3789.

41 Davies, D.G., Parsek, M.R., Pearson, J.P., Iglewski, B.H., Costerton, J.W. and Greenberg, E.P. (1998) The involvement of cell-to-cell signals in the development of a bacterial biofilm. *Science*, **280**, 295–298.

42 Lee, M.S. and Morrison, D.A. (1999) Identification of a new regulator in *Streptococcus pneumoniae* linking quorum sensing to competence for genetic transformation. *J. Bacteriol.*, **181**, 5004–5016.

43 Bassler, B.L., Wright, M. and Silverman, M.R. (1994) Multiple signalling systems controlling expression of luminescence in *Vibrio harveyi*: sequence and function of genes encoding a second sensory pathway. *Mol. Microbiol.*, **13**, 273–286.

44 Chen, X., Schauder, S., Potier, N., Van Dorsselaer, A., Pelczer, I., Bassler, B.L., and Hughson, F.M. (2002) Structural identification of a bacterial quorum-sensing signal containing boron. *Nature*, **415**, 545–549.

45 Neiditch, M.B., Federle, M.J., Miller, S.T., Bassler, B.L. and Hughson, F.M. (2005) Regulation of LuxPQ receptor activity by the quorum-sensing signal autoinducer-2. *Mol. Cell*, **18**, 507–518.

46 Freeman, J.A. and Bassler, B.L. (1999) A genetic analysis of the function of LuxO, a two-component response regulator involved in quorum sensing in *Vibrio harveyi*. *Mol. Microbiol.*, **31**, 665–677.

47 Neiditch, M.B., Federle, M.J., Pompeani, A.J., Kelly, R.C., Swem, D.L., Jeffrey, P.D., Bassler, B.L. and Hughson, F.M. (2006) Ligand-induced asymmetry in histidine sensor kinase complex regulates quorum sensing. *Cell*, **126**, 1095–1108.

48 Milburn, M.V., Prive, G.G., Milligan, D.L., Scott, W.G., Yeh, J., Jancarik, J., Koshland, D.E. Jr. and Kim, S.H. (1991) Three-dimensional structures of the ligand-binding domain of the bacterial aspartate receptor with and without a ligand. *Science*, **254**, 1342–1347.

49 Yeh, J.I., Biemann, H.P., Pandit, J., Koshland, D.E. and Kim, S.H. (1993) The three-dimensional structure of the ligand-binding domain of a wild-type bacterial chemotaxis receptor. Structural comparison to the cross-linked mutant forms and conformational changes upon ligand binding. *J. Biol. Chem.*, **268**, 9787–9792.

50 Yeh, J.I., Biemann, H.P., Prive, G.G., Pandit, J., Koshland, D.E. Jr. and Kim, S.H. (1996) High-resolution structures of the ligand binding domain of the wild-type bacterial aspartate receptor. *J. Mol. Biol.*, **262**, 186–201.

51 Biemann, H.P. and Koshland, D.E. Jr. (1994) Aspartate receptors of *Escherichia coli* and *Salmonella typhimurium* bind ligand with negative and half-of-the-sites cooperativity. *Biochemistry*, **33**, 629–634.

52 Chervitz, S.A. and Falke, J.J. (1996) Molecular mechanism of transmembrane signaling by the aspartate receptor: a model. *Proc. Natl. Acad. Sci. USA* **93**, 2545–2550.

53 Ottemann, K.M., Xiao, W., Shin, Y.K. and Koshland, D.E. Jr. (1999) A piston model for transmembrane signaling of the aspartate receptor. *Science*, **285**, 1751–1754.

54 Chervitz, S.A. and Falke, J.J. (1995) Lock on/off disulfides identify the transmembrane signaling helix of the aspartate receptor. *J. Biol. Chem.*, **270**, 24043–24053.

55 Miller, A.S. and Falke, J.J. (2004) Side chains at the membrane–water interface modulate the signaling state of a transmembrane receptor. *Biochemistry*, **43**, 1763–1770.

56 Draheim, R.R., Bormans, A.F., Lai, R.Z. and Manson, M.D. (2005) Tryptophan residues flanking the second transmembrane helix (TM2) set the signaling state of the Tar chemoreceptor. *Biochemistry*, **44**, 1268–1277.

57 Draheim, R.R., Bormans, A.F., Lai, R.Z. and Manson, M.D. (2006) Tuning a bacterial chemoreceptor with protein–membrane interactions. *Biochemistry*, **45**, 14655–14664.

58 Yao, V.J. and Spudich, J.L. (1992) Primary structure of an archaebacterial transducer, a methyl-accepting protein associated with sensory rhodopsin I. *Proc. Natl. Acad. Sci. USA*, **89**, 11915–11919.

59 Hulko, M., Berndt, F., Gruber, M., Linder, J.U., Truffault, V., Schultz, A., Martin, J. et al. (2006) The HAMP domain structure implies helix rotation in transmembrane signaling. *Cell*, **126**, 929–940.

60 Starrett, D.J. and Falke, J.J. (2005) Adaptation mechanism of the aspartate receptor: electrostatics of the adaptation subdomain play a key role in modulat-

ing kinase activity. *Biochemistry*, **44**, 1550–1560.

61 Coleman, M.D., Bass, R.B., Mehan, R.S. and Falke, J.J. (2005) Conserved glycine residues in the cytoplasmic domain of the aspartate receptor play essential roles in kinase coupling and on–off switching. *Biochemistry*, **44**, 7687–7695.

62 Winston, S.E., Mehan, R. and Falke, J.J. (2005) Evidence that the adaptation region of the aspartate receptor is a dynamic four-helix bundle: cysteine and disulfide scanning studies. *Biochemistry*, **44**, 12655–12666.

63 Jin, T. and Inouye, M. (1994) Transmembrane signaling. Mutational analysis of the cytoplasmic linker region of Taz1-1, a Tar–EnvZ chimeric receptor in *Escherichia coli. J. Mol. Biol.*, **244**, 477–481.

64 Aravind, L. and Ponting, C.P. (1999) The cytoplasmic helical linker domain of receptor histidine kinase and methyl-accepting proteins is common to many prokaryotic signalling proteins. *FEMS Microbiol. Lett.*, **176**, 111–116.

65 Ponting, C.P., Schultz, J., Milpetz, F., and Bork, P. (1999) SMART: identification and annotation of domains from signalling and extracellular protein sequences. *Nucleic Acids Res.*, **27**, 229–232.

66 Anantharaman, V., Balaji, S., and Aravind, L. (2006) The signaling helix: a common functional theme in diverse signaling proteins. *Biol. Direct.*, **1**, 25.

67 DeLano, W.L. (2002) *The PyMOL Molecular Graphics System*, DeLano Scientific, San Carlos, CA.

12
Sensory Transport Proteins
Reinhard Krämer

12.1
Introduction

Translocation of solutes across membranes is the core catalytic function of membrane transport proteins. Due to the recently growing number of available high-resolution structures for carrier proteins, their mechanism of action seems to become increasingly accessible for a molecular understanding. This chapter, however, does not deal with the catalytic function of membrane transport proteins, but with additional functions beyond solute transport that are related to the perception of certain stimuli, the conversion of these stimuli into biochemical signals, and the transduction of these signals into useful cellular responses. This process has to be discriminated from the capacity of many transport proteins for metabolic regulation (i.e., the direct modulation of transport activity by binding of substrate(s) or effectors).

A number of examples of sensory transport proteins are discussed here, which are able to either transfer a particular signal across the membrane to another compound of a cellular signal transduction cascade or to transfer information related to an external stimulus, most commonly some type of stress, as a signal to the metabolic network of the cell, thereby leading to meaningful physiological answers. Consequently, examples of two different classes of sensory transporters are dealt with. (i) Transport proteins that, in addition to transporting a particular substrate, are able to sense the availability of this substrate and to transduce this information to the regulatory networks of the cell, which in turn leads to cellular responses, in general at the level of gene transcription. Some cases of this type of sensory transport proteins in fact represent a mechanism of "transport sensing" in *sensu stricto*, which means a mechanism by which the actual state of activity of a carrier system is sensed directly and used as an information for the cell's metabolism. Mainly two kinds of transport systems are discussed in this subsection – transport proteins coupled to two-component systems for further signal transduction (see also Chapter 8) and sensing mediated by compounds related to phosphotransferase systems (PTSs) (see also Chapters 18–20). (ii) The second class comprises transport

Bacterial Signaling. Edited by Reinhard Krämer and Kirsten Jung
Copyright © 2010 WILEY-VCH Verlag GmbH & Co. KGaA, Weinheim
ISBN: 978-3-527-32365-4

proteins that directly sense external stimuli related to stress situations and which then, in turn, regulate their activity in order to achieve adaptation of the physiological situation of the cell to this particular stress condition. In this section, mainly the response of transport systems to both hyperosmotic (osmosensory uptake systems) and hypo-osmotic stress (mechanosensitive channels) are discussed.

The latter example again makes clear that it is not easy to strictly discriminate between sensing and regulation in the case of transport systems. To keep the concept simple, this chapter thus neither includes modulation of transport systems by binding of effector ligands to additional regulatory sites on transport proteins, for example, regulation of carrier proteins by pH (proton binding), other cations (e.g., Ca^{2+}), or nucleotides (e.g., ATP, ADP, NADH, NADPH). Also direct effector/activity relations (e.g., gating of ligand-gated channels by binding of the appropriate ligand) are not interpreted here as transmembrane signaling.

On the other hand, a certain overlap with other chapters of this book cannot be avoided. This becomes most obvious for two paradigms of sensory and regulatory cellular networks in bacteria, namely PTS systems with their multiple regulatory connections to other functional elements in the cell including other transport systems, as well as two-component regulation systems.

12.2
Sensing of Transport Activity

Transport proteins that are able to sense their own state of activity and to transduce this information as a signal to the cell are the paradigm of an ideal sensory transport system. The presence of a particular substrate of regulatory interest for the cell is monitored not via a specialized receptor protein, as, for example, in the case of attractant or repellent receptors related to chemotaxis (see Chapter 9) or by a membrane bound receptor kinase of a two-component system (see Chapter 9), but by the transporter itself. The functional concept behind this mechanism requires that the catalytic action of the transporter directly generates a particular signal that can then be transduced to downstream elements of a coupled signal transduction cascade leading to further cellular action. This requirement of direct (conformational) coupling is also a critical point of this hypothesis, as will become clear in the following, since the initial event – the direct coupling of a useful signal to the catalytic activity of a transporter – is not understood in many cases, at least not at the molecular level.

The examples that probably come closest to a molecular understanding of how this signal directly related to the action of a transport system is generated are represented by a number of signaling networks originating from or connected to the action of PTS as well as ATP-binding cassette (ABC)-type transport systems. Related information to these concepts is also described in the chapters on cAMP signaling (Chapter 20), intracellular phosphotransfer (Chapter 19), and trigger enzymes (Chapter 18). These chapters focus on the regulatory aspects of these networks; in the present chapter, however, we concentrate on the mechanistic aspects of sensing.

12.2.1
Escherichia coli Maltose System and the Global Regulator Mlc

The first sensory network to be discussed is the control of the maltose regulon in *E. coli*, which is one of the best-studied examples of this kind and thus also referred to in other chapters of this book for various aspects. In this rather complex mechanism, a central aspect is binding of two transcription factors, MalT and Mlc, to transport systems in dependence of the state of activity of the respective carriers (Figure 12.1). The regulatory network involves shuttling of the two transcription factors, MalT and Mlc, respectively, between their (regulatory) promoter-binding sites at the DNA and their docking sites at particular subunits of two transport systems, the MalK protein of the ABC-type maltose and maltodextrin uptake system, on the one hand, and the EIIBC^{Glc} subunit of the glucose PTS, on the other (see Chapters 19 and 20 for details on the architecture and function of bacterial PTS systems). MalT is a transcriptional

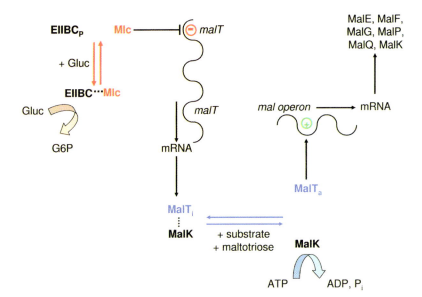

Figure 12.1 Transport sensing and regulation in the maltose system in *E. coli* (for details, see text). MalT is a transcription activator for *mal* genes. It is regulated both on the level of activity and transcription (area shaded light blue). Activity regulation of MalT depends on the presence of its inducer maltotriose and on sequestration to the ATPase subunit MalK of the ABC-type maltose/ maltodextrin uptake system. MalT in its inactive state (MalT$_i$) is firmly bound only to the idle MalK protein, not being active in ATP hydrolysis coupled to transport. When MalK is active in transport, MalT is released, binds maltotriose (MalT$_a$), and binds to the promoter of the *mal* operon triggering transcription of the *mal* genes. Transcription of the *malT* gene itself depends on the absence of the transcriptional inhibitor Mlc from the respective promoter site (area shaded light red). When the glucose PTS actively takes up glucose, the EIIBC component of the glucose PTS becomes dephosphorylated and sequesters the Mlc protein, which leads to transcription of the *malT* gene.

activator of the *mal* operon(s), which includes the genes responsible for maltose and maltodextrin metabolism, transport and regulation, whereas Mlc is a global repressor of gene transcription in *E. coli* [1]. In the default situation, the transcriptional activator MalT is firmly bound to the MalK subunit of the maltose transport system. During transport, the MalK subunit energizes maltose transport via ATP hydrolysis. If the MalK subunit is now hydrolyzing ATP (i.e., if the maltose transporter actively takes up its substrate, maltose or maltodextrin), MalT is released from its binding site at MalK, shuttles to the respective promoter site(s), and activates gene transcription of the *mal* operon(s), provided the inducer, maltotriose, a product of maltodextrin metabolism is present and bound to MalT [1, 2]. In this scenario, the maltose transport system is a sensory transporter since the catalytic action of the MalK subunit of this transporter generates a conformational signal that is transduced into the transcriptional activation by MalT. Although the three-dimensional structure of MalK is known [3, 4], as well as the binding region of MalT at MalK [5], the correct understanding of the conformational aspects of MalT–MalK interaction has not yet been achieved at the molecular level; it is clear, however, that ATP binding to MalK and subsequent hydrolysis, respectively, lead to significant changes in the conformation of the MalK subunit [6]. It should not be neglected that, in addition to being regulated by MalT, the maltose system is subject to catabolite repression (see Chapter 19). This aspect of the regulatory control is mediated by the cAMP–C-reactive protein complex, which has to bind to the *mal* operon(s) for full activation of *mal* gene transcription. This leads to a direct connection to the PTS system, since the concentration of cAMP is controlled by PTS [7, 8] (for details, see Chapters 19 and 20).

There is, however, also a direct connection to PTS via the Mlc protein. The global repressor Mlc controls a large number of genes, among them genes for several sugar transporters, for the glucose PTS, central components of the PTS phosphorylation cascade, as well as the *malT* gene. Similar to MalT, Mlc also shuttles between its docking site at the $EIIBC^{Glc}$ subunit of the glucose PTS and the respective promoter sites at the DNA [1, 9–11]. The core principle of this mechanism now is based on the fact that Mlc is only sequestered to the dephosphorylated form of $EIIBC^{Glc}$. As explained in Chapters 19 and 20 in detail, a high degree of phosphorylation of PTS components indicates an idle transporter, whereas dephosphorylated components are indicative of a transport-engaged PTS. Hence, when the glucose PTS is actively taking up glucose and phosphorylating it to glucose-6-phosphate, $EIIBC^{Glc}$ will be predominantly in the dephosphorylated state, Mlc is firmly bound, and transcription of the target genes of Mlc is released, among them the *malT* gene. Thus, MalT is subject to a dual control, both on the level of activity (see previous paragraph) and transcription by Mlc (Figure 12.1). As for MalT, the regulatory state of Mlc is controlled by the state of activity of the corresponding transport system, in this case the glucose PTS. For PTS systems, the structural information related to the sensory function is easier to rationalize. It is directly related to the phosphorylation state of the EIIBC component of the glucose PTS. Fortunately, for this sensory system a high-resolution structure of a tetrameric Mlc in complex with four molecules of $EIIB^{Glc}$, the cytoplasmic domain of the $EIIBC^{Glc}$ component of PtsG is available [12]. This structure provides a detailed view of the protein–protein interface and, at the same time, suggests a mechanism based on steric hindrance that alters the binding surface and thus the probability of interaction in dependence of the phosphorylation state of EIIB^{Glc}. Taken together, the mechanistic design of PTS transport systems renders them ideally suited for transport sensing and subsequent signal transduction. Conceptually, this is based on the fact that most components of the PTS system are chemically distinguishable in dependence of their state of activity, due to the difference in phosphorylation. In contrast to many other transport systems, these multiple aspects of PTS systems (transport, sensing, and regulation) were recognized long ago and have been described in detail before (Postma et al., 1994) [7].

Moreover, as a true bacterial paradigm of sensory transport systems, PTS systems provide us with further examples of direct activity sensing. Glucose-dependent induction mediated by the Mlc protein was shown to use the phosphorylation state of the EIIBC^{Glc} component as a signal for regulation. A number of other components of the phosphorylation chain of PTS systems are also used to transmit the information of PTS engaged in taking up glucose to a surprisingly large number of other signaling pathways. This is dealt with in more detail in Chapters 19 and 20. Among them are the processes of inducer exclusion and catabolite repression via adenylate cyclase activation, which are both connected to the phosphorylation state of the EIIA^{Glc} component of the PTS. The EI component is connected to chemotaxis (see Chapter 11) and the Hpr protein to glycogen metabolism (see Chapter 20). Further connections of components of particular PTS systems are observed for dihydroxy acetone (DHA) metabolism as well as nitrogen regulation. This broad use of individual components of the PTS for sensory and regulatory purposes indicates that (i) the phosphorylation state of PTS components, which is directly coupled to the transport activity of individual PTS systems, is a very useful signal for the cell in order to obtain information about the external availability of particular substrates (e.g., glucose), and (ii) sensing and regulation is an intrinsic rather than a side aspect of this kind of transport system (PTS). The latter conclusion is in line with the hypothesis that the primordial function of PTS may have been regulation and not transport.

12.2.2
E. coli Uhp System

Another class of transport proteins, capable of direct transport activity sensing, directly feed into two-component regulatory systems, thus adding an additional component to the regulatory network. Consequently, these types of systems have thus been named three-component systems (see also Chapter 8). The archetypical example, known and investigated since more than 20 years, is the regulation of the *uhpABCT* operon in *E. coli*. From the structural point of view, UhpAB is a two-component system consisting of the membrane-bound sensor kinase UhpB and the soluble response regulator UhpA [13, 14]. The expression of the *uhp* operon is induced by glucose-6-phosphate and the inducer must be present in the external medium. UhpC and UhpT are structurally closely related proteins both belonging to the large class of transporters of the major facilitator superfamily. Surprisingly, although structurally resembling a secondary transporter, the UhpC protein was found to function as a constitutively expressed sensor protein binding extracellular glucose-6-phosphate. Upon binding of the inducer, UhpC is supposed to change its conformation and to interact with the sensor kinase UhpB, thus triggering expression of the *uphT* gene via UhpA [15, 16]. The transporter UhpT then takes up glucose-6-phosphate. It was hypothesized that a primordial gene encoding the glucose-6-phosphate transporter was duplicated and one copy was then modified to a sensor protein, thereby loosing its transport function.

On the basis of insertion mutagenesis into *uhpC* and *uhpB*, it was suggested that the transmembrane domains of the two membrane proteins interact, so that binding of glucose-6-phosphate to UhpC would result in autophosphorylation of UhpB and further signal transduction via UhpA [13]. Despite the fact that UhpC was interpreted to be a transporter-derived sensor only, this scheme would, on the basis of the transporter-like structure of UhpC, fit into the concept of activity-related transport sensing. More recently, this view was modified by the finding that in fact UhpC is still able to transport glucose-6-phosphate, albeit with much lower activity than UhpT [17]. This result seems to put the Uhp system even closer into the class of transport activity sensing systems; however, a molecular analysis shed some doubt on this interpretation. By a series of site-specific mutations it was shown that transport and sensing functions of UhpT can individually and independently be modified. Thus, transport of glucose-6-phosphate by UhpC is not required for sensing and the sensing function is not necessarily coupled to transport [18]. More precisely, the affinity of UhpC for glucose-6-phosphate as an inducer could be differentiated from its affinity for the same sugar phosphate as a transport substrate. Consequently, the glucose-6-phosphate-binding sites responsible for transport and sensing do not seem to be identical, which would contradict the simple model of transport activity sensing, in which the conformational action of the transport protein upon binding and/or translocation of the transport substrate would represent the information for the signal transduction cascade coupled to this transporter with sensory function.

It should be noted here that there is a wealth of information on transport activity sensing by transporter like sensors, also called nutrient sensors, for glucose as well as amino acids transport systems in yeast. The glucose sensors Snf3 and Rgt2, and the amino acid sensor Ssy1, are the prototypes of this category of nutrient sensors in yeast [19, 20]. Upon binding of their respective ligands at the cell surface, these proteins activate signaling pathways in the cytoplasm leading to transcriptional regulation of genes encoding transport systems for these nutrients.

12.2.3
Dicarboxylic Acid Uptake in *E. coli* and Rhizobia

Growing evidence is accumulating for a similar type of coupling of transporter and two-component systems for dicarboxylic acid uptake in bacteria. In this case, the sensory part is a carrier with its major function in dicarboxylate transport. In Rhizobia (*Sinorhizobium meliloti* and *Rhizobium leguminosarum*), dicarboxylic acids are taken up by the Dct system, which consists of a secondary transporter

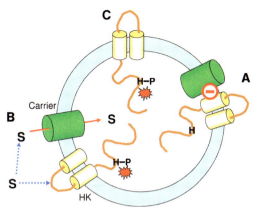

Figure 12.2 Transport sensing by dicarboxylate transporters in *E. coli* and in rhizobia (hypothesis). (A) In the absence of the substrate dicarboxylate (S), the transport protein (green cylinder) interacts with the histidine kinase of the dicarboxylate sensing two-component system (yellow cylinders). This protein–protein interaction leads to the inhibition of signal transfer to the cognate response regulator of the two-component system. (B) When substrate is present, it interacts both with the periplasmic loop of the sensor kinase and with the carrier protein. As a consequence, the transporter dissociates from the histidine kinase releasing inhibition and the signal is transduced to the response regulator. (C) In a mutant in which the transporter is deleted, the two-component system is not inhibited and constantly activates its response regulator leading to constitutive transcription of the respective target genes.

DctA and a two-component system DctBD [21, 22]. Succinate, for example, binds to the periplasmic domain of the membrane-bound histidine kinase DctB [23] and induces *dctA* expression via the response regulator DctD. Surprisingly, deletion of the transporter DctA causes constitutive expression of the *dctA* gene, thus indicating that DctA is involved in dicarboxylate sensing by DctB, although biochemical evidence for a functional interaction is still missing (Figure 12.2).

A related system, Dcu, has been studied in more detail in *E. coli*. It consists of the two-component system DcuSR and the dicarboxylate/succinate antiporter DcuB, one of the three C4-dicarboxylate carriers in this organism [24, 25]. DcuSR controls genes of fumarate respiration in *E. coli*, including the *dcuB* gene. Expression of *dcuB* is strongly stimulated by addition of fumarate, which binds to the periplasmic domain of the histidine kinase DcuS [26, 27]. Similar as in the example described above, deletion of *dcuB* causes constitutive expression of the genes controlled by DcuSR, thus indicating that DcuB plays an important role in the signaling network responding to C4-dicarboxylates. Recently, it was shown by site-directed mutation that the transport function of DcuB and its function in sensing and/or regulation can be uncoupled, thus suggesting that different sites in DcuB are supposed to be responsible for the two different functions [28]. Consequently, we end up with a transporter DcuB that has a dual function, substrate transport and substrate sensing, and thus fits into the concept of transport sensing; however, it is not entirely clear whether and how these two functions are structurally related or functionally coupled. In principle, at least two explanations are possible, (i) the transport and the regulatory binding sites, respectively, at DcuB for dicarboxylates are different and the regulatory function of DcuB depends on binding of dicarboxylates to the regulatory binding site only or (ii) the amino acid residues shown to be related to the regulatory function of DcuB actually represent the interaction site with DcuS. The latter explanation would mean that binding of substrate to the transport binding site and/or substrate translocation would trigger a conformational change around the amino acids shown to be involved in regulation via the DcuSR system.

12.2.4
LysP/CadC System in *E. coli*

Recently, another instructive example of this kind of conformational coupling has been described – the LysP/CadC couple in *E. coli*. The CadBA system is supposed to be involved in the response of *E. coli* to acid stress, and combines the lysine decarboxylase CadA and the lysine/cadverine antiporter CadB. The expression of *cadBA* is controlled by CadC – a membrane-bound transcriptional activator. Expression of *cadBA* now depends on the presence of external lysine and low pH. The involvement of this system in acid stress response may be explained by the fact that CadC, in addition to the sensing mechanism described here, was also suggested to participate in signal transduction via proteolytic cleavage in response to low external pH [29]. Concerning lysine as an external stimulus, recent genetic and biochemical analysis of this system demonstrated that lysine does not directly interact with CadC, but rather triggers CadC activity via LysP, a secondary uptake carrier [30]. A series of mutations in CadC provided strong evidence for a physical interaction between LysP and CadC being essential for LysP-dependent signal transduction. Consequently, it was suggested that the conformational state of LysP in dependence of its function as a lysine transporter, or, in other words, in dependence of the presence of external lysine, controls binding of LysP to CadC and thus CadC activity. This situation again can be interpreted in the framework of transport activity sensing; in the absence of lysine, LysP binds to and thereby inactivates CadC. When lysine is available, LysP becomes active and concomitantly will change its conformation. As a consequence, LysP will dissociate from CadC and thus release CadC for activation of the *cadBA* operon.

12.2.5
Ammonium Signaling

A completely different way to couple the action of a transporter to a signal transduction cascade is represented by the combination of bacterial ammonium transporters and interacting PII proteins. Ammonia is translocated by Amt proteins in many bacteria and there is strong evidence that Amt proteins function, beside substrate (ammonia) translocation, also in sensing ammonium [31]. Amt proteins are an integral part in the bacterial nitrogen control network. In *E. coli* and many other bacteria, the intracellular nitrogen status is monitored by GlnK – a small trimeric signal transduction protein of the PII family. GlnK regulates the activity of many other proteins in the cell [32]. It has been shown that GlnK binds to the *E. coli* AmtB ammonium transporter, which is, like GlnK, also a trimer in its functional state [33]. GlnK activity is regulated by uridylylation, and only the deuridylylated form, which indicates a surplus of cellular nitrogen availability, binds to AmtB and thereby blocks ammonia uptake [34]. AmtB, however, is also actively involved in this process, since it has been shown that the transporter must be active (i.e., ammonium must be present) for efficient deuridylylation and sequestration of GlnK [35]. Consequently, AmtB can be interpreted in this process to function as an ammonium sensor.

It should be added here that there is strong evidence for the yeast ammonium transporter Mep2 playing a dual role in ammonium transport and ammonium sensing, too. The two functions can experimentally be separated by site-directed mutation and the latter function is connected to intracellular signal transduction pathways [36].

12.2.6
Further Transport Systems with Substrate Sensing Function

There are further systems and mechanisms which are related to transport sensing. In some cases, periplasmic binding proteins of ABC-type transport systems may serve, beside their core function in binding transport substrates and delivering them to the membrane bound subunits of ABC systems, also in sensing these substrates and transducing a signal to downstream regulatory pathways. An example is the binding protein BctC of the tripartite tricarboxylate transporter BctCBA of *Bordetella pertussis*. BctC serves as an integral part of the signaling cascade of the citrate sensitive BctDE two-component system. Furthermore, signal transduction between particular channels in the outer membrane of Gram-negative bacteria with interacting transport systems of the inner membrane, a mechanism discussed in Chapter 7, is in principle also related to transporters with sensing functions.

Finally, we would like to mention another example of substrate sensing by a transporter, or of regulation dependent on substrate concentration, which has been studied to great detail at the molecular level – the Na^+/H^+ antiporter NhaA from *E. coli* [37]. In this case, protons are both substrate and regulatory effector. The binding sites, however, for protons as a substrate and as an effector, respectively, are different, and proton binding to the effector site is a regulatory event and not coupled to signaling. Consequently, NhaH and related proteins are not discussed here.

12.3
Stress Sensing by Transport Proteins

The second class of sensory transport systems discussed in this chapter is fundamentally different from that described above in its mechanism of sensing and regulation. The physiological steady state of a living cell is constantly challenged by external stress factors, for example, by varying pH, by changing osmolarity and/or salt concentration, by heat, or by chill, to mention only the most frequent types of stress. The adapted reaction to various types of stress is a pivotal capacity of the bacterial cell and is essential for survival. A number of receptor systems participate in the response to these stimuli (see also Chapter 8) and in some cases, transport systems are involved, too. The best-studied example is the response to osmotic stress. Consequently, whereas in the previous section of this chapter, carrier proteins were discussed, which are able to transduce information about their own state of activity (transport sensing), the type of transporters described in this section is characterized by mechanisms that render the carrier protein sensitive to external stimuli.

Bacteria frequently encounter osmotic stress, in particular soil bacteria. Hypo-osmotic stress occurs when the external osmolality is decreasing (e.g., after rainfall), whereas hyperosmotic stress is a consequence of an increase of external osmolality (e.g., upon sunshine). The mechanical impact on bacterial cells originates from the (passive) physical response of massive water influx into (hypo-osmotic conditions) or efflux out of the cell (hyperosmotic conditions) in response to osmotic stress. The cell counteracts these life-threatening situations by the action of fast efflux systems – mechanosensitive channels in the case of hypo-osmotic stress or by efficient, energy-coupled uptake of compatible solutes mediated by transporters in response to hyperosmotic stress.

12.3.1
Mechanosensitive Channels

Mechanosensitive channels function as emergency valves, triggering fast efflux of solutes upon hypo-osmotic stress, to prevent the cell from rupture caused by massive water influx as a consequence of decreased external osmolality. There are several types of mechanosensitive channels, and two of them, MscL and MscS, have been characterized in detail both in terms of structure and function. The three-dimensional structure of both channels has been solved by X-ray crystallography [38, 39] and their dynamic properties have been investigated by spectroscopic techniques [40, 41]. MscL functions as a homopentamer, whereas MscS is a homoheptamer. Mechanosensitive channels embedded in the phospholipid bilayer are able to directly sense the membrane tension both in intact cells and in artificial bilayers, which develops upon the presence of hypo-osmotic stress. In response to changed membrane tension, mechanosensitive channels open and close by an iris-like movement of their transmembrane domains [40]. When inserted into liposomes (artificial membranes) mechanosensitive channels retain their mechanosensitive properties; this indicates that all aspects of membrane sensing and regulation of channel functions reside in the protein itself.

How is the stimulus (i.e., membrane tension) related to osmotic stress, perceived by mechanosensitive channels in mechanistic terms? Application of membrane interacting agents was instrumental for answering this question. Insertion of particular amphipathic molecules (e.g., lyso-phospholipids) as well as local anesthetics was able to trigger the action of mechanosensitive channels (channel gating) even in the absence of membrane tension. These amphipaths are known to insert preferentially in one leaflet of the bilayer and thus may alter the curvature of the bilayer membrane [42]. A similar effect on gating was observed when mechanosensitive channels were inserted into membranes with nonphysiological bilayer thickness based on the presence of short (less than 18 hydrocarbons) or long (greater than 18 hydrocarbons) fatty acid side-chains [43]. These two kinds of influences originating directly from the surrounding bilayer were investigated on a structural level using spectroscopic techniques and were found to result in structural changes of mechanosensitive channels [43]. On the basis of this concept it was concluded that the function of mechanosensitive channels, also called mechanotransduction, is directly controlled by the surrounding bilayer, and that gating stimulated by osmotic stress is defined by local and global asymmetries in the transbilayer tension profile of the membrane.

12.3.2
Osmosensory Uptake Systems

In the response to hyperosmotic stress, not channels but transporters are involved. There is a wealth of information on a variety of osmosensory transport systems in bacteria and three of them are studied to significant molecular detail with respect to their response to osmotic stress – ProP from *E. coli*, OpuA from *Lactococcus lactis*, and BetP from *Corynebacterium glutamicum* [44–46]. The sensing mechanisms of these systems share a significant functional similarity. Sensing and regulation seems to be triggered primarily by changes of the cytoplasmic conditions (e.g., ion strength, solute concentration) in response to hyperosmotic stress, rather than by sensing changes in the external surrounding of the cell directly. This is in contrast to most two-component sensory systems, where frequently external loops of the respective histidine kinases are the sites where external stimuli are perceived. Furthermore, in the case of these three osmosensory transporters, specialized, terminally located domains of the membrane proteins seem to be critically involved in the mechanisms of stress sensing and intramolecular signal transduction [44].

Owing to the fact that only for one of these transporters a high-resolution structure and thus information on the molecular level is available, we will focus on BetP from *C. glutamicum* in the rest of this section. BetP is a secondary transport system that catalyzes uptake of glycine betaine into cells of *C. glutamicum* coupled to the inward movement of two Na^+ ions. BetP is a homotrimer, each individual subunit is a membrane protein of 595 amino acid residues organized into 12 transmembrane segments and two terminal domains of about 55 amino acid residues each facing the cytoplasmic side (Figure 12.3). Its transport activity is essentially dependent on the presence of hyperosmotic stress – transport is zero without external stimulus and becomes active within less than 1 s upon onset of osmotic stress. BetP can be heterologously expressed in *E. coli* cells, as well as reconstituted in functionally active form in proteoliposomes; it retains both its transport and its sensory function in an artificial surrounding. This argues for the fact that BetP comprises all these functions without the need of further cofactors. Mainly in experiments using proteoliposomes, at least three factors necessary for and cooperating in stimulus perception by BetP under physiological conditions have been identified: (i) the presence and structural integrity of the C- and N-terminal domains of BetP, (ii) a negatively charged

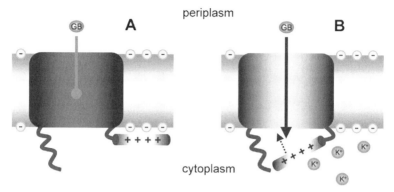

Figure 12.3 Osmosensing and signaling by the betaine transporter BetP from *C. glutamicum* (hypothesis). In the absence of osmotic stress, the C-terminal domain of BetP is immobile and probably fixed near the membrane surface. Upon hyperosmotic stress and concomitant increase in the cytoplasmic K^+ concentration, the C-terminal domain becomes mobile, interacts with the cytoplasmic surface of BetP (N-terminal domain and internal loops) and activates betaine transport.

membrane surface in the surrounding, and (iii) the rise in cytoplasmic K^+ in a concentration range of roughly 200–300 mM upon hyperosmotic stress [45, 47]. Very recently, a fourth factor to be added to this list was identified but not yet elucidated to great extent: (iv) an appropriate lipid composition of the surrounding bilayer membrane with respect to the fatty acid composition of the phospholipids [48].

How do these parameters and modules work together in the sensing process of BetP? The current model suggests that the C-terminal domain of BetP in the default condition (i.e., in the absence of osmotic stress) is fixed somewhere, probably near the membrane surface and/or the N-terminal domain (Figure 12.3). In this state, the pathway for betaine and concomitant Na^+ transport in the membrane body of BetP is blocked. Upon an increase of the external osmolality, the cytoplasmic volume (correctly the free cellular water) is decreased because of water flowing out of the cell. This leads to an increase in the concentration of all internal solutes. The elevated internal K^+ concentration now is a stimulus for BetP and is somehow sensed by the C-terminal domain. As a response, the C-terminal domain undergoes a significant conformational change. This event is then transduced to the membrane part of BetP, most probably by direct interaction of the C-terminal domain with internal loops of the transporter, and the transport pathway becomes active [47]. Interestingly, recent data on the three-dimensional structure of the protein add further information to the process of sensing and regulation by BetP [49]. It turned out that the C-terminal domain of one BetP protomer within the functional trimer is stretching out and is located on top of the cytoplasmic surface of the neighboring BetP protomer when BetP is in the occluded state with bound substrate. Since the functional contribution of the C-terminal domain to sensing and regulation of BetP in dependence of osmotic stress has been experimentally proven, the whole process can putatively be interpreted by a mechanism in which the C-terminal domain of each protomer of BetP is able to silence and activate, respectively, the neighboring protomer, with respect to betaine transport. This hypothesis has to be proven by investigating the dynamic behavior of different parts of BetP during the transport and the activation/deactivation cycle.

Currently, an uncertainty about a particular aspect in the signaling mechanisms of these transporters remains, which was discussed above with respect to transport sensing, too. There is growing evidence that also these transporters perceive stimuli directly originating from the surrounding hydrophobic phase of the membrane as a monitor for stress situations. This type of sensory interaction was shown to be valid in particular for mechanotransduction by mechanosensitive channels (see Section 12.3.1). There are several arguments in favor of this concept in the case of BetP: (i) membrane-active amphipaths were shown to be able to trigger BetP activation [50], (ii) an alteration of membrane fluidity by low temperature leads to BetP activation [51], and (iii) the activation profile of BetP was shown to depend on the composition of the surrounding lipid bilayer [48]. Although not yet fully conclusive, a dual-sensing mechanism at least for BetP has to be considered, by which different stimuli from both the hydrophilic (solute concentration in the cytoplasm) and from the hydrophobic surroundings are integrated by the transporter, leading to a meaningful answer of the carrier in terms of an adapted stress response.

As already noted, the compatible solute carriers ProP from *E. coli* and OpuA from *L. lactis* resemble BetP in their mechanism of sensing and regulation to a significant extent [44, 46]. On the other hand, however, besides the fact that the respective domains of the these transporters involved in sensing are different, a C-terminal domain with a coiled-coil structure in the case of ProP [52] and a CBS domain for OpuA [53], interestingly, also the suggested stimulus is different. Whereas the stimulus reflecting the changed state of the cytoplasm was found to be restricted to K^+ in the case of BetP [54–56], it was reported to be the ionic strength for OpuA [53, 57] and the hydration state of the transporter for ProP [46].

There are certainly a large number of additional transport systems that respond to internal or external stimuli; however, in most cases, as explained above, these responses are more closely related to metabolic regulation, namely activity regulation in response to effector binding (e.g., nucleotide binding, such as ATP, NADH, etc.), and will consequently not be discussed in this chapter.

12.4
Conclusions and Perspective

Sensory transport proteins cover a relatively broad field of mechanisms by which they perceive environmental or metabolic stimuli and how they transduce these stimuli into useful signals for the bacterial cell. The first type of mechanism described in this chapter representing a direct connection between transport proteins and signaling networks involves monitoring of the transporter's state of activity as the actual information that is transduced. Three examples of mechanisms fitting into this concept have been described here – signaling via components of the PTS, via the energy coupling subunit of primary, ABC-type transporters, and via conformational interaction of secondary transporters. This list combines all three major classes of bacterial transport mechanisms, primary and secondary transport as well as PTS systems, which may indicate that this type of signaling is both important and widely distributed in solute transport.

The first class of examples, PTS-related signal transduction, is relatively well-defined since the actual information that is monitored (i.e., the state of phosphorylation of components of PTS systems in dependence of their transport activity) is understood in detail. The second example, represented by the MalK/MalT couple of the maltodextrin ABC transporter, very likely depends on conformational changes of MalK, supposed to be related to the ATPase function of the MalK subunit and thus to transport activity of the transporter. Finally, also some secondary carriers show functional coupling between the activity state of the carrier with sensory function and a component of the corresponding signal transduction pathway (e.g., histidine kinases). Although biochemical studies strongly suggest conformational coupling also in this case, details on the molecular level are not available yet. Unfortunately, detailed structural information of the extent and kind of conformational changes occurring during the catalytic cycle of carriers is not known in general, and is in particular not yet available for secondary carriers. Correlation of conformational changes of a carrier during substrate translocation with binding properties of signal transducing proteins will be an interesting topic for future structural and functional investigations.

Two types of transporters sensing environmental stimuli have been described in Section 12.3 – mechanosensitive channels and osmosensory transporters. Both systems, although being extremely different both in structure and function (passive channels versus active transporters), perceive their stimuli indirectly via an impact of the respective stress conditions to cellular compartments or structures. Whereas in Section 12.2, signaling by conformational transfer via specific protein–protein interactions is the point of interest, in the examples described in Section 12.3, the core interest is in the molecular understanding of the mechanics of sensing and signal transduction by particular domains of these proteins. A combination of high-resolution structure determination and a variety of biophysical methods elucidating the dynamic behavior of these membrane proteins will provide us with novel insights into these nanomachines.

References

1 Böhm, A. and Boos, W. (2004) Gene regulation in prokaryotes by subcellular relocalization of transcription factors. *Curr. Opin. Microbiol.*, **7**, 151–156.

2 Panagiotidis, C.H., Boos, W., and Shuman, H.A. (1998) The ATP-binding cassette subunit If the maltose transporter MalK antagonizes MalT, the activator of the *Escherichia coli* mal regulon. *Mol. Microbiol.*, **30**, 535–546.

3 Diederichs, K., Diez, J., Greller, G., Müller, C., Breed, J., Schnell, C., Vonrhein, C., Boos, W., and Welte, W. (2000) Crystal structure of MalK, the ATPase subunit of the trehalose/maltose BC transporter of the archaeon *Thermococcus litoralis*. *EMBO J.*, **19**, 5951–5961.

4 Oldham, M.L., Khare, D., Quiocho, F.A., Davidson, A.L., and Chen, J. (2007) Crystal structure of a catalytic intermediate of the maltose transporter. *Nature*, **450**, 515–521.

5 Joly, N., Böhm, A., Boos, W., and Richet, E. (2004) MalK, the ATP-binding cassette component of the *Escherichia coli* maltodextrin transporter, inhibits the transcriptional activator MalT by antagonizing inducer binding. *J. Biol. Chem.*, **279**, 33123–33130.

6 Davidson, A.L., Dassa, E., Orelle, C., and Chen, J. (2008) Structure, function and evolution of bacterial ATP-binding cassette systems. *Microbiol. Mol. Biol. Rev.*, **72**, 317–364.

7 Postma, P.W., Lengeler, J.S., and Jacobson, G.R. (1993) Phosphoenolpyruvate: carbohydrate phosphotransferase systems of bacteria. *Microbiol. Rev.*, **57**, 543–594.

8 Görke, B. and Stülke, J. (2008) Carbon catabolite repression in bacteria: many ways to make the most out of nutrients. *Nat. Rev. Microbiol.*, **6**, 613–624.

9 Tanaka, Y., Kimata, K., and Aiba, H. (2000) A novel regulatory role of glucose transporter of *Escherichia coli*: membrane sequestration of a global repressor Mlc. *EMBO J.*, **19**, 5344–5352.

10 Nam, T.-W., Cho, S.-H., Shin, D., Kim, J.-H., Jeong, J.-Y., Lee, J.-H., Roe, J.-H., Peterkofsky, A., Kang, S.-O., Ryu, S., and Seok, Y.-J. (2001) The *Escherichia coli* glucose transporter enzyme IIBC^{Glc} recruits the global repressor Mlc. *EMBO J.*, **20**, 491–498.

11 Shin, D., Lim, S., Seok, Y.-J., and Ryu, S. (2001) Heat shock RNA polymerase is involved in the transcription of *mlc* and crucial for induction of the Mlc regulon by glucose in *Escherichia coli*. *J. Biol. Chem.*, **276**, 25871–25875.

12 Nam, T.-W., Jung, H.I., An, Y.J., Park, Y.-H., Lee, S.H., Seok, Y.-J., and Cha, S.-S. (2008) Analyses of Mlc–IIBGlc interaction and a plausible mechanism of Mlc inactivation by membrane sequestration. *Proc. Natl. Acad. Sci. USA*, **105**, 3751–3756.

13 Island, M.D., Wei, B.Y., and Kadner, R.J. (1992) Structure and function of the *uhp* genes for the sugar phosphate transport system in *E. coli* and *Salmonella typhimurium*. *J. Bacteriol.*, **174**, 2754–2762.

14 Kadner, R.J., Island, M.D., Dahl, J.J., and Webber, C.A. (1994) A transmembrane signalling complex controls transcription of the Uhp sugar phosphate transport system. *Res. Microbiol.*, **145**, 381–387.

15 Island, M.D. and Kadner, R.J. (1993) Interplay between the membrane-associated UhpB and UhpC regulatory proteins. *J. Bacteriol.*, **175**, 5028–5034.

16 Kadner, R.J., Webber, C.A., and Island, M.D. (1993) The family of organophosphate transport proteins includes a transmembrane regulatory protein. *J. Bioenerg. Biomembr.*, **25**, 637–645.

17 Schwöppe, C., Winkler, H.H., and Neuhaus, H.E. (2002) Properties of the glucose-6-phosphate transporter from *Chlamydia pneumoniae* (HPTcp) and the glucose-6-phosphate sensor from *Escherichia coli* (UhpC). *J. Bacteriol.*, **184**, 2108–2115.

18 Schwöppe, C., Winkler, H.H., and Neuhaus, H.E. (2003) Connection of transport and sensing by UhpC, the sensor for external glucose-6-phosphate in *Escherichia coli*. *Eur. J. Biochem.*, **270**, 1450–1457.

19 Lalonde, S., Boles, E., Hellman, H., Barker, L., Patrick, J.W., Frommer, W.B., and Ward, J.M. (1999) The dual function of sugar carriers: transport and sugar sensing. *Plant Cell*, **11**, 707–726.

20 Boles, E. and Andre, B. (2004) Role of transporter-like sensors in glucose and amino acid signalling in yeast. *Top. Curr. Genet.*, **9**, 121–154.

21 Yarosh, O.K., Charles, T.C., and Finan, T.M. (1989) Analysis of C_4-dicarboxylate transport genes in *Rhizobium meliloti*. *Mol. Microbiol.*, **3**, 813–823.

22 Reid, C.J. and Poole, P.S. (1998) Roles of DctA and DctB in signal detection by the dicarboxylic acid transport system of *Rhizobium leguminosarum*. *J. Bacteriol.*, **180**, 2660–2669.

23 Zhou, Y.F., Nan, B., Nan, J., Ma, Q., Panjikar, S., Liang, Y.H., Wang, Y., and Su, X.D. (2008) C4-dicarboxylate sensing mechanism revealed by the crystal structures of DctB sensor domain. *J. Mol. Biol.*, **383**, 49–61.

24 Sic, S., Andrews, S.C., Unden, G., and guest, J.R. (1994) *Escherichia coli* possesses two homologous anaerobic C4-dicarboxylate membrane transporters (DcuA and DcuB) distince from the aerobic dicarboxylate transport system (Dct). *J. Bacteriol.*, **176**, 6470–6478.

25 Unden, G. and Kleefeld, A. (2004) C4-dicarboxylate degradation in aerobic and anaerobic growth, in *EcoSal – Escherichia coli and Salmonella: Cellular and Molecular Biology* (ed. R. Curtiss), ASM Press, Washington, DC.

26 Kneuper, H., Janausch, I.G., Vijayan, V., Zweckstetter, M., Bock, V., Griesinger, C., and Unden, G. (2005) The nature of the stimulus and of the fumarate binding site of the fumarate sensor DcuS of *Escherichia coli*. *J. Biol. Chem.*, **280**, 20596–20603.

27 Cheung, J. and Hendrickson, W.A. (2008) Crystal structures of C4-dicarboxylate ligand complexes with sensor domains of histidine kinases DcuS and DctB. *J. Biol. Chem.*, **283**, 30256–30265.

28 Kleefeld, A., Ackermann, B., Bauer, J., Krämer, J., and Unden, G. (2009) The fumarate/succinate antiporter DcuB of *Escherichia coli* is a bifunctional protein with sites for regulation of DcuS-dependent gene expression. *J. Biol. Chem.*, **284**, 256–275.

29 Lee, H.Y., Kom, J.H., Bang, I.S., and Park, Y.K. (2008) The membrane-bound transcriptional regulator CadC is activated by proteolytic cleavage in response to acid stress. *J. Bacteriol.*, **190**, 5120–5126.

30 Tetsch, L., Koller, C., Haneburger, I., and Jung, K. (2007) The membrane-integrated transcriptional activator CadC of *Escherichia coli* senses lysine indirectly via the interaction with the lysine permease LysP. *Mol. Microbiol.*, **67**, 570–583.

31 Javelle, A., Lupo, D., Li, X.-D., Merrick, M., Chami, M., Ripoche, P., and Winkler, F. (2007) Structural and mechanistic aspects of Amt/Rh proteins. *J. Struct. Biol.*, **158**, 474–481.

32 Arcondeguy, T., Jack, R., and Merrick, M. (2001) P(II) signal transduction proteins, pivotal players in microbial nitrogen control. *Microbiol. Mol. Biol. Rev.*, **65**, 80–105.

33 Khademi, S., O'Connel, J., Remis, J., Robles-Colmenares, Y., Miercke, L.J., and Stroud, R.M. (2004) Mechanism of ammonia transport by Amt/MEP/Rh: structure of AmtB at 1.35 A. *Science*, **305**, 1587–1594.

34 Conroy, M.J., Durand, A., Lupo, D., Li, X.-D., Bullough, P.A., Winkler, F., and Merrick, M. (2007) The crystal structure of the *Escherichia coli* AmtB–GlnK complex reveals how GlnK regulates the ammonia channel. *Proc. Natl. Acad. Sci. USA*, **104**, 1213–1228.

35 Javelle, A., Severi, E., Thornton, J., and Merrick, M. (2004) Ammonium sensing in *Escherichia coli*. Role of the ammonium transporter AmtB and AmtB–GlnK complex formation. *J. Biol. Chem.*, **279**, 8530–8538.

36 van Nuland, A., Vondormael, P., Donaton, M., Alenquer, M., Lourenco, A., Quintino, E., Versele, M., and Thevelein, J.M. (2006) Ammonium permease-based sensing mechanism for rapid ammonium activation of the protein kinase A pathway in yeast. *Mol. Microbiol.*, **59**, 1485–1505.

37 Padan, E. (2008) The enlightening encounter between structure and function in the NhaA Na^+–H^+ antiporter. *Trends Biochem. Sci.*, **33**, 435–443.

38 Chang, G., Spencer, R., Lee, A., Barclay, M., and Rees, D. (1998) Structure of the MscL homologue from *Mycobacterium tuberculosis*: a gated mechanosensitive ion channel. *Science*, **282**, 2220–2236.

39 Bass, R.B., Strop, P., Barclay, M., and Rees, D. (2002) Crystal structure of *Escherichia coli* MscS, a voltage-modulated and mechanosensitive channel. *Science*, **298**, 1582–1587.

40 Betanzos, M., Chiang, C.-S., Guy, H.R., and Sukharev, S. (2002) A large iris-like expansion of a mechanosensitive channel protein induced by membrane tension. *Nat. Struct. Biol.*, **9**, 704–710.

41 Perozo, E. (2006) Gating prokaryotic mechanosensitive channels. *Nat. Rev. Mol. Cell. Biol.*, **7**, 109–119.

42 Martinac, B., Adler, J., and Kung, C. (1990) Mechanosensitive ion channels of *E. coli* activated by amphipaths. *Nature*, **348**, 261–263.

43 Perozo, E., Kloda, A., Cortes, D.M., and Martinac, B. (2002) Physical properties underlying the transduction of bilayer deformation forces during mechanosensitive channel gating. *Nat. Struct. Biol.*, **9**, 696–703.

44 Poolman, B., Spitzer, J.J., and Wood, J.M. (2004) Bacterial osmosensing: roles of membrane structure and electrostatics in lipid–protein and protein–protein interactions. *Biochim. Biophys. Acta*, **1666**, 88–104.

45 Morbach, S. and Krämer, R. (2005) Structure and function of the betaine uptake system BetP of *Corynebacterium glutamicum*: strategies to sense osmotic and chill stress. *J. Mol. Microbiol. Biotechnol.*, **10**, 143–153.

46 Wood, J.M. (2006) Osmosensing by bacteria. *Sci. STKE*, **357**, 43–50.

47 Ott, V., Koch, J., Späte, K., Morbach, S., and Krämer, R. (2008) Regulatory properties and interaction of the C- and N-terminal domains of BetP, an osmoregulated betaine transporter from *Corynebacterium glutamicum*. *Biochemistry*, **47**, 12208–12218.

48 Özcan, N., Ejsing, C.S., Shevchenko, A., Lipski, A., Morbach, S., and Krämer, R. (2007) Osmolality, temperature, and membrane lipid composition modulate the activity of betaine transporter BetP in *Corynebacterium glutamicum*. *J. Bacteriol.*, **189**, 7485–7496.

49 Ressl, S., Terwisscha van Scheltinga, A., Vonrhein, C., Ott, V., and Ziegler, C. (2009) Molecular basis of transport and regulation of the Na^+/betaine symporter BetP. *Nature*, **458**, 47–52.

50 Rübenhagen, R., Rönsch, H., Jung, H., Krämer, R., and Morbach, S. (2000) Osmosensor and osmoregulator properties of the betaine carrier BetP from *Corynebacterium glutamicum* in proteoliposomes. *J. Biol. Chem.*, **275**, 735–741.

51 Özcan, N., Krämer, R., and Morbach, S. (2005) Chill activation of compatible solute transporters in *Corynebacterium glutamicum* at the level of transport activity. *J. Bacteriol.*, **187**, 4752–4759.

52 Hillar, A., Culham, D.E., Vernikovska, Y.I., Wood, J.M., and Boggs, J.M. (2005) Formation of an antiparallel, intermolecular coiled coil is associated with *in vivo* dimerization of osmosensor and osmoprotectant transporter ProP in *Escherichia coli*. *Biochemistry*, **44**, 10170–10180.

53 Biemans-Oldehinkel, E., Mahmood, N.A., and Poolman, B. (2006) A sensor for intracellular ionic strength. *Proc. Natl. Acad. Sci. USA*, **103**, 10624–10629.

54 Rübenhagen, R., Morbach, S., and Krämer, R. (2001) The osmoreactive betaine carrier BetP from *Corynebacterium glutamicum* is a sensor for cytoplasmic K^+. *EMBO J.*, **20**, 5412–5420.

55 Schiller, D., Rübenhagen, R., Krämer, R., and Morbach, S. (2004) The C-terminal domain of the betaine carrier BetP of *Corynebacterium glutamicum* is directly involved in sensing K^+ as an osmotic stimulus. *Biochemistry*, **43**, 5583–5591.

56 Schiller, D., Krämer, R., and Morbach, S. (2004) Cation specificity of osmosensing by the betaine carrier BetP from *Corynebacterium glutamicum*. *Biochemistry*, **43**, 5583–5591.

57 Biemans-Oldehinkel, E. and Poolman, B. (2003) On the role of the two extracytoplasmic substrate-binding domains in the ABC transprter OpuA. *EMBO J.*, **22**, 5983–5993.

13
Regulated Intramembrane Proteolysis in Bacterial Transmembrane Signaling
Thomas Wiegert

13.1
Introduction

Many transmembrane signaling pathways are characterized by the proteolytic cleavage of a membrane-spanning regulatory protein in the plane of the membrane, a mechanism called regulated intramembrane proteolysis (RIP), which has been well investigated for eukaryotic models [1–4]. Here, RIP promotes the controlled release of membrane-anchored proteins such as transcription factors like the mammalian sterol regulatory element-binding protein or epidermal growth factor ligands, but also is causative for the cleavage of the β-amyloid precursor protein and the production of amyloidogenic Aβ peptides, which are major risk factors of Alzheimer's disease. In the past decade emerging evidence has revealed that RIP also plays a prominent role in a variety of important and highly controlled bacterial transmembrane signaling processes in stress response, sporulation, cell division, cell cycle regulation, quorum sensing (QS), pheromone and toxin production, and biofilm formation.

The pivotal step in RIP is catalyzed by a family of peptidases named intramembrane cleaving proteases (I-CLiPs) that catalyze cleavage of transmembrane domains of substrate proteins [5]. In virtually all cases of transmembrane signaling by RIP, this intramembrane cleavage is initiated by a preceding proteolytic processing in the extracytoplasmic part of the substrate protein, also referred to as site-1 proteolysis, which makes the substrate competent for site-2 proteolysis – the intramembrane cleavage step in RIP. Moreover, additional cytoplasmic proteases can be involved, so that in summary an extracytoplasmic signal triggers a concerted proteolytic cascade in three cellular compartments with participation of at least three different proteases to transmit information and elicit cellular responses (Figure 13.1A).

Signals that trigger RIP and the type of responding site-1 proteases (S1Ps) seem to be specific for respective processes, less variation can be found for the type of site-2 proteases (S2Ps). A clear classification of bacterial transmembrane signaling pathways that involve RIP is difficult because mechanisms are diverse and often uniquely described for one bacterial species (Figure 13.1). Several examples

Bacterial Signaling. Edited by Reinhard Krämer and Kirsten Jung
Copyright © 2010 WILEY-VCH Verlag GmbH & Co. KGaA, Weinheim
ISBN: 978-3-527-32365-4

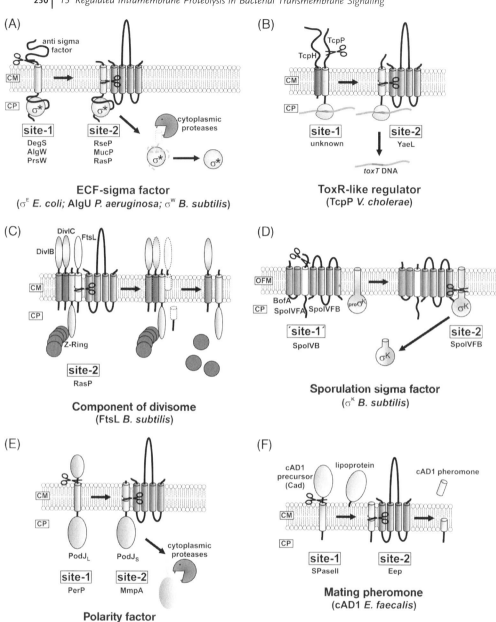

Figure 13.1 Bacterial transmembrane signaling processes that involve RIP. CM, cytoplasmic membrane; CP, cytoplasm. For explanations, see text.

are well investigated for the regulation of alternative sigma factors of the extracytoplasmic function (ECF) family. Activation of ECF sigma factors through RIP of the corresponding transmembrane anti-sigma factor seems to represent a more general mechanism (Figure 13.1A). Here, current knowledge on bacterial I-CLiPs and transmembrane signaling pathways in bacteria using RIP is summarized.

13.2
Bacterial I-CLiPs

I-CLiPs described so far are multispanning membrane proteins and belong to three mechanistic groups. They are divided into four families, the S2P zinc metalloproteases, the presenilins and signal peptide peptidase families of aspartyl proteases, and the rhomboid family of serine proteases [5]. For bacteria, only members of the S2P and rhomboid families have been found in a great variety of species. Almost all described bacterial transmembrane signaling pathways that involve RIP make use of I-CLiPs of the S2P family. Although a lot of biochemical and structural analyses have been performed on *Escherichia coli* GlpG [6–13], the only bacterial rhomboid protease with a defined substrate and function is *Providencia stuartii* AarA, which mediates QS [14].

S2Ps possess the classical consensus HExxH motif of a metalloprotease and a conserved sequence (LDG) with an aspartate as the third coordinating residue of the catalytic zinc atom. Rhomboids contain a GxSG motif that is characteristic of serine proteases of the chymotrypsin/trypsin/elastase family. The crystal structures of the rhomboid GlpG of *E. coli* [7, 8, 12] and *Haemophilus influenzae* [15], and of the S2P of the archaebacterial species *Methanocaldococcus jannaschii* [16], have been resolved recently, which helped to unravel some of the big questions in intramembrane proteolysis – how hydrolysis of a transmembrane segment in the lipid bilayer can take place, how hydrophilic water molecules enter the active site, and how transmembrane substrate proteins gain access [17]. The active sites are located within a folded, proteinaceous domain and sequestered from the surrounding lipids. GlpG from *E. coli* holds a water-filled cavity where the active-site residue serine is placed and it opens to the extracellular side. In contrast, *M. jannaschii* S2P has a polar channel that allows water entry to the catalytic zinc atom. Both structures point to two different conformations of the molecules – a closed conformation and an open conformation that would allow lateral substrate entry. For both types of I-CLiPs, a mechanism of gating by transmembrane helices was proposed (Figure 13.2A); however, S2P helices assigned for gating are not conserved and are absent in *E. coli* RseP, for example [18].

The mechanism of substrate recognition of bacterial I-CLiPs remains unanswered, and might be diverse and complex. Substrate requirements were determined for the S2P RseP of *E. coli*. The enzyme is able to cleave a variety of different transmembrane segments unrelated in sequence, provided that the transmembrane region contains residues of low helical propensity [19], which act to stabilize the

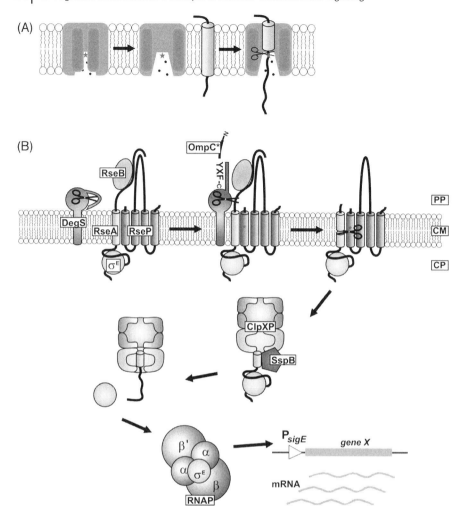

Figure 13.2 RIP catalyzed by S2P family I-CLiPs that contain a catalytically active zinc. (A) Gating mechanism by transmembrane helices that allows the substrate protein to enter the proteinaceous pore [16]. Water molecules are represented as dots, the catalytically active zinc as a star. (B) Regulation of E. coli σ^E by RIP. CM, cytoplasmic membrane; CP, cytoplasm; PP, periplasm. For explanations, see text.

substrate–enzyme interaction [18]. The site-1 proteolysis step that removes the bulk extracytoplasmic part of the substrate protein is crucial for recognition by the S2P; however, RseP is able to attack its substrate protein without the site-1 truncation step under certain conditions [20, 21]. For *Bacillus subtilis* RasP two unrelated *in vivo* substrates, the anti-sigma factor RsiW and the cell division protein FtsL [22, 23], were found that display sequence conservation in the cytoplasmic part that may be involved in substrate recognition [23].

13.3
Regulation of ECF Sigma Factors by RIP

The sigma factor is an essential component of bacterial RNA polymerase (RNAP), responsible for promoter recognition and initiation of transcription. Apart from the housekeeping sigma factor, alternative sigma factors become activated only under certain conditions to promote expression of a certain set of genes. A distinct class of alternative sigma factors that are involved in the regulation of extracytoplasmic functions was recognized in 1994 [24]. Genes encoding these ECF sigma factors are often cotranscribed with one or more negative regulators, which in many cases include a transmembrane anti-sigma factor that sequesters the sigma factor from interaction with RNAP [25]. In response to an environmental signal, the sigma factor is released, and, in almost all cases where the underlying mechanisms was solved, through RIP of the anti-sigma factor (Figure 13.1A).

13.3.1
Regulation of the *E. coli* σ^E-Dependent Envelope Stress Response

The most prominent and best investigated member of the ECF family of sigma factors regulated by RIP is *E. coli* σ^E (reviewed in [4, 26, 27]). About 20 promoters are σ^E controlled [28], among them the promoter of its own gene (*rpoE*), one of the four promoters of *rpoH* (*rpoH3*) encoding the heat shock sigma factor σ^{32}, promoters of four genes acting directly on the folding of *E. coli* envelope proteins (*dsbC*, *fkpA*, *skp*, and *surA*), the promoter of the periplasmic protease *degP* gene, and the promoter of the I-CLiP gene *rseP*. The *rpoE* gene encoding σ^E is cotranscribed with *rseA*, *rseB*, and *rseC*. RseA is a transmembrane anti-sigma factor with a type II topology (N-in, C-out), whereas RseB is a soluble periplasmic protein that interacts with the extracytoplasmic part of RseA (Figure 13.2B). RseC seems to exert no direct effect on modulating the σ^E-mediated stress response [29, 30]. σ^E forms a complex with the anti-sigma factor RseA and binds to it with approximately 300-fold greater affinity than to core RNAP [31]. To release σ^E for interaction with RNAP, RseA is sequentially cleaved by the proteases DegS [32], RseP (RseP was formerly named YaeL or EcfE) [33, 34], and cytoplasmic proteases like ClpXP [35] upon a specific signal, which activates the σ^E-dependent extracytoplasmic stress response. DegS is a transmembrane serine protease with the catalytic domain facing the periplasm. It was shown that the catalytic site of DegS is covered by its own C-terminal PDZ domain, inhibiting the proteolytic activity [36–38]. The DegS PDZ domain binds to C-terminal residues (consensus sequence YXF-COOH) of non-native outer membrane porins like OmpC that may accumulate in the periplasm upon envelope stress like a heat shock, denoted as the OMP signal. The OMP signal thereby uncovers the catalytic site of DegS that becomes active to cleave RseA in a site-specific manner [36]. RseA periplasmically truncated by DegS becomes the substrate for the membrane-embedded RseP zinc metalloprotease of the S2P family of I-CLiPs, the catalytic domain of which lies within the cytoplasmic membrane [33, 34]. Site-2 clipped RseA (RseA-S2) that is released to the cytoplasm still sequesters σ^E from interaction with RNAP unless it is completely removed by cytoplasmic proteases. RseA-S2 ends with the amino acid sequence VAA at its C-terminus (Figure 13.2B), which represents a C-motif 1 for recognition by the ClpXP AAA$^+$ protease [39]. Therefore, site-2 cleavage of RseA exposes a cryptic proteolytic tag that resembles the C-terminal SsrA-tag. The VAA-tag is recognized by the SspB adapter protein, and delivers site-2 clipped RseA to proteolysis by ClpXP and probably other proteases [35, 40]. The ATP-dependent unfolding activity of ClpX tears RseA-S2 apart from σ^E that now is free to bind to core RNAP and to initiate transcription of σ^E-controlled genes of the σ^E regulon.

Two other aspects of RIP of RseA have been substantially addressed. (i) It was shown that one of the two periplasmic PDZ domains of RseP prevents premature intramembrane proteolysis of full-length RseA, probably through interaction of the RseP PDZ domain with a glutamine-rich region of the extracytoplasmic RseA domain [20]. (ii) Attempts were made to define the role of RseB, which was known to negatively modulate σ^E induction [29]. RseB inhibits proteolysis by DegS *in vitro* by binding tightly to a conserved region near the C-terminus of the poorly structured RseA periplasmic domain, but the RseA sequences that mediate DegS recognition and RseB binding do not overlap directly. In addition, RseB inhibition is independent from the OMP signal that activates DegS (i.e., binding of the C-termini of porins to the PDZ domain of DegS [41]). Removing both RseB and DegS partially relieves the inhibitory mechanisms that prevent RseP cleavage of intact RseA. Therefore, by turning off RseP-initiated proteolysis of RseA, RseB and DegS make the system both sensitive to the OMP signal, and insensitive (robust) to variations in the absolute levels of DegS and RseP [21]. The *in vivo* role of RseB and probable inputs into the envelope-stress response that regulate RseB activity remain to be elucidated. However, the crystal structure of RseB has been resolved recently [42, 43] and it was proposed that RseB detects mislocalized lipoproteins.

13.3.2
σ^E Homologous Systems in Gram-Negative Pathogenic Bacteria

The investigation of stress responses in pathogenic bacteria is of particular interest because these regulatory mechanisms enable them to survive host defense mechanisms. A significant role of σ^E homologous envelope stress responses has been determined for the enterobacterial pathogens *Vibrio cholerae* [44], *Salmonella enterica* sv. *Typhimurium* [45, 46], *Haemophilus influenza* [47], and *Pseudomonas aeruginosa* [48, 49]; however, in most cases the role of RIP has not been investigated directly. σ^E of *V. cholera* and *S. enterica* sv. *Typhimurium* is induced by cationic antimicrobial peptides and confers resistance to them. For *V. cholerae*, it was shown that the induction is dependent on the outer membrane protein OmpU that contains a C-terminal YDF motif [50, 51].

P. aeruginosa is an opportunistic pathogen that is one of the major causes of morbidity in cystic fibrosis (CF). It is able to produce large amounts of the polysaccharide alginate that causes the mucoid colony morphology of the bacterium. It is believed that alginate production is responsible for biofilm formation that enables the bacterium to survive host defenses and antibiotic treatment. Transcription of the large alginate biosynthetic operon *algD–algA* is controlled by the ECF sigma factor AlgU (also designated as AlgT) and its anti-sigma factor MucA, and clinical isolates from CF patients of alginate overproducing strains of *P. aeruginosa* are usually mutated in *mucA* [52]. Recent work showed that the AlgU/MucA sigma factor/anti-sigma factor system is homologous to *E. coli* σ^E/RseA and is controlled by RIP by the same components (Figures 13.1A and 13.2B) – AlgW (DegS), MucP (RseP), and MucB (RseB). In addition, a small secretory protein (MucE) that might represent an outer membrane protein was identified that, upon overexpression, induced mucoidy of *P. aeruginosa*. Like *E. coli* OmpC, its C-terminal amino acid residues are believed to activate AlgW through interaction with its PDZ domain, although there are some variations in the recognition motif [53].

In *Bordetella bronchiseptica*, a pathogen of humans and animals that colonizes the respiratory tract, an S2P (HurP) is essential for heme-dependent induction of the gene for an outer membrane heme receptor (BhuR) and downstream genes. HurP, which is able to complement *V. cholerae* YaeL, is believed to function in RIP of HurR, an anti-sigma factor of the ECF sigma factor HurI [54].

13.3.3
Regulation of the *Bacillus subtilis* σ^W Regulon

σ^W is one the seven ECF sigma factors encoded by the Gram-positive model bacterium *B. subtilis*. It is autoregulated and controls about 30 promoters. σ^W constitutes an antibiosis regulon that confers intrinsic immunity against different antimicrobial agents [55]. It is induced by a variety of signals like vancomycin treatment [56], antimicrobial peptides [57, 58], phage infection, and most probably in an artificial manner by alkaline shock [59]. The σ^W anti-sigma factor RsiW undergoes RIP in a manner similar to *E. coli* RseA (Figure 13.1A), under participation of the S2P family I-CLiP RasP (formerly YluC) [22]. Site-2 clipped RsiW also presents a proteolytic tag at its C-terminus that is recognized mainly by ClpXP [60]. A substantial difference to *E. coli* σ^W/RseA was discovered for the site-1 proteolysis step that renders RsiW a substrate for RasP. A participation of DegS homologs of *B. subtilis* could not be proven and two different experimental approaches identified the *prsW* gene (formerly *ypdC*) as the determinant of site-1 proteolysis, also in reconstituted *E. coli* systems [61, 62]. PrsW displays similarities to a superfamily of multispanning membrane proteins with the yeast RCE1 CAAX-prenyl endopeptidase as the founding and sole member with so far known function. It was suggested that these proteins that are present in a variety of bacterial species, but not, for example, in *E. coli*, are metalloproteases that are potentially involved in protein and/or peptide modification and secretion [63]. The molecular signal that activates the site-1 proteolytic step and other regulatory factors that might be involved in degradation of RsiW are unknown to that time point.

13.3.4
Possible Role of RIP in Regulation of other ECF Sigma Factors

The cell envelope of the Gram-positive bacterium *Mycobacterium tuberculosis* is highly hydrophobic due to mycolic acids and it contains many unique lipids that are implicated in pathogenesis. Analysis of a knockout of the S2P I-CLiP *Rv2869c* revealed that the protease has an impact on cell envelope composition and the transcriptional regulation of multiple lipid biosynthetic and lipid degrading genes. The *Rv2869c* minus strain is highly impaired for growth and persistence in infected lungs of mice [64]. Although regulatory pathways that are controlled by *Rv2869c* have not yet been solved, it is conceivable that one or more of the 10 ECF sigma factors *M. tuberculosis* encodes [65] will be regulated by RIP through *Rv2869c*. Good candidates are σ^L and σ^K, both of which are inactivated by transmembrane anti-sigma factors [66, 67].

The Gram-negative bacterium *Myxococcus xanthus* has a complex lifecycle. Under starvation conditions cells swarm together to build mounds and intricate fruiting bodies, and it is also able to form heat resistant spores. One of its features is the light-induced synthesis of carotenoids as a protection mechanism against damage caused by light, regulated by the *carQRS* operon. It was shown that CarQ is an ECF sigma factor [68] and CarR the corresponding transmembrane anti-sigma factor [69]. CarR becomes unstable upon exposure of the cells to light and it was speculated that it is degraded in analogy to *E. coli* RseA [69].

13.4
Regulation of ToxR-Like Transcriptional Regulators via RIP

TcpP and ToxR of *V. cholerae* are bitopic membrane proteins with a cytoplasmic winged helix–turn–helix domain of the DNA-binding domain of OmpR-like response regulators of bacterial two-component systems. Since they do not interact with a sensor histidine kinase and do not contain a phosphoryl acceptor domain, they represent the simplest form of a bacterial transmembrane signaling system, named the ToxR-like family of transcriptional regulators. TcpP and ToxR are associated with the effector proteins TcpH and ToxS, respectively. TcpP is stabilized from proteolytic degradation through association with TcpH, which is a membrane anchored extracytoplasmic protein (Figure 13.1B). TcpP/TcpH and ToxR/ToxS cooperate to form a membrane localized complex that activates the transcription of the *toxT* gene. ToxT itself regulates transcription of the genes encoding cholera toxin and the toxin-coregulated pilus [70–72]. In a screen to search for mutations that restore TcpP function in the absence of TcpH, the *yaeL* gene was found, encoding the homolog of the *E. coli* RseP I-CLiP. In a *yaeL* minus strain, a truncated most probably site-1 clipped form of TcpP was detected that still activates *toxT* transcription. DegS is not involved in the initial site-1 proteolytic step [73]. It is noteworthy that YaeL (RseP) of *V. cholerae* regulates two different signaling processes involved in pathogenesis – the σ^E-mediated stress response (see Section 13.3.2) and the production of cholera toxin through the TcpP system.

CadC is the other prominent member of the ToxR-like family of transcriptional regulators. Under conditions of both low external pH and presence of lysine in the medium it activates the *cadBA* operon that encodes a lysine decarboxylase (CadA) and a lysine-cadaverine antiporter (CadB) by direct binding to the *cadBA* promoter. CadC of *E. coli* is held in an inactive state in the absence of lysine by interaction of its transmembrane domain with the lysine permease LysP. When lysine becomes available for LysP, CadC is released and becomes susceptible for activation by low pH [74]. For *S. enterica* sv. *Typhimurium* it was shown that acid stress causes activation through proteolytic processing of CadC by an unknown protease. However, the C-terminally truncated form of CadC remains bound to the cytoplasmic membrane, making it unlikely that RIP is involved in regulation of CadC activity [75].

13.5
Involvement of RIP in Regulation of Bacterial Cell Division and Differentiation

13.5.1
Involvement of RIP in Timing of Cell Division in *B. subtilis*

B. subtilis FtsL is a bitopic membrane protein that is essential for the initiation of septation during central and sporulation specific asymmetric cell division. It is part of the divisome that includes the FtsZ ring and it is thought to interact with two other proteins, DivIC and DivIB (Figure 13.1C). Depletion of FtsL results in the arrest of septation and rapid increase in cell length [76], and is followed closely by degradation of DivIC. FtsL itself is highly unstable and a role of DivIB for its stabilization especially at higher temperatures was reported [77]. The fact that *ftsL* transcription is regulated by the DNA replication initiation protein led to the assumption that FtsL is an important regulator of timing of cell division that delays cell division under conditions of impaired DNA replication [78]. Swapping and truncation of the cytoplasmic domain resulted in FtsL stabilization and did not affect its function. Deletion of the gene of the S2P I-CLiP RasP (YluC) abolished FtsL proteolysis and diminished cell length because cell division occurred faster. FtsL is degraded by RasP in a reconstituted *E. coli* system and amino acid residues in the cytoplasmic part of FtsL in the vicinity of the transmembrane domain that are also present in RsiW seem to be crucial for recognition by RasP [23]. In conclusion, RIP of FtsL regulates cell division in *B. subtilis*. Mechanisms that may modulate FtsL proteolysis are unknown, and whether FtsL is subjected to a site-1 proteolytic step in its extracellular domain has to be answered in the future (Figure 13.1C).

13.5.2
Activation of the Alternative Sporulation Sigma Factor σ^K of *B. subtilis*

In a complex and tightly regulated program *B. subtilis* is able to form stress-resistant spores in response to starvation. The cell divides asymmetrically and a small cell, which is the forespore, and a large cell, called the mother cell, is formed.

Further development of both cells is completely different and mainly controlled by four alternative sigma factors, but there is mutual communication to ensure that gene expression is coordinated [79–81]. The mother cell engulfs the forespore which then is surrounded by two membranes – the inner forespore membrane (IFM), which derives from the forspore cell, and the outer forespore membrane (OFM), which originates from the mother cell. Completion of engulfment activates σ^G in the forespore, initiating a transmembrane signaling pathway that activates σ^K in the mother cell. This so-called σ^K checkpoint is critical for efficient completion of endospore formation [82]. σ^K is synthesized as an inactive OFM anchored precursor (pro-σ^K) that is released by the activity of the I-CLiP SpoIVFB [83, 84], which is produced by the mother cell and localized in the OFM (Figure 13.1C). Unlike other transmembrane signaling processes described so far, RIP of pro-σ^K does not require a site-1 proteolysis step, but is negatively regulated by the two OFM proteins BofA and SpoIVFA that form a multimeric complex together with SpoIVFB [85, 86]. SpoVFA is thought to serve as a platform to bring the inhibitory protein BofA in contact to SpoIVFB [86, 87]. A quasi-site-1 processing step is performed by the σ^G-controlled SpoIVB serine protease that is synthesized in the forespore, and secreted into the space between the IFM and OFM (Figure 13.1D). SpoIVB cleaves the extracellular domain of SpoIVFA at multiple sites [88–90], which is a conformational signal that activates SpoIVFB to cleave pro-σ^K and release σ^K to the cytoplasm of the mother cell where it interacts with RNAP (Figure 13.1D). A second serine protease called CtpB also affects pro-σ^K processing [91]. It was proposed that the SpoIVB serine protease triggers pro-σ^K processing indirectly in a second mechanism by cleaving and activating CtpB as a second serine protease that in turn targets the extracellular domain of SpoIVFA [90]. Another report suggested that CtpB targets BofA [92]. Recently it was shown that SpoIVB and CtpB act independently but that SpoIVB has to be present that CtpB can access SpoIVFA for cleavage [93].

13.5.3
Regulation of the Cell Polarity Determinant PodJ of *Caulobacter crescentus*

In its cell cycle the α-proteobacterium *C. crescentus* executes an orchestrated cell differentiation program involving two different cell types: a motile swarmer cell and a sessile stalked cell. The chemotactic swarmer cell that has pili and a single flagellum at one cell pole is not able to initiate DNA replication until it converts to a stalked cell by replacing the flagellum and pili by a stalk with holdfast at its tip to adhere to surfaces. DNA replication and assembly of pili and the flagellum at the pole opposite to the stalk precede an asymmetric cell division that forms a new swarmer cell. The polarity determinant protein PodJ effects a spatial control in that developmental program. The *podJ* gene encodes a 974-amino-acid protein with one transmembrane domain of type II topology (Figure 13.1E). It is exclusively expressed in the predivisional cell and its product PodJ$_L$ localizes to the incipient swarmer pole to recruit factors for pili biogenesis. During cell division, the periplasmic domain of PodJ$_L$ is proteolytically removed to form PodJ$_S$ that stays attached to the flagellated pole, and that is necessary for chemotaxis and later for holdfast formation [94, 95]. The protease performing this site-1 proteolytic step was identified as PerP (Figure 13.1E) – a protein with a putative signal sequence or membrane anchor at its N-terminus and a conserved aspartic protease motif in its periplasmic domain [96]. PodJ$_S$ is finally cleared from the cells by the action of the S2P family protease MmpA during the swarmer-to-stalked transition, releasing its N-terminus into the cytosol that is most probably further degraded (Figure 13.1E). Interestingly, the *mmpA* knockout strain can be complemented with *E. coli rseP* [97]. In the absence of PerP or MmpA, PodJ isoforms loose their asymmetric distribution in the cells; however, no obvious phenotype of these mutants can be discerned for unknown reasons. Timing of RIP of PodJ seems to be performed by transcriptional regulation of *perP* expression. Strains with constitutively expressed *perP* are unable to form pili [96].

13.6
Involvement of RIP in Cell–Cell Communication

13.6.1
Production of Peptide Sex Pheromones in *Enterococcus faecalis*

E. faecalis is a Gram-positive opportunistic human pathogen harboring mobile genetic elements that provide antibiotic resistance and virulence determinants. One of these elements is pAD1, encoding a hemolysin and the resistance to ultraviolet light. pAD1 is a highly conjugative plasmid that mediates a mating response between plasmid-free cells that produce a peptide sex pheromone, and pAD1-containing cells that produce a plasmid-encoded surface protein important for initiating contact between the donor and recipient cell. The pheromone called cAD1 is an octapeptide that is transported into the pAD1-containing cells and binds to a repressor protein of mating-responsive genes on pAD1, thereby activating the mating response. Self-induction of conjugation functions among pAD1-containing cells is prevented by the partial shutdown of pheromone production and by the production of the pAD1-encoded inhibitor octapeptide iAD1 that competitively inhibits pAD1 function [98]. It was shown that production of the cAD1 octapeptide and of several other pheromones depends on a S2P-familiy I-CLiP that was named Eep (Figure 13.1F) [99]. Genome sequencing revealed that the octapeptide sequence is part of a signal sequence of a surface lipoprotein (Cad) of unknown function; the same is true for other peptide sex pheromones [100]. Indirect evidence suggested that pre-Cad is secreted and then processed by the lipoprotein signal peptidase SPase II and that the remaining signal peptide is cleaved by Eep to produce the octapeptide pheromone (Figure 13.1F). Moreover, production of the plasmid-encoded precursor of the iAD1 inhibitory octapeptide that resembles a signal sequence is dependent on Eep [101]. Most recently it was shown that Eep recognizes amino acids N-terminal to the *E. faecalis* pheromone cCF10 in the signal sequence of the lipoprotein precursor CcfA [102].

13.6.2
Rhomboid-Mediated QS in *Providencia stuartii*

For the Gram-negative human pathogen *P. stuartii* it was shown that the rhomboid I-CLiP AarA is responsible for the production of an unknown extracellular signal that mediates QS [103]. In a search originated to identify rhomboids of the related bacterium *Proteus mirabilis*, the *tatA* gene was found as a multicopy suppressor of the *P. stuartii aarA* knockout phenotype. TatA is a bitopic protein of type I topology (C-in, N-out) that is part of the TAT protein secretion machinery. The TatA protein of *P. stuartii* possesses an N-terminal extension of seven amino acids when compared with *E. coli* or *P. mirabilis* TatA. This extension is clipped of by the AarA rhomboid I-CLiP, which activates TatA and TAT-dependent protein translocation, and enables secretion of the QS signal from the cell [14]. Although this mechanism of intramembrane proteolysis of TatA is indirect with respect to transmembrane signaling and there is no indication for regulation, it is the first report of a bacterial rhomboid substrate, and may be the first indication that I-CLiP function in bacteria is even more diverse and involved in many more important processes than we can imagine today.

13.7
Conclusions

We are just beginning to understand the importance of RIP in transmembrane signaling and are still far from having a complete picture of its relevance in diverse regulatory processes in bacteria. Intramembrane proteases are believed to be suited as good targets for new antimicrobial drugs; therefore, RIP in bacteria is an interesting and prospering field for future investigations.

References

1 Brown, M.S., Ye, J., Rawson, R.B., and Goldstein, J.L. (2000) Regulated intramembrane proteolysis: a control mechanism conserved from bacteria to humans. *Cell*, **100**, 391–398.

2 Urban, S. and Freeman, M. (2002) Intramembrane proteolysis controls diverse signaling pathways throughout evolution. *Curr. Opin. Genet. Dev.*, **12**, 512–518.

3 Wolfe, M.S. and Kopan, R. (2004) Intramembrane proteolysis: theme and variations. *Science*, **305**, 1119–1123.

4 Ehrmann, M. and Clausen, T. (2004) Proteolysis as a regulatory mechanism. *Annu. Rev. Genet.*, **38**, 709–724.

5 Weihofen, A. and Martoglio, B. (2003) Intramembrane-cleaving proteases: controlled liberation of proteins and bioactive peptides. *Trends Cell Biol.*, **13**, 71–78.

6 Maegawa, S., Ito, K., and Akiyama, Y. (2005) Proteolytic action of GlpG, a rhomboid protease in the *Escherichia coli* cytoplasmic membrane. *Biochemistry*, **44**, 13543–13552.

7 Wang, Y., Zhang, Y., and Ha, Y. (2006) Crystal structure of a rhomboid family intramembrane protease. *Nature*, **444**, 179–180.

8 Wu, Z., Yan, N., Feng, L., Oberstein, A., Yan, H., Baker, R.P., Gu, L., Jeffrey, P.D.,

Urban, S., and Shi, Y. (2006) Structural analysis of a rhomboid family intramembrane protease reveals a gating mechanism for substrate entry. *Nat. Struct. Mol. Biol.*, **13**, 1084–1091.

9 Akiyama, Y. and Maegawa, S. (2007) Sequence features of substrates required for cleavage by GlpG, an *Escherichia coli* rhomboid protease. *Mol. Microbiol.*, **64**, 1028–1037.

10 Maegawa, S., Koide, K., Ito, K., and Akiyama, Y. (2007) The intramembrane active site of GlpG, an *E. coli* rhomboid protease, is accessible to water and hydrolyses an extramembrane peptide bond of substrates. *Mol. Microbiol.*, **64**, 435–447.

11 Baker, R.P., Young, K., Feng, L., Shi, Y., and Urban, S. (2007) Enzymatic analysis of a rhomboid intramembrane protease implicates transmembrane helix 5 as the lateral substrate gate. *Proc. Natl. Acad. Sci. USA*, **104**, 8257–8262.

12 Ben-Shem, A., Fass, D., and Bibi, E. (2007) Structural basis for intramembrane proteolysis by rhomboid serine proteases. *Proc. Natl. Acad. Sci. USA*, **104**, 462–466.

13 Urban, S. and Shi, Y. (2008) Core principles of intramembrane proteolysis: comparison of rhomboid and site-2 family proteases. *Curr. Opin. Struct. Biol.*, **4**, 432–441.

14 Stevenson, L.G., Strisovsky, K., Clemmer, K.M., Bhatt, S., Freeman, M., and Rather, P.N. (2007) Rhomboid protease AarA mediates quorum-sensing in *Providencia stuartii* by activating TatA of the twin-arginine translocase. *Proc. Natl. Acad. Sci. USA*, **104**, 1003–1008.

15 Lemieux, M.J., Fischer, S.J., Cherney, M.M., Bateman, K.S., and James, M.N. (2007) The crystal structure of the rhomboid peptidase from *Haemophilus influenzae* provides insight into intramembrane proteolysis. *Proc. Natl. Acad. Sci. USA*, **104**, 750–754.

16 Feng, L., Yan, H., Wu, Z., Yan, N., Wang, Z., Jeffrey, P.D., and Shi, Y. (2007) Structure of a site-2 protease family intramembrane metalloprotease. *Science*, **318**, 1608–1612.

17 Urban, S. and Shi, Y. (2008) Core principles of intramembrane proteolysis: comparison of rhomboid and site-2 family proteases. *Curr. Opin. Struct. Biol.*, **18**, 432–441.

18 Koide, K., Ito, K., and Akiyama, Y. (2008) Substrate recognition and binding by RseP, an *Escherichia coli* intramembrane protease. *J. Biol. Chem.*, **283**, 9562–9570.

19 Akiyama, Y., Kanehara, K., and Ito, K. (2004) RseP (YaeL), an *Escherichia coli* RIP protease, cleaves transmembrane sequences. *EMBO J.*, **23**, 4434–4442.

20 Kanehara, K., Ito, K., and Akiyama, Y. (2003) YaeL proteolysis of RseA is controlled by the PDZ domain of YaeL and a Gln-rich region of RseA. *EMBO J.*, **22**, 6389–6398.

21 Grigorova, I.L., Chaba, R., Zhong, H.J., Alba, B.M., Rhodius, V., Herman, C., and Gross, C.A. (2004) Fine-tuning of the *Escherichia coli* σ^E envelope stress response relies on multiple mechanisms to inhibit signal-independent proteolysis of the transmembrane anti-sigma factor, RseA. *Genes Dev.*, **18**, 2686–2697.

22 Schöbel, S., Zellmeier, S., Schumann, W., and Wiegert, T. (2004) The *Bacillus subtilis* σ^W anti-sigma factor RsiW is degraded by intramembrane proteolysis through YluC. *Mol. Microbiol.*, **52**, 1091–1105.

23 Bramkamp, M., Weston, L., Daniel, R.A., and Errington, J. (2006) Regulated intramembrane proteolysis of FtsL protein and the control of cell division in *Bacillus subtilis*. *Mol. Microbiol.*, **62**, 580–591.

24 Lonetto, M.A., Brown, K.L., Rudd, K.E., and Buttner, M.J. (1994) Analysis of the *Streptomyces coelicolor sigE* gene reveals the existence of a subfamily of eubacterial RNA polymerase σ factors involved in the regulation of extracytoplasmic functions. *Proc. Natl. Acad. Sci. USA*, **91**, 7573–7577.

25 Helmann, J.D. (2002) The extracytoplasmic function (ECF) sigma factors. *Adv. Microbiol. Physiol.*, **46**, 47–110.

26 Ades, S.E. (2004) Control of the alternative sigma factor σ^E in *Escherichia coli*. *Curr. Opin. Microbiol.*, **7**, 157–162.

27 Alba, B.M. and Gross, C.A. (2004) Regulation of the *Escherichia coli*

27 σ^E-dependent envelope stress response. *Mol. Microbiol.*, **52**, 613–619.

28 Dartigalongue, C., Missiakas, D., and Raina, S. (2001) Characterization of the *Escherichia coli* σ^E regulon. *J. Biol. Chem.*, **276**, 20866–20875.

29 Missiakas, D., Mayer, M.P., Lemaire, M., Georgopoulos, C., and Raina, S. (1997) Modulation of the *Escherichia coli* σ^E (RpoE) heat-shock transcription-factor activity by the RseA, RseB and RseC proteins. *Mol. Microbiol.*, **24**, 355–371.

30 De Las, P.A., Connolly, L., and Gross, C.A. (1997) The σ^E-mediated response to extracytoplasmic stress in *Escherichia coli* is transduced by RseA and RseB, two negative regulators of σ^E. *Mol. Microbiol.*, **24**, 373–385.

31 Campbell, E.A., Tupy, J.L., Gruber, T.M., Wang, S., Sharp, M.M., Gross, C.A., and Darst, S.A. (2003) Crystal structure of *Escherichia coli* σ^E with the cytoplasmic domain of its anti-sigma RseA. *Mol. Cell*, **11**, 1067–1078.

32 Ades, S.E., Connolly, L.E., Alba, B.M., and Gross, C.A. (1999) The *Escherichia coli* σ^E-dependent extracytoplasmic stress response is controlled by the regulated proteolysis of an anti-sigma factor. *Genes Dev.*, **13**, 2449–2461.

33 Alba, B.M., Leeds, J.A., Onufryk, C., Lu, C.Z., and Gross, C.A. (2002) DegS and YaeL participate sequentially in the cleavage of RseA to activate the σ^E-dependent extracytoplasmic stress response. *Genes Dev.*, **16**, 2156–2168.

34 Kanehara, K., Ito, K., and Akiyama, Y. (2002) YaeL (EcfE) activates the σ^E pathway of stress response through a site-2 cleavage of anti-σ^E, RseA. *Genes Dev.*, **16**, 2147–2155.

35 Flynn, J.M., Levchenko, I., Sauer, R.T., and Baker, T.A. (2004) Modulating substrate choice: the SspB adaptor delivers a regulator of the extracytoplasmic-stress response to the AAA$^+$ protease ClpXP for degradation. *Genes Dev.*, **18**, 2292–2301.

36 Walsh, N.P., Alba, B.M., Bose, B., Gross, C.A., and Sauer, R.T. (2003) OMP peptide signals initiate the envelope-stress response by activating DegS protease via relief of inhibition mediated by its PDZ domain. *Cell*, **113**, 61–71.

37 Wilken, C., Kitzing, K., Kurzbauer, R., Ehrmann, M., and Clausen, T. (2004) Crystal structure of the DegS stress sensor: how a PDZ domain recognizes misfolded protein and activates a protease. *Cell*, **117**, 483–494.

38 Hasselblatt, H., Kurzbauer, R., Wilken, C., Krojer, T., Sawa, J., Kurt, J., Kirk, R., Hasenbein, S., Ehrmann, M., and Clausen, T. (2007) Regulation of the σ^E stress response by DegS: how the PDZ domain keeps the protease inactive in the resting state and allows integration of different OMP-derived stress signals upon folding stress. *Genes Dev.*, **21**, 2659–2670.

39 Flynn, J.M., Neher, S.M., Kim, Y.-I., Sauer, R.T., and Baker, T.A. (2003) Proteomic discovery of cellular substrates of the ClpXP protease reveals five classes of ClpX-recognition signals. *Mol. Cell*, **11**, 671–683.

40 Chaba, R., Grigorova, I.L., Flynn, J.M., Baker, T.A., and Gross, C.A. (2007) Design principles of the proteolytic cascade governing the σ^E-mediated envelope stress response in *Escherichia coli*: keys to graded, buffered, and rapid signal transduction. *Genes Dev.*, **21**, 124–136.

41 Cezairliyan, B.O. and Sauer, R.T. (2007) Inhibition of regulated proteolysis by RseB. *Proc. Natl. Acad. Sci. USA*, **104**, 3771–3776.

42 Kim, D.Y., Jin, K.S., Kwon, E., Ree, M., and Kim, K.K. (2007) Crystal structure of RseB and a model of its binding mode to RseA. *Proc. Natl. Acad. Sci. USA*, **104**, 8779–8784.

43 Wollmann, P. and Zeth, K. (2007) The structure of RseB: a sensor in periplasmic stress response of *E. coli*. *J. Mol. Biol.*, **372**, 927–941.

44 Kovacikova, G. and Skorupski, K. (2002) The alternative sigma factor σ^E plays an important role in intestinal survival and virulence in *Vibrio cholerae*. *Infect. Immun.*, **70**, 5355–5362.

45 Humphreys, S., Stevenson, A., Bacon, A., Weinhardt, A.B., and Roberts, M. (1999) The alternative sigma factor, σ^E, is critically important for the virulence of *Salmonella typhimurium*. *Infect. Immun.*, **67**, 1560–1568.

46 Testerman, T.L., Vazquez-Torres, A., Xu, Y., Jones-Carson, J., Libby, S.J., and Fang, F.C. (2002) The alternative sigma factor σ^E controls antioxidant defences required for *Salmonella* virulence and stationary-phase survival. *Mol. Microbiol.*, **43**, 771–782.

47 Craig, J.E., Nobbs, A., and High, N.J. (2002) The extracytoplasmic sigma factor, final σ^E, is required for intracellular survival of nontypeable *Haemophilus influenzae* in J774 macrophages. *Infect. Immun.*, **70**, 708–715.

48 Martin, D.W., Schurr, M.J., Yu, H., and Deretic, V. (1994) Analysis of promoters controlled by the putative sigma factor AlgU regulating conversion to mucoidy in *Pseudomonas aeruginosa*: relationship to σ^E and stress response. *J. Bacteriol.*, **176**, 6688–6696.

49 DeVries, C.A. and Ohman, D.E. (1994) Mucoid-to-nonmucoid conversion in alginate-producing *Pseudomonas aeruginosa* often results from spontaneous mutations in *algT*, encoding a putative alternate sigma factor, and shows evidence for autoregulation. *J. Bacteriol.*, **176**, 6677–6687.

50 Crouch, M.L., Becker, L.A., Bang, I.S., Tanabe, H., Ouellette, A.J., and Fang, F.C. (2005) The alternative sigma factor sigma is required for resistance of *Salmonella enterica* serovar *Typhimurium* to antimicrobial peptides. *Mol. Microbiol.*, **56**, 789–799.

51 Mathur, J., Davis, B.M., and Waldor, M.K. (2007) Antimicrobial peptides activate the *Vibrio cholerae* σ^E regulon through an OmpU-dependent signaling pathway. *Mol. Microbiol.*, **63**, 848–858.

52 Ramsey, D.M. and Wozniak, D.J. (2005) Understanding the control of *Pseudomonas aeruginosa* alginate synthesis and the prospects for management of chronic infections in cystic fibrosis. *Mol. Microbiol.*, **56**, 309–322.

53 Qiu, D., Eisinger, V.M., Rowen, D.W., and Yu, H.D. (2007) Regulated proteolysis controls mucoid conversion in *Pseudomonas aeruginosa*. *Proc. Natl. Acad. Sci. USA*, **104**, 8107–8112.

54 King-Lyons, N.D., Smith, K.F., and Connell, T.D. (2007) Expression of *hurP*, a gene encoding a prospective site 2 protease, is essential for heme-dependent induction of *bhuR* in *Bordetella bronchiseptica*. *J. Bacteriol.*, **189**, 6266–6275.

55 Helmann, J.D. (2006) Deciphering a complex genetic regulatory network: the *Bacillus subtilis* σ^W protein and intrinsic resistance to antimicrobial compounds. *Sci. Prog.*, **89**, 243–266.

56 Cao, M., Wang, T., Ye, R., and Helmann, J.D. (2002) Antibiotics that inhibit cell wall biosynthesis induce expression of the *Bacillus subtilis* σ^W and σ^M regulons. *Mol. Microbiol.*, **45**, 1267–1276.

57 Pietiäinen, M., Gardemeister, M., Mecklin, M., Leskela, S., Sarvas, M., and Kontinen, V.P. (2005) Cationic antimicrobial peptides elicit a complex stress response in *Bacillus subtilis* that involves ECF-type sigma factors and two-component signal transduction systems. *Microbiology*, **151**, 1577–1592.

58 Butcher, B.G. and Helmann, J.D. (2006) Identification of *Bacillus subtilis* σ^W dependent genes that provide intrinsic resistance to antimicrobial compounds produced by bacilli. *Mol. Microbiol.*, **60**, 765–782.

59 Wiegert, T., Homuth, G., Versteeg, S., and Schumann, W. (2001) Alkaline shock induces the *Bacillus subtilis* σ^W regulon. *Mol. Microbiol.*, **41**, 59–71.

60 Zellmeier, S., Schumann, W., and Wiegert, T. (2006) Involvement of Clp protease activity in modulating the *Bacillus subtilis* σ^W stress response. *Mol. Microbiol.*, **61**, 1569–1582.

61 Heinrich, J. and Wiegert, T. (2006) YpdC determines site-1 degradation in regulated intramembrane proteolysis of the RsiW anti-sigma factor of *Bacillus subtilis*. *Mol. Microbiol.*, **62**, 566–579.

62 Ellermeier, C.D. and Losick, R. (2006) Evidence for a novel protease governing regulated intramembrane proteolysis and resistance to antimicrobial peptides in *Bacillus subtilis*. *Genes Dev.* **20**, 1911–1922.

63 Pei, J. and Grishin, N.V. (2001) Type II CAAX prenyl endopeptidases belong to a novel superfamily of putative

membrane-bound metalloproteases. *Trends Biochem. Sci.*, **26**, 275–277.

64 Makinoshima, H. and Glickman, M.S. (2005) Regulation of *Mycobacterium tuberculosis* cell envelope composition and virulence by intramembrane proteolysis. *Nature*, **436**, 406–409.

65 Cole, S.T., Brosch, R., Parkhill, J., Garnier, T., Churcher, C., Harris, D., Gordon, S.V., Eiglmeier, K., Gas, S., Barry, C.E. III *et al.* (1998) Deciphering the biology of *Mycobacterium tuberculosis* from the complete genome sequence. *Nature*, **393**, 537–544.

66 Hahn, M.Y., Raman, S., Anaya, M., and Husson, R.N. (2005) The *Mycobacterium tuberculosis* extracytoplasmic-function sigma factor SigL regulates polyketide synthases and secreted or membrane proteins and is required for virulence. *J. Bacteriol.*, **187**, 7062–7071.

67 Said-Salim, B., Mostowy, S., Kristof, A.S., and Behr, M.A. (2006) Mutations in *Mycobacterium tuberculosis Rv0444c*, the gene encoding anti-SigK, explain high level expression of MPB70 and MPB83 in *Mycobacterium bovis*. *Mol. Microbiol.*, **62**, 1251–1263.

68 Martinez-Argudo, I., Ruiz-Vazquez, R.M., and Murillo, F.J. (1998) The structure of an ECF-sigma-dependent, light-inducible promoter from the bacterium *Myxococcus xanthus*. *Mol. Microbiol.*, **30**, 883–893.

69 Browning, D.F., Whitworth, D.E., and Hodgson, D.A. (2003) Light-induced carotenogenesis in *Myxococcus xanthus*: functional characterization of the ECF sigma factor CarQ and antisigma factor CarR. *Mol. Microbiol.*, **48**, 237–251.

70 Krukonis, E.S., Yu, R.R., and DiRita, V.J. (2000) The *Vibrio cholerae* ToxR/TcpP/ToxT virulence cascade: distinct roles for two membrane-localized transcriptional activators on a single promoter. *Mol. Microbiol.*, **38**, 67–84.

71 Crawford, J.A., Krukonis, E.S., and DiRita, V.J. (2003) Membrane localization of the ToxR winged-helix domain is required for TcpP-mediated virulence gene activation in *Vibrio cholerae*. *Mol. Microbiol.*, **47**, 1459–1473.

72 Krukonis, E.S. and DiRita, V.J. (2003) DNA binding and ToxR responsiveness by the wing domain of TcpP, an activator of virulence gene expression in *Vibrio cholerae*. *Mol. Cell*, **12**, 157–165.

73 Matson, J.S. and DiRita, V.J. (2005) Degradation of the membrane-localized virulence activator TcpP by the YaeL protease in *Vibrio cholerae*. *Proc. Natl. Acad. Sci. USA*, **102**, 16403–16408.

74 Tetsch, L., Koller, C., Haneburger, I., and Jung, K. (2008) The membrane-integrated transcriptional activator CadC of *Escherichia coli* senses lysine indirectly via the interaction with the lysine permease LysP. *Mol. Microbiol.*, **67**, 570–583.

75 Lee, Y.H., Kim, J.H., Bang, I.S., and Park, Y.K. (2008) The membrane-bound transcriptional regulator CadC is activated by proteolytic cleavage in response to acid stress. *J. Bacteriol.*, **190**, 5120–5126.

76 Daniel, R.A., Harry, E.J., Katis, V.L., Wake, R.G., and Errington, J. (1998) Characterization of the essential cell division gene *ftsL* (*yIID*) of *Bacillus subtilis* and its role in the assembly of the division apparatus. *Mol. Microbiol.*, **29**, 593–604.

77 Daniel, R.A. and Errington, J. (2000) Intrinsic instability of the essential cell division protein FtsL of *Bacillus subtilis* and a role for DivIB protein in FtsL turnover. *Mol. Microbiol.*, **36**, 278–289.

78 Goranov, A.I., Katz, L., Breier, A.M., Burge, C.B., and Grossman, A.D. (2005) A transcriptional response to replication status mediated by the conserved bacterial replication protein DnaA. *Proc. Natl. Acad. Sci. USA*, **102**, 12932–12937.

79 Kroos, L. and Yu, Y.T. (2000) Regulation of sigma factor activity during *Bacillus subtilis* development. *Curr. Opin. Microbiol.*, **3**, 553–560.

80 Errington, J. (2003) Regulation of endospore formation in *Bacillus subtilis*. *Nat. Rev. Microbiol.*, **1**, 117–126.

81 Hilbert, D.W. and Piggot, P.J. (2004) Compartmentalization of gene expression during *Bacillus subtilis* spore formation. *Microbiol. Mol. Biol. Rev.*, **68**, 234–262.

82 Cutting, S., Oke, V., Driks, A., Losick, R., Lu, S., and Kroos, L. (1990) A forespore checkpoint for mother cell gene expression during development in *B. subtilis. Cell*, **62**, 239–250.

83 Rudner, D.Z., Fawcett, P., and Losick, R. (1999) A family of membrane-embedded metalloproteases involved in regulated proteolysis of membrane-associated transcription factors. *Proc. Natl. Acad. Sci. USA*, **96**, 14765–14770.

84 Yu, Y.T. and Kroos, L. (2000) Evidence that SpoIVFB is a novel type of membrane metalloprotease governing intercompartmental communication during *Bacillus subtilis* sporulation. *J. Bacteriol.*, **182**, 3305–3309.

85 Resnekov, O., Alper, S., and Losick, R. (1996) Subcellular localization of proteins governing the proteolytic activation of a developmental transcription factor in *Bacillus subtilis. Genes Cells*, **1**, 529–542.

86 Rudner, D.Z. and Losick, R. (2002) A sporulation membrane protein tethers the pro-σ^K processing enzyme to its inhibitor and dictates its subcellular localization. *Genes Dev.*, **16**, 1007–1018.

87 Zhou, R. and Kroos, L. (2004) BofA protein inhibits intramembrane proteolysis of pro-σ^K in an intercompartmental signaling pathway during *Bacillus subtilis* sporulation. *Proc. Natl. Acad. Sci. USA*, **101**, 6385–6390.

88 Wakeley, P.R., Dorazi, R., Hoa, N.T., Bowyer, J.R., and Cutting, S.M. (2000) Proteolysis of SpoIVB is a critical determinant in signalling of pro-σ^K processing in *Bacillus subtilis. Mol. Microbiol.*, **36**, 1336–1348.

89 Dong, T.C. and Cutting, S.M. (2003) SpoIVB-mediated cleavage of SpoIVFA could provide the intercellular signal to activate processing of pro-σ^K in *Bacillus subtilis. Mol. Microbiol.*, **49**, 1425–1434.

90 Campo, N. and Rudner, D.Z. (2006) A branched pathway governing the activation of a developmental transcription factor by regulated intramembrane proteolysis. *Mol. Cell*, **23**, 25–35.

91 Pan, Q., Losick, R., and Rudner, D.Z. (2003) A second PDZ-containing serine protease contributes to activation of the sporulation transcription factor σ^K in *Bacillus subtilis. J. Bacteriol.*, **185**, 6051–6056.

92 Zhou, R. and Kroos, L. (2005) Serine proteases from two cell types target different components of a complex that governs regulated intramembrane proteolysis of pro-σ^K during *Bacillus subtilis* development. *Mol. Microbiol.*, **58**, 835–846.

93 Campo, N. and Rudner, D.Z. (2007) SpoIVB and CtpB are both forespore signals in the activation of the sporulation transcription factor σ^K in *Bacillus subtilis. J. Bacteriol.*, **189**, 6021–6027.

94 Viollier, P.H., Sternheim, N., and Shapiro, L. (2002) Identification of a localization factor for the polar positioning of bacterial structural and regulatory proteins. *Proc. Natl. Acad. Sci. USA*, **99**, 13831–13836.

95 Hinz, A.J., Larson, D.E., Smith, C.S., and Brun, Y.V. (2003) The *Caulobacter crescentus* polar organelle development protein PodJ is differentially localized and is required for polar targeting of the PleC development regulator. *Mol. Microbiol.*, **47**, 929–941.

96 Chen, J.C., Hottes, A.K., McAdams, H.H., McGrath, P.T., Viollier, P.H., and Shapiro, L. (2006) Cytokinesis signals truncation of the PodJ polarity factor by a cell cycle-regulated protease. *EMBO J.*, **25**, 377–386.

97 Chen, J.C., Viollier, P.H., and Shapiro, L. (2005) A membrane metalloprotease participates in the sequential degradation of a *Caulobacter* polarity determinant. *Mol. Microbiol.*, **55**, 1085–1103.

98 Dunny, G.M. and Leonard, B.A. (1997) Cell–cell communication in grampositive bacteria. *Annu. Rev. Microbiol.*, **51**, 527–564.

99 An, F.Y., Sulavik, M.C., and Clewell, D.B. (1999) Identification and characterization of a determinant (*eep*) on the *Enterococcus faecalis* chromosome that is involved in production of the peptide sex pheromone cAD1. *J. Bacteriol.*, **181**, 5915–5921.

100 Clewell, D.B., An, F.Y., Flannagan, S.E., Antiporta, M., and Dunny, G.M.

(2000) Enterococcal sex pheromone precursors are part of signal sequences for surface lipoproteins. *Mol. Microbiol.*, **35**, 246–247.

101 An, F.Y. and Clewell, D.B. (2002) Identification of the cAD1 sex pheromone precursor in *Enterococcus faecalis*. *J. Bacteriol.*, **184**, 1880–1887.

102 Chandler, J.R. and Dunny, G.M. (2008) Characterization of the sequence specificity determinants required for processing and control of sex pheromone by the intramembrane protease Eep and the plasmid-encoded protein PrgY. *J. Bacteriol.*, **190**, 1172–1183.

103 Rather, P.N., Ding, X., Baca-DeLancey, R.R., and Siddiqui, S. (1999) *Providencia stuartii* genes activated by cell-to-cell signaling and identification of a gene required for production or activity of an extracellular factor. *J. Bacteriol.*, **181**, 7185–7191.

14
Protein Chemical and Electron Paramagnetic Resonance Spectroscopic Approaches to Monitor Membrane Protein Structure and Dynamics –
Methods Chapter
Daniel Hilger and Heinrich Jung

14.1
Introduction

Site-directed protein labeling techniques alone or in combination with spectroscopic methods such as fluorescence or electron paramagnetic resonance (EPR) spectroscopy are widely used to probe protein structure and structural alterations relevant for function [1–3]. The respective approaches are frequently applied to membrane proteins, since the determination of the three-dimensional structures of this important group of proteins is still highly challenging despite recent advances [4–6]. Due to their hydrophobic nature and, in part, high conformational flexibility, membrane proteins are often very difficult to crystallize. Furthermore, X-ray structures reflect only one conformational state and in many cases other conformational intermediates are hard to obtain via crystallography. Nuclear magnetic resonance spectroscopy is a powerful technology to characterize these dynamics as well as the structure and interactions of proteins at atomistic detail in near-physiological milieus [7–9]. However, its application is restricted with respect to the size of the molecule that can be analyzed. By contrast, there is no size limitation for site-directed protein-labeling techniques and proteins can be analyzed in their native environment, such as the membrane. While the latter techniques can in principle provide information on all levels of structural organization of a protein, they are neither thought to replace structure determination (e.g., by crystallization) nor do they become dispensable when a high-resolution structure is available. Instead, high-resolution structures may serve as guides for labeling approaches, which in turn may provide complementary information, for example, on the structure of yet unresolved protein regions, sites of ligand binding, regulatory mechanisms, membrane topography of a protein, and most importantly protein dynamics.

In this chapter, we discuss labeling techniques relying on site-directed cysteine modification. We start with an overview of applications of the cysteine chemistry for evaluating protein structure and dynamics. This overview is given since cysteine labeling *per se* can provide information on all levels of structural organization without requiring advanced biophysical technology. Subsequently, we specifically focus on site-directed spin labeling in conjunction with EPR spectroscopy. The utilization of this combinatory approach for the investigation of static and dynamic properties of biomacromolecules is discussed.

14.2
Cysteine Chemistry

14.2.1
General Considerations

Cysteine, an amino acid of average size and relatively low side-chain polarity, is frequently found in the hydrophobic interior of soluble proteins or as a constituent of catalytically active sites [10]. In membrane proteins cysteine is usually located in transmembrane domains. The thiolate anion resulting from deprotonation of the thiol group (typical pK_a 8.3) represents the most powerful nucleophile available from the naturally occurring amino acids. This property makes cysteine amenable to highly specific labeling under mild reaction conditions (e.g., neutral pH, room temperature). Modification of cysteine most likely proceeds via a nucleophilic addition or displacement reaction [1]. The following groups of cysteine-modifying reagents are most commonly used: alkylating reagents including alkyl halides, haloacetamides, and maleimides form a very stable thioether bond with cysteine. The alkylation reaction can be quenched at desired timepoints by adding thiol compounds like dithiothreitol or β-mercaptoethanol. Labeled protein and quencher-reagent adducts can easily be separated by dialysis or gel filtration, or in the case of vesicular systems by centrifugation. Alkylation of cysteine is highly specific when protein labeling is carried out at a pH below 8.0 (e.g., between pH 6.5 and 8) which prevents modification of amino groups as present, for example, in lysine (typical pK_a of the side-chain group 10.8). Alternatively, highly selective cysteine modification can be achieved via formation of mixed disulfides. For this purpose, methane thiosulfonate (MTS) derivatives are currently most frequently used (Figure 14.1) [11]. If necessary, the reaction can be stopped by bovine serum albumin (BSA) (e.g., in a vesicular system when labeled target and BSA can be easily separated by centrifugation), but not by reducing agents. Finally, cysteine is specifically modified by organic mercurial compounds (e.g., *p*-hydroxymercuribenzoate) and metal ions, such as silver, gold, mercury, and zinc [1, 12].

The unique reactivity of cysteine has prompted investigators to use site-specific mutagenesis to place cysteine at particular positions in a protein for the subsequent attachment of probe molecules [1, 6, 11, 13–15]. The following prerequisites of cysteine labeling need to be fulfilled. When native cysteine interferes with the labeling reaction and ligand protection is not applicable, a functional cysteine-free protein variant needs to be generated. With respect to target function, alanine, serine, or valine is usually tolerated best in place of cysteine; thereby, the choice of the

Fluoresceine-5-maleimide
MTS-CH$_3$ (MMTS)
MTS-CH$_2$CH$_2$SO$_3^-$ (MTSES)
MTS-CH$_2$CH$_2$N$^+$(CH$_3$)$_3$ (MTSET)

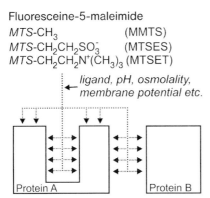

Figure 14.1 Schematic presentation of the cysteine accessibility approach. Scanning of the accessibility of site-specifically introduced cysteine residues to thiol reagents identifies solvent-exposed positions in a protein or protein complex. Depending on the size and chemical nature of the thiol reagent, information on the existence and location of hydrophilic cavities or channels in a protein may be obtained. Furthermore, different conformational states of a protein induced, for example, by ligand binding or alteration of physicochemical parameters may be detected via an altered cysteine accessibility pattern. Inhibition of cysteine modification by ligand combined with ligand-affinity analyses may help to identify residues lining a ligand-binding crevice. Protein–protein interaction sites can be determined by comparing cysteine accessibility patterns of participating proteins in the monomeric and homo- or hetero-oligomeric state. Alternatively, a monomeric protein partner is labeled with a bulky thiol reagent, and the effect of labeling on subsequent oligomer formation and complex activity is analyzed to obtain information on crucial interaction sites. (See text for details).

replacement depends on space and polarity of the local environment of the native cysteine. Subsequently, protein variants are engineered that contain one or two cysteine residues at the desired positions. Although cysteine substitutions are usually well tolerated by proteins, it is necessary to test protein function in order to get an idea whether the overall three-dimensional structure is likely to be similar to the wild-type. Furthermore, cysteine residues supposed to be labeled need to be kept reduced in order to prevent oxidation of the thiol group (e.g., cysteic acid or disulfide formation). This can be achieved by preincubation with dithiothreitol or β-mercaptoethanol; however, these thiol compounds need to be removed prior to the labeling reaction or the labeling reagent must be added in concentrations exceeding the thiol compounds. Alternatively, tris(2-carboxymethyl)phosphine (TCEP) can be applied as reducing reagent [16]. TCEP does not need to be removed before labeling with alkylating reagents as it does not contain a thiol group. For use with vesicular systems it should be noted that TCEP is membrane impermeable.

A simple method to detect cysteine labeling measures the effect of the modification on protein activity. Here, negative results have to be interpreted with caution as they reflect either toleration of the label at the respective site or no labeling. Along this line, partial inhibition of activity can be due to partial labeling and/or partial toleration of the label. The success of the labeling reaction can also be determined by titrating unmodified cysteine residues after the labeling reaction. Traditionally, free cysteine is quantified via mixed disulfide formation with Ellman's reagent (5,5'-dithiobis(2-nitrobenzoic acid), $\lambda_{max} = 412$ nm, $\varepsilon = 14\,150$ M^{-1} cm^{-1} in 0.1 M phosphate buffer, neutral pH, $\varepsilon = 13\,700$ M^{-1} cm^{-1} in 6 M guanidinium hydrochloride or 8 M urea) [17, 18]. To overcome steric and electrostatic constraints of the Ellman's assay, mediators (e.g., cystamine) are added to the assay or Ellman's reagent is replaced by smaller and uncharged 4,4'-dithiodipyridine ($\lambda_{max} = 324$ nm, $\varepsilon = 21\,400$ M^{-1} cm^{-1} in 0.1 M phosphate buffer, neutral pH, $\varepsilon = 18\,500$ M^{-1} cm^{-1} in 5 M guanidinium hydrochloride). Either way, thiol concentrations down to 0.6 nmol/ml can be determined [17]. To detected completely buried thiol groups, it may still be necessary to denature the protein, for example, by treatment with 6 M guanidinium hydrochloride or 8 M urea. However, it should be noted that integral membrane proteins may resist all trials of complete unfolding. Apart from indirect methods, cysteine modification can be directly detected by Western blot analysis (e.g., when a cysteine-specific biotin derivative is used [19]) or by utilizing radioactive or spectroscopically active (e.g., fluorescent or paramagnetic) thiol reagents. The spectroscopic methods usually require protein purification before detection. Time courses of labeling of purified protein can continuously be followed utilizing fluorescent thiol reagents (e.g., 4-acetamido-4'-maleimidylstilbene-2,2'-disulfonate, N-(1-pyrene)maleimide, monobromobimane) that have no or very low fluorescence until they react with thiols to form fluorescent adducts (for further information see, for example, the *Handbook of Fluorescent Probes and Research Products* online addition: http://probes.invitrogen.com/handbook/). In addition, radioactive or spectroscopically active thiol reagents represent an alternative to Ellman's reagent and can in turn be used for titrating unmodified cysteine residues after labeling with hardly detectable compounds. Finally, labeling can be directly detected by mass spectrometry – a method that also works for large hydrophobic membrane proteins [20]. Further information and protocols for site-directed cysteine labeling can be found elsewhere [1, 5, 6, 11].

14.2.2
Applications of Cysteine Chemistry

14.2.2.1 Cysteine Accessibility Analyses
Cysteine accessibility analyses probe the differential reactivity of native or site-specifically introduced cysteine residues in proteins with thiol reagents of different size, polarity, and charge. These analyses discriminate between solvent-accessible and buried positions, thereby providing information on protein topology (Figure 14.1). For data interpretation it should be considered that not only steric constrains, but also the local environment of the targeted cysteine residues influences reactivity. In this context, it has been shown that the local electrostatic potential modulates the reactivity of cysteine in a protein probably via influencing formation and stability of the thiolate anion [21]. Along this line, cysteine modification is inhibited when the thiol group is exposed to an apolar environment as found, for example, in a phospholipid bilayer [22]. Furthermore, as static and dynamic properties of a protein influence the reactivity of individual cysteine residues, analysis of cysteine accessibility is often not an all-or-nothing approach. Instead, data

interpretation requires determination of precise labeling kinetics in order to discriminate between different locations. Furthermore, it is usually necessary to scan a series of successive positions in the domain of interest to obtain a convincing picture of its structural characteristics and dynamics. This is also necessary since an individual cysteine substitution may perturb the local structure.

In practice, cysteine accessibility analyses are often used to elucidate the topography of membrane proteins (e.g., determination of the boarders and orientation of transmembrane domains and connecting loops) (e.g., [13, 14, 19, 23, 24]). A location on the outside or inside of a vesicular system can be determined by comparing the results of cysteine modification with membrane impermeant (e.g., [2-(trimethylammonium)ethyl]-MTS ([2-(trimethylammonium)ethyl] methane thiosulfonate MTSET), (2-sulfonatoethyl)-MTS ((2-sulfonatoethyl) methane thiosulfonate MTSES)) and permeant (e.g., N-ethylmaleimide, methyl-MTS (methyl methane thiosulfonate (MMTS)) thiol reagents (Figure 14.1). The comparative analysis may be further substantiated by including right-side-out and inside-out or detergent-permeabilized membrane vesicles. It should be noted that the membrane permeability of a thiol reagents depends on the specific reaction conditions used for labeling. For example, following Fick's first law, the diffusive flux of a solute is proportional to its concentration gradient. As a consequence, whether a specific thiol reagent is able to modify cysteine located on the opposite side of the membrane crucially depends on the applied concentration of the reagent and the time of incubation. Alteration of these parameters may render a "membrane impermeant" into a "membrane permeant" reagent as demonstrated, for example, for biotin maleimide labeling of human P-glycoprotein [19].

Cysteine accessibility analyzes can also be used to map residues lining a hydrophilic cavity or channel in a protein as demonstrated for various receptors and transporters (e.g., [11, 22, 25, 26]). Positively charged MTSET and ethylammonium MTS and negatively charged MTSES are well suited for this purpose (Figure 14.1). They are very hydrophilic with a relative solubility in water: octanol greater than 2500 : 1 [27]. The largest, MTSET, fits into a cylinder 6 Å in diameter and 10 Å long; thus, the reagents are approximately the same size as low-molecular-weight ligands such as amines or amino acids [22]. This approach can be used to locate a ligand binding-site crevice in a protein. Labeling of cysteine lining the binding site should alter binding irreversibly and, reciprocally, ligand (substrates or competitive inhibitors) should retard the reaction with the thiol reagent [25]. Furthermore, secondary structure elements can be identified by analyzing the cysteine accessibility pattern of a domain located on a protein surface or lining a hydrophilic cavity or channel in a protein. If an ideal α-helix with 3.6 amino acids per turn is formed, then every third or fourth amino acid should be accessible from the solvent phase, with adjacent residues exposed depending on the width of the accessible stripe. In right-handed coiled-coil structures the pitch is about 3.9–4.0 and in left-handed coiled-coils 3.5 residues per turn. If the domain forms a β-strand, than every other residue might be exposed [11]. For example, mapping of highly exposed and buried positions in the adaptation region of the bacterial aspartate receptor fitted best to heptad repeat in a coiled-coil four-helix bundle [28].

Different functional states and underlying conformational alterations induced by ligand binding or changes of physical or chemical parameters (e.g., membrane potential, temperature, pH) may result in changes in the reactivity of substituted cysteines. These alterations can be assessed by comparing rates of reaction of cysteine with polar thiol reagents (e.g., MTS reagents) under the conditions of interest (Figure 14.1). For example, stimulation of reactivity by addition of a ligand most likely reflects a ligand-induced conformational alteration that moves the respective cysteine from a more buried (or apolar) into a more water-exposed environment. Inhibition of reactivity by a ligand may be due to either a conformational alteration or direct steric hindering when the thiol group is placed into a ligand-binding site crevice. Here, analysis of the effect of cysteine substitution and modification on ligand affinities facilitates data interpretation.

Finally, protein–protein interactions can be analyzed by an *in vitro* scanning approach termed protein interactions by cysteine modification (PICM) [29]. Here, cysteine residues are engineered on the surface of a protein of known structure and a bulky thiol reagent (e.g., fluoresceine-5-maleimide, tetramethylrhodamine-5-maleimide) is coupled to each cysteine residue. Next, the effect of both cysteine substitution and bulky probe attachment on complex assembly and activity is determined. Bulky probe coupling at an essential interaction site is expected to disrupt complex assembly and/or activity, while probe coupling outside the interaction site typically has little or no effect [29]. PICM has been applied to the bacterial chemotaxis pathway, where it has mapped out essential interactions sites between the chemoreceptor Tar and the histidine kinase CheA (e.g., [30]).

14.2.2.2 Proximity Relationships in Proteins by Cysteine Cross-Linking

Cysteine cross-linking provides information on proximity relationships in proteins and protein complexes. Estimation of intramolecular proximity relationships requires the introduction of pairs of cysteine residues at positions of interest; intermolecular distances are assessed by cross-linking cysteine residues individually introduced at the surface of interacting proteins. Cysteine at neighboring positions in a tertiary or quaternary structure may form disulfide bonds (ideal length 2.05 Å [10]) under oxidizing conditions. The latter conditions are found in the periplasm of Gram-negative bacteria, or they can be induced by adding the "zero-length cross-linker" $Cu(1,10-phenanthroline)_3$ or I_2 [6, 31]. Larger distances between two cysteine residues can be bridged by homobifunctional cross-linkers such as 1,6-bis-(maleimido)hexane (3.47–15.64 Å) or more rigid N,N'-p-phenylenedimaleimide (9.2–12.29 Å) [32]. Successful cross-linking can be detected by changes in the electrophoretic mobility of the target protein (or protein complex) in sodium dodecylsulfate–polyacrylamide gel electrophoresis. However, detection of intramolecular cross-linking in membrane protein is more demanding as the protein usually runs in diffuse bands. Here, safe detection of cross-linking may require introduction of a specific protease site between the cysteine positions, and cleavage of the protein after cross-linking (e.g., [33]). In addition, single cysteine variants should be tested in control experiments to discriminate intramolecular from potential intermolecular cross-linking. Furthermore, it should be considered that cross-linking has the tendency to underestimate distances. This is in most part due to the conformational flexibility of proteins that may cause a transient proximity of two positions sufficient for cross-linking.

Disulfide scanning has very successfully been used to map subunit interphases of various receptor oligomers (e.g., bacterial chemotaxis receptors, G-protein-coupled receptors) and to relate changes of intersubunit contacts to distinct functional states of the receptors (e.g., [6, 25, 28, 34–36]).

14.3
Site-Directed Spin Labeling and EPR Spectroscopy

14.3.1
Why EPR Spectroscopy?

The aforementioned site-directed cysteine labeling studies can provide information on all levels of structural organization of proteins without any notable and expensive technical requirements. This raises the question why a technical demanding biophysical approach such as EPR (also known as electron spin resonance) becomes a rapidly expanding field in characterizing the structure and dynamics of biomacromolecules, even when it addresses similar problems as classical biochemical approaches. Solid reasons for this are (i) the small size and low perturbing nature of the label used for EPR, (ii) the variety of different EPR techniques that can be accomplished with only a single type of reporter group, and (iii) the high sensitivity of EPR towards small changes in the microenvironment of the label. These properties can be used for qualitative *and* quantitative analyzes of different parameters specifying tertiary contacts and topographic localization of the label, the polarity and proticity in the immediate environment of the label, and the accessibility of the label to water. Furthermore, accurate distances between introduced reporter groups can be determined. To get a general idea of the potential of EPR spectroscopy, different EPR methods are briefly depicted and sample applications are provided.

14.3.2
Site-Directed Spin Labeling of Proteins

EPR spectroscopy is based on the measurement of the resonant absorption of microwave radiation by unpaired electrons in a strong magnetic field. However, with the exception of radical enzymes (e.g., metalloproteases), the vast majority of soluble and membrane proteins do not possess unpaired electrons and are not accessible to EPR analyses without introducing paramagnetic centers in form of stable radicals. Site-directed spin labeling (SDSL) is a well-established method to achieve the incorporation of paramagnetic spin labels in proteins. This approach basically follows the cysteine substitution approach described above; thereby, introduced thiol groups are modified with a specific nitroxide spin label [37]

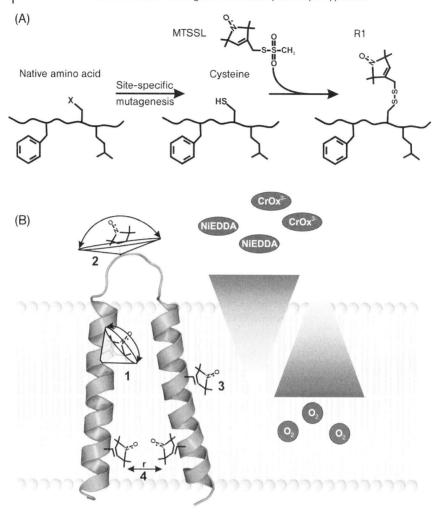

Figure 14.2 Strategy and applications of SDSL. (A) Schematic representation of the principle of SDSL of proteins. Shown is the reaction of the MTS spin label with the sulfhydryl group of cysteine to generate the spin label side-chain R1. (B) Overview of different structural information that can be obtained by EPR spectroscopy (modified from [2]). Analysis of spin label side-chain mobility reveals information on tertiary interaction with high (1) or low (2) restrictions of spin label dynamics. (3) The collision frequency of the spin label with paramagnetic quenchers of different polarity provides information of the spin label location with respect to the protein/water/membrane boundaries. The triangles represent the concentration gradients of polar reagents $CrOx^{3-}$ or NiEDDA and apolar oxygen, respectively. (4) Distance measurements in the range between 0.8 and 8 nm allows the determination of secondary, tertiary and quaternary structures. (See text for details).

(Figure 14.2A). Due to its small size and poor perturbing nature, (1-oxyl-2,2,5,5-tetramethylpyrroline-3-methyl)-MTS is the most commonly used thiol specific nitroxide spin label [38, 39]. In direct comparison to the chromophores used for fluorescence analyses, the molecular volume of the MTS spin label side-chain (R1) is much smaller, resembling the molecular volume of a large native amino acid. In addition, R1 is flexible, and capable of participating in polar and hydrophobic interactions, and therefore native structure and functions are retained in the majority of labeled proteins [2, 39, 40].

The EPR spectrum of protein-immobilized spin label yields structural information based on nitroxide side-chain mobility [41], its accessibility to collisions with lipid- or water-soluble quenchers [42], and the polarity of its immediate surrounding or distances between the nitroxide and a second paramagnetic center in the protein [43–45] (Figure 14.2B).

14.3.3
Information on Protein Structure and Dynamics Based on Spin Label Dynamics

The continuous-wave (cw) EPR spectrum of singly spin-labeled proteins measured at room temperature reveals information on the reorientational motion of the bound nitroxide side-chain, which is dependent on the local structure and dynamics of the protein [41]. The spin label motion results from rotational diffusion of the entire protein, the intrinsic flexibility of the backbone, and rotational isomerization about internal bonds of the side-chain [37, 46]. For high-molecular-weight and membrane-reconstituted proteins, the contributions from protein rotary diffusion are negligible with respect to the time range of EPR sensitivity [2]. Hence, the EPR spectra of spin-labeled proteins are considered to be affected mainly by the dynamics of the spin label side-chain and local backbone fluctuations. The amplitudes, rates, and geometry of the reorientational motion of the nitroxide side-chain are generally analyzed in terms of the spin label mobility that is reflected in the EPR spectral line shape [39]. The mobility can be quantified by two semiempirical mobility parameters: the line width of the central line of cw X-band EPR spectra (ΔH_0) (Figure 14.3A) and the spectral breadth, represented by the so-called second moment $<H^2>$ [39, 47]. With these parameters, correlations between the mobility of the spin label side-chain and protein topography have been established in soluble helical proteins T4 lysozyme

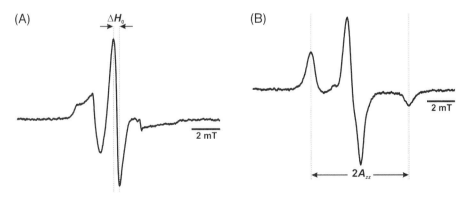

Figure 14.3 EPR spectra of a singly spin-labeled protein: (A) 293 K (room temperature) spectrum and (B) 170 K spectrum. Spectral parameters are indicated.

and annexin 12 [39, 48], as well as for the membrane protein bacteriorhodopsin [2]. At helix surface sites or loop regions, weak interactions between the nitroxide and other parts of the protein result in a high mobility of the spin label side-chain (Figure 14.2B). In this case, ΔH_0 and $<H^2>$ of the cw EPR spectrum are characterized by small values. On the contrary, at tertiary contacts or buried sites, strong van der Waals interactions with adjacent side-chains or backbone atoms restrict the mobility of the spin label leading to an increase of ΔH_0 and $<H^2>$ [39] (Figure 14.2B). Apart from the discrimination between surface-exposed positions with high conformational freedom and buried, conformationally restricted sites, a sequential analysis of a series of spin-labeled protein variants also allows the detection of secondary structure elements [49–52]. Furthermore, time-resolved measurements of spin label dynamics can follow conformational changes up to the 100 μs range [53].

14.3.3.1 Example

Both applications have been used to investigate the structure and dynamics of the transducer protein NpHtrII associated with the phototaxis receptor sensory rhodopsin II (NpSRII) from Natronomonas pharaonis [54]. The nitroxide mobilities of several spin-labeled residues at the C-terminal end of the second transmembrane domain (TMII) of the transducer NpHtrII were determined and correlated to the position of the analyzed amino acids in the protein primary structure. The plot of the $<H^2>$ versus residue number shows a periodic alternation of the mobility in this region of TMII, which is characteristic for α-helices in tertiary contacts. Close tertiary interactions were shown to be formed between TMII of NpHtrII and helix F of NpSrII [54]. In addition, to investigate conformational changes in these transmembrane domains during signal transduction, time-resolved EPR measurements were performed. Upon light activation of the receptor, positions in helix F and TMII show a transient increase in the mobility that in turn was attributed to a rotational motion of both domains. These conformational changes are thought to be of importance for the signal transfer from the receptor to the cytoplasmic signal domain of NpHtrII [54].

14.3.4
Information on Protein Structure and Dynamics Based on Spin Label Accessibility

The accessibility of the nitroxide to polar and apolar paramagnetic reagents can provide information on the secondary and tertiary structure of proteins as well as on the location of the nitroxide side-chain with respect to the protein/water/membrane boundaries [42]. The accessibility of the spin label is characterized by the collision frequency of the nitroxide with paramagnetic exchange reagents, and depends on the product of the local concentration and the translational diffusion coefficient of the paramagnetic probe. The most commonly used reagents are water-soluble paramagnetic Ni(II)-ethylenediamine diacetate (NiEDDA) or chromium oxalate ($CrOx^{3-}$) and molecular oxygen. In water/membrane systems, polar metal complexes preferentially partition into the aqueous phase, whereas apolar oxygen can penetrate the lipid bilayer [37] (Figure 14.2B). The relation between the collision frequency of the spin label side-chains with NiEDDA or $CrOx^{3-}$ and molecular oxygen can be used to determine whether a side-chain is exposed to water or to the lipid bilayer. Spin labels that are buried in the protein interior, however, are accessible neither to the polar nor apolar paramagnetic probe [37]. Quantification of the collision frequencies by room temperature EPR power saturation measurements in combination with nitroxide scanning represents a valuable method for analyzing the structure and topology of soluble and membrane proteins [37, 55, 56].

14.3.4.1 Example

Nitroxide accessibility analysis performed on the Na^+/proline transporter PutP of *Escherichia coli* identified an additional transmembrane domain (TMII in the revised model) that was not seen in previous membrane topology predictions [51]. For this purpose, SDSL was performed at series of positions in PutP, and ratios of collision frequencies of the spin label with apolar oxygen and polar $CrOx^{3-}$ were determined. High ratios were taken as indication for a location within the membrane, whereas low ratios were assumed to originate from positions in loop regions exposed to an aqueous environment. Furthermore, boundaries of the transmembrane domain were characterized by intermediate collision frequencies with $CrOx^{3-}$. Together, the results revealed that parts of an originally proposed periplasmic loop region form a transmembrane domain (TMII), and defined the boundaries of the second and third transmembrane domains and the connecting loop [51].

14.3.5
Polarity and Proticity in the Spin Label Microenvironment

The analysis of the solvent accessibility of the nitroxide side-chain by power saturation measurements is supplemented by the determination of the polarity and proticity of the spin label microenvironment. This kind of information can be used to monitor, for example, the water density at specific internal or external sites of the protein and its variation upon oligomerization or conformational alterations [2, 57]. In EPR spectra the environmental polarity and proticity of the nitroxide side-chain are revealed by the hyperfine tensor element A_{zz} (Figure 14.3B) and **g** tensor component g_{xx} of the spin label that can be regarded as polarity indexes [58]. Unpolar environments of the spin label are generally characterized by high g_{xx} and low A_{zz} values, whereas polar environments by low g_{xx} and high A_{zz} values [2, 58]. In addition, based on different sensitivities of the tensor components towards the participation of the nitroxide group in hydrogen bonding, a plot of g_{xx} and against A_{zz} provides information on the protic and aprotic properties of the microenvironment [59]. While the tensor component g_{xx} is most sensitive to differences in proticity, the tensor component A_{zz} shows a higher sensitivity to differences in polarity [58]. While cw X-band EPR spectroscopy of spin-labeled proteins in frozen solution is restricted to obtain the hyperfine tensor element A_{zz}, high-field EPR techniques allows additionally the determination of the tensor element g_{xx} of nitroxides with high accuracy due to the enhanced spectral resolution of rigid-limit spectra of disordered samples [60–62].

14.3.5.1 Example

High-field (95-GHz) EPR spectroscopy was used to study the polarity in the environment of residues along the putative proton channel of bacteriorhodopsin [59]. The g_{xx} values obtained for the studied positions varied depending on the distance between the spin-labeled residues and the cytoplasmic surface. Positions located in loops on the cytoplasmic and extracellular side of the membrane showed the smallest g_{xx} values and, therefore, exhibited the highest polarity in their environment, which is concordant with accessibility to water. The g_{xx} values of nitroxides increased towards the middle of the membrane and reached a maximum at positions near the protein bound retinal. The corresponding low polarity in this region suggests a hydrophobic barrier that has to be overcome by the proton during its translocation through the proton channel [59].

14.3.6
Intra- and Intermolecular Distances by Double Spin Labeling and Interspin Distance Measurements

The attachment of two spin labels at specific sites within proteins allows the determination of the interspin distance via EPR spectroscopy in the range of approximately 0.5–8 nm [2, 44] (Figure 14.2B). The distance measurements are based on the dipole–dipole coupling between spins that is proportional to the inverse cube of the distance r^{-3} [63]. Up to interspin distances of 2 nm, the dipolar interaction leads to considerable broadening of the cw EPR spectrum when motional averaging effects are absent. This is the case for measurements of protein samples at temperature below 200 K [64] or under conditions of high viscosity [65]. The distance determination within the range of 0.8–2 nm can be accomplished by a variety of approaches including peak height ratios, second moments, deconvolution based methods, and fitting algorithms of cw EPR spectra [2]. For distances less than 0.8 nm, the influence of exchange interaction between the nitroxides on the EPR lineshape increases due to partial overlap of the nitrogen π-orbitals. In this distance range (0.45–1.1 nm) the intensities of the half-field transitions provide a reliable distance measure, which are not sensitive to exchange interaction [66]. Above interspin distances of 2 nm, cw EPR is not applicable anymore to separate the dipolar broadening contribution from the natural line width in the EPR spectrum. However, pulse EPR techniques such as double electron–electron resonance (DEER) [67, 68] and double-quantum coherence [69] can access a distance range between 1.6 and 8 nm, and are thus powerful applications for the determination of intra- and intermolecular distances in proteins [70–72].

14.3.6.1 Example
Intermolecular distance measurements by four-pulse DEER have been applied to the pH-dependent Na^+/H^+ antiporter NhaA of *E. coli* to determine the structure of the presumably physiological dimer [73] that could not be obtained in a previous highly

resolved X-ray structure [74]. For structure elucidation, nine distance distributions between singly spin-labeled NhaA molecules in the range between 2 and 6 nm were used in combination with restraint-driven rigid-body transformations [75] to arrange the two component molecules against each other. The intrinsic imprecision of SDSL EPR caused by its reliance on labels with a size of around 1 nm were overcome by explicit modeling of spin label conformations with rotamer libraries. By this technique a dimer model was obtained with a resolution that is limited only by the resolution of the X-ray structure of the monomer [73]. Furthermore, four-pulse DEER measurements of NhaA under different pH values suggested a monomer–dimer equilibrium that depends moderately on pH. Thereby, raising the pH from 5.8 to 8 resulted in a significant increase of the degree of oligomerization and *vice versa* [72].

14.4
Conclusions

In general, the understanding of protein function increases with expanding knowledge of the structure and dynamics. The protein chemical and SDSL EPR approaches described above can provide such information, even when classical techniques for structural determination fail or if the structure is known, but no information on the dynamics of the explored system is available. By applying these methods, protein topology and topography, secondary structure elements, protein–protein interactions, proximity relationships in proteins and protein complexes, and the location of binding-site crevices can be explored in terms of static or dynamic aspects. Thereby, site-directed protein labeling techniques are relatively inexpensive, require only small amounts of protein, and can be applied in native environments like lipid bilayers with no significant upper limit on the size of the protein.

For correct assignment of experimental results to structural sites in proteins or protein complexes, only a limited number of labels can be introduced into the investigated system at one time. The relatively small number of resulting structural constraints implicates that probe techniques alone do not allow the determination of protein structures at atomistic resolution. However, sparse constraints can aid *de novo* protein modeling approaches to generate highly accurate, atomic-detail protein models. In addition, as the wealth of experimental protein structures will further grow and, hence, the application of comparative modeling will increase, site-directed labeling techniques will also become more important for experimental evaluation of homology derived structural models.

Acknowledgment

This work was financially supported by the Deutsche Forschungsgemeinschaft (Ju333/3-2, Ju333/4-2, and Exc114-1).

References

1 Lundblad, R.L. (2005) The modification of cysteine, in *Chemical Reagents for Protein Modification*, 3rd edn, CRC Press, Boca Raton, FL, pp. 139–191.

2 Bordignon, E. and Steinhoff, H.-J. (2007) Membrane protein structure and dynamics studied by site-directed spin labeling ESR, in *ESR Spectroscopy in Membrane Biophysics* (eds M.A. Hemminga and L.J. Berliner), Springer, New York, pp. 129–164.

3 Heyduk, T. (2002) Measuring protein conformational changes by FRET/LRET. *Curr. Opin. Biotechnol.*, **13**, 292–296.

4 Lacapère, J.J., Pebay-Peyroula, E., Neumann, J.M., and Etchebest, C. (2007) Determining membrane protein structures: still a challenge! *Trends Biochem. Sci.*, **32**, 259–270.

5 Guan, L. and Kaback, H.R. (2007) Site-directed alkylation of cysteine to test solvent accessibility of membrane proteins. *Nat. Protoc.*, **2**, 2012–2017.

6 Bass, R.B., Butler, S.L., Chervitz, S.A., Gloor, S.L., and Falke, J.J. (2007) Use of site-directed cysteine and disulfide chemistry to probe protein structure and dynamics: applications to soluble and transmembrane receptors of bacterial chemotaxis. *Methods Enzymol.*, **423**, 25–51.

7 Wüthrich, K. (1986) *NMR of Proteins and Nucleic Acids*, John Wiley & Sons, Inc., New York.

8 Kay, L.E. (2005) NMR studies of protein structure and dynamics. *J. Magn. Reson.*, **173**, 193–207.

9 Rhee, J.E., Sheng, W., Morgan, L.K., Nolet, R., Liao, X., and Kenney, L.J. (2008) Amino acids important for DNA recognition by the response regulator OmpR. *J. Biol. Chem.*, **283**, 8664–8677.

10 Creighton, T.E. (1992) *Proteins, Structure and Molecular Principles*, 2nd edn, Freeman, New York.

11 Karlin, A. and Akabas, M.H. (1998) Substituted-cysteine accessibility method. *Methods Enzymol.*, **293**, 123–145.

12 Horn, R. (1998) Explorations of voltage-dependent conformational changes using cysteine scanning. *Methods Enzymol.*, **293**, 145–155.

13 Frillingos, S., Sahin-Toth, M., Wu, J., and Kaback, H.R. (1998) Cys-scanning mutagenesis: a novel approach to structure function relationships in polytopic membrane proteins. *FASEB J.*, **12**, 1281–1299.

14 Jung, H., Rübenhagen, R., Tebbe, S., Leifker, K., Tholema, N., Quick, M., and Schmid, R. (1998) Topology of the Na^+/proline transporter of *Escherichia coli*. *J. Biol. Chem.*, **273**, 26400–26407.

15 Jung, K., Jung, H., Wu, J., Prive, G.G., and Kaback, H.R. (1993) Use of site-directed fluorescence labeling to study proximity relationships in the lactose permease of *Escherichia coli*. *Biochemistry*, **32**, 12273–12278.

16 Getz, E.B., Xiao, M., Chakrabarty, T., Cooke, R., and Selvin, P.R. (1999) A comparison between the sulfhydryl reductants tris(2-carboxyethyl)phosphine and dithiothreitol for use in protein biochemistry. *Anal. Biochem.*, **273**, 73–80.

17 Riener, C.K., Kada, G., and Gruber, H.J. (2002) Quick measurement of protein sulfhydryls with Ellman's reagent and with 4,4′-dithiodipyridine. *Anal. Bioanal. Chem.*, **373**, 266–276.

18 Riddles, P.W., Blakeley, R.L., and Zerner, B. (1983) Reassessment of Ellman's reagent. *Methods Enzymol.*, **91**, 49–60.

19 Loo, T.W. and Clarke, D.M. (1995) Membrane topology of a cysteine-less mutant of human P-glycoprotein. *J. Biol. Chem.*, **270**, 843–848.

20 Whitelegge, J.P., le Coutre, J., Lee, J.C., Engel, C.K., Prive, G.G., Faull, K.F., and Kaback, H.R. (1999) Toward the bilayer proteome, electrospray ionization-mass spectrometry of large, intact transmembrane proteins. *Proc. Natl. Acad. Sci. USA*, **96**, 10695–10698.

21 Britto, P.J., Knipling, L., and Wolff, J. (2002) The local electrostatic environment determines cysteine reactivity of tubulin. *J. Biol. Chem.*, **277**, 29018–29027.

22 Javitch, J.A. and Susan, G.A. (1998) Probing structure of neurotransmitter transporters by substituted-cysteine accessibility method. *Methods Enzymol.*, **296**, 331–346.

23 Sorgen, P.L., Hu, Y., Guan, L., Kaback, H.R., and Girvin, M.E. (2002) An approach to membrane protein structure without crystals. *Proc. Natl. Acad. Sci. USA*, **99**, 14037–14040.

24 Danielson, M.A., Bass, R.B., and Falke, J.J. (1997) Cysteine and disulfide scanning reveals a regulatory alpha-helix in the cytoplasmic domain of the aspartate receptor. *J. Biol. Chem.*, **272**, 32878–32888.

25 Javitch, J.A., Shi, L., and Liapakis, G. (2002) Use of the substituted cysteine accessibility method to study the structure and function of G protein-coupled receptors. *Methods Enzymol.*, **343**, 137–156.

26 Raba, M., Baumgartner, T., Hilger, D., Klempahn, K., Hartel, T., Jung, K., and Jung, H. (2008) Function of transmembrane domain IX in the Na^+/proline transporter PutP. *J. Mol. Biol.*, **382**, 884–893.

27 Akabas, M.H., Stauffer, D.A., Xu, M., and Karlin, A. (1992) Acetylcholine receptor channel structure probed in cysteine-substitution mutants. *Science*, **258**, 307–310.

28 Winston, S.E., Mehan, R., and Falke, J.J. (2005) Evidence that the adaptation region of the aspartate receptor is a dynamic four-helix bundle: cysteine and disulfide scanning studies. *Biochemistry*, **44**, 12655–12666.

29 Bass, R.B., Miller, A.S., Gloor, S.L., Falke, J.J., Melvin, I., Simon, B.R.C., and Alexandrine, C. (2007) The PICM chemical scanning method for identifying domain–domain and protein–protein interfaces: applications to the core signaling complex of *E. coli* chemotaxis. *Methods Enzymol.*, **423**, 3–24.

30 Miller, A.S., Kohout, S.C., Gilman, K.A., and Falke, J.J. (2006) CheA Kinase of bacterial chemotaxis: chemical mapping of four essential docking sites. *Biochemistry*, **45**, 8699–8711.

31 Lai, W., Hazelbauer, G.L., Melvin, I., Simon, B.R.C., and Alexandrine, C. (2007) Analyzing transmembrane chemoreceptors using *in vivo* disulfide formation between introduced cysteines. *Methods Enzymol.*, **423**, 299–316.

32 Green, N.S., Reisler, E., and Houk, K.N. (2001) Quantitative evaluation of the lengths of homobifunctional protein cross-linking reagents used as molecular rulers. *Protein Sci.*, **10**, 1293–1304.

33 Hilger, D., Böhm, M., Hackmann, A., and Jung, H. (2008) Role of Ser-340 and Thr-341 in transmembrane domain IX of the Na^+/proline transporter PutP of *Escherichia coli* in ligand binding and transport. *J. Biol. Chem.*, **283**, 4921–4929.

34 Guo, W., Shi, L., Filizola, M., Weinstein, H., and Javitch, J.A. (2005) Crosstalk in G protein-coupled receptors: changes at the transmembrane homodimer interface determine activation. *Proc. Natl. Acad. Sci. USA*, **102**, 17495–17500.

35 Hazelbauer, G.L., Falke, J.J., and Parkinson, J.S. (2008) Bacterial chemoreceptors: high-performance signaling in networked arrays. *Trends Biochem. Sci.*, **33**, 9–19.

36 Swain, K.E. and Falke, J.J. (2007) Structure of the conserved HAMP domain in an intact, membrane-bound chemoreceptor: a disulfide mapping study. *Biochemistry*, **46**, 13684–13695.

37 Hubbell, W.L., Gross, A., Langen, R., and Lietzow, M.A. (1998) Recent advances in site-directed spin labeling of proteins. *Curr. Opin. Struct. Biol.*, **8**, 649–656.

38 Berliner, L.J., Grünwald, J., Hankovszky, H.O., and Hideg, K. (1982) A novel reversible thiol-specific spin label: papain active site labeling and inhibition. *Anal. Biochem.*, **119**, 450–455.

39 McHaourab, H.S., Lietzow, M.A., Hideg, K., and Hubbell, W.L. (1996) Motion of spin-labeled side chains in T4 lysozyme. Correlation with protein structure and dynamics. *Biochemistry*, **35**, 7692–7704.

40 Borbat, P.P., Mchaourab, H.S., and Freed, J.H. (2002) Protein structure determination using long-distance constraints from double-quantum coherence ESR: study of T4 lysozyme. *J. Am. Chem. Soc.*, **124**, 5304–5314.

41 Columbus, L. and Hubbell, W.L. (2002) A new spin on protein dynamics. *Trends Biochem. Sci.*, **27**, 288–295.

42 Altenbach, C., Greenhalgh, D.A., Khorana, H.G., and Hubbell, W.L. (1994) A collision gradient method to determine the

immersion depth of nitroxides in lipid bilayers: application to spin-labeled mutants of bacteriorhodopsin. *Proc. Natl. Acad. Sci. USA*, **91**, 1667–1671.

43 Bordignon, E., Klare, J.P., Doebber, M., Wegener, A.A., Martell, S., Engelhard, M., and Steinhoff, H.-J. (2005) Structural analysis of a HAMP domain: the linker region of the phototransducer in complex with sensory rhodopsin II. *J. Biol. Chem.*, **280**, 38767–38775.

44 Jeschke, G. and Polyhach, Y. (2007) Distance measurements on spin-labelled biomacromolecules by pulsed electron paramagnetic resonance. *Phys. Chem. Chem. Phys.*, **9**, 1895–1910.

45 Borbat, P.P. and Freed, J.H. (2007) Measuring distances by pulsed dipolar ESR spectroscopy: spin-labeled histidine kinases. *Methods Enzymol.*, **423**, 52–116.

46 Feix, J.B. and Klug, C.S. (1998) Site-directed spin labeling of membrane proteins and peptide-membrane interactions, in *Biological Magnetic Resonance 14. Spin Labeling: The Next Millenium* (ed. L.J. Berliner), Plenum, New York, p. 253.

47 Slichter, C.P. (1992) *Principles of Magnetic Resonance*, Springer, Berlin.

48 Isas, J.M., Langen, R., Haigler, H.T., and Hubbell, W.L. (2002) Structure and dynamics of a helical hairpin and loop region in annexin 12: a site-directed spin labeling study. *Biochemistry*, **41**, 1464–1473.

49 Perozo, E., Cortes, D.M., and Cuello, L.G. (1998) Three-dimensional architecture and gating mechanism of a K^+ channel studied by EPR spectroscopy. *Nat. Struct. Biol.*, **5**, 459–469.

50 Salwinski, L. and Hubbell, W.L. (1999) Structure in the channel forming domain of colicin E1 bound to membranes: the 402–424 sequence. *Protein Sci.*, **8**, 562–572.

51 Wegener, C., Tebbe, S., Steinhoff, H.J., and Jung, H. (2000) Spin labeling analysis of structure and dynamics of the Na^+/proline transporter of *Escherichia coli*. *Biochemistry*, **39**, 4831–4837.

52 Dong, J., Yang, G., and McHaourab, H.S. (2005) Structural basis of energy transduction in the transport cycle of MsbA. *Science*, **308**, 1023–1028.

53 Steinhoff, H.J., Mollaaghababa, R., Altenbach, C., Hideg, K., Krebs, M., Khorana, H.G., and Hubbell, W.L. (1994) Time-resolved detection of structural changes during the photocycle of spin-labeled bacteriorhodopsin. *Science*, **266**, 105–107.

54 Wegener, A.A., Klare, J.P., Engelhard, M., and Steinhoff, H.J. (2001) Structural insights into the early steps of receptor-transducer signal transfer in archaeal phototaxis. *EMBO J.*, **20**, 5312–5319.

55 Doebber, M., Bordignon, E., Klare, J.P., Holterhues, J., Martell, S., Mennes, N., Li, L., Engelhard, M., and Steinhoff, H.-J. (2008) Salt-driven equilibrium between two conformations in the HAMP domain from *Natronomonas pharaonis*: the language of signal transduction? *J. Biol. Chem.*, **283**, 28691–28701.

56 Cuello, L.G., Cortes, D.M., and Perozo, E. (2004) Molecular architecture of the K_vAP voltage-dependent K^+ channel in a lipid bilayer. *Science*, **306**, 491–495.

57 Finiguerra, M.G., Blok, H., Ubbink, M., and Huber, M. (2006) High-field (275 GHz) spin-label EPR for high-resolution polarity determination in proteins. *J. Magn. Reson.*, **180**, 197–202.

58 Plato, M., Steinhoff, H.J., Wegener, C., Törring, J.T., Savitsky, A., and Möbius, K. (2002) Molecular orbital study of polarity and hydrogen bonding effects on the g and hyperfine tensors of site directed NO spin labeled bacteriorhodopsin. *Mol. Phys.*, **100**, 3711–3721.

59 Steinhoff, H.J., Savitsky, A., Wegener, C., Pfeiffer, M., Plato, M., and Mobius, K. (2000) High-field EPR studies of the structure and conformational changes of site-directed spin labeled bacterio-rhodopsin. *Biochim. Biophys. Acta*, **1457**, 253–262.

60 Prisner, T.F., van der Est, A., Bittl, R., Lubitz, W., Stehlik, D., and Möbius, K. (1995) Time-resolved W-band (95 GHz) EPR spectroscopy of Zn-substituted reaction centers of *Rhodobacter sphaeroides* R-26. *Chem. Phys.*, **194**, 361–370.

61 Huber, M. and Törring, J.T. (1995) High-field EPR on the primary electron donor cation radical in single crystals of heterodimer mutant reaction centers of

photosynthetic bacteria – first characterization of the G-tensor. *Chem. Phys.*, **194**, 379–385.

62 Möbius, K., Savitsky, A., Wegener, C., Plato, M., Fuchs, M., Schnegg, A., Dubinskii, A.A., Grishin, Y.A., Grigor'ev, I.A., Kühn, M., Duché, D., Zimmermann, H., and Steinhoff, H.-J. (2005) Combining high-field EPR with site-directed spin labeling reveals unique information on proteins in action. *Magn. Reson. Chem.*, **43**, S4–S19.

63 Jeschke, G. (2002) Distance measurements in the nanometer range by pulse EPR. *Chem. Phys. Chem.*, **3**, 927–932.

64 Rabenstein, M.D. and Shin, Y.K. (1995) Determination of the distance between two spin labels attached to a macromolecule. *Proc. Natl. Acad. Sci. USA*, **92**, 8239–8243.

65 Altenbach, C., Oh, K.J., Trabanino, R.J., Hideg, K., and Hubbell, W.L. (2001) Estimation of inter-residue distances in spin labeled proteins at physiological temperatures: experimental strategies and practical limitations. *Biochemistry*, **40**, 15471–15482.

66 Eaton, S.S., More, K.M., Sawant, B.M., and Eaton, G.R. (1983) Use of the ESR half-field transition to determine the interspin distance and the orientation of the interspin vector in systems with two unpaired electrons. *J. Am. Chem. Soc.*, **105**, 6560–6567.

67 Milov, A.D., Salikhov, K.M., and Shirov, M.D. (1981) Application of ELDOR in electron-spin echo for paramagnetic center space distribution in solids. *Fiz. Tverd. Tela*, **23**, 975–982.

68 Pannier, M., Veit, S., Godt, A., Jeschke, G., and Spiess, H.W. (2000) Dead-time free measurement of dipole–dipole interactions between electron spins. *J. Magn. Reson.*, **142**, 331–340.

69 Borbat, P.P. and Freed, J.H. (1999) Multiple-quantum ESR and distance measurements. *Chem. Phys. Lett.*, **313**, 145–154.

70 Sen, K.I., Logan, T.M., and Fajer, P.G. (2007) Protein dynamics and monomer–monomer interactions in AntR activation by electron paramagnetic resonance and double electron–electron resonance. *Biochemistry*, **46**, 11639–11649.

71 Park, S.Y., Borbat, P.P., Gonzalez-Bonet, G., Bhatnagar, J., Pollard, A.M., Freed, J.H., Bilwes, A.M., and Crane, B.R. (2006) Reconstruction of the chemotaxis receptor-kinase assembly. *Nat. Struct. Mol. Biol.*, **13**, 400–407.

72 Hilger, D., Jung, H., Padan, E., Wegener, C., Vogel, K.P., Steinhoff, H.J., and Jeschke, G. (2005) Assessing oligomerization of membrane proteins by four-pulse DEER: pH-dependent dimerization of NhaA Na^+/H^+ antiporter of *E. coli*. *Biophys. J.*, **89**, 1328–1338.

73 Hilger, D., Polyhach, Y., Padan, E., Jung, H., and Jeschke, G. (2007) High-resolution structure of a Na^+/H^+ antiporter dimer obtained by pulsed EPR distance measurements. *Biophys. J.*, **93**, 3675–3683.

74 Hunte, C., Screpanti, E., Venturi, M., Rimon, A., Padan, E., and Michel, H. (2005) Structure of a Na^+/H^+ antiporter and insights into mechanism of action and regulation by pH. *Nature*, **435**, 1197–1202.

75 Sompornpisut, P., Liu, Y.S., and Perozo, E. (2001) Calculation of rigid-body conformational changes using restraint-driven Cartesian transformations. *Biophys. J.*, **81**, 2530–2546.

Part III
Intracellular Signaling

Introduction

Kirsten Jung, Michael Y. Galperin, and Reinhard Krämer

The bacterial signal transduction machinery senses changes in extracellular and/or intracellular parameters, and transmits these signals to various cellular components, eliciting adaptive changes in bacterial physiology, metabolism, and cell behavior. Most of these processes involve ordered sequences of biochemical reactions inside the cell, which are preferentially carried out by enzymes or mediated by second messengers and result typically in alterations of gene expression. Principles of intracellular signaling and signal transduction are comprehensively described in Part III, namely:

(1) Histidine kinase/response regulator systems that perceive intracellular stimuli and regulate a variety of processes, including gene expression and post-translational modification of proteins (Chapters 15 to 17).
(2) Transcriptional regulators that concurrently function as sensors (Chapters 16 and 17).
(3) Trigger enzymes with dual functions that catalyze enzymatic reactions and interfere in various ways with regulatory processes (Chapters 18 and 19).
(4) The catabolite repression pathway that includes sensor adenylate cyclases and regulates the transcription of various operons (including those responsible for flagellar biogenesis) through modulation of the cellular levels of cAMP. The phosphotransferase system regulates several independent processes, including chemotaxis inducer exclusion of alternative substrates, and the activity of soluble adenylate cyclase that controls the cellular cAMP levels (Chapters 19 and 20).
(5) The c-di-GMP-dependent signaling pathways that include a variety of sensor diguanylate cyclases and c-di-GMP-specific phosphodiesterases, and control bacterial motility, biofilm formation, and other processes (Chapters 15 and 21).
(6) The hormone-like effector molecules (p)ppGpp that generally promote survival under unfavorable conditions (Chapter 22).
(7) RNAs with versatile sensory and regulatory potential (Chapter 23).

Bacterial Signaling. Edited by Reinhard Krämer and Kirsten Jung
Copyright © 2010 WILEY-VCH Verlag GmbH & Co. KGaA, Weinheim
ISBN: 978-3-527-32365-4

(8) Signaling through serine/threonine protein kinases and serine/threonine protein phosphatases that regulates transcription and the activity of various metabolic enzymes (Chapter 24).
(9) Regulated proteolysis that plays an important role to control the level of critical components in signal transduction pathways (Chapter 25).

Gene expression is altered at the end of most signal transduction pathways. Chapter 26 summarizes modern high-throughput methods to identify gene targets and regulatory interactions.

15
Protein Domains Involved in Intracellular Signal Transduction
Michael Y. Galperin

15.1
Introduction

The current understanding of bacterial signal transduction is a result of the concerted effort of many scientists all over the world who have genetically characterized, cloned, expressed, purified, and finally biochemically characterized components of the signaling machinery from a variety of bacteria. However, owing to the complexity of bacterial signaling systems, experimental approaches alone could not provide a holistic view of the general organization of the signal transduction machinery even in the relatively well-studied model organisms, such as *Escherichia coli* and *Bacillus subtilis*. Such a view could have emerged only after the completion of the respective genome-sequencing projects, which allowed enumeration of all (predicted) signal transduction systems encoded in these genomes. The results of these studies were nothing short of shocking, revealing that even *E. coli* and *B. subtilis* encode a large number of signaling systems that had not been identified in any previous experiments [1]. Even now, 12 years after the completion of the *E. coli* K-12 genome, signals sensed by at least seven histidine kinases (AtoS, RstB, YehU, YpdA, YfhK, YedV, and YjoN) remain either unknown or poorly defined [2].

Further analyses of these genomes revealed additional proteins that were likely to participate in signal transduction. Thus, out of 29 uncharacterized proteins with conserved protein domains of unknown function DUF1 and DUF2 (also known as, respectively, GGDEF and EAL domains) encoded in the *E. coli* genome, six had a typical membrane receptor architecture with an N-terminal periplasmic domain, anchored by two transmembrane helices and followed by a long cytoplasmic domain. Analyzes of other bacterial genomes left no doubt that, in addition to the well-studied sensory histidine kinases and methyl-accepting chemotaxis proteins (MCPs), bacteria utilize other classes of receptor proteins, including membrane-anchored adenylate cyclases, diguanylate cyclases (c-di-GMP synthetases) and c-di-GMP-specific phosphodiesterases (c-di-GMP hydrolases), serine/threonine protein kinases, and protein phosphatases [3, 4], not to mention the membrane-associated phosphotranslocases of the phosphoenolpyruvate-dependent sugar phosphotransferase system (PTS) that were shown to serve as chemotaxis sensors for their sugar substrates. These findings reveal a much more complex signaling network than had been envisaged before the onset of genomics [5, 6]. They also indicate that, in full accordance with our prediction made at the very beginning of the genome era [7], the new paradigm of genome-based microbiology was effectively replacing the old paradigm of studying one gene (or one phenotype) at a time.

As I have stated previously, one of the goals of bacterial genome analysis should be the ability to reconstruct microbial signaling pathways and predict responses of any given microorganism to various environmental factors, based solely on the DNA sequence of its genome [3, 4]. Unfortunately, signals perceived by the great majority of cellular receptor proteins are still unknown and will remain that way for the foreseeable future. The exact signaling mechanisms and signal outputs are often obscure as well. However, experimental studies of signal transduction in various model organisms have led to the gradual accumulation of data on specific signaling domains. Structural characterization of signaling domains in the course of structural genomics initiative provides additional insights into the potential ligand-binding sites and ligand specificity of the respective domains. That is why computational analysis of signaling proteins has been, and still remains, an easy and effective way of investigating mechanisms of signal transduction and of generating hypotheses that could drive subsequent experimental analysis. In this chapter, I provide a listing of the most widespread protein domains involved in bacterial signal transduction and briefly review the available information on their functions.

15.2
Computational Analysis of Signaling Domains

Computational analysis of bacterial signal transduction systems in diverse bacteria relies on a key common feature of all those systems, the modular structure of their component proteins. Despite their great variety, most signal transduction proteins are composed of the same set of structural components, referred to as conserved protein domains. Protein domains are often defined as discrete structural units of proteins that can be found in different contexts, but generally convey conserved or related functions [8]. As a result, uncovering the domain organization of a given protein often provides a way towards understanding its function. That is what makes comparative analysis of signal transduction components systems encoded in various bacterial genomes so effective – recognition of the previously characterized domains paves the way to the identification of new signaling domains, which, in turn, leads to their experimental characterization.

Identification of known domains is easily achieved through the comparison of the protein sequences against common protein domain databases, such as Pfam (http://pfam.sanger.ac.uk/), SMART (http://smart.embl.de), CDD (http://www.ncbi.nlm.nih.gov/Structure/cdd/cdd.shtml, or COG (http://www.ncbi.nlm.nih.

gov/COG) [9–12]. As a result, protein entries in popular protein sequence databases, such as UniProt [13] and the National Center for Biotechnology Information (NCBI) RefSeq database [14], display domain content of the respective protein as part of the record and provide direct links to the domain database entries. NCBI RefSeq provides links to the CDD database, whereas UniProt entries are linked to a variety of databases, including Pfam, SMART, ProDom, and PROSITE, and to the integrated protein domain database, InterPro (http://www.ebi.ac.uk/interpro), which combines the data from a variety of databases [15]. For each of its domains, Pfam offers a possibility to view the taxonomic distribution of domain representatives and download all protein sequences with the given domain from a given taxonomic node. InterPro offers an even more convenient view of the domain distribution in the major phylogenetic groups, but provides somewhat lesser download capabilities.

In addition to the general protein domain resources, there are several tools that are specifically dedicated to the analysis of signal transduction. The popular MiST database (http://genomics.ornl.gov/mist/ [16]), created by Luke Ulrich and Igor Zhulin at the Oak Ridge National Laboratory in Oak Ridge, Tennessee, lists the key signal transduction proteins encoded in completely sequenced bacterial genomes and provides domain structures of these proteins. The author of this chapter maintains the Signaling Census web site, http://www.ncbi.nlm.nih.gov/Complete_Genomes/SignalCensus.html [4], which provides counts of signal transduction proteins, including serine/threonine protein kinases, in representative microbial genomes with links to NCBI protein database, UniProt, Pfam, and CDD entries. A similarly organized Response Regulator Census web site, http://www.ncbi.nlm.nih.gov/Complete_Genomes/RRcensus.html [17], provides counts of response regulator proteins encoded in completely sequenced bacterial genomes and classifies them based on their output domain architecture (i.e., the nature of their output domains).

15.3
Intracellular Sensory Domains

Sensory (signal input) domains of bacterial receptors are as diverse as the signals they perceive. A bacterial genome often (albeit not always, see below) encodes only a single copy of any given domain. That is where the abundance of sequenced genomes becomes most helpful, as comparison of related domain sequences from diverse bacteria allows identification of the conserved and variable regions within the domain, prediction of its secondary structure, and other features. Still, computational analysis of a new domain is usually insufficient to predict its binding specificity and, hence, the biological function. That requires experimental studies, which typically come much later. As a result, the list of well-characterized sensory domains is relatively short and includes (see relevant section for nomenclature derivation): PAS, GAF, BLUF, globin coupled sensor GCS, HNOB, hemerythrin, KdpD, and phytochrome domains (Table 15.1).

Table 15.1 Widespread intracellular sensory domains.

Domain name	Example (amino acid range)	Binding specificity	Pfam entry Name	Pfam entry No.[a]	CDD entry	PDB entry	References
PAS	P50466 (30–115)	heme, flavin, ATP	PF00989	9681	cl02459	2phy	[24]
GAF	P19323 (202–354)	cGMP	PF01590	7549	cl00853	1f5m	[33]
BLUF	Q53119 (16–108)	FAD	PF04940	220	cl04855	2byc	[43]
GCS	O07621 (32–178)	heme (O_2)	Pfam-B _13616	170[a]	cd01068	1or4	[50]
HNOB	Q02153 (1–172)	heme (NO)	PF07700	217	cl06647	1xbn	[53]
Hr	Q60AX2 (11–70)	oxygen	PF00814	2164	cl08548	2awc	[57]
KdpD	P21865 (20–230)	unknown (turgor-sensing)	PF02702	309	cl03665	2r8r	[64]
PHY	Q3IUZ1 (367–552)	tetrapyrrole	PF00360	1467	cl02849	2vea	[68]

a) The number of protein entries containing the respective domain according to the Pfam database [9] release 23.0 (19 August 2008); the number of bacterial receptor proteins with the globin sensor domain was calculated from the results of a BLAST search against NCBI protein database (28 December 2008).

15.3.1
PAS Domain

The PAS domain (see also Chapters 8 and 11) is one of the most widespread protein domains encoded in bacterial and archaeal genomes with more than 10 000 representatives in the NCBI protein database (Table 15.1). It was originally described in eukaryotes and got its name from the three proteins where it was first recognized: the *Drosophila melanogaster* period (Per) clock protein, human aryl hydrocarbon receptor nuclear translocator (ARNT), and *Drosophila* single-minded (Sim) protein [18, 19]. It was subsequently described as the flavin-binding domain in the *E. coli* chemotaxis receptor Aer and found in a number of other bacterial proteins [20–22]. The so-called PAS fold (a variant of the profilin fold [23]) has a relatively simple structure consisting of a six-stranded antiparallel β-sheet flanked by five α-helices. Its ligand-binding pocket allows the PAS domain to accommodate a variety of small molecules and bind them in several different fashions. The best-studied variants are those binding heme, flavin (flavin mononucleotide (FMN) and flavin adenine dinucleotide (FAD)), and *p*-hydroxycinnamic acid [24]. Remarkably, the heme is bound noncovalently [25], *p*-hydroxycinnamic acid is covalently bound to a cysteine residue through a thioester linkage [26], whereas FMN is bound noncovalently but forms a covalent adduct with a conserved cysteine residue as part of the signaling mechanism [27]. Accordingly, different PAS families employ different sensing mechanisms, although in each case the external signal causes a conformational change in the PAS domain that is used to transmit the signal to the C-terminally located signal transduction domains [28]. In heme-containing PAS domains, the presence of oxygen affects the position of the PAS-bound heme molecule, causing a change in the general conformation of the PAS domain [29]. Similarly, in FAD-containing PAS domains, conformational changes are triggered by changes in the redox state of the flavin ligand that lead to reorganization of hydrogen bonds within the flavin-binding pocket [30]. FMN-containing PAS domains (often referred to as LOV (light, oxygen, and voltage) domains) respond to redox changes or light by forming a covalent bond between FMN and the protein [31]. In the bacterial blue-light photoreceptor photoactive yellow protein, conformational changes in the PAS domain are triggered by light absorption, which causes cis/trans isomerization of the *p*-hydroxycinnamic acid chromophore. PAS domains could have other potential ligands with additional mechanisms of signal transduction. For example, one of the three PAS domains of the *B. subtilis* histidine kinase KinA reportedly binds ATP [32].

15.3.2
GAF Domain

The GAF domain is second, after PAS, most common ligand-binding domain in bacteria (Table 15.1). It was originally described as a common domain in cGMP-specific phosphodiesterases, *a*denylyl cyclases, and *F*hlA protein (hence GAF), capable of binding cGMP and phytochromes [33], and is believed to function as a molecular switch regulating the activity of downstream domains [34, 35]. In spite of limited sequence similarity, PAS and GAF domains share the same profilin fold and have somewhat similar three-dimensional structures [36], suggesting that they might have common ancestry. Like PAS, GAF domains bind a variety of ligands, including cGMP, cAMP, heme, bilin, and other tetrapyrroles [37–40]. Two recent papers suggested that GAF domains might play additional roles in the cell. One of them reported specific binding of sodium ions to the GAF domains of cyanobacterial adenylate cyclases, which leads to the inhibition of the downstream adenylate cyclase domains [41]. The other work found that a stand-alone GAF domain protein, product of the *Escherichia coli yebR* gene, had free methionine-(*R*)-sulfoxide reductase activity [42]. These papers show that GAF domains might have additional functions that beyond simple ligand binding and suggest that these might bring further surprises.

15.3.3
BLUF Domain

The BLUF domain (see also Chapter 10) was first described as a FAD-containing photosensor domain in the transcriptional antirepressor AppA, which controls photosynthetic gene expression in the purple bacterium *Rhodobacter sphaeroides* in response to light and oxygen by binding to the transcriptional repressor PpsR, and in the photo-activated adenylate cyclase from the flagellate protist *Euglena gracilis* [43]. The domain name comes from the abbreviation of "sensor of *b*lue *l*ight *u*sing *F*AD," although BLUF was reported to bind riboflavin and FMN equally well [44]. Blue-light absorption by the FAD appears to cause reorientation of several residues in the chromophore-surrounding pocket of the BLUF domain, affecting the interface between the BLUF and the output domain [45–47]. A recent computational analysis suggested participation of BLUF-containing proteins in a variety of light-regulated signaling pathways [48].

15.3.4
GCS Domain

GCS (globin-coupled sensor) domains are heme-binding protein domains homologous to vertebrate hemoglobins and myoglobins. The first GCS domain was described in the aerotactic transducer HemAT from *B. subtilis* that couples an oxygen-sensing GCS domain to an MCP domain [49, 50]. GCS domains were subsequently discovered in association with other transmitter modules, including diguanylate cyclase (GGDEF) domains [51]. GCS-bound heme has been shown to bind oxygen, CO, and NO with comparable affinities, suggesting that *in vivo* GCS domains serve primarily as oxygen sensors. Remarkably, in *E. coli*, the *dos* (*yddU*) gene, which encodes an oxygen-regulated c-di-GMP-specific phosphodiesterase Dos (*d*irect *o*xygen *s*ensor [29]) is cotranscribed with the *yddV* gene, which encodes a diguanylate cyclase of GCS-xxx-GGDEF domain architecture [52]. Thus, although both genes of the *yddVU* operon regulate cellular c-di-GMP levels in response to oxygen, they sense oxygen by two entirely different protein domains.

15.3.5
HNOB Domain

The HNOB domain got its name from the "*h*eme *NO* *b*inding" domain found in eukaryotic soluble guanylate cyclases [53]. It was later renamed the H-NOX (*h*eme *n*itric *o*xide and *o*xygen binding) domain because some members of this domain family turned out to specifically bind oxygen [54], whereas other members have an extremely high affinity to NO [55]. HNOB domains are usually found either in the stand-alone form or as the N-terminal sensory domains of MCPs and adenylate (and/or guanylate) cyclases [53, 54]. Stand-alone HNOB domains are often encoded in the same operons with histidine kinases and diguanylate cyclases, suggesting potential interactions of these proteins [53]. The current database also includes a protein from *Magnetococcus* sp. MC-1 that combines the HNOB domain with the c-di-GMP-binding PilZ domain. The HNOB-bound heme adopts a highly distorted structure that is believed to be important for signaling [54], which has been shown to include significant conformational changes in the HNOB structure [56].

15.3.6
Hr Domain

Hr (hemerythrin) is a nonheme di-iron oxygen-transporting protein domain initially discovered in invertebrate animals. A homologous oxygen-binding Hr was described in the chemotaxis receptor DcrH from the sulfate-reducing bacterium *Desulfovibrio vulgaris* [57]. Binding of molecular O_2 to the diferrous active site of *D. vulgaris* Hr was shown to trigger a conformational change in the N-terminal loop of this protein domain, which could play a role in the transduction of the sensory signal [58]. Genome sequence analyzes subsequently found Hr domains a variety of other bacteria and archaea [59–61], suggesting that Hr serves as a widespread oxygen-sensing domain.

15.3.7
KdpD Domain

The KdpD domain (see also Chapter 8) forms the N-terminal signal input domain of the osmolarity-sensing histidine kinase KdpD, which gave it its name. This sensor histidine kinase consists of two cytoplasmic fragments, 1–400 and 500–894, connected by a 100-amino-acid integral membrane fragment that consists of four transmembrane segments connected by short loops. The membrane fragment does not appear to serve any specific function beyond anchoring [62]. In contrast, the N-terminal cytoplasmic fragment was shown to play a key part in signaling [63, 64]. This 400-amino-acid fragment consists of two separate protein domains, a 210-amino-acid KdpD domain and a 125-amino-acid domain of the universal stress protein (USP) family. The USP domain is a short potentially ATP-binding domain has been shown to improve cell survival rate under a variety of stress conditions; the mechanism of its action remains unknown [65, 66]. The KdpD domain apparently serves as a sensory domain of the whole protein [63]. In all available protein sequences, the KdpD domain exists as part of the KdpD–USP pair, either in stand-alone proteins, or fused to histidine kinases, or, in two recently sequenced proteins, to the PTS EIIA protein domain. The recently solved crystal structure of the KdpD domain (PDB: 2r8r) shows a dimeric structure with each domain composed of a five-stranded β-sheet surrounded by eight α-helices. An important contribution towards understanding the exact mechanisms of sensing of K^+ limitation and salt stress by KdpD has been a recent study that showed that the USP domain of the KdpD–USP pair directly interacts with its soluble homolog, UspC [67]. UspC can also interact with the KdpE/DNA complex, therefore controlling expression of the KdpD-regulated genes [64].

15.3.8
PHY Domain

The PHY (phytochrome) (see also Chapter 10) domain [68] is found in a variety of bacterial and plant photoreceptors, sensing changes in red and far-red illumination. It is sometimes found in a stand-alone form but is usually associated with PAS and GAF domains, forming a PAS–GAF–PHY combination, which provides sensory modules for histidine kinases and diguanylate cyclases, serine/threonine protein kinases and serine/threonine protein phosphatases. Recent data show that the PHY domain is structurally related to PAS and GAF domains and interacts with them to form the phytochrome sensory modules [69, 70].

15.4
Intracellular Signal-Transducing and Output Domains

Given that the principal intracellular signal transduction pathways are discussed in detail elsewhere in this book, I only briefly list the key protein domains that are involved in these pathways (Table 15.2).

15.4.1
Two-Component Signal Transduction

The two-component signal transduction machinery includes several well-conserved domains (see also Chapter 8), such as the *h*istidine kinase-type *ATPase* catalytic (HATPase_c) domain and the receiver (phosphoacceptor) REC domain, and several diverse domains, including various histidine kinase dimerization domains (HisKA, HisKA_2, HisKA_3, HWE_HK) and various signal output domains, see [17, 71, 72] and Table 15.2. It is important to note that diversity of response regulator output domains (DNA-binding, RNA-binding, enzymatic, and enzymatically inactive protein-binding domains) allows the two-component machinery to regulate most, if not all, other signal transduction systems in the bacterial cell.

15.4.2
Chemotaxis

The most notable domain in the chemotaxis signal transduction machinery is the signal-transducing domain of chemotaxis sensors (MCPs) (see also Chapter 11) that consists of two long α-helices and a connecting turn fragment, which interacts with CheW-like domain of the histidine kinase CheA through its conserved sequence motif QTNLLALNAAIEAARAGExGRGFAVVAxEVRxLA [73]. Although a great majority of MCPs are transmembrane proteins, some are cytoplasmic, including the well-characterized *B. subtilis* aerotactic sensor HemAT [49]. Sequence comparisons revealed several principal classes of MCPs differing in the overall structure, interaction with CheA, and phylogenetic distribution [74]. Despite their names, MCPs do not necessarily depend on methylation for their function and can serve as chemotactic sensors in organisms that do not encode either MCP methylase CheR or demethylase CheB (Table 15.2). Rather, methylation and demethylation of strategically located glutamic acid residues change the charge distribution along the α-helices, which affects the packing of these α-helices against each other and, in turn, the interaction between MCP and CheA. The CheB methylesterase domain is often fused to the receiver domain in chemotaxis-specific response regulators, which adds yet another level of regulation of the chemotaxis signaling mechanism.

Table 15.2 Common types of intracellular signal-transducing and output domains.

Domain name	Example (amino acid range)	Function	Pfam entry Name	Pfam entry No.[a]	CDD entry	PDB entry	References
Two-component signal transduction system							
HisKA	P0AEJ4 (231–290)	phosphocarrier, His	PF00512	18 204	cl00080	1joy	[71]
HATPase	P0AEJ4 (333–436)	phosphorylation of HisKA	PF02518	33 607	cl00075	1bxd	[71]
HPT	P07363 (1–105)	phosphocarrier, His	PF01627	2669	cl00086	2a0b	[96, 97]
REC	P0AA16 (1–120)	phosphocarrier, Asp	PF00072	33 605	cd00156	2che	[71, 98]
HAMP	P0AEJ4 (160–229)	dimerization	PF00672	14 209	cl01054	1h2s	[99, 100]
AAA	Q06065 (145–366)	ATPase, binding to σ^{54}	PF00158	5590	cd00009	1ojl	[101, 102]
BTAD	P25941 (119–263)	regulatory	PF03704	374	—	2fez	[103]
HTH GerE (LuxR_C)	P0AF28 (150–215)	DNA binding	PF00196	7183	cl10457	1rnl	[104]
wHTH (trans_reg_C)	P0AA16 (138–232)	DNA binding	PF00486	9615	cl00209	1opc	[105]
LytTR	P0AO17 (149–238)	DNA binding	PF04397	1358	cl04498	3d6w	[106, 107]
HTH_AraC	O31517 (314–360)	DNA binding	PF00165	15 055	cl02815	1bl0	[108, 109]
Spo0A_C	P06534 (152–267)	DNA binding	PF08769	115	cl08493	1fc3	[110]
HTH_Fis	Q06065 (413–454)	DNA binding	PF02954	5423	cl01091	1ety	[111]
ANTAR	P10932 (135–190)	RNA binding	PF03861	526	cl04297	1qo0	[112, 113]
Chemotaxis							
MCP	P02942 (296–518)	dimerization, CheA binding	PF00015	7762	—	1qu7	[73, 114]
HAMP	P0AEJ4 (160–229)	dimerization	PF00672	14 209	cl01054	1h2s	[99, 100]
CheR	P07364 (92–285)	MCP methylation	PF03705 PF01739	993	COG1352	1af7	[115]
CheB	P07330 (150–345)	MCP demethylation	PF01339	933	cl03170	1chd	[116]
CheW	P07363 (509–640)	binding to MCP, CheY	PF01584	2710	cl00256	1k0s	[117]

Table 15.2 (Continued)

Domain name	Example (amino acid range)	Function	Pfam entry Name	Pfam entry No.[a]	CDD entry	PDB entry	References
PTS, catabolite repression							
PTS EI	P08839 (1–545)	autophosphorylation, phosphotransfer to HPr	PF05524 PF00391 PF02896	489	COG1080	2hwg	[118, 119]
PTS_HPr	P0AA04 (5–85)	phosphocarrier, His	PF00381	1122	cl00206	1cm2	[120]
PTS_EIIA	P69783 (18–150)	phosphocarrier, His	PF00358	895	cl00162	1f3q	[121]
PTS_EIIB	P69786 (399–474)	phosphocarrier, His	PF00367	1440	cl00164	1iba	[122]
PTS_EIIC	P69786 (1–397)	phosphotransfer	PF02378	2932	cl00557	—	[123]
ACyc_1	P00936 (1–830)	cAMP synthesis	PF01295	219	cl01168	—	[124]
ACyc_3	P40137 (221–410)	cAMP synthesis	PF00211	2201	cl00925	1wc1	[125, 126]
CpdA	P0AEW4 (15–250)	cAMP hydrolysis	PF00149 PF08413	160	COG1409	2hy1	[127, 128]
c-di-GMP-mediated signaling							
GGDEF	O87374 (144–300)	c-di-GMP synthesis	PF01590	10 126	cl00291	1w25	[129, 130]
EAL	O87374 (319–560)	c-di-GMP hydrolysis	PF00990	5861	cl00290	2bas	[131, 132]
HD-GYP	Q4UU85 (195–370)	c-di-GMP hydrolysis	—	1445[a]	COG2206	—	[133, 134]
PilZ	P76010 (112–230)	c-di-GMP binding	PF07238	1574	cl01260	1ywu	[83, 135]
FleQ	Q51460 (5–125)	c-di-GMP binding	PF06490	116	cl05812	—	[87]
Phosphoserine/phosphothreonine signaling							
S_TKc	P0A5S4 (10–275)	protein phosphorylation (Ser, Thr)	PF00069	32 109	cd00180	1mru	[136]
PP2C	P71588 (9–230)	pSer, pThr hydrolysis	PF00481	2142	cl00120	1txo	[137]
FHA	P66799 (285–385)	pThr, pTyr binding	PF00498	2559	cl00062	2fez	[138, 139]

a) See footnote to Table 15.1.

15.4.3
Sugar: PTS

The phosphoenolpyruvate-dependent sugar: PTS (see also Chapters 19 and 20) is uniquely organized so that the PTS-mediated regulation is performed by the same protein components that participate in the phosphotransfer cascade (EI–HPr–EIIA–EIIB–EIIC–sugar), which leads to the phosphorylative uptake of its sugar substrates [75, 76]. It should be noted there are several distinct sugar-specific PTS EII components whose roles in regulation are not entirely clear; Table 15.2 lists only the glucose-specific EII variants [77, 78]. In addition, PTS has been shown to regulate class I adenylate cyclase (*E. coli*-like); activities of the far more widespread class III adenylate cyclases appear to be regulated by the adjacent sensory domains [79–81]. So far, adenylate cyclase activity and cAMP-dependent catabolite repression have been demonstrated only in bacteria. Homologs of the *Aeromonas hydrophila* adenylate cyclase [82], often referred to as thermophilic or class IV adenylate cyclase, are encoded in many archaea [4], but neither of these archaeal enzymes has been experimentally characterized.

15.4.4
c-di-GMP-Mediated Signaling

Despite significant progress in recent years, detailed mechanisms of the c-di-GMP-mediated signaling remain obscure (see also Chapter 21). A dedicated c-di-GMP-binding domain, PilZ, has been identified [83] and experimentally demonstrated to undergo conformational changes in response to c-di-GMP binding [84]. Still, it is not clear what the targets of PilZ action are and how these targets are affected by the conformational changes in PilZ. Other potential targets for c-di-GMP action include inactivated GGDEF and EAL domains [85, 86], and the FleQ domain [87], a divergent variant of the REC domain. In neither of these cases, however, are the mechanisms of c-di-GMP known at this time.

15.4.5
Serine/threonine Protein Phosphorylation Signaling System

Reversible protein phosphorylation on serine, threonine, or tyrosine residues is a key regulatory mechanism in eukaryotic cells. In the past several years, serine/threonine protein kinases have been recognized in a variety of prokaryotic cells (see also Chapter 24); they are the only (known) type of sensor proteins encoded in many archaeal genomes [4]. Despite their abundance, serine/threonine protein kinases are still poorly studied and their targets remain mostly unknown. In the several organisms where serine/threonine phosphorylation has been investigated in more detail, it was found to be part of global regulation of gene expression [88–91]. Recent proteomic studies of *E. coli*, *B. subtilis*, and *Lactococcus lactis* revealed numerous proteins phosphorylated on serine or threonine residues, including certain enzymes of central metabolism [92–94]. These findings suggest that this kind of regulation could be extremely important in a wide variety of prokaryotes.

Table 15.3 Diversity of the sensor – output domain combinations.

Sensor domain[a]	Signaling system					
	Histidine kinase	Chemotaxis (MCP)	cAMP-mediated	c-di-GMP-mediated	pSer, pThr	Other
PAS	P10955	Q5V4W2	Q7WWV9	P77334	A6W4X7	Q55178
GAF	Q8CK17	B3E5K3	Q72T23	A4XM27	Q8CK17	A0JZT1
BLUF	—	—	A7BT71	P75990	—	A0Z3A0
GCS	Q1D4K2	O07621	B4WU25	Q5P4K8	A0DTZ6	B1WNW3
HNOB	—	Q97E73	Q5YLC2	A0L531	—	Q9A451
PHY	A5P3T1	—	Q9RZA4	Q2IY35	A6W9I3	Q84XX0

a) Sensor domains are as in Table 15.1. Each cell lists an example (UniProt entry) of a protein that combines the respective sensor and signal transduction domains.

15.5
Diversity of Intracellular Signaling Pathways

An important property of the bacterial signal transduction systems is the variety of combinations between a relatively limited number of key sensor domains (Table 15.1) and signal transduction domains (Table 15.2). Almost any sensor domain can be found in combination in almost any signal transduction domain (Table 15.3). It means that signals perceived by these sensors can be processed through different signal transduction pathways [95] and elicit different cellular responses, sometimes at the level of gene expression, sometimes at the level of enzymatic activity, sometimes at the level of the whole-cell response (e.g., chemotaxis, phototaxis, biofilm formation). From the evolutionary point of view, it appears that different bacteria may utilize different adaptation strategies in response to the same environmental – or intracellular – stimuli. Obviously, there still remain many puzzles in the organization of bacterial signal transduction systems and detailed experimental analysis of the constituent domains remains a promising venue towards a better understanding of bacterial adaptation mechanisms.

Acknowledgments

This work was supported by the Intramural Research Program of the National Institutes of Health at the National Library of Medicine.

References

1 Fabret, C., Feher, V.A., and Hoch, J.A. (1999) Two-component signal transduction in *Bacillus subtilis*: how one organism sees its world. *J. Bacteriol.*, **181**, 1975–1983.

2 Galperin, M.Y. (2009) Sensory transduction network of *E. coli*, in *Systems Biology and Biotechnology of Escherichia coli* (ed. S.Y. Lee), Springer, Berlin, pp. 133–148.

3 Galperin, M.Y. (2004) Bacterial signal transduction network in a genomic perspective. *Environ. Microbiol.*, **6**, 552–567.
4 Galperin, M.Y. (2005) A census of membrane-bound and intracellular signal transduction proteins in bacteria: bacterial IQ, extroverts and introverts. *BMC Microbiol.*, **5**, 35.
5 Galperin, M.Y., Nikolskaya, A.N., and Koonin, E.V. (2001) Novel domains of the prokaryotic two-component signal transduction systems. *FEMS Microbiol. Lett.*, **203**, 11–21.
6 Kennelly, P.J. (2002) Protein kinases and protein phosphatases in prokaryotes: a genomic perspective. *FEMS Microbiol. Lett.*, **206**, 1–8.
7 Koonin, E.V. and Galperin, M.Y. (1997) Prokaryotic genomes: the emerging paradigm of genome-based microbiology. *Curr. Opin. Genet. Dev.*, **7**, 757–763.
8 Bateman, A. and Ponting, C.P. (2003) Identifying domains, repeats and motifs from protein sequences, in *Frontiers in Computational Genomics* (eds M.Y. Galperin and E.V. Koonin), Caister Academic Press, Wymondham, pp. 123–144.
9 Finn, R.D., Tate, J., Mistry, J., Coggill, P.C., Sammut, S.J., Hotz, H.R., Ceric, G., Forslund, K., Eddy, S.R., Sonnhammer, E.L. *et al.* (2008) The Pfam protein families database. *Nucleic Acids Res.*, **36**, D281–D288.
10 Letunic, I., Doerks, T., and Bork, P. (2009) SMART 6: recent updates and new developments. *Nucleic Acids Res.*, **37**, D229–D232.
11 Marchler-Bauer, A., Anderson, J.B., Chitsaz, F., Derbyshire, M.K., DeWeese-Scott, C., Fong, J.H., Geer, L.Y., Geer, R.C., Gonzales, N.R., Gwadz, M. *et al.* (2009) CDD: specific functional annotation with the Conserved Domain Database. *Nucleic Acids Res.*, **37**, D205–D210.
12 Tatusov, R.L., Galperin, M.Y., Natale, D.A., and Koonin, E.V. (2000) The COG database: a tool for genome-scale analysis of protein functions and evolution. *Nucleic Acids Res.*, **28**, 33–36.
13 Wu, C.H., Apweiler, R., Bairoch, A., Natale, D.A., Barker, W.C., Boeckmann, B., Ferro, S., Gasteiger, E., Huang, H., Lopez, R. *et al.* (2006) The Universal Protein Resource (UniProt): an expanding universe of protein information. *Nucleic Acids Res.*, **34**, D187–D191.
14 Pruitt, K.D., Tatusova, T., Klimke, W., and Maglott, D.R. (2009) NCBI Reference Sequences: current status, policy and new initiatives. *Nucleic Acids Res.*, **37**, D32–D36.
15 Hunter, S., Apweiler, R., Attwood, T.K., Bairoch, A., Bateman, A., Binns, D., Bork, P., Das, U., Daugherty, L., Duquenne, L. *et al.* (2009) InterPro: the integrative protein signature database. *Nucleic Acids Res.*, **37**, D211–D215.
16 Ulrich, L.E. and Zhulin, I.B. (2007) MiST: a microbial signal transduction database. *Nucleic Acids Res.*, **35**, D386–D390.
17 Galperin, M.Y. (2006) Structural classification of bacterial response regulators: diversity of output domains and domain combinations. *J. Bacteriol.*, **188**, 4169–4182.
18 Crews, S.T., Thomas, J.B., and Goodman, C.S. (1988) The *Drosophila* single-minded gene encodes a nuclear protein with sequence similarity to the *per* gene product. *Cell*, **52**, 143–151.
19 Hoffman, E.C., Reyes, H., Chu, F.F., Sander, F., Conley, L.H., Brooks, B.A., and Hankinson, O. (1991) Cloning of a factor required for activity of the Ah (dioxin) receptor. *Science*, **252**, 954–958.
20 Zhulin, I.B., Taylor, B.L., and Dixon, R. (1997) PAS domain S-boxes in Archaea, Bacteria and sensors for oxygen and redox. *Trends Biochem. Sci.*, **22**, 331–333.
21 Bibikov, S.I., Biran, R., Rudd, K.E., and Parkinson, J.S. (1997) A signal transducer for aerotaxis in *Escherichia coli*. *J. Bacteriol.*, **179**, 4075–4079.
22 Ponting, C.P. and Aravind, L. (1997) PAS: a multifunctional domain family comes to light. *Curr. Biol.*, **7**, R674–R677.
23 Andreeva, A., Howorth, D., Chandonia, J.M., Brenner, S.E., Hubbard, T.J., Chothia, C., and Murzin, A.G. (2008) Data growth and its impact on the SCOP database: new developments. *Nucleic Acids Res.*, **36**, D419–D425.
24 Taylor, B.L. and Zhulin, I.B. (1999) PAS domains: internal sensors of oxygen,

24. redox potential, and light. *Microbiol. Mol. Biol. Rev.*, **63**, 479–506.
25. Gilles-Gonzalez, M.A. and Gonzalez, G. (2005) Heme-based sensors: defining characteristics, recent developments, and regulatory hypotheses. *J. Inorg. Biochem.*, **99**, 1–22.
26. Genick, U.K., Soltis, S.M., Kuhn, P., Canestrelli, I.L., and Getzoff, E.D. (1998) Structure at 0.85 A resolution of an early protein photocycle intermediate. *Nature*, **392**, 206–209.
27. Crosson, S. and Moffat, K. (2001) Structure of a flavin-binding plant photoreceptor domain: insights into light-mediated signal transduction. *Proc. Natl. Acad. Sci. USA*, **98**, 2995–3000.
28. Brudler, R., Gessner, C.R., Li, S., Tyndall, S., Getzoff, E.D., and Woods, V.L. Jr. (2006) PAS domain allostery and light-induced conformational changes in photoactive yellow protein upon I2 intermediate formation, probed with enhanced hydrogen/deuterium exchange mass spectrometry. *J. Mol. Biol.*, **363**, 148–160.
29. Delgado-Nixon, V.M., Gonzalez, G., and Gilles-Gonzalez, M.A. (2000) Dos, a heme-binding PAS protein from *Escherichia coli*, is a direct oxygen sensor. *Biochemistry*, **39**, 2685–2691.
30. Key, J., Hefti, M., Purcell, E.B., and Moffat, K. (2007) Structure of the redox sensor domain of *Azotobacter vinelandii* NifL at atomic resolution: signaling, dimerization, and mechanism. *Biochemistry*, **46**, 3614–3623.
31. Crosson, S., Rajagopal, S., and Moffat, K. (2003) The LOV domain family: photo-responsive signaling modules coupled to diverse output domains. *Biochemistry*, **42**, 2–10.
32. Stephenson, K. and Hoch, J.A. (2001) PAS-A domain of phosphorelay sensor kinase A: a catalytic ATP-binding domain involved in the initiation of development in *Bacillus subtilis. Proc. Natl. Acad. Sci. USA*, **98**, 15251–15256.
33. Aravind, L. and Ponting, C.P. (1997) The GAF domain: an evolutionary link between diverse phototransducing proteins. *Trends Biochem. Sci.*, **22**, 458–459.
34. Martinez, S.E., Beavo, J.A., and Hol, W.G. (2002) GAF domains: two-billion-year-old molecular switches that bind cyclic nucleotides. *Mol. Interv.*, **2**, 317–323.
35. Hurley, J.H. (2003) GAF domains: cyclic nucleotides come full circle. *Sci. STKE*, **2003**, PE1.
36. Ho, Y.S., Burden, L.M., and Hurley, J.H. (2000) Structure of the GAF domain, a ubiquitous signaling motif and a new class of cyclic GMP receptor. *EMBO J.*, **19**, 5288–5299.
37. Ikeuchi, M. and Ishizuka, T. (2008) Cyanobacteriochromes: a new superfamily of tetrapyrrole-binding photoreceptors in cyanobacteria. *Photochem. Photobiol. Sci.*, **7**, 1159–1167.
38. Cornilescu, G., Ulijasz, A.T., Cornilescu, C.C., Markley, J.L., and Vierstra, R.D. (2008) Solution structure of a cyanobacterial phytochrome GAF domain in the red-light-absorbing ground state. *J. Mol. Biol.*, **383**, 403–413.
39. Martinez, S.E., Heikaus, C.C., Klevit, R.E., and Beavo, J.A. (2008) The structure of the GAF A domain from phosphodiesterase 6C reveals determinants of cGMP binding, a conserved binding surface, and a large cGMP-dependent conformational change. *J. Biol. Chem.*, **283**, 25913–25919.
40. Handa, N., Mizohata, E., Kishishita, S., Toyama, M., Morita, S., Uchikubo-Kamo, T., Akasaka, R., Omori, K., Kotera, J., Terada, T. et al. (2008) Crystal structure of the GAF-B domain from human phosphodiesterase 10A complexed with its ligand, cAMP. *J. Biol. Chem.*, **283**, 19657–19664.
41. Cann, M. (2007) A subset of GAF domains are evolutionarily conserved sodium sensors. *Mol. Microbiol.*, **64**, 461–472.
42. Lin, Z., Johnson, L.C., Weissbach, H., Brot, N., Lively, M.O., and Lowther, W.T. (2007) Free methionine-(R)-sulfoxide reductase from *Escherichia coli* reveals a new GAF domain function. *Proc. Natl. Acad. Sci. USA*, **104**, 9597–9602.
43. Gomelsky, M. and Klug, G. (2002) BLUF: a novel FAD-binding domain involved in sensory transduction in microorganisms. *Trends Biochem. Sci.*, **27**, 497–500.
44. Laan, W., Bednarz, T., Heberle, J., and Hellingwerf, K.J. (2004) Chromophore

composition of a heterologously expressed BLUF-domain. *Photochem. Photobiol. Sci.*, **3**, 1011–1016.

45 Jung, A., Reinstein, J., Domratcheva, T., Shoeman, R.L., and Schlichting, I. (2006) Crystal structures of the AppA BLUF domain photoreceptor provide insights into blue light-mediated signal transduction. *J. Mol. Biol.*, **362**, 717–732.

46 Grinstead, J.S., Avila-Perez, M., Hellingwerf, K.J., Boelens, R., and Kaptein, R. (2006) Light-induced flipping of a conserved glutamine sidechain and its orientation in the AppA BLUF domain. *J. Am. Chem. Soc.*, **128**, 15066–15067.

47 Masuda, S., Tomida, Y., Ohta, H., and Takamiya, K. (2007) The critical role of a hydrogen bond between Gln63 and Trp104 in the blue-light sensing BLUF domain that controls AppA activity. *J. Mol. Biol.*, **368**, 1223–1230.

48 Singh, A.H., Doerks, T., Letunic, I., Raes, J., and Bork, P. (2009) Discovering functional novelty in metagenomes: examples from light-mediated processes. *J. Bacteriol.*, **191**, 32–41.

49 Hou, S., Larsen, R.W., Boudko, D., Riley, C.W., Karatan, E., Zimmer, M., Ordal, G.W., and Alam, M. (2000) Myoglobin-like aerotaxis transducers in Archaea and Bacteria. *Nature*, **403**, 540–544.

50 Hou, S., Freitas, T., Larsen, R.W., Piatibratov, M., Sivozhelezov, V., Yamamoto, A., Meleshkevitch, E.A., Zimmer, M., Ordal, G.W., and Alam, M. (2001) Globin-coupled sensors: a class of heme-containing sensors in Archaea and Bacteria. *Proc. Natl. Acad. Sci. USA*, **98**, 9353–9358.

51 Freitas, T.A., Hou, S., and Alam, M. (2003) The diversity of globin-coupled sensors. *FEBS Lett.*, **552**, 99–104.

52 Mendez-Ortiz, M.M., Hyodo, M., Hayakawa, Y., and Membrillo-Hernandez, J. (2006) Genome-wide transcriptional profile of *Escherichia coli* in response to high levels of the second messenger 3′,5′-cyclic diguanylic acid. *J. Biol. Chem.*, **281**, 8090–8099.

53 Iyer, L.M., Anantharaman, V., and Aravind, L. (2003) Ancient conserved domains shared by animal soluble guanylyl cyclases and bacterial signaling proteins. *BMC Genomics*, **4**, 5.

54 Pellicena, P., Karow, D.S., Boon, E.M., Marletta, M.A., and Kuriyan, J. (2004) Crystal structure of an oxygen-binding heme domain related to soluble guanylate cyclases. *Proc. Natl. Acad. Sci. USA*, **101**, 12854–12859.

55 Nioche, P., Berka, V., Vipond, J., Minton, N., Tsai, A.L., and Raman, C.S. (2004) Femtomolar sensitivity of a NO sensor from Clostridium botulinum. *Science*, **306**, 1550–1553.

56 Ma, X., Sayed, N., Beuve, A., and van den Akker, F. (2007) NO and CO differentially activate soluble guanylyl cyclase via a heme pivot-bend mechanism. *EMBO J.*, **26**, 578–588.

57 Xiong, J., Kurtz, D.M. Jr., Ai, J., and Sanders-Loehr, J. (2000) A hemerythrin-like domain in a bacterial chemotaxis protein. *Biochemistry*, **39**, 5117–5125.

58 Isaza, C.E., Silaghi-Dumitrescu, R., Iyer, R.B., Kurtz, D.M. Jr., and Chan, M.K. (2006) Structural basis for O_2 sensing by the hemerythrin-like domain of a bacterial chemotaxis protein: substrate tunnel and fluxional N terminus. *Biochemistry*, **45**, 9023–9031.

59 Kao, W.C., Wang, V.C., Huang, Y.C., Yu, S.S., Chang, T.C., and Chan, S.I. (2008) Isolation, purification and characterization of hemerythrin from *Methylococcus capsulatus* (Bath). *J. Inorg. Biochem.*, **102**, 1607–1614.

60 French, C.E., Bell, J.M., and Ward, F.B. (2008) Diversity and distribution of hemerythrin-like proteins in prokaryotes. *FEMS Microbiol. Lett.*, **279**, 131–145.

61 Bailly, X., Vanin, S., Chabasse, C., Mizuguchi, K., and Vinogradov, S.N. (2008) A phylogenomic profile of hemerythrins, the nonheme diiron binding respiratory proteins. *BMC Evol. Biol.*, **8**, 244.

62 Heermann, R., Fohrmann, A., Altendorf, K., and Jung, K. (2003) The transmembrane domains of the sensor kinase KdpD of *Escherichia coli* are not essential for sensing K^+ limitation. *Mol. Microbiol.*, **47**, 839–848.

63 Heermann, R., Altendorf, K., and Jung, K. (2000) The hydrophilic N-terminal

63. domain complements the membrane-anchored C-terminal domain of the sensor kinase KdpD of *Escherichia coli*. *J. Biol. Chem.*, **275**, 17080–17085.

64. Heermann, R., Altendorf, K., and Jung, K. (2003) The N-terminal input domain of the sensor kinase KdpD of *Escherichia coli* stabilizes the interaction between the cognate response regulator KdpE and the corresponding DNA-binding site. *J. Biol. Chem.*, **278**, 51277–51284.

65. Nystrom, T. and Neidhardt, F.C. (1994) Expression and role of the universal stress protein, UspA, of *Escherichia coli* during growth arrest. *Mol. Microbiol.*, **11**, 537–544.

66. Zarembinski, T.I., Hung, L.W., Mueller-Dieckmann, H.J., Kim, K.K., Yokota, H., Kim, R., and Kim, S.H. (1998) Structure-based assignment of the biochemical function of a hypothetical protein: a test case of structural genomics. *Proc. Natl. Acad. Sci. USA*, **95**, 15189–15193.

67. Heermann, R., Fohrmann, A., Altendorf, K., and Jung, K. (2009) The universal stress protein UspC scaffolds the KdpD/KdpE signaling cascade of *Escherichia coli* under salt stress. *J. Mol. Biol.*, **386**, 134–148.

68. Davis, S.J., Vener, A.V., and Vierstra, R.D. (1999) Bacteriophytochromes: phytochrome-like photoreceptors from non-photosynthetic eubacteria. *Science*, **286**, 2517–2520.

69. Essen, L.O., Mailliet, J., and Hughes, J. (2008) The structure of a complete phytochrome sensory module in the Pr ground state. *Proc. Natl. Acad. Sci. USA*, **105**, 14709–14714.

70. Yang, X., Kuk, J., and Moffat, K. (2008) Crystal structure of *Pseudomonas aeruginosa* bacteriophytochrome: photoconversion and signal transduction. *Proc. Natl. Acad. Sci. USA*, **105**, 14715–14720.

71. Grebe, T.W. and Stock, J.B. (1999) The histidine protein kinase superfamily. *Adv. Microb. Physiol.*, **41**, 139–227.

72. Gao, R., Mack, T.R., and Stock, A.M. (2007) Bacterial response regulators: versatile regulatory strategies from common domains. *Trends Biochem. Sci.*, **32**, 225–234.

73. Zhulin, I.B. (2001) The superfamily of chemotaxis transducers: from physiology to genomics and back. *Adv. Microb. Physiol.*, **45**, 157–198.

74. Alexander, R.P. and Zhulin, I.B. (2007) Evolutionary genomics reveals conserved structural determinants of signaling and adaptation in microbial chemoreceptors. *Proc. Natl. Acad. Sci. USA*, **104**, 2885–2890.

75. Postma, P.W., Lengeler, J.W., and Jacobson, G.R. (1993) Phosphoenolpyruvate: carbohydrate phosphotransferase systems of bacteria. *Microbiol. Rev.*, **57**, 543–594.

76. Deutscher, J., Francke, C., and Postma, P.W. (2006) How phosphotransferase system-related protein phosphorylation regulates carbohydrate metabolism in bacteria. *Microbiol. Mol. Biol. Rev.*, **70**, 939–1031.

77. Reizer, J., Reizer, A., Merrick, M.J., Plunkett, G., Rose, D.J., and Saier, M.H. (1996) Novel phosphotransferase-encoding genes revealed by analysis of the *Escherichia coli* genome: a chimeric gene encoding an Enzyme I homologue that possesses a putative sensory transduction domain. *Gene*, **181**, 103–108.

78. Tchieu, J.H., Norris, V., Edwards, J.S., and Saier, M.H. Jr. (2001) The complete phosphotranferase system in *Escherichia coli*. *J. Mol. Microbiol. Biotechnol.*, **3**, 329–346.

79. Wolfgang, M.C., Lee, V.T., Gilmore, M.E., and Lory, S. (2003) Coordinate regulation of bacterial virulence genes by a novel adenylate cyclase-dependent signaling pathway. *Dev. Cell*, **4**, 253–263.

80. Lory, S., Wolfgang, M., Lee, V., and Smith, R. (2004) The multi-talented bacterial adenylate cyclases. *Int. J. Med. Microbiol.*, **293**, 479–482.

81. Ohmori, M. and Okamoto, S. (2004) Photoresponsive cAMP signal transduction in cyanobacteria. *Photochem. Photobiol. Sci.*, **3**, 503–511.

82. Sismeiro, O., Trotot, P., Biville, F., Vivares, C., and Danchin, A. (1998) *Aeromonas hydrophila* adenylyl cyclase 2: a new class of adenylyl cyclases with thermophilic properties and sequence similarities to proteins from hyperthermophilic archaebacteria. *J. Bacteriol.*, **180**, 3339–3344.

83 Amikam, D. and Galperin, M.Y. (2006) PilZ domain is part of the bacterial c-di-GMP binding protein. *Bioinformatics*, **22**, 3–6.

84 Benach, J., Swaminathan, S.S., Tamayo, R., Handelman, S.K., Folta-Stogniew, E., Ramos, J.E., Forouhar, F., Neely, H., Seetharaman, J., Camilli, A. *et al.* (2007) The structural basis of cyclic diguanylate signal transduction by PilZ domains. *EMBO J.*, **26**, 5153–5166.

85 Kazmierczak, B.I., Lebron, M.B., and Murray, T.S. (2006) Analysis of FimX, a phosphodiesterase that governs twitching motility in Pseudomonas aeruginosa. *Mol. Microbiol.*, **60**, 1026–1043.

86 Lee, V.T., Matewish, J.M., Kessler, J.L., Hyodo, M., Hayakawa, Y., and Lory, S. (2007) A cyclic-di-GMP receptor required for bacterial exopolysaccharide production. *Mol. Microbiol.*, **65**, 1474–1484.

87 Hickman, J.W. and Harwood, C.S. (2008) Identification of FleQ from *Pseudomonas aeruginosa* as a c-di-GMP-responsive transcription factor. *Mol. Microbiol.*, **69**, 376–389.

88 Hanlon, W.A., Inouye, M., and Inouye, S. (1997) Pkn9, a Ser/Thr protein kinase involved in the development of *Myxococcus xanthus*. *Mol. Microbiol.*, **23**, 459–471.

89 Umeyama, T., Lee, P.C., and Horinouchi, S. (2002) Protein serine/threonine kinases in signal transduction for secondary metabolism and morphogenesis in Streptomyces. *Appl. Microbiol. Biotechnol.*, **59**, 419–425.

90 Saskova, L., Novakova, L., Basler, M., and Branny, P. (2007) Eukaryotic-type serine/threonine protein kinase StkP is a global regulator of gene expression in *Streptococcus pneumoniae*. *J. Bacteriol.*, **189**, 4168–4179.

91 Ulijasz, A.T., Falk, S.P., and Weisblum, B. (2009) Phosphorylation of the RitR DNA-binding domain by a Ser–Thr phosphokinase: implications for global gene regulation in the streptococci. *Mol. Microbiol.*, **71**, 382–390.

92 Macek, B., Mijakovic, I., Olsen, J.V., Gnad, F., Kumar, C., Jensen, P.R., and Mann, M. (2007) The serine/threonine/tyrosine phosphoproteome of the model bacterium *Bacillus subtilis*. *Mol. Cell Proteomics*, **6**, 697–707.

93 Macek, B., Gnad, F., Soufi, B., Kumar, C., Olsen, J.V., Mijakovic, I., and Mann, M. (2008) Phosphoproteome analysis of *E. coli* reveals evolutionary conservation of bacterial Ser/Thr/Tyr phosphorylation. *Mol. Cell. Proteomics*, **7**, 299–307.

94 Soufi, B., Gnad, F., Jensen, P.R., Petranovic, D., Mann, M., Mijakovic, I., and Macek, B. (2008) The Ser/Thr/Tyr phosphoproteome of *Lactococcus lactis* IL1403 reveals multiply phosphorylated proteins. *Proteomics*, **8**, 3486–3493.

95 Zhulin, I.B., Nikolskaya, A.N., and Galperin, M.Y. (2003) Common extracellular sensory domains in transmembrane receptors for diverse signal transduction pathways in bacteria and Archaea. *J. Bacteriol.*, **185**, 285–294.

96 Matsushika, A. and Mizuno, T. (1998) The structure and function of the histidine-containing phosphotransfer (HPt) signaling domain of the *Escherichia coli* ArcB sensor. *J. Biochem.*, **124**, 440–445.

97 Xu, Q. and West, A.H. (1999) Conservation of structure and function among histidine-containing phosphotransfer (HPt) domains as revealed by the crystal structure of YPD1. *J. Mol. Biol.*, **292**, 1039–1050.

98 Volz, K. (1993) Structural conservation in the CheY superfamily. *Biochemistry*, **32**, 11741–11753.

99 Aravind, L. and Ponting, C.P. (1999) The cytoplasmic helical linker domain of receptor histidine kinase and methyl-accepting proteins is common to many prokaryotic signalling proteins. *FEMS Microbiol. Lett.*, **176**, 111–116.

100 Williams, S.B. and Stewart, V. (1999) Functional similarities among two-component sensors and methyl-accepting chemotaxis proteins suggest a role for linker region amphipathic helices in transmembrane signal transduction. *Mol. Microbiol.*, **33**, 1093–1102.

101 Morett, E. and Segovia, L. (1993) The sigma 54 bacterial enhancer-binding protein family: mechanism of action and phylogenetic relationship of their functional domains. *J. Bacteriol.*, **175**, 6067–6074.

102 Sallai, L. and Tucker, P.A. (2005) Crystal structure of the central and C-terminal domain of the sigma^{54}-activator ZraR. *J. Struct. Biol.*, **151**, 160–170.

103 Yeats, C., Bentley, S., and Bateman, A. (2003) New knowledge from old: *in silico* discovery of novel protein domains in *Streptomyces coelicolor*. *BMC Microbiol.*, **3**, 3.

104 Baikalov, I., Schroder, I., Kaczor-Grzeskowiak, M., Grzeskowiak, K., Gunsalus, R.P., and Dickerson, R.E. (1996) Structure of the *Escherichia coli* response regulator NarL. *Biochemistry*, **35**, 11053–110561.

105 Martinez-Hackert, E. and Stock, A.M. (1997) The DNA-binding domain of OmpR: crystal structures of a winged helix transcription factor. *Structure*, **5**, 109–124.

106 Nikolskaya, A.N. and Galperin, M.Y. (2002) A novel type of conserved DNA-binding domain in the transcriptional regulators of the AlgR/AgrA/LytR family. *Nucleic Acids Res.*, **30**, 2453–2459.

107 Sidote, D.J., Barbieri, C.M., Wu, T., and Stock, A.M. (2008) Structure of the *Staphylococcus aureus* AgrA LytTR domain bound to DNA reveals a beta fold with an unusual mode of binding. *Structure*, **16**, 727–735.

108 Gallegos, M.T., Schleif, R., Bairoch, A., Hofmann, K., and Ramos, J.L. (1997) Arac/XylS family of transcriptional regulators. *Microbiol. Mol. Biol. Rev.*, **61**, 393–410.

109 Rhee, S., Martin, R.G., Rosner, J.L., and Davies, D.R. (1998) A novel DNA-binding motif in MarA: the first structure for an AraC family transcriptional activator. *Proc. Natl. Acad. Sci. USA*, **95**, 10413–10418.

110 Lewis, R.J., Krzywda, S., Brannigan, J.A., Turkenburg, J.P., Muchova, K., Dodson, E.J., Barak, I., and Wilkinson, A.J. (2000) The trans-activation domain of the sporulation response regulator Spo0A revealed by X-ray crystallography. *Mol. Microbiol.*, **38**, 198–212.

111 Yuan, H.S., Finkel, S.E., Feng, J.A., Kaczor-Grzeskowiak, M., Johnson, R.C., and Dickerson, R.E. (1991) The molecular structure of wild-type and a mutant Fis protein: relationship between mutational changes and recombinational enhancer function or DNA binding. *Proc. Natl. Acad. Sci. USA*, **88**, 9558–9562.

112 Pearl, L., O'Hara, B., Drew, R., and Wilson, S. (1994) Crystal structure of AmiC: the controller of transcription antitermination in the amidase operon of *Pseudomonas aeruginosa*. *EMBO J.*, **13**, 5810–5817.

113 Shu, C.J. and Zhulin, I.B. (2002) ANTAR: an RNA-binding domain in transcription antitermination regulatory proteins. *Trends Biochem. Sci.*, **27**, 3–5.

114 Szurmant, H. and Ordal, G.W. (2004) Diversity in chemotaxis mechanisms among the bacteria and archaea. *Microbiol. Mol. Biol. Rev.*, **68**, 301–319.

115 Djordjevic, S., Goudreau, P.N., Xu, Q., Stock, A.M., and West, A.H. (1998) Structural basis for methylesterase CheB regulation by a phosphorylation-activated domain. *Proc. Natl. Acad. Sci. USA*, **95**, 1381–1386.

116 West, A.H., Martinez-Hackert, E., and Stock, A.M. (1995) Crystal structure of the catalytic domain of the chemotaxis receptor methylesterase. CheB. *J. Mol. Biol.*, **250**, 276–290.

117 Bilwes, A.M., Alex, L.A., Crane, B.R., and Simon, M.I. (1999) Structure of CheA, a signal-transducing histidine kinase. *Cell*, **96**, 131–141.

118 Liao, D.I., Silverton, E., Seok, Y.J., Lee, B.R., Peterkofsky, A., and Davies, D.R. (1996) The first step in sugar transport: crystal structure of the amino terminal domain of enzyme I of the *E. coli* PEP: sugar phosphotransferase system and a model of the phosphotransfer complex with HPr. *Structure*, **4**, 861–872.

119 Teplyakov, A., Lim, K., Zhu, P.P., Kapadia, G., Chen, C.C., Schwartz, J., Howard, A., Reddy, P.T., Peterkofsky, A., and Herzberg, O. (2006) Structure of phosphorylated enzyme I, the phosphoenolpyruvate: sugar phosphotransferase system sugar translocation signal protein. *Proc. Natl. Acad. Sci. USA*, **103**, 16218–16223.

120 Napper, S., Delbaere, L.T., and Waygood, E.B. (1999) The aspartyl replacement of the active site histidine in histidine-containing protein, HPr, of the *Escherichia coli* phosphoenolpyruvate: sugar

phosphotransferase system can accept and donate a phosphoryl group. Spontaneous dephosphorylation of acyl-phosphate autocatalyzes an internal cyclization. *J. Biol. Chem.*, **274**, 21776–21782.

121 Feese, M.D., Comolli, L., Meadow, N.D., Roseman, S., and Remington, S.J. (1997) Structural studies of the *Escherichia coli* signal transducing protein IIAGlc: implications for target recognition. *Biochemistry*, **36**, 16087–16096.

122 Cai, M., Williams, D.C. Jr., Wang, G., Lee, B.R., Peterkofsky, A., and Clore, G.M. (2003) Solution structure of the phosphoryl transfer complex between the signal-transducing protein IIAGlucose and the cytoplasmic domain of the glucose transporter IICBGlucose of the *Escherichia coli* glucose phosphotransferase system. *J. Biol. Chem.*, **278**, 25191–25206.

123 Saier, M.H. Jr. and Reizer, J. (1992) Proposed uniform nomenclature for the proteins and protein domains of the bacterial phosphoenolpyruvate: sugar phosphotransferase system. *J. Bacteriol.*, **174**, 1433–1438.

124 Barzu, O. and Danchin, A. (1994) Adenylyl cyclases: a heterogeneous class of ATP-utilizing enzymes. *Prog. Nucleic Acid Res. Mol. Biol.*, **49**, 241–283.

125 Katayama, M. and Ohmori, M. (1997) Isolation and characterization of multiple adenylate cyclase genes from the cyanobacterium *Anabaena* sp. strain PCC 7120. *J. Bacteriol.*, **179**, 3588–3593.

126 Coudart-Cavalli, M.P., Sismeiro, O., and Danchin, A. (1997) Bifunctional structure of two adenylyl cyclases from the myxobacterium *Stigmatella aurantiaca*. *Biochimie*, **79**, 757–767.

127 Richter, W. (2002) $3',5'$-Cyclic nucleotide phosphodiesterases class III: members, structure, and catalytic mechanism. *Proteins*, **46**, 278–286.

128 Shenoy, A.R., Capuder, M., Draskovic, P., Lamba, D., Visweswariah, S.S., and Podobnik, M. (2007) Structural and biochemical analysis of the Rv0805 cyclic nucleotide phosphodiesterase from *Mycobacterium tuberculosis*. *J. Mol. Biol.*, **365**, 211–225.

129 Hecht, G.B. and Newton, A. (1995) Identification of a novel response regulator required for the swarmer-to-stalked-cell transition in *Caulobacter crescentus*. *J. Bacteriol.*, **177**, 6223–6229.

130 Chan, C., Paul, R., Samoray, D., Amiot, N.C., Giese, B., Jenal, U., and Schirmer, T. (2004) Structural basis of activity and allosteric control of diguanylate cyclase. *Proc. Natl. Acad. Sci. USA*, **101**, 17084–17089.

131 Tal, R., Wong, H.C., Calhoon, R., Gelfand, D., Fear, A.L., Volman, G., Mayer, R., Ross, P., Amikam, D., Weinhouse, H. et al. (1998) Three *cdg* operons control cellular turnover of cyclic di-GMP in *Acetobacter xylinum*: genetic organization and occurrence of conserved domains in isoenzymes. *J. Bacteriol.*, **180**, 4416–4425.

132 Schmidt, A.J., Ryjenkov, D.A., and Gomelsky, M. (2005) Ubiquitous protein domain EAL encodes cyclic diguanylate-specific phosphodiesterase: enzymatically active and inactive EAL domains. *J. Bacteriol.*, **187**, 4774–4781.

133 Galperin, M.Y., Natale, D.A., Aravind, L., and Koonin, E.V. (1999) A specialized version of the HD hydrolase domain implicated in signal transduction. *J. Mol. Microbiol. Biotechnol.*, **1**, 303–305.

134 Ryan, R.P., Fouhy, Y., Lucey, J.F., Crossman, L.C., Spiro, S., He, Y.W., Zhang, L.H., Heeb, S., Camara, M., Williams, P. et al. (2006) Cell–cell signaling in *Xanthomonas campestris* involves an HD-GYP domain protein that functions in cyclic di-GMP turnover. *Proc. Natl. Acad. Sci. USA*, **103**, 6712–6717.

135 Ryjenkov, D.A., Simm, R., Romling, U., and Gomelsky, M. (2006) The PilZ domain is a receptor for the second messenger c-di-GMP: the PilZ domain protein YcgR controls motility in enterobacteria. *J. Biol. Chem.*, **281**, 30310–30314.

136 Young, T.A., Delagoutte, B., Endrizzi, J.A., Falick, A.M., and Alber, T. (2003) Structure of *Mycobacterium tuberculosis* PknB supports a universal activation mechanism for Ser/Thr protein kinases. *Nat. Struct. Biol.*, **10**, 168–174.

137 Pullen, K.E., Ng, H.L., Sung, P.Y., Good, M.C., Smith, S.M., and Alber, T.

(2004) An alternate conformation and a third metal in PstP/Ppp, the *M. tuberculosis* PP2C-Family Ser/Thr protein phosphatase. *Structure*, **12**, 1947–1954.

138 Durocher, D., Henckel, J., Fersht, A.R., and Jackson, S.P. (1999) The FHA domain is a modular phosphopeptide recognition motif. *Mol. Cell*, **4**, 387–394.

139 Pallen, M., Chaudhuri, R., and Khan, A. (2002) Bacterial FHA domains: neglected players in the phospho-threonine signalling game? *Trends Microbiol.*, **10**, 556–563.

16
Sensing of Oxygen by Bacteria
Gottfried Unden, Martin Müllner, and Florian Reinhart

16.1
Introduction

Molecular oxygen (O_2) has adverse effects on bacteria. It functions as an electron acceptor and substrate for catabolism for aerobic, facultatively anaerobic, and microaerobic bacteria. In anaerobic bacteria, or in microaerobic and aerobic bacteria under hyperoxic conditions, O_2 induces oxidative stress response. Therefore, facultatively anaerobic and microaerobic bacteria measure the presence of O_2 to adapt metabolism. In facultative bacteria, O_2 represses anaerobic respiration and fermentation, which are less efficient with respect to molar ATP yield [1–3]. The bacteria measure the O_2 tension in order to coordinate expression of alternative pathways and to protect O_2-sensitive enzymes from damage. There are, however, bacteria that perform reactions of anaerobic catabolism like denitrification in parallel with aerobic respiration [4, 5].

Bacteria contain O_2 sensors for responding to the presence of O_2. Regulation occurs mostly at the transcriptional level, but sensors like Aer and WhiB3 control aerotaxis and metabolism at the protein level. The sensors react either directly with O_2, or indirectly by interaction with enzymes or metabolites of aerobic metabolism. Many bacteria contain more than one O_2 sensor that can function in different ways and regulate different target genes. Use of two sensors allows sensing by different modes and coordination of related pathways. Direct sensors control expression in response to the actual O_2 concentration, whereas indirect sensors to respond to the metabolic situation in response to O_2, but other parameters like carbon or [H] supply as well. The combination of direct and indirect sensing allows an efficient adaptation to various aspects of metabolic requirements. Many genes of facultatively anaerobic or microaerobic metabolism are regulated by more than one O_2 senor. Thus, the *cyd* operon of *Escherichia coli* encoding the microaerobic quinol oxidase *bd* is transcriptionally regulated by the sensors FNR (fumarate/nitrate reductase regulator) and ArcBA, resulting in maximal expression under microaerobic, and partial repression under aerobic and anaerobic conditions [6]. *Bacillus subtilis* and *Staphylococcus*, for example, contain three different O_2 sensors for direct (FNR_{Bs}, Rex, and NreB) or indirect (ResE/ResD and SrrA/SrrB) sensing.

Bacterial Signaling. Edited by Reinhard Krämer and Kirsten Jung
Copyright © 2010 WILEY-VCH Verlag GmbH & Co. KGaA, Weinheim
ISBN: 978-3-527-32365-4

This chapter describes sensors that respond to O_2. Sensors reacting with reactive oxygen species (O_2^-, H_2O_2 and peroxides) that induce oxidative stress response will be mentioned only briefly (for a review, see [7]). These include the sensor SoxR with a $[2Fe-2S]^{1+}$ cluster that is oxidized to $[2Fe-2S]^{2+}$ in the presence of O_2^- [8]. Aconitase of some bacteria is bifunctional and has in addition to the enzymatic a regulatory function, and senses oxidative stress and iron starvation [9]. In *E. coli*, the catalytic [4Fe–4S] cluster of the enzyme (in particular of isoenzyme AcnB) is decomposed under oxidative stress (H_2O_2 or O_2^-) and iron starvation. The cofactor-less apoAcnB modulates the expression of superoxide dismutase (*sodA* gene) by controlling translation, presumably by binding to specific structures in the mRNA. The H_2O_2 and peroxide stress sensor OxyR is oxidized by peroxides at cysteine residues to the sulfenic acid (R-SOH) or disulfide state that affects DNA-binding of the transcriptional activator [10–12]. In the sensor OhrR from *B. subtilis*, a single cysteine residue is oxidized by organic hydroperoxides to the sulfenic acid that affects the regulatory activity of OhrR [13].

16.2
O_2 as a Signal

O_2 can interact with the sensors by oxidation of cofactors (e.g., oxidative degradation of $[4Fe-4S]^{2+}$ of FNR), or by specific binding to sensors (e.g., FixL from *Bradyrhizobium*), or indirectly by reaction with oxygenases or oxidases, which then produce a signal that is perceived by the sensor. Due to the significance of regulation by O_2 and the chemical versatility of O_2, many and largely different sensing and signaling devices are found in bacteria.

Air-saturated water contains about 210 μM O_2, but in many biotopes the available O_2 concentration is much lower, since O_2 supply by diffusion without convection is slow for distances below 100 μm compared to O_2 consumption rates by respiring bacteria (Figure 16.1). Many bacteria grow in an environment with permanently or temporarily limiting O_2 concentration and have to cope with microaerobic conditions (1–10 μM or mbar of O_2). The regulatory O_2 tensions are in the same range (typically 1–5 μM O_2, corresponding to approximately 0.5–2.5% air saturation) [14]. Obviously the switch points characterize a physiological *in situ* situation, but do not represent the K_d values of the sensors for O_2.

The direct O_2 sensors are located in the cytoplasm or are attached to the membrane and sense O_2 from the cytoplasm, whereas indirect sensors are often linked to electron transfer chains in the cytoplasmic membrane or perceive the signal from the membrane. The diffusion rate of O_2 in water and the membrane is very high over the small distances (micrometer range) of bacterial cells, compared to the O_2 consumption by respiration (Figure 16.1B) [3]. Therefore the cytoplasmic O_2 tension is comparable to the extracellular, even under microaerobic conditions. The rapid O_2 supply by diffusion therefore supplies sufficient O_2 under aerobic and microaerobic conditions for regulatory and even catabolic processes, like growth on aromatic compounds that depend on the function of cytoplasmic mono- or dioxygenases [14–16].

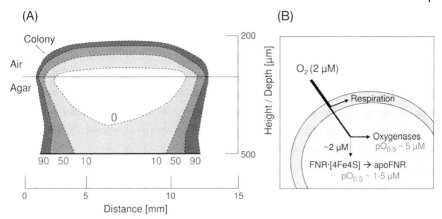

Figure 16.1 Oxygen gradients in surface colonies of bacteria (A) and diffusion of O_2 over short distances into bacterial cytoplasm (B). Scheme (A) gives the contour map of O_2 concentration in a surface colony grown an agar that can be determined with microelectrodes. The contour lines give the O_2 content (% of air saturation). Scheme (B) shows the amount of O_2 which can be estimated to diffuse over short distance ($\leq 1\,\mu m$) into the bacterial cells amounting to approximately $7\,\mu mol\ O_2/min/mg$ protein under microaerobic conditions ($2\,\mu M$ or 2 mbar O_2 in the medium). The diffusion rate compares to O_2 consumption of about $0.5\,\mu mol\ O_2/min/mg$ protein by respiration of *E. coli* under the same conditions. Therefore O_2 diffuses rapidly to the cytoplasm and the cytoplasmic equals the extracellular O_2 concentration under aerobic and microaerobic conditions, and supplies O_2 for regulatory and catabolic purpose in the cytoplasm. The switch point ($pO_{0.5}$ or half-maximal activity) for regulation by FNR and for cytoplasmic oxygenases is 1–5 and 5–10 $\mu M\ O_2$, respectively. For details, see text and [3, 15, 16].

16.3
Direct O_2 Sensors

Many O_2 sensors interact directly with O_2 (Table 16.1). Heme B, $[4Fe–4S]^{2+}$ clusters, or $FADH_2$ are used for reaction with O_2 that can be mere binding of O_2 or oxidation and disintegration of the cofactor. The reactions trigger conformational changes to transmit the signal to the output domains of the sensors. Heme B and $[4Fe–4S]^{2+}$ are used by various types of sensors (sensor kinases, enzymes, and gene regulators). The mode of signal transduction and output differs greatly, which can be activity control of the kinase domain in sensor kinases, of enzyme activity, of DNA binding, or interaction with a downstream regulatory protein.

16.3.1
Heme B-Containing Sensors

16.3.1.1 FixL
Expression of the O_2-sensitive nitrogenase of diazotrophic bacteria is carefully controlled in response to O_2. Most bacteria perform N_2 fixation under anoxic conditions or protect the enzyme from the deleterious effects of O_2. Oxygen sensors

Table 16.1 Bacterial sensors directly interacting with O_2.

Sensor (type)/bacteria	Stimulus	Cofactor/reaction with O_2	Signal chain and output
FixL (sensor kinase)/ Sinorhizobium meliloti	O_2	heme B/binding of O_2	FixL/FixJ two-component system (inactivation of kinase by O_2); gene activation of *nif* and *fix* genes
Dos (enzyme)/ Escherichia coli	O_2	heme B/binding of O_2 or oxidation	PAS domain with heme B controls phosphodiesterase activity; inactivation of Dos by O_2
HemAT (chemoreceptor)/ Bacillus subtilis	O_2	heme B	receptor complex HemAT–CheW–CheA; phosphorylation of CheY; stimulation of CheA and smooth swimming in the presence of O_2
FNR (gene regulator)/ E. coli	O_2	$[4Fe-4S]^{2+}$/cluster degradation	dimer/monomer conversion and gene activation by FNR_2 (anoxic form)
NreB (sensor kinase)/ Staphylococcus carnosus	O_2	$[4Fe-4S]^{2+}$/cluster degradation	NreB/NreC two-component system (inactivation of kinase by O_2); activation of nitrate respiration by NreC$\sim$P (anoxic conditions)
WhiB3 (sensor)/ Mycobacterium tuberculosis	O_2	$[4Fe-4S]$	control of central metabolism
NifL/Azotobacter vinelandii	O_2 (MKH_2)[a]	$FADH_2/FADH_2 + O_2$ $\rightarrow FAD + H_2O_2$	NifL is antiactivator of NifA (binding of NifA by oxidized NifL); activation of *nif* and *fix* genes by NifA (anaerobic conditions)

a) *Klebsiella*.
Reactivation of O_2-inactivated sensors in not clear for many of the sensors. Binding of O_2 to heme-containing sensors (FixL, Dos, HemAT) is reversible under anoxic conditions. Regeneration of $[4Fe-4S]^{2+}$-containing sensors under anoxic conditions after cluster disassembly has been suggested. The $FADH_2$-binding sensor NifL of *Klebsiella* is reactivated at the membrane by menaquinol (MKH_2) after O_2 inactivation. The quoted bacteria contain well-characterized examples for the sensor. See text for references.

that control expression of N_2 fixation genes represent well-defined examples for O_2 sensing. Many of the sensors respond directly to O_2. In *Sinorhizobium meliloti*, O_2-responsive regulation is complex [17]. The two-component system FixL/FixJ activates the expression of the *fix* and *nif* genes under anoxic conditions by a

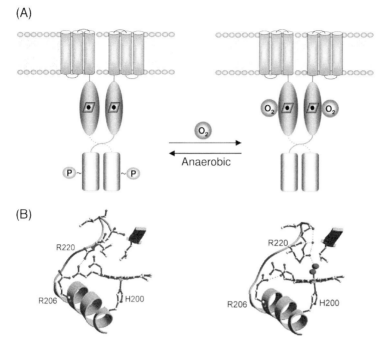

Figure 16.2 FixL as a direct O_2 sensor using heme B for O_2 sensing. (A) FixL consists of an N-terminal membrane domain followed by a PAS and the kinase domain in the cytoplasm. Binding of O_2 to heme B in the PAS domain converts the kinase to the inactive state, and prevents phosphorylation of the kinase and of the response regulator FixJ. Phosphorylated FixJ transcriptionally activates expression of the *fix* and *nif* genes. (B) Heme B-binding site in detail. Binding of O_2 at the sixth coordination site of heme B lifts the Fe ion in the tetrapyrrol plane, which affects the ligation of the Fe(II) and propionate residue of heme by the His200 and Arg206/220 residues in the region of the linker helix of the PAS domain. The shift is transferred to the kinase domain which inhibits the activity in oxy-FixL. (Part (B) with permission from [20]).

regulatory cascade. FixL/FixJ control the expression of the NifA and FixF transcription factors, which in turn activate the expression of the *fix* and *nif* genes.

FixL is inserted by three or four transmembrane helices in the membrane, which are followed by a large cytoplasmic region with a Per/Arnt/Sim (PAS) domain and the histidine kinase domains (Figure 16.2). The PAS domain carries heme B which is fivefold coordinated by the protein and has a free sixth coordination site for binding O_2 [18–22]. In the deoxy state, the central Fe(II) of the heme is in the high-spin state and the kinase is active. By binding of O_2 to the heme, Fe(II) is transferred from the high-spin to the low-spin state (Figure 16.2B). Oxygen binding lifts the Fe(II) in the porphyrin ring and affects binding of the heme to the PAS domain. The heme interacts via the Fe(II) and the propionate residues with histidine and arginine residues in the linker region, and the preceding loop of the PAS domain. The structural changes at the heme upon O_2 binding are transmitted by the arginine and histidine residues to the linker region and the kinase domain that follows the PAS domain. The shift causes inactivation of the kinase in oxy-FixL and of the signal cascade, and the expression of the *nif* and *fix* genes comes to an end.

16.3.1.2 Dos

The Dos (direct oxygen sensor) protein of *E. coli* is a homotetrameric enzyme with phosphodiesterase activity. The activity of the enzyme is controlled by O_2 [23], but the physiological role of the O_2-dependent diesterase activity is not clear. Dos binds heme B in a N-terminal PAS domain like FixL, but the Fe(II) of heme B is sixfold coordinated. The heme B of Dos is able to react with O_2 by binding the molecule which displaces the sixth ligand, a methionine residue [24–26]. Alternatively, O_2 oxidizes the Fe(II) to Fe(III). The diesterase is active only in the deoxy (or reduced Fe(II)) state. Transfer of Dos to the Fe(III) and the oxy form causes a conformational switch which is linked to the loss of the esterase activity.

16.3.1.3 HemAT

B. subtilis is able to perform aerotaxis in media containing O_2 gradients. In chemotaxis, positive or negative stimuli are transmitted by chemoreceptors to a cytoplasmic signal transduction system that controls rotation of the flagellar motor. *B. subtilis* uses homologs of the *E. coli* chemotaxis proteins for signal transduction in the cytoplasm, including CheW, histidine kinase CheA, and response regulator CheY. The chemoreceptor for sensing O_2 in the aerotaxis of *B. subtilis* is the HemAT protein [27]. HemAT is a homodimer and binds heme in a globin-like domain that has only limited amino acid sequence similarity to hemoglobin-like proteins. The oxygenated form of HemAT shows structural changes which are induced by O_2 binding. The oxy-HemAT dimer is more symmetric and allows formation of the HemAT–CheW–CheA receptor complex. Binding of O_2 stimulates kinase activity of CheA and formation of phosphorylated CheY that controls motor switching. Oxygen (which functions as an attractant) therefore stimulates kinase activity and induces smooth swimming in increasing O_2 gradients.

16.3.2
$[4Fe-4S]^{2+}$-Containing Sensors

16.3.2.1 FNR$_{Ec}$

The O_2 sensor FNR of *E. coli* controls the expression of a multitude of genes, mainly of anaerobic respiration [2, 3, 28]. It is a member of the cAMP receptor protein (CRP)/FNR family of transcriptional activators with a helix–turn–helix DNA-binding domain. In the anaerobic state FNR binds a $[4Fe-4S]^{2+}$ cluster by four essential cysteine residues [7, 29, 30]. FNR $[4Fe-4S]^{2+}$ is a homodimer that activates gene expression in a mode similar to CRP [31]. FNR responds very sensitively to O_2 concentrations as low as 1 µM (or 1 mbar) in the medium or cytoplasm (corresponding to 0.5% air saturation) [14]. Presence of O_2 causes degradation of the $[4Fe-4S]^{2+}$ cluster, with $[3Fe-4S]^{1+}$ and $[2Fe-2S]^{2+}$ as the intermediates (Figure 16.3) [32–35]. The final product is apoFNR with the formal loss of two Fe^{3+}, two Fe^{2+}, and four

(A) FNR

(B) NreB

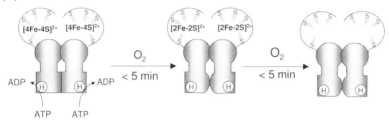

Figure 16.3 FNR and NreB as direct [4Fe–4S]$^{2+}$-based O$_2$ sensors. Scheme (A) shows the reaction of dimeric anaerobic FNR that binds [4Fe–4S]$^{2+}$ and the products formed by reaction of the FeS cluster with O$_2$. The half-maximal O$_2$ concentrations (pO$_{0.5}$ values) and reaction times for the conversion of [4Fe–4S]FNR to [2Fe–2S]FNR and apoFNR are given. The times are approximately the same *in vivo* and *in vitro*. Only dimeric FNR binds to target promoters. Cluster insertion in FNR requires cysteine desulfurase IscS. (B) Response of the sensor kinase NreB of *Staphylococcus carnosus* to O$_2$. The scheme shows the response of the FeS cluster to O$_2$ and the effect on the kinase activity. For details, see text.

S^{2-}. Apo-(cluster-less)FNR is found in aerobically grown *E. coli* or after exposure of anaerobic FNR to O$_2$ *in vitro* [36, 37]. The degradation of [4Fe–4S]$^{2+}$ starts with the oxidation of one Fe(II) in the cluster and release of one Fe(II) ([4Fe–4S]$^{2+}$ + O$_2$ [3Fe–4S]$^{1+}$ + O$_2^-$ + Fe^{2+}) [38–40]. The [3Fe–4S]$^{1+}$ cluster is not stable and rapidly degrades to [2Fe–2S]$^{2+}$ by spontaneous release of one Fe(III) and two sulfide ions. FNR with [2Fe–2S]$^{2+}$ is rather stable in the absence of O$_2$. Conversion of [2Fe–2S]FNR to apoFNR is triggered by O$_2$ or O$_2^-$ [36]. The conversion of the (cubic) [4Fe–4S]$^{2+}$ to the (planar) [2Fe–2S]$^{2+}$ and the loss of the cluster are supposed to induce significant structural rearrangements that affect the dimerization helix of FNR, resulting in monomerization [32]. The monomeric forms ([2Fe–2S]FNR and apoFNR) are no longer able to bind to DNA and to function as gene regulator.

ApoFNR is the main form of FNR in aerobically grown bacteria [36, 37]. [2Fe–2S] FNR might be of physiological significance when O$_2$ is present only intermittently or at low O$_2$ tensions. Under such conditions [2Fe–2S]FNR could allow rapid reversal to the anaerobic form. It is not known, however, whether apoFNR or [2Fe–2S]FNR are recycled to [4Fe–4S]FNR in significant amounts. There are indications for recycling of O$_2$ inactivated FNR; on the other hand, is apoFNR more sensitive to proteolysis.

16.3.2.2 FNR$_{Bs}$

B. subtilis contains a variant of FNR that is also a member of the CRP/FNR family [41]. FNR$_{Bs}$ activates the expression of genes of nitrate/nitrite respiration under anoxic conditions. FNR$_{Bs}$ from anaerobic bacteria contains a [4Fe–4S]$^{2+}$ cluster which activates expression of FNR$_{Bs}$-dependent genes. Unlike FNR from *E. coli*, the cysteine residues binding the FeS cluster are at the C-terminal end of the protein close to the DNA-binding site [42]. The region contains three conserved cysteine residues; the fourth ligand of the cluster is not known. Reaction with O$_2$ causes loss of the cluster [42]. [4Fe–4S]FNR$_{Bs}$ and apoFNR$_{Bs}$ are dimers, but only the former functions in gene activation. It was suggested that cluster degradation changes the arrangement of the DNA-binding site and disturbs DNA-binding and gene activation.

16.3.2.3 NreB

The cytoplasmic NreB/NreC two-component system activates the expression of nitrate/nitrite respiration of the bacteria under anoxic conditions [43]. NreB/NreC is found in *Staphylococcus* strains, *Bacillus clausii*, and *Geobacillus* spp. The sensor kinase NreB binds [4Fe–4S]$^{2+}$ in a PAS domain by four conserved cysteine residues (Figure 16.3B) [44, 45]. The [4Fe–4S]$^{2+}$-containing form is found in anaerobically grown bacteria. It has high kinase activity, resulting in phosphorylation of the response regulator NreC that binds to target promoters and activates gene expression. Upon reaction with air the cluster is degraded with a half-life time of about 2.5 min to a [2Fe–2S]$^{2+}$-containing form. [2Fe–2S]NreB is also not stable and within 5 min most of the [4Fe–4S]NreB is converted to apoNreB. Exposure to air decreases the content of [4Fe–4S]$^{2+}$ and kinase activity of NreB with similar half-life times. Therefore, the PAS domain of NreB uses a [4Fe–4S]$^{2+}$ cofactor for O$_2$ sensing in a mode similar to that of FNR of *E. coli*, and even the mode and rate of [4Fe–4S]$^{2+}$ cluster decomposition is similar to FNR of *E. coli*.

16.3.2.4 WhiB3

Mycobacterium tuberculosis requires O$_2$ for replication, but can persist in the host for very long periods without O$_2$, if adaptation to O$_2$ depletion occurs in an ordered mode or gradually. The regulator WhiB3 plays an important role in acquiring persistence and survival of the bacteria during nutrient and O$_2$ starvation [46]. WhiB3 contains a FeS cluster that appears to be of the [4Fe–4S]$^{2+}$ type. The cluster responds to O$_2$ by degradation. WhiB3 is supposed to cause a metabolic switch, but the level and targets where WhiB3 interacts are not known.

16.3.2.5 [4Fe–4S]$^{2+}$ as a Universal Cofactor for O$_2$ Sensing

[4Fe–4S]$^{2+}$ clusters are widely used cofactors for sensing of O$_2$. The cofactor is found in unrelated proteins, like the adjuncts to the transcriptional regulators FNR and FNR$_{Bs}$, the PAS domain of NreB, and the WhiB3 protein. FNR and FNR$_{Bs}$ contain short N- or C-terminal, extensions to the CRP core that carry most of the essential cysteine residues. The proteins all contain three or four conserved cysteine residues for binding of the [4Fe–4S] cluster, but the sequences and the spacing of the cysteine residues show no clear similarities for FNR (Cys(20)-X_2-Cys-X_5-Cys-X_{92}-Cys), FNR_{Bs} (Cys(227)-X_2-Cys-X_4-Cys), NreB (Cys(C59)-X_2-Cys-X_{11}-Cys-X_2-Cys), and WhiB3 (C(23)-X_{29}-C-X_2-C-X_5-C). It appears that the clusters have been acquired by unrelated proteins and that the sensory [4Fe–4S]$^{2+}$ clusters can be adopted in different modes. The sensors all respond to O_2 by disassembly of the cluster. The mode of signal processing and output, however, varies largely. The same cofactor controls dimerization (FNR), kinase activity (NreB), DNA-binding (FNR_{Bs}), and the functioning of other proteins (WhiB3). The labile [4Fe–4S] cluster of aconitase (Cys(C710)-X_{58}-Cys-X_2-Cys) appears to be different in terms of its sensitivity to H_2O_2 and O_2^-.

16.3.3
FAD-Containing Sensors

FAD/$FADH_2$ is found as a cofactor in various sensors responding to O_2. NifL from *Azotobacter vinelandii* is supposed to react directly with O_2, whereas other FAD-containing sensors are indirect O_2 sensors.

16.3.3.1 NifL
The NifL/NifA system of *A. vinelandii* controls synthesis of nitrogenase in response to O_2, which for its part regulates the activity of the transcriptional regulator NifA [47]. NifL is a homotetramer and each monomer consists of two N-terminal PAS domains that are connected by a linker region to the C-terminal output domain. The output domain shares sequence similarity with histidine kinases, but signal transfer to NifA occurs only by protein contacts without phosphorylation. NifA is a σ^{54}-dependent transcriptional activator.

The function of NifL is controlled by the redox state of FAD (FAD/$FADH_2$; $E_0' = -226$ mV in the protein) [48]. NifL of anaerobic bacteria contains $FADH_2$ and does not interact with NifA, which is then able to activate expression of the nitrogen fixation (*nif*) genes. Upon oxidation of the $FADH_2$, NifL binds NifA and functions as an anti-activator preventing expression of *nif* genes. The physiological electron donor for reduction of the $FADH_2$ in NifL of *A. vinelandii* is not clear. NifL from *Klebsiella*, however, binds under O_2 limitation to the membrane and the FAD is reduced by the respiratory quinones [49].

The $FADH_2$ of *A. vinelandii* NifL is buried in a cavity, and channels allow the access of O_2 and the release of the reduced product, H_2O_2 [50]. The oxidation and subsequent protonation of FAD initiates conformational changes in the FAD region, which is transmitted to the interface of the PAS domains in the tetramer. According to this scheme NifL is a direct O_2 sensor, but the reduction of the FAD might be linked to the respiratory chain.

Table 16.2 Bacterial sensors responding indirectly to the presence of O_2.

Sensor (type)/bacteria	Direct signal	Sensory device in sensor/reaction of sensor	Signal chain and output
ArcB (sensor kinase)/ *Escherichia coli*	Q/QH_2	Cys-SH of ArcB/2Cys-SH + Q → Cys-S-S-Cys + QH_2	ArcB/ArcA two-component system (inactivation by O_2); repression of aerobic catabolism under N_2
Aer (chemoreceptor)/ *E. coli*	electron transport	FAD/oxidation/ reduction	receptor complex Aer–CheW–CheA; phosphorylation of CheY and control flagellar motor (aerotaxis)
RegB/PrrB (sensor kinase)/*Rhodobacter*	Q/QH_2 or oxidase cbb_3	Cys(265)-SH, binding site for Q or cbb_3	RegB/RegA (Prrb/PrrA) two-component system; activation of genes (photosynthesis, anaerobic ET)
Rex (repressor)/ *Streptomyces, Bacillus*	NADH: NAD^+	NADH: NAD binding	inactivation of repressor by NADH; gene repression at low NADH, high NAD^+

All sensor systems respond indirectly to the presence of O_2. The membrane-bound sensor kinases ArcB, RegB, and PrrB, and the chemoreceptor Aer, perceive their direct signal from the aerobic electron transport, and respond either by a redox reaction or by binding of an intermediate (Q, or a subunit of oxidase cbb_3). The quoted bacteria contain well-characterized examples for the sensor. For details, see text and Figure 16.4. Q, ubiquinone; QH_2, ubiquinol; ET, electron transport.

16.4
Indirect O_2 Sensors

Indirect O_2 sensors respond to components of the respiratory chain or to cytoplasmic factors that change in a characteristic mode in response to O_2 availability (Table 16.2).

16.4.1
Electron Transport-Linked Sensors

16.4.1.1 ArcB/ArcA

The ArcB/ArcA two-component system of *E. coli* plays an important role in the regulation of facultative anaerobic metabolism. ArcB (Figure 16.4) is inserted in the membrane by two transmembrane helices that are followed by a cytoplasmic leucine zipper and a PAS domain [51]. ArcB is a hybrid kinase containing a catalytic kinase domain with conserved His292, a receiver domain with conserved Asp576 and a phosphotransfer domain with a second conserved histidine residue (His717). Under anoxic conditions the protein is autophosphorylated followed by intramolecular phosphate transfer (H292 → D576 → H717) and phosphorylation of ArcA. ArcA ∼ P binds to target promoters and controls gene expression. Under oxic conditions

(A) ArcB-ArcA

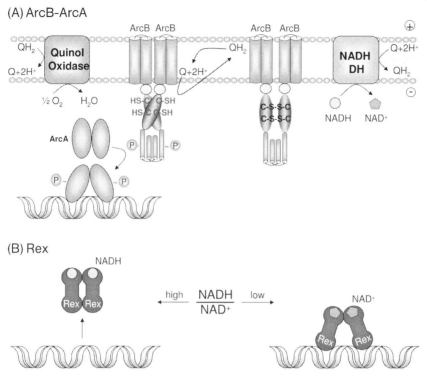

(B) Rex

Figure 16.4 ArcB/ArcA and Rex as indirect O_2 sensors. (A) ArcB/ArcA: the sensor kinase ArcB is a dimer and consists of two transmembrane helices, followed by a leucine zipper, PAS, and composite kinase domains. Residues Cys180 and Cys241 in the cytoplasmic PAS domain respond to the redox state of the respiratory quinones. During aerobic respiration the quinone: quinol ratio is high, resulting in the oxidation of the cysteine residues. Formation of the intermolecular disulfides (Cys180–Cys180 and Csy241–Cys241) inhibits kinase activity. (B) The complex of the repressor Rex with NAD^+ binds to promoter DNA and represses the expression of the target genes. In the absence of O_2 and high carbon or [H] supply, the NADH: NAD^+ ratio increases. NADH binds with high affinity to Rex, leading to the release of Rex from DNA. The Rex · NADH complex has a closed conformation which is no longer able to bind to DNA.

when the kinase is inactive, ArcB functions as an ArcA $\sim$ P phosphatase. ArcB senses O_2 indirectly by the redox state of the respiratory chain [52]. Oxidized quinones, which prevail under aerobic conditions, oxidize two cysteine residues (Cys180 and Cys241) of the PAS domain in two steps (Figure 16.4A). The kinase is inactive under aerobic conditions when intermolecular Cys180–Cys180 and Cys241–Cys241 disulfides are formed. The disulfides are supposed to cause domain rearrangement affecting kinase activity. It is not clear how the membrane integral quinone is able to oxidize the peripheral Cys180 and Cys241. However, cysteine residues in the periplasmic domain of the membrane bound protein disulfide reductase DsbB [53] can be reduced by membrane integral quinol; oxidation of the cysteine residues of ArcB could be achieved in a similar way.

16.4.1.2 Aer

Aer regulates swimming behavior of *E. coli* in gradients of O_2 (aerotaxis) and of other nutrients [54]. Aer is an atypical chemoreceptor that is fixed to the membrane. It carries, in contrast to typical chemoreceptors, the receptor site in the cytoplasm. Aer is fixed by two transmembrane helices to the membrane. The receptor domain is located N-terminal to the transmembrane helices at the cytoplasmic side of the membrane. It consists of a PAS domain that binds FAD noncovalently [55, 56]. It is supposed that respiration or redox changes in the electron transport chain oxidize and reduce the FAD, which then controls the function of the PAS domain [57]. The signal transmission module is C-terminal to the membrane anchor. It is composed of a HAMP domain (found in *h*istidine kinases, *a*denylyl cyclases, *m*ethyl-accepting chemotaxis proteins, and *p*hosphatases) and a signaling domain typical for chemoreceptors. The HAMP domain represents the output module that is regulated by the sensing (or input) PAS domain and transmits the signal to the signaling domain in the cytoplasm. The C-terminal signaling domain is arranged in a complex with the adapter protein CheW and the chemotaxis histidine kinase CheA. The signaling domain controls in the complex the activity of the kinase CheA and thus phosphorylation of the response regulator CheY. Phosphorylated CheY regulates switching of the flagellar motor.

16.4.1.3 PrrB/PrrA and RegB/RegA

Rhodobacter capsulatus contains the two-component system RegB/RegA for controlling the expression of photosynthesis, respiration, and related genes in response to O_2 [58, 59]. A closely homologous system is represented by PrrB/PrrA of *Rhodobacter sphaeroides* [60]. The RegB/PrrB histidine kinases consist of an N-terminal membrane integral domain with six transmembrane helices and a C-terminal kinase domain. The domain functions as a kinase in the anaerobic (default) state and as a phosphatase under oxic conditions.

The sensor kinases perceive the signal ("presence of O_2") from the respiratory chain. Alternative models imply different signaling molecules and modes of signal transduction. One model suggests that during function in aerobic respiration cytochrome oxidase cbb_3 produces an inhibitory signal and converts the kinase to the phosphorylase mode [60]. Under anoxic conditions the kinase is active, resulting in the expression of the Reg/Prr controlled genes. Alternative data suggest that the quinones of the respiratory chain supply the signal controlling RegB/PrrB function. At high O_2 tension, quinone prevails, and is suggested to bind to a motif in the membrane region of RegB and thereby inhibit the kinase activity [58]. Alternatively it was suggested that a conserved cysteine residue Cys265 in RegB is oxidized under high Q: QH_2 ratios as in aerobic respiration [59]. This leads to an intermolecular disulfide and a RegB/RegB dimer with decreased kinase activity.

16.4.1.4 ResE/ResD and SrrA/SrrB

The ResE/ResD ("regulation of respiration") two-component system represents a major system for anaerobic gene regulation in *B. subtilis* [61, 62]. The sensor kinase

ResE consists of two transmembrane helices that are separated by a large extra-cytoplasmic loop. The second transmembrane helix is followed by HAMP, PAS, and kinase domains. ResE/ResD controls the expression of anaerobic catabolic genes, including the fnr_{Bs} gene. The mechanism of signal perception by ResE is unknown. The SrrB/SrrA ("*Staphylococcus* respiratory response") two-component system is highly similar to ResE/ResD and controls aerobic/anaerobic shift of metabolism, including induction of fermentative enzymes repression of citric acid cycle genes/enzymes under anaerobic conditions, and synthesis of virulence factors of *S. aureus* [63, 64]. It is supposed that both systems respond indirectly to O_2 and operate in addition to direct (FNR_{Bs}/NreB/NreC) and indirect (Rex) O_2 sensors.

16.4.2
NADH-Linked Systems

16.4.2.1 Rex

Gram-positive bacteria like *Streptomyces*, *Staphylococcus*, and *Bacillus* contain the repressor Rex ("redox") which responds to cellular NADH: NAD^+ ratios [65, 66]. Rex controls the expression of genes encoding respiratory enzymes, dehydrogenases, and other enzymes of redox metabolism. Rex is wide-spread in Gram-positive bacteria. In bacteria the NADH: NAD^+ ratios show characteristic variations in response to O_2 that allows rapid oxidation of NADH by respiration. Under oxic conditions with low NADH: NAD^+ ratio, Rex represses transcription of genes.

Rex is composed of an N-terminal helix–turn–helix DNA-binding and a C-terminal NADH-binding domain with a characteristic Rossmann fold [66]. At low NADH: NAD^+ ratios dimeric Rex in complex with NAD^+ binds to operator sites and represses target genes (Figure 16.4B). Rex binds NADH with much higher affinity than NAD. Binding of NADH shifts the domain arrangement in Rex and brings the DNA-binding domains in close contact, which disturbs DNA binding. After release of the protein, the promoter becomes accessible to transcription. The NADH: NAD^+ ratio is affected in addition to the presence of O_2 by many metabolic parameters like availability of suitable sources of [H] or carbon, which makes the NADH: NAD^+ ratio and Rex versatile in sensing many metabolic situations in addition to the presence of O_2.

Acknowledgments

Work in the authors laboratory was supported by Deutsche Forschungsgemeinschaft.

References

1 Gunsalus, R.P. (1992) Control of electron flow in *Escherichia coli*: coordinated transcription of respiratory pathway genes. *J. Bacteriol.*, **174**, 7069–7074.

2 Guest, J.R. (1995) The Leeuwenhoek Lecture, 1995. Adaptation to life without oxygen. *Philos. Trans. R. Soc. Lond. B. Biol. Sci.*, **350**, 189–202.

3 Unden, G., Becker, S., Bongaerts, J., Holighaus, G., Schirawski, J. et al. (1995) O_2-sensing and O_2-dependent gene regulation in facultatively anaerobic bacteria. *Arch. Microbiol.*, **164**, 81–90.

4 Körner, H. and Zumft, W.G. (1989) Expression of denitrification enzymes in response to the dissolved oxygen level and respiratory substrate in continuous culture of *Pseudomonas stutzeri*. *Appl. Environ. Microbiol.*, **55**, 1670–1676.

5 Su, J.J., Liu, B.Y., and Liu, C.Y. (2001) Comparison of aerobic denitrification under high oxygen atmosphere by *Thiosphaera pantotropha* ATCC 35512 and *Pseudomonas stutzeri* SU2 newly isolated from the activated sludge of a piggery wastewater treatment system. *J. Appl. Microbiol.*, **90**, 457–462.

6 Cotter, P.A., Melville, S., Albrecht, J., and Gunsalus, R.P. (1997) Aerobic regulation of cytochrome *d* oxidase (cydAB) operon expression in *Escherichia coli*: roles of Fnr and ArcA in repression and activation. *Mol. Microbiol*, **25**, 605–615.

7 Green, J. and Paget, M.S. (2004) Bacterial redox sensors. *Nat. Rev. Microbiol.*, **2**, 954–966.

8 Hidalgo, E., Bollinger, J.M., Bradley, J.M., Walsh, C.T., and Demple, B.J. (1995) Binuclear [2Fe–2S] clusters in the *Escherichia coli* SoxR protein and role of the metal centers in transcription. *J. Biol. Chem.*, **270**, 20908–20914.

9 Tang, Y., Guest, J.R., Artymiuk, P.J., Read, R.C., and Green, J. (2004) Post-transcriptional regulation of bacterial motility by aconitase proteins. *Mol. Microbiol.*, **51**, 1817–1826.

10 Storz, G., Tartaglia, L.A., and Ames, B.N. (1990) The OxyR regulon. *Science*, **248**, 189–194.

11 Zheng, M., Åslund, F., and Storz, G. (1998) Activation of the OxyR transcription factor by reversible disulfide bond formation. *Science*, **279**, 1718–1721.

12 Kim, S.O., Merchant, K., Nudelman, R., Beyer, W.F., Keng, T. et al. (2002) OxyR: a molecular code for redox-related signaling. *Cell*, **109**, 383–396.

13 Fuangthong, M. and Helmann, J.D. (2002) The OhrR repressor senses organic hydroperoxides by reversible formation of a cysteine-sulfenic acid derivative. *Proc. Natl. Acad. Sci. USA*, **99**, 6690–6695.

14 Becker, S., Holighaus, G., Gabrielczyk, T., and Unden, G. (1996) O_2 as the regulatory signal for FNR-dependent gene regulation in *Escherichia coli*. *J. Bacteriol.*, **178**, 4515–4521.

15 Arras, T., Schirawski, J., and Unden, G. (1998) Availability of O_2 as a substrate in the cytoplasm of bacteria under aerobic and microaerobic conditions. *J. Bacteriol.*, **180**, 2133–2136.

16 Unden, G. and Schirawski, J. (1997) The oxygen-responsive transcriptional regulator FNR of *Escherichia coli*: the search for signals and reactions. *Mol. Microbiol.*, **25**, 205–210.

17 Fischer, H.M. (1994) Genetic regulation of nitrogen fixation in rhizobia. *Microbiol. Rev.*, **58**, 352–386.

18 Gong, W., Hao, B., and Chan, M.K. (2000) New mechanistic insights from structural studies of the oxygen-sensing domain of *Bradyrhizobium japonicum* FixL. *Biochemistry*, **39**, 3955–3962.

19 Gilles-Gonzalez, M.A., Ditta, G.S., and Helinski, D.R. (1991) A haemoprotein with kinase activity encoded by the oxygen sensor of *Rhizobium meliloti*. *Nature*, **350**, 170–172.

20 Gilles-Gonzalez, M.A. and Gonzalez, G. (2005) Heme-based sensors: defining characteristics, recent developments, and regulatory hypotheses. *J. Inorg. Biochem.*, **99**, 1–22.

21 Gilles-Gonzalez, M.A. and Gonzalez, G. (1993) Regulation of the kinase activity of heme protein FixL from the two-component system FixL/FixJ of *Rhizobium meliloti*. *J. Biol. Chem.*, **268**, 16293–16297.

22 Gilles-Gonzalez, M.A. and Gonzalez, G. (2004) Signal transduction by heme-containing PAS-domain proteins. *J. Appl. Physiol.*, **20054**, **96**, 774–783.

23 Delgado-Nixon, V.M., Gonzalez, G., and Gilles-Gonzalez, M.A. (2000) Dos, a heme-binding PAS protein from *Escherichia coli*, is a direct oxygen sensor. *Biochemistry*, **39**, 2685–2691.

24 Gonzalez, G., Dioum, E.M., Bertolucci, C.M., Tomita, T., Ikeda-Saito, M. et al. (2002) Nature of the displaceable heme-axial residue in the *EcDos* protein, a heme-

based sensor from *Escherichia coli*. *Biochemistry*, **41**, 8414–8421.

25 Park, H., Suquet, C., Satterlee, J.D., and Kang, C. (2004) Insights into signal transduction involving PAS domain oxygen-sensing heme proteins from the X-ray crystal structure of *Escherichia coli* Dos heme domain (*Ec*DosH). *Biochemistry*, **43**, 2738–2746.

26 Kurokawa, H., Lee, D.S., Watanabe, M., Sagami, I., Mikami, B. *et al.* (2004) A redox-controlled molecular switch revealed by the crystal structure of a bacterial heme PAS sensor. *J. Biol. Chem.*, **279**, 20186–20193.

27 Zhang, W. and Phillips, G.N. (2003) Structure of the oxygen sensor in *Bacillus subtilis*: signal transduction of chemotaxis by control of symmetry. *Structure*, **11**, 1097–1110.

28 Salmon, K., Hung, S.P., Mekjian, K., Baldi, P., Hatfield, G.W. *et al.* (2003) Global gene expression profiling in *Escherichia coli* K12. The effects of oxygen availability and FNR. *J. Biol. Chem.*, **278**, 29837–29855.

29 Kiley, P.J. and Beinert, H. (1998) Oxygen sensing by the global regulator, FNR: the role of the iron–sulfur cluster. *FEMS Microbiol. Rev.*, **22**, 341–352.

30 Kiley, P.J. and Beinert, H. (2003) The role of Fe–S proteins in sensing and regulation in bacteria. *Curr. Opin. Microbiol.*, **6**, 181–185.

31 Wing, H.J., Green, J., Guest, J.R., and Busby, S.J. (2000) Role of activating region 1 of *Escherichia coli* FNR protein in transcription activation at class II promoters. *J. Biol. Chem.*, **275**, 29061–29065.

32 Lazazzera, B.A., Beinert, H., Khoroshilova, N., Kennedy, M.C., and Kiley, P.J. (1996) DNA binding and dimerization of the Fe–S-containing FNR protein from *Escherichia coli* are regulated by oxygen. *J. Biol. Chem.*, **271**, 2762–2768.

33 Khoroshilova, N., Beinert, H., and Kiley, P. (1995) Association of a polynuclear iron–sulfur center with a mutant FNR protein enhances DNA binding. *Proc. Natl. Acad. Sci. USA*, **92**, 2499–2503.

34 Khoroshilova, N., Popescu, C., Münck, E., Beinert, H., and Kiley, P.J. (1997) Iron–sulfur cluster disassembly in the FNR protein of *Escherichia coli* by O_2: [4Fe–4S] to [2Fe–2S] conversion with loss of biological activity. *Proc. Natl. Acad. Sci. USA*, **94**, 6087–6092.

35 Green, J., Bennett, B., Jordan, P., Ralph, E.T., Thomson, A.J., and Guest, J.R. (1996) Reconstitution of the [4Fe–4S] cluster in FNR and demonstration of the aerobic-anaerobic transcription switch *in vitro*. *Biochem. J.*, **316**, 887–892.

36 Sutton, V.R., Stubna, A., Patschkowski, T., Münck, E., Beinert, H. *et al.* (2004) Superoxide destroys the $[2Fe–2S]^{2+}$ cluster of FNR from *Escherichia coli*. *Biochemistry*, **43**, 791–798.

37 Reinhart, F., Achebach, S., Koch, T., and Unden, G. (2008) Reduced apo-fumarate nitrate reductase regulator (apoFNR) as the major form of FNR in aerobically growing *Escherichia coli*. *J. Bacteriol.*, **190**, 879–886.

38 Crack, J.C., Green, J., Le Brun, N.E., and Thomson, A.J.J. (2006) Detection of sulfide release from the oxygen-sensing [4Fe–4S] cluster of FNR. *Biol. Chem.*, **281**, 18909–18913.

39 Crack, J.C., Le Brun, N.E., Thomson, A.J., Green, J., and Jervis, A.J. (2008) Reactions of nitric oxide and oxygen with the regulator of fumarate and nitrate reduction, a global transcriptional regulator, during anaerobic growth of *Escherichia coli*. *Methods Enzymol.*, **437**, 191–209.

40 Crack, J.C., Green, J., Cheesman, M.R., Le Brun, N.E., and Thomson, A.J. (2007) Superoxide-mediated amplification of the oxygen-induced switch from [4Fe–4S] to [2Fe–2S] clusters in the transcriptional regulator FNR. *Proc. Natl. Acad. Sci. USA*, **104**, 2092–2097.

41 Cruz Ramos, H., Boursier, L., Moszer, I., Kunst, F., Danchin, A., and Glaser, P. (1995) Anaerobic transcription activation in *Bacillus subtilis*: identification of distinct FNR-dependent and -independent regulatory mechanisms. *EMBO J.*, **14**, 5984–5994.

42 Reents, H., Gruner, I., Harmening, U., Böttger, L.H., Layer, G. *et al.* (2006) *Bacillus subtilis* Fnr senses oxygen via a [4Fe–4S] cluster coordinated by three cysteine residues without change in the

oligomeric state. *Mol. Microbiol.*, **60**, 1432–1445.

43 Fedtke, I., Kamps, A., Krismer, B., and Götz, F. (2002) The nitrate reductase and nitrite reductase operons and the *narT* gene of *Staphylococcus carnosus* are positively controlled by the novel two-component system NreBC. *J. Bacteriol.*, **184**, 6624–6634.

44 Kamps, A., Achebach, S., Fedtke, I., Unden, G., and Götz, F. (2004) Staphylococcal NreB: an O_2-sensing histidine protein kinase with an O_2-labile iron-sulphur cluster of the FNR type. *Mol. Microbiol.*, **52**, 713–723.

45 Müllner, M., Hammel, O., Mienert, B., Schlag, S., Bill, E., and Unden, G. (2008) A PAS domain with an oxygen labile $[4Fe-4S]^{2+}$ cluster in the oxygen sensor kinase NreB of *Staphylococcus carnosus*. *Biochemistry*, **47**, 13921–13932.

46 Singh, A., Guidry, L., Narasimhulu, K.V., Mai, D., Trombley, J. et al. (2007) *Mycobacterium tuberculosis* WhiB3 responds to O_2 and nitric oxide via its [4Fe-4S] cluster and is essential for nutrient starvation survival. *Proc. Natl. Acad. Sci. USA*, **104**, 111562–111567.

47 Martinez-Argudo, I., Little, R., Shearer, N., Johnson, P., and Dixon, R. (2004) The NifL–NifA system: a multidomain transcriptional regulatory complex that integrates environmental signals. *J. Bacteriol.*, **186**, 601–610.

48 Macheroux, P., Hioll, S., Austin, S., Eydmann, T., Jones, T. et al. (1998) Electron donation to the flavoprotein NifL, a redox-sensing transcriptional regulator. *Biochem. J.*, **332**, 413–419.

49 Thummer, R., Klimmek, O., and Schmitz, R.A. (2007) Biochemical studies of *Klebsiella pneumoniae* NifL reduction using reconstituted partial anaerobic respiratory chains of *Wolinella succinogenes*. *J. Biol. Chem.*, **282**, 12517–12526.

50 Key, J., Hefti, M., Purcell, E.B., and Moffat, K. (2007) Structure of the redox sensor domain of *Azotobacter vinelandii* NifL at atomic resolution: signaling, dimerization, and mechanism. *Biochemistry*, **46**, 3614–3623.

51 Georgellis, D., Kwon, O., and Lin, E.C. (2001) Quinones as the redox signal for the Arc two-component system of bacteria. *Science*, **292**, 2314–2316.

52 Malpica, R., Franco, B., Rodriguez, C., Kwon, O., and Georgellis, D. (2004) Identification of a quinone-sensitive redox switch in the ArcB sensor kinase. *Proc. Natl. Acad. Sci. USA*, **101**, 13318–13323.

53 Bader, M., Muse, W., Ballou, D.P., Gassner, C., and Bardwell, J.C. (1999) Oxidative protein folding is driven by the electron transport system. *Cell*, **98**, 217–227.

54 Taylor, B.L. (2007) Aer on the inside looking out: paradigm for a PAS–HAMP role in sensing oxygen, redox and energy. *Mol. Microbiol.*, **65**, 1415–1424.

55 Bibikov, S.I., Biran, R., Rudd, K.E., and Parkinson, J.S. (1997) A signal transducer for aerotaxis in *Escherichia coli*. *J. Bacteriol.*, **179**, 4075–4079.

56 Repik, A., Rebbapragada, A., Johnson, M.S., Haznedar, J.O., Zhulin, I.B. et al. (2000) PAS domain residues involved in signal transduction by the Aer redox sensor of *Escherichia coli*. *Mol. Microbiol.*, **36**, 806–816.

57 Edwards, J.C., Johnson, M.S., and Taylor, B.L. (2006) Differentiation between electron transport sensing and proton motive force sensing by the Aer and Tsr receptors for aerotaxis. *Mol. Microbiol.*, **62**, 823–837.

58 Swem, L.R., Gong, X., Yu, C.A., and Bauer, C.E. (2006) Identification of a ubiquinone-binding site that affects autophosphorylation of the sensor kinase RegB. *J. Biol. Chem.*, **281**, 6768–6775.

59 Swem, L.R., Kraft, B.J., Swem, D.L., Setterdahl, A.T., Masuda, S. et al. (2003) Signal transduction by the global regulator RegB is mediated by a redox-active cysteine. *EMBO J.*, **22**, 4699–4708.

60 Kim, Y.-J., Ko, I.-J., Lee, J.-M., Kang, H.-Y., Kim, Y.M. et al. (2007) Dominant role of the cbb_3 oxidase in regulation of photosynthesis gene expression through the PrrBA system in *Rhodobacter sphaeroides* 2.4.1. *J. Bacteriol.*, **189**, 5617–5625.

61 Nakano, M.M. and Hulett, F.M. (1997) Adaptation of *Bacillus subtilis* to oxygen limitation. *FEMS Microbiol. Letters*, **157**, 1–7.

62 Geng, H., Zuber, P., and Nakano, M.M. (2007) Regulation of respiratory genes by ResD–ResE signal transduction system in *Bacillus subtilis*. *Methods Enzymol.*, **422**, 448–464.

63 Yarwood, J.M., McCormick, J.K., and Schlievert, P.M. (2001) Identification of a novel two-component regulatory system that acts in global regulation of virulence factors of *Staphylococcus aureus*. *J. Bacteriol.*, **183**, 1113–1123.

64 Throup, J.P., Zappacosta, F., Lunsford, R.D., Annan, R.S., Lonsdale, J.T. *et al.* (2001) The srhSR gene pair from *Staphylococcus aureus*: genomic and proteomic approaches to the identification and characterization of gene function. *Biochemistry*, **40**, 10392–10401.

65 Brekasis, D. and Paget, M.S. (2003) A novel sensor of NADH/NAD$^+$ redox poise in *Streptomyces coelicolor* A3$_2$. *EMBO J.*, **22**, 4856–4865.

66 Wang, E., Bauer, M.C., Rogstam, A., Linse, S., Logan, D.T., and von Wachenfeld, C. (2008) Structure and functional properties of the *Bacillus subtilis* transcriptional repressor Rex. *Mol. Microbiol.*, **69**, 466–478.

17
Microbial Sensor Systems for Dihydrogen, Nitric Oxide, and Carbon Monoxide
Rainer Cramm and Bärbel Friedrich

17.1
Introduction

The transition of the atmosphere to an aerobic environment about 2000 million years ago resulted in the ubiquitous use of the gas molecular oxygen as a universal terminal electron acceptor in the respiratory chains of almost all higher organisms. However, life began in the absence of free molecular oxygen and many recent microorganisms still use gases other than oxygen as electron acceptors.

Gases are also used as transmitter molecules for inter- or intracellular signaling purposes. The gaseous plant hormone ethylene has long been known to be involved in a large number of developmental processes [1]. Notably, the first two-component receptor to be discovered in plants was the ethylene receptor ETR1 (*e*thylene *r*esistant *1*) from *Arabidopsis thaliana*. In vertebrates it is now well established that nitric oxide and carbon monoxide serve as second messenger molecules, and are linked to neurotransmitter and vasodilatory actions, and hydrogen sulfide has been recognized recently as a third gaseous messenger [2]. Notably, all of these gases are cytotoxic at high concentrations.

Gaseous second messenger molecules appear to be lacking in prokaryotes. Yet, many bacteria and Archaea possess gas-sensing systems since gases take part in metabolic processes. For example, dihydrogen, methane, carbon monoxide, dihydrogen sulfide, nitric oxide, nitrous oxide, and dinitrogen are consumed or produced in energy metabolism. Interestingly, specific sensor systems have been discovered for only three gases: dihydrogen, nitric oxide, and carbon monoxide. All biological gas sensors known to date are metalloproteins, with the metal center being the active site for interaction with the effector gas. Dihydrogen is sensed by a binuclear center similar to [NiFe]-hydrogenase enzymes [3]. For nitric oxide and carbon monoxide, different sensor proteins exist [4]. Each of them contains at least one out of only three principal types of metallocenters: heme, nonheme iron, and iron–sulfur clusters.

Genome sequencing of hundreds of microbial genomes have made available a huge number of protein sequences for bioinformatic analyses. Several molecular

Bacterial Signaling. Edited by Reinhard Krämer and Kirsten Jung
Copyright © 2010 WILEY-VCH Verlag GmbH & Co. KGaA, Weinheim
ISBN: 978-3-527-32365-4

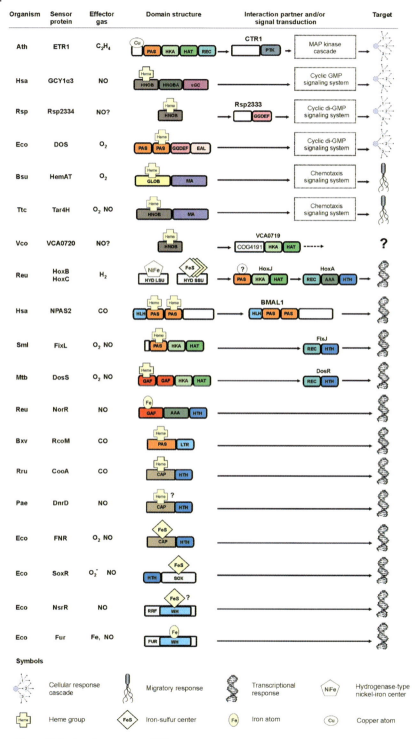

Figure 17.1 (caption see page 309)

building blocks, so-called domains, were discovered in the past and can be retrieved by searching public databases like COG [5], SMART [6], and Pfam [7]. In fact, all sensor proteins discussed in this chapter exhibit more or less defined domain structures that are illustrated in Figure 17.1. While many sensor proteins directly bind DNA and thus act as transcription factors, others appear to take part in signaling cascades consisting of two or more components. Direct interaction partners of sensor proteins have been included in Figure 17.1 wherever possible and appropriate.

17.2
Sensing of Molecular Hydrogen

Many bacterial species, Archaea and lower eukaryotes utilize the oxidative cleavage of molecular hydrogen (H_2) into electrons and protons for the generation of energy. Alternatively some anaerobic microorganisms evolve gaseous hydrogen by reduction of protons to release excess reducing power. Both reactions (i.e., H_2 oxidation and H_2 production) are catalyzed by a group of enzymes designated hydrogenases [8]. This section focuses on the regulation of hydrogenase synthesis by H_2; more specifically, on the question how an organism recognizes gaseous H_2 in its environment, and how this signal is converted and transmitted to the transcription apparatus in the cell.

Figure 17.1 Schematic overview of gas-sensing proteins. Sensors are grouped by their cognate signal transduction systems and their domain structure.
Domain abbreviations: AAA, ATPases associated with a wide variety of cellular activities; CAP, effector domain of the CAP (CRP) family of transcription factors; EAL, candidate diguanylate phosphodiesterase catalytic domain; FUR, domain specific for the regulator Fur; GAF, ligand-binding domain initially found in phytochromes and cGMP-specific phosphodiesterases; GGDEF, diguanylate cyclase catalytic domain; GLOB, globin domain present in globin-coupled-sensors; HAT, histidine kinase-like ATPase; HKA, dimerization and phosphoacceptor domain of histidine kinases; HLH, helix–loop–helix domain found in specific DNA-binding proteins that act as transcription factors; HNOB, nitric oxide and dioxygen-binding heme domain initially found in eukaryotic soluble guanylate cyclases; HNOBA, domain found associated with the HNOB domain; HTH, helix–turn–helix DNA-binding domain; HYD LSU, domain specific for large subunits of [NiFe]-hydrogenases; HYD SSU, domain specific for small subunits of [NiFe]-hydrogenases; LTR, LytTr-type DNA-binding domain; MA, methyl-accepting chemotaxis sensory transducer domain; PAS, multipurpose ligand binding domain found in Archaea, Eubacteria and Eukarya; REC, receiver domain that is phosphorylated by histidine kinases; RRF, domain specific for regulators of the Rrf2 family; sGC, guanylate cyclization catalytic domain (cGMP forming); SOX, domain specific for the regulator SoxR; WH, winged-helix DNA binding domain.
Organism abbreviations: Ath, *Arabidopsis thaliana*; Bsu, *Bacillus subtilis*; Bxv, *Burkholderia xenovorans*; Eco, *Escherichia coli*; Hsa, *Homo sapiens*; Mtb, *Mycobacterium tuberculosis*; Pae, *Pseudomonas aeruginosa*; Reu, *Ralstonia eutropha*; Rru, *Rhodospirillum rubrum*; Rsp, *Rhodobacter sphaeroides*; Sml, *Sinorhizobium meliloti*; Ttc, *Thermoanaerobacter tencongensis*; Vco, *Vibrio cholerae*.

17.2.1
Environmental Signals that Direct Hydrogenase Control

Most of the H_2 in natural habitats is produced by bacterial fermentation and is immediately consumed by anaerobic microorganisms that perform methanogenesis, anaerobic respiration, or anaerobic photosynthesis. Thus, only scarce amounts of H_2 escape into the oxic environment where phylogenetically diverse predominantly facultatively H_2-utilizing bacterial species thrive. Complex regulatory circuits instrumental in H_2 metabolism are usually observed in bacteria that are capable of shifting from aerobic to anaerobic growth mode and/or use alternative energy sources in addition to H_2, such as inorganic and organic compounds or light. Well-studied control systems are: (i) transcriptional regulation of multiple hydrogenases in *Escherichia coli* by oxygen and the fermentation intermediate formate, (ii) control of an H_2 uptake hydrogenase in the phototrophic bacterium *Rhodobacter capsulatus* by H_2 and the internal redox status of the cell, and (iii) coordinate regulation of two complex hydrogenase gene clusters in the facultatively chemolithoautotrophic β-proteobacterium *Ralstonia eutropha* H16 by H_2 and global substrate availability [8].

17.2.2
Hydrogen-Activating Proteins in Nature

At first glance the reversible cleavage of H_2 into protons and electrons ($H_2 \leftrightarrow 2H^+ + 2e^-$) appears to be a simple reaction. In reality, chemical as well as biological activation of H_2 relies on the participation of efficient catalysts. Three distinct, phylogenetically unrelated classes of hydrogenases, grouped by their metal content, have been found in nature: (i) the [FeFe]-hydrogenases that contain a di-iron H_2-activating site conduct primarily H_2 evolution, (ii) the [NiFe]-hydrogenases that host a heterodinuclear Ni–Fe-active site are predominantly engaged in H_2 uptake, and (iii) the [Fe]-hydrogenases that accommodate a monoiron-containing organic cofactor function as H_2-forming methylenetetrahydromethanopterine dehydrogenases in a small group of methanogenic Archaea [9].

Regulation of hydrogenase biosynthesis by H_2 has so far been observed only in organisms containing [NiFe]-hydrogenases. Therefore this group of proteins is discussed in more depth. X-ray diffraction data of heterodimeric [NiFe]-hydrogenases from *Desulfovibrio* species [10], now taken as prototypical [NiFe]-hydrogenase, revealed that the Ni–Fe cofactor is deeply buried in the large approximately 60-kDa subunit (Figure 17.2A). The nickel is coordinated to the protein via four thiol groups provided by cysteine residues; two of these are bridging ligands that also coordinate the iron [11]. Fourier transform infrared spectroscopy revealed that the iron coordination sphere is completed by three diatomic ligands – one CO and two CN^- [12]. An electronic coupling exists between the Ni–Fe site and the nearest (proximal) Fe–S cluster located in the small electron-transferring subunit approximately 14 Å away. Hydrophobic tunnels that may channel the substrate between the protein surface and the active site as well as a proton-conducting channel were identified inside the hydrogenase [11]. Although the amino acid environments directly surrounding both the active site and the proximal Fe–S cluster are highly similar in all [NiFe]-

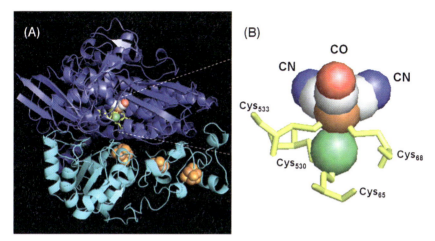

Figure 17.2 Structure of a standard [NiFe]-hydrogenase. (A) The large subunit (blue) carries the NiFe active site. The small subunit (green) accommodates three [Fe–S] clusters. (B) The active site is composed of Ni (green), Fe (red), two CN^- and one CO. The metals are coordinated by cysteine ligands in reference to the structure of *Desulfovibrio gigas* [10].

hydrogenases, this class of proteins displays extensive diversity with regard to its modular composition, biochemical features and physiological functions [13, 14]. All [NiFe]-hydrogenases undergo post-translational maturation that involves at least six auxiliary proteins. These so-called Hyp proteins mediate the assembly of the Ni–Fe cofactor into the large subunit precursor [15]. Thus, hydrogenase biosynthesis requires substantial resources of a living cell.

17.2.3
What Makes the H_2 Signaling Hydrogenase Different from the Energy-Providing Hydrogenase?

H_2 signaling regulatory hydrogenases (RHs) do not sustain growth on H_2. They have been discovered in relatively few microorganisms including the symbiotic nitrogen-fixing *Bradyrhizobium japonicum*, the phototrophic bacterium *Rhodobacter capsulatus*, and the aerobic hydrogen bacterium *R. eutropha* [16]. What makes these H_2 signaling RHs so unique?

Structural comparison of RH proteins with their energy-providing hydrogenase counterparts reveals three key features that are characteristic of H_2-sensing hydrogenases. (i) Lack of a membrane-targeting signal peptide and (ii) absence of a hydrophobic membrane-anchoring region. These observations are consistent with an intracellular location of the H_2 sensor. Instead of a hydrophobic membrane anchor the RH proteins carry an approximately 55-amino-acid C-terminal extension at the small subunit that is typical for regulatory hydrogenases, suggesting a crucial role of this domain in the H_2-sensing process. (iii) Assembly of the Ni–Fe active site into RH proteins also involves assistance of the Hyp proteins. However, proteolytic processing of the large subunit that normally completes the insertion of the Ni–Fe cofactor does not occur in the maturation of RH-like proteins [16].

The H_2 signaling proteins from both *R. eutropha* and *R. capsulatus* were shown to catalyze H_2 oxidation as well as H_2 evolution using artificial dyes as electron mediators [14, 16]. The data unambiguously demonstrate that H_2-sensing hydrogenases can be considered as "real" hydrogenase catalysts. A genetically designed overexpression system combined with protein fusion affinity tags provided the basis for obtaining high quantities of the otherwise poorly expressed regulatory protein. Biochemical studies using purified RH samples from *R. eutropha* identified a number of characteristic features. (i) Compared to standard [NiFe]-hydrogenases that display a specific activity of approximately 30–80 U/mg protein, the activity of RH proteins is approximately two orders of magnitude lower [17]. (ii) The electron paramagnetic resonance (EPR) spectrum of the RH lacks two EPR signals (Ni-A, Ni-B) from the nickel-based unpaired spin [18]. These signals are characteristic for the oxidized, catalytically inactive resting states of the standard [NiFe]-hydrogenase. In this case conversion of the enzyme to the active state, which implies the reductive removal of oxygen from the Ni–Fe site, may take hours (Ni-A) or minutes (Ni-B) [18]. The fact that the RH instantaneously reacts with H_2 [17] is compatible with its function of a sensing protein that needs to be on alert the moment H_2 occurs in the environment. (iii) Unlike the majority of [NiFe]-hydrogenases the RH activity is insensitive to oxygen, carbon monoxide, and also acetylene [17]. Mutant analyses and spectroscopic data are in favor of a common composition of the Ni–Fe active site (Figure 17.2B) in the RH comprising four cofactor-coordinating cysteine residues [19] and three diatomic ligands bound to the iron [17]. The RH represents the first example for which the presence of a hydride (H^-) in the bridge between nickel and iron could be demonstrated [20].

To elucidate the molecular basis of the exquisite O_2 resistance the protein environment surrounding the Ni–Fe-active site of H_2-sensing proteins has been further explored. Sequence alignment and modeling in reference to the standard X-ray structure uncovered a substitution of two highly conserved amino acids (a valine and a leucine residue) by the bulkier residues isoleucine and a phenylalanine in the RH-like proteins. The position of these specific amino acids was predicted to be located at the end of the gas tunnel near the Ni–Fe active site [11]. This discovery led to the hypothesis that the larger amino acid residues might limit the size of the tunnel and in turn might prevent O_2 from accessing the active site, whereas still permitting the smaller H_2 molecule to enter the catalytic center. This concept was approved by replacing the respective residues in the H_2 sensors of *R. eutropha* and *R. capsulatus* using site-directed mutagenesis. In fact, the resulting mutant proteins now equipped with valine and leucine at the target positions proved to be O_2-sensitive and behaved in this respect similarly to the standard [NiFe]-hydrogenases [21, 22]. These observations nicely demonstrate how a protein protects itself against O_2 inactivation, thereby acquiring the ability to respond without any lag period to the availability of H_2.

17.2.4
H_2 Signaling Cascade

The H_2 signaling cascade in *R. eutropha* and *R. capsulatus* is based on a bacterial two-component regulatory system consisting of the NtrC-like response regulator HoxA

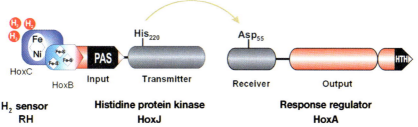

Figure 17.3 Regulatory components involved in H_2 signal transduction.

(HupR in *R. capsulatus*), its cognate histidine protein kinase HoxJ (HupT), and the H_2 sensor RH (HoxBC or HupUV) (Figure 17.3). In *R. eutropha*, the structural and maturation genes for an energy-linked membrane-bound [NiFe]-hydrogenase (MBH) are arranged together with the four regulatory components in a single operon that maps on megaplasmid pHG1 (Figure 17.4). Transcription of the MBH operon is initiated at a strong σ^{54}-dependent promoter whose proper function depends on the activation by HoxA. Two weak internal promoters that are recognized by the major σ^{70} of RNA polymerase are active in the absence of HoxA [23].

Insights into the function of the H_2 signaling cascade were obtained by the analysis of mutants carrying nonpolar deletions in the four regulatory genes [24]. The effect of the individual mutations on transcription was monitored measuring β-galactosidase activity expressed from a MBH–*lacZ* operon fusion. A deletion in the output domain of HoxA, consisting of a nucleotide-binding site and a DNA-interacting region, leads to a total loss of MBH gene transcription, pointing to a principal role of the response regulator HoxA (Figure 17.3). Unexpectedly a knockout of a conserved aspartate residue in the receiver module of HoxA, which is the target of phosphorylation, has just the opposite effect – MBH gene transcription is significantly upregulated even in the absence of H_2. Likewise mutational inactivation of the transmitter module in HoxJ (Figure 17.3), which is essential for autophosphorylation of the kinase, also leads to constitutive hydrogenase operon expression. On the other hand, mutants with deletions in the input domain of HoxJ completely fail to express the MBH operon. A similarly negative phenotype is obtained by a knockout of the RH genes [24]. These results suggest that the response regulator HoxA is inactive in

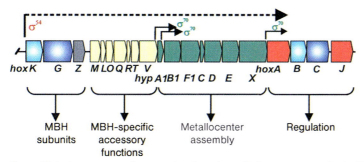

Figure 17.4 Arrangement of structural and regulatory hydrogenase genes in the MBH operon.

the phosphorylated form – an observation that gained support from data obtained with the HupR homolog of R. capsulatus [14]. This is interesting with regard to orthodox two-component regulatory systems where the phosphorylated response regulator normally represents the active configuration. From the data the conclusion is drawn that H_2 in concert with the RH and the input module of HoxJ counteract the negative effect exerted by phosphorylation of HoxA.

17.2.5
H_2 Sensor Complex in Action

Unlike in classical two-component systems, the external regulatory signal H_2 is not directly received by the input domain of the cognate histidine protein kinase, but needs an additional signal transmitter module. Nature has copied the structural and biochemical elements of the Ni–Fe center to design a specific H_2-sensing protein that recognizes the gaseous molecule and, in turn, generates a regulatory signal. This notion raises a number of intriguing questions:

- Does signal transduction imply a direct interaction of the regulatory hydrogenase with the protein kinase?
- Is binding of H_2 to the sensor protein sufficient for emitting a signal or does H_2 signaling imply a redox reaction?
- If RH and kinase HoxJ directly interact in signal transduction, how does H_2 affect the kinase activity and subsequently modification of the response regulator HoxA?

Experimental results clearly point to a direct interaction of the RH and the kinase HoxJ since *in vitro* studies uncovered a large protein complex consisting of four HoxJ, two HoxB, and two HoxC subunits. The C-terminal peptide extension of the small RH subunit HoxB is essential for both the formation of the super complex and the regulatory function of the RH as deduced from the phenotypic characterization of RH mutants deleted in this region. Although the truncated version of the RH has lost its ability to form the $(HoxBC)_2$ oligomer, it is still able to form the HoxBC heterodimer and to catalyze H_2 oxidation. Moreover, interaction of the H_2 sensor with HoxJ relies on the input module of the kinase that comprises a Per/Arnt/Sim (PAS) domain [25]. PAS domains have a function in protein–protein interaction, and may be involved in sensing light and various redox signals. In several cases PAS domains are associated with cofactors such as heme groups, flavins, or Fe–S clusters, or they simply transmit the redox signal via conformational changes [26].

To shed light on the question, if H_2 signaling implies a redox reaction, mutants with site-specific alterations in the RH of R. eutropha were analyzed by using assays that discriminate between the capacity of binding H_2 to the catalytic site (*para/ortho* H_2 conversion), of splitting H_2 in the absence of external electron acceptors (D_2/H^+ exchange), or of catalyzing the entire H_2 oxidation cycle by including a redox dye (e.g., methylene blue) as electron acceptor [27]. Exchange of a key glutamate residue (HoxC-E13) in the vicinity of the active site of the RH leads to a knockout of the regulatory and the H_2-activating functions, whereas H_2-binding ability is maintained in the mutant protein. This observation strongly suggests that H_2 has to undergo at least a proton-exchange reaction, prior to signal emission, leading to the formation of a hydride (H$^-$). A significant decline of the methylene blue-dependent H$_2$ oxidation rate by the RH is observed after substitution of a conserved aspartate residue (HoxC-D15). This exchange, however, scarcely affects the regulatory potential of the RH, indicating that H$_2$ signal transmission does not require an enhanced H$_2$ turnover.

17.2.6
Concluding Remarks and Perspectives

In the absence of H$_2$ the bacterium *R. eutropha* is smart enough to save its cellular resources for the facultative use of organic carbon and energy sources instead of initiating the synthesis of a nonprofitable highly sophisticated biosynthetic pathway that is necessary to build an active [NiFe]-hydrogenase. Under H$_2$ deprivation transcription of the hydrogenase genes is repressed since the response regulator HoxA is not in the appropriate conformation to activate σ^{54} RNA polymerase. Accumulating evidence suggests that unlike in orthodox two-component systems, the nonphosphorylated form of HoxA activates transcription, whereas modification of HoxA by phosphorylation exerts a negative effect on hydrogenase gene transcription. The phosphorylation status of HoxA is controlled by the H$_2$ signaling complex consisting of a high molecular oligomer of the H$_2$-sensing regulatory [NiFe]-hydrogenase RH and the histidine protein kinase HoxJ. In the presence of H$_2$ the signaling complex counteracts phosphorylation of HoxA, either by inhibiting the kinase activity of HoxJ or by inducing a phosphatase, most likely in conjunction with HoxA. The underlying molecular mechanism is still an open question. As illustrated in a model (Figure 17.5), it is known, however, that interaction of RH and HoxJ relies on individual features assigned to the C-terminus of the hydrogenase small subunit and the N-terminal region of the input domain of HoxJ. Signal reception by the RH does not only involve binding of H$_2$ to the Ni–Fe active site, but H$_2$ has to undergo at least an initial redox reaction before signal transmission occurs. It is not yet clear if the Fe–S clusters assigned to the small RH subunit are the only players in the electron transfer or if an additional organic cofactor participates in this process as proposed previously [17]. It is difficult to trap such a cofactor in highly purified RH-HoxJ samples [16]. Thus, experiments are underway to improve the quality of native complex preparations. Moreover, the roles of the PAS domain and of highly conserved key amino acids in the input domain of HoxJ in the process of H$_2$ signaling wait to be further explored.

17.3
Sensing of Nitric Oxide and Carbon Monoxide

The diatomic gases nitric oxide (NO) and carbon monoxide (CO) share similar chemical properties and biological functions. Both show a general affinity for metal atoms and thus are widely used as molecular probes (e.g., for spectroscopic

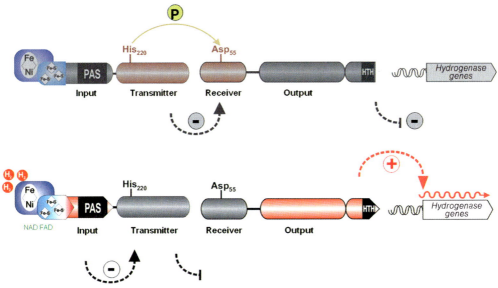

Figure 17.5 Model of H_2 sensing. (A) Hydrogenase gene transcription is repressed. (B) Hydrogenase gene transcription is activated by H_2 signal transduction.

investigations of metal centers). Given this unspecific behavior, it is notoriously difficult to decide whether the capacity of a protein to react with CO or NO *in vitro* is of physiological significance *in vivo*. Furthermore, even exogenous addition of these molecules or artificial donor substances to cell cultures may lead to "collateral" reactions that never occur in the natural habitat of these organisms. This is a particular problem for O_2-sensing proteins, since most of them react to some extent with NO and/or CO. On the other hand, such artificial effects do not preclude the possibility that certain gas sensors may have more than one effector molecule. In fact, in some cases the physiological relevance of such a promiscuous behavior has been well established. In this chapter, sensor proteins that appear to be dedicated to a specific response, rather that having multiple effector molecules, are termed "primary" sensors. Proteins listed as "secondary" sensors take part in multiple regulation circuits and, in some cases, the physiological relevance of the binding of alternative ligands is still a matter of debate.

17.3.1
Primary Sensors for NO

In mammals, NO is involved in a vast array of biological functions [28]. NO is produced from arginine by inducible and constitutive isoforms of NO synthases that are involved in neurotransmission, control of the vascular tone, and immune defense. The latter function relies on the deleterious effects of NO when present above micromolar concentrations. NO inhibits many metalloenzymes, elicits lipid peroxidation, and mutates DNA. Given the ubiquity and toxicity of NO, it is not surprising that cells have developed specialized systems to control NO-dependent reactions. In eukaryotes, the main receptors for NO are the soluble guanylate cyclases, which come in different isoforms [29]. Soluble guanylate cyclases are heterodimeric heme enzymes that are inactive in the absence of NO. NO binding to the heme displaces an endogenous histidine ligand, thereby triggering a conformational change that leads to synthesis of the signal molecule cGMP. cGMP binds to target proteins like protein kinases, ion channels, and phosphodiesterases, resulting in an array of different cell-specific downstream outputs.

Several bacteria and also some fungi are able to denitrify, and hence exposed to NO formed endogenously as a product of nitrite reduction by nitrite reductase [30]. NO is also liberated during the oxidation of ammonia by nitrifying bacteria, and affects the production of ammonia mono-oxygenase and formation of biofilms [31]. It has been known for many years that enterobacteria like *E. coli* produce significant amounts of NO during oxygen-limited growth on nitrate [32]. Other microorganisms may encounter NO that is produced in their environment, either by other cells or by chemical reactions (e.g., by decomposition of nitrite). Furthermore, NO defense is a challenging task for pathogenic bacteria that have to cope with NO produced by the immune system of the host [33].

17.3.1.1 NorR-Type NO-Sensing Regulators

NorR was initially discovered in the denitrifier *R. eutropha* as a NO-responsive transcriptional activator for expression of a dicistronic operon consisting of *norA*, encoding a di-iron protein, and *norB*, encoding the respiratory NO reductase [34]. Genes for NorR orthologs are present in the genomes of several denitrifying and nondenitrifying β- and γ-proteobacteria [35]. The protein has been studied on the biochemical and the physiological level in *R. eutropha* and the ammonifier *E. coli*. In *E. coli*, NorR activates expression of the genes for the flavorubredoxin NorV and its cognate reductase NorW. Flavorubredoxins are soluble NO reductases widely distributed among proteobacteria and Firmicutes that are unrelated to respiratory NO reductases, but are of similar physiological importance for detoxification of NO [36]. In other bacteria, predicted targets of *norR* include genes for flavohemoglobin and hybrid cluster protein [35], both of which have been implicated in defense against NO. It is interesting to note that NorR and all of its target genes are usually located in direct vicinity. A similar correlation has not been observed for any of the NO-sensing regulator described below; it is not known, however, if the placement of *norR* in cis is the consequence of a mechanistic constraint. In *R. eutropha*, coproduction of NorA and NorB may represent a strategy for balanced production of NO, since NorA has been shown to bind NO [37]. NO binding by NorA results in a feedback loop, since competition of NorR and NorA for NO impairs activation of NorR, and in turn decreases transcriptional activation of the genes for NorA and NorB. Orthologs of NorA (YtfE, ScdA, DnrN) in *E. coli*, *Staphylococcus aureus* and *Neisseria gonorrhoeae* have been shown to facilitate iron–sulfur cluster assembly in the presence of NO.

NorR is a distant relative of the NifA/NtrC family of regulators that interact with the alternative sigma factor σ^{54} [38]. In contrast to NifA and NtrC, however, which are the response regulators of classical two-component systems, NorR is a one-component system that directly senses the effector molecule NO. NorR shows a tripartite domain structure of an N-terminal GAF domain (a common domain in cGMP-specific phosphodiesterases, *a*denylyl cyclases, and *F*hlA protein), a central ATPase domain for interaction with σ^{54}, and a C-terminal DNA-binding domain. It is assumed that, in the absence of NO, NorR is inactive due to an interaction of the signaling domain with the central domain which blocks ATP hydrolysis and hence open complex formation. Upon binding of NO the blockade is relieved and NorR becomes active. NorR binds NO at a monoiron center that is coordinated by the GAF domain [39, 40]. Three aspartates, an arginine, and a cysteine have been identified by mutational analysis as the likely iron ligands in both *R. eutropha* NorR and *E. coli* NorR [40, 41].

17.3.1.2 NsrR-Type NO Sensing Regulators

NsrR, another primary NO sensor, is encoded in the genomes of many proteobacteria, bacilli, and some actinobacteria [35]. Responsiveness of NsrR-controlled genes towards NO was verified experimentally in *E. coli*, two *Neisseria* species, and *Bacillus subtilis* [42]. Interestingly, there is a considerable overlap between the regulons predicted for NsrR and NorR, respectively. For example, each of them may control expression of genes for flavohemoglobin, flavorubredoxin, and hybrid cluster protein Hcp in different species. Nevertheless NorR and NsrR may fulfill different physiological roles, because several β- and γ-proteobacterial genomes contain genes for both of them [35].

NsrR belongs to the Rrf2 protein family, which comprises a number of putative transcriptional regulators from diverse bacteria. A close relative of NsrR is the *E. coli* regulator IscR that controls expression of genes for the ISC iron–sulfur cluster assembly machinery [43]. IscR coordinates a [2Fe–2S] cluster and acts as a repressor on most of its target genes. Most members of the IscR regulon are derepressed in the presence of apoIscR, suggesting a feedback mechanism triggered by assembly of the IscR [2Fe–2S] cluster by the ISC machinery. Given the sequence similarity between NsrR and IscR, it has been postulated that an iron–sulfur cluster is responsible for NO-sensing by NsrR [42]. In such a scenario, holoNsrR would repress its target genes until the protein is inactivated by destruction or modification of its iron–sulfur cluster by NO or NO-derived compounds. Indeed, NsrR-regulated genes can also be derepressed by iron starvation, indicating that iron is an important component of active NsrR.

17.3.1.3 NO-Sensing Regulators Containing CAP Domains

In many proteobacteria, sensing of NO is mediated by proteins belonging to the CAP superfamily of transcriptional regulators [44]. These proteins show a tripartite structure of an N-terminal sensor module, a dimerization domain, and a C-terminal DNA-binding domain. The sensor module contains a CAP domain that is the defining element of the family. The best-studied member of the CAP superfamily is the cAMP receptor proteinCRP (also known as *c*atabolite *a*ctivator *p*rotein, hence

"CAP") of E. coli [45]. While CRP binds cAMP, CAP domains of other proteins may ligate a metallocenter for sensing effector molecules. Prominent examples are the heme-containing CO sensor CooA [46] (see Section 17.3.2.1) and the fumarate/nitrate reductase regulator (FNR) protein of E. coli that uses an iron–sulfur cluster to sense O_2 [47]. While FNR has also been implicated in NO sensing (see Section 17.3.3.3), other members of the CAP superfamily exist that appear to be more specialized NO sensors. All of the latter proteins lack the conserved cysteine residues that coordinate the iron–sulfur cluster in FNR and thus must rely on another mechanism for gas detection.

Based on the sequence of the recognition helix in the DNA-binding domain, NO-sensing CAP domain proteins fall into at least three different subgroups termed DNR, NnrR, and HcpR [35]. While DNR proteins are found in β- and γ-proteobacteria, NnrR proteins can be attributed to α-proteobacteria. HcpR proteins are found predominantly in δ-proteobacteria and some cyanobacteria, and genes for HcpR-related proteins have been found in the genomes from clostridia and some CFB (*c*ytophaga-*f*lavobacter-*b*acteroidetes) group bacteria.

Recently, the three-dimensional structure of the sensor domain of DNR from P. aeruginosa was solved at 2.1-Å resolution [48]. DNR adopts a similar general folding structure as previously described for CRP and CooA. The crystallized protein did not contain a cofactor. However, the presence of a hydrophobic cavity well suited for accommodation of a cofactor was observed. Indeed, apoDNR supplemented with heme does bind CO and NO, indicating that DNR, in its active state *in vivo*, contains heme like its relative CooA.

17.3.2
Primary Sensors for CO

Like NO, CO is an important vasodilating agent in higher eukaryotes [49]. Both NO and CO inhibit the hypoxic induction of vascular endothelial growth factor by exerting an indirect effect on the DNA binding activity of the transcription factor hypoxia-inducible factor-1 (HIF-1). HIF-1 plays a crucial role in cellular adaptation to low oxygen. In eukaryotes, CO is produced primarily from the degradation of heme by the heme oxygenase enzyme system. Interestingly, heme oxygenase was found in close proximity to soluble guanylate cyclase in the brain, suggesting that CO may act as an alternative physiological regulator of NO-dependent soluble guanylate cyclases [50]. However, to date the best candidate eukaryotic CO receptor is the neuronal PAS domain protein 2 (NPAS2) – a transcription factor that regulates the mammalian circadian rhythm [51]. NPAS2 contains two PAS domains adjacent to an N-terminal dimerization domain. The PAS tandem arrangement is reminiscent of the bacterial O_2 sensors (e.g., FixJ from rhizobia or Dos (*d*irect *o*xygen *s*ensor) from E. coli); however, in contrast to these proteins, both PAS domains of NPAS2 carry heme groups.

Numerous metabolically and phylogenetically distinct prokaryotes use CO as an energy or carbon source [52, 53]. Obligate anaerobes, including methanogens, sulfate reducers, and acetogens, use the bifunctional nickel-dependent CO dehydrogenases/acetyl-CoA synthase complex for cell carbon synthesis from CO_2 in the Wood–Ljungdahl pathway. Aerobic carboxidotrophic bacteria like *Oligotropha carboxidovorans* contain molybdenum-dependent CO dehydrogenases (CODH) that produce CO_2 and two electrons that are channeled into the respiratory chain. CO oxidation is typically coupled to CO_2 fixation via the Calvin–Benson–Bassham reductive pentose phosphate cycle. In contrast, some facultative or obligate anaerobes like *Rhodospirillum rubrum* and *Carboxydothermus hydrogenoformans* contain monofunctional nickel CODHs that are involved in energy generation by oxidation of CO. Electrons are delivered to CO-insensitive hydrogenases that are suggested to function as energy-conserving proton pumps. While the bifunctional nickel-containing enzymes form CO as an intermediate and are formed independently of exogenous CO, the formation of CODHs from both aerobic and anaerobic CO-oxidizing bacteria is CO-dependent. This response is mediated by specific CO-sensing regulator proteins that control the expression of genes for CO oxidation and thus are considered as primary CO sensors.

17.3.2.1 CooA – A CO Sensor of Anaerobic Carboxidotrophs

Expression of the genes for a monofunctional Ni-CODH and a CO-insensitive hydrogenase in *R. rubrum* is controlled by the transcriptional activator CooA [46]. CooA belongs to the CAP/CRP superfamily and thus is distantly related to the oxygen sensor FNR and the DnrD/NNR group of NO sensors [54]. In contrast to FNR, which forms homodimers only in its active state, CooA dimerizes regardless of redox potential or the presence of CO. Each of the N-terminal CAP domains of the CooA homodimer coordinates a B-type heme. The CooA dimer can adopt three different states. In the inactive oxidized state, heme iron is ligated by a cysteine and by a proline residue of the opposing monomer (yielding a six-coordinate low-spin state). Under reducing conditions, which are required for the activity of the CODH, Fe(III)-CooA is reduced to Fe(II)-CooA that is able to bind CO. Not much is known about the transition of Fe(III)-CooA to Fe(III)-CooA under physiological conditions and, moreover, the redox partner of CooA in the cell is unknown.

In the reduced state, the cysteine heme ligand is replaced by histidine. Since the heme of CooA is six-coordinate and low-spin in both the reduced and the CO-bound states, CO must displace an axial ligand from the heme. Displacement of the proline ligand by CO results in movement of the heme, accompanied by a conformational change that is translated to the DNA-binding domain. In this active state, CooA binds to specific DNA sites upstream of the promoter of its target genes and activates expression by recruiting RNA polymerase. Given that CO can bind to virtually all heme-containing proteins, nature's choice of CooA as a heme-based CO sensor systems is probably not surprising. Nevertheless, compared to other heme-containing biosensors (e.g., FixL, see Section 17.3.3.3), CooA is an unusually specific system. For example, inactive Fe(III)-CooA is formed by auto-oxidation upon reaction of Fe(II)-CooA with O_2 [55]. NO fails to activate Fe(II)-CooA since it displaces both of the heme ligands. Other well-known heme-binding molecules like azide an cyanide do not bind to CooA.

17.3.2.2 RcoM – A CO Sensor of Aerobic Carboxidotrophs

CooA is not expected to occur in aerobic carboxidotrophic bacteria because the CooA/CODH system works at low redox potentials below -300 mV. In several organisms that lack a *cooA* gene, a gene adjacent to the *coo* operon codes for another CO-dependent transcriptional activator, RcoM [56]. RcoM was characterized from *Burkholderia xenovorans* strain LB400 – a well-studied degrader of polychlorinated biphenyls that is able to oxidize CO under both aerobic and anaerobic conditions. In fact, *B. xenovorans* contains two paralogous proteins, RcoM1 and RcoM2, both of which have been shown to act as CO dependent transcriptional activators *in vivo*.

Like in CooA, a heme group is used as a prosthetic group for CO-sensing by RcoM. However, RcoM shows a different domain structure of an N-terminal PAS domain and a C-terminal LytTR DNA-binding domain. Thus, RcoM is more related to the CO-responsive mammalian NPAS2 and the oxygen sensor Dos from *E. coli*. In RcoM2 of *B. xenovorans*, a histidine and a cysteine residue within the PAS domain have been identified as the heme iron axial ligands in oxidized Fe(III)-RcoM. While histidine remains bound to the iron through changes of the redox state, cysteine is replaced by a neutral ligand in Fe(II)-RcoM and CO in Fe(II)-RcoM-CO, respectively.

17.3.3
Hypothetical or Secondary Sensors Systems for NO and/or CO

17.3.3.1 Eukaryotic-Style NO Sensing in Prokaryotes

In prokaryotes, a family of proteins was found to contain a heme domain with considerable similarity to the heme domain of the well-characterized NO sensor of eukaryotes – soluble guanylate cyclase. This newly identified domain was termed HNOB (heme NO binding). HNOB domains may be fused to methyl-accepting (MA) domains, suggesting that such proteins are the sensors components of chemotactic response systems. Notably, the principle layout of these sensor proteins resemble a group of O_2 sensor proteins termed HemAT involved in chemotaxis in Archaea and bacteria (e.g., HemAT of *B. subtilis*) [57]. HemATs contain a heme-binding globin domain in place of a HNOB domain. HNOB–MA fusion proteins have been characterized from the strict anaerobes *Clostridium botulinum* (SonO) and *Thermoanaerobacter tencongensis* (Tar4H). The crystal structure of the HNOB domain of the *T. tengcongensis* chemotaxis receptor has been determined at 1.8-Å resolution. The overall folding scaffold confirmed the similarity of this domain to soluble guanylate cyclases. Since a bound dioxygen molecule was present in the distal heme pocket, renaming of HNOB to H-NOX (heme NO and/or O_2 binding) was suggested to emphasize the fact that this domain can bind to both of the two gases.

There are also examples for proteins carrying only a HNOB domain. Genes for such "stand-alone" proteins are found in direct vicinity of genes encoding possible transmitters of a signal transduction chain. The putative interaction partner may carry, among others, histidine kinase domains (e.g., VCA0720 in *Vibrio cholerae*) or GGDEF domains systems (e.g., Rsp2334 in *Rhodobacter sphaeroides*). GGDEF domains have been implicated in c-di-GMP signaling. While the formation of a heme iron–NO complex has been verified for several HNOB proteins, neither the formation of a functional downstream signaling chain nor an essential physiological role for NO sensing by HNOB proteins has yet been proven experimentally.

17.3.3.2 NO Sensing by Fur, SoxR, and OxyR

NO can react with O_2, peroxides, and superoxides to give various reactive nitrogen species (RNS) including nitrogen dioxide, dinitrogen trioxide, and peroxynitrite that elicit formation of transition metal-NO adducts and nitrosylation of proteins [58, 59]. In fact, RNS are presumed to be much more cytotoxic than NO itself. Since RNS are predominantly formed when oxidatively stressed cells encounter NO, one might expect some cross-talk between regulatory systems for nitrosative and oxidative stress. Indeed, the main peroxide and superoxide sensors of *E. coli*, OxyR and SoxR, have been shown to respond to nitrosative stress [60]. Furthermore, NO can damage certain iron–sulfur proteins like mammalian ferrochelatase [61] and aconitase [62], which may increase the amount of so-called free iron. This scenario implies a linkage between nitrosative stress and iron homeostasis.

NO Sensing by Fur In *E. coli*, the expression of genes involved in iron acquisition and utilization is controlled by the ferric uptake regulator Fur [63]. In excess iron, the iron-sensing site of Fur is occupied by an Fe^{2+} ion. Iron-substituted Fur protein binds to specific DNA sequences in the promoter region of its target genes, leading to repression of gene expression. Iron limitation leads to derepression of the Fur regulon. Notably, exposure of Fur and the homologous PerR of *B. subtilis* to NO results in inactive proteins (see [42] and references therein). For Fur, a ferrous dinitrosyl complex was characterized that did inhibit DNA-binding *in vitro*. Derepression of the Fur regulon, at least under certain growth conditions, was demonstrated for *E. coli* and *Staphylococcus aureus*.

NO Sensing by SoxR Exposure of *E. coli* to superoxide results in activation of the transcriptional activator SoxR [60]. SoxR belongs to the MerR family of regulators that is defined by an N-terminal DNA-binding domain, a central dimerization region, and a C-terminal sensory domain [64]. The sensory domain of inactive SoxR carries a $[2Fe–2S]^{1+}$ cluster. One-electron oxidation to the $[2Fe–2S]^{2+}$ state by superoxide activates SoxR, while loss of this cluster does not affect protein folding, DNA binding, or promoter affinity. Activated SoxR triggers expression of the gene for SoxS – a transcriptional regulator for about 100 genes that are involved in the oxidative stress response. There is evidence that SoxR does also act as an NO sensor [65]. In contrast to superoxide, NO activates the SoxR protein by direct formation of a dinitrosyliron species, leading to an NO-dependent induction of the oxidative stress response.

Sensing of Nitrosative Stress by OxyR Another major regulator of the oxidative stress response of *E. coli* is the peroxide sensor OxyR [60]. Activated OxyR induces the expression of genes encoding, for example, catalase, alkyl hydroperoxide reductase, glutathione reductase, and the small regulatory RNA *oxyS*. OxyR is believed to be activated by oxidation of a cysteine residue that forms a disulfide bridge in the inactive protein. This mode of regulation is found in several redox sensing systems from both prokaryotes and eukaryotes. For example, the transcription factors nuclear factor-γB, activator protein-1, and p53 contain reactive thiols in their DNA-binding regions, the modification of which alters DNA association *in vitro* [66]. Although OxyR is probably not activated by NO gas itself, it has been shown that the critical cysteine can be nitrosylated *in vitro* by S-nitrosothiol compounds. Nitrosylating compounds may be also formed *in vivo* during nitrosative stress imposed by NO. The physiological impact of this reaction has been addressed by microarray experiments that did reveal a weak effect of the nitrosylating compound GSNO on the expression of members of the OxyR regulon.

17.3.3.3 Detecting Multiple Diatomic Gases: Sensors Responding to O_2, CO, and NO

NO Sensing by FNR The oxygen sensor FNR of *E. coli* contains a CAP domain that coordinates a $[4Fe–4S]^{2+}$ cluster [47]. This cluster is degraded by oxygen, resulting in FNR inactivation. *In vitro*, FNR reacts also with NO, leading to formation of a dinitrosyl-iron-cysteine complex with the iron–sulfur cluster. Both oxygen-damaged FNR and NO-modified FNR have been shown to be impaired in DNA binding [44]. Anaerobic exposure of *E. coli* cells to NO affected the expression level of both FNR-activated and FNR-repressed genes. The latter group of genes include the gene for Hmp, a flavohemoglobin that detoxifies NO by forming nitrate from the oxidation of NO with O_2 [67]. However, expression of the *hmp* gene is subjected to a complex interplay of the regulators FNR, MetR, NsrR, and Fur, all of which respond to NO or nitrosylation compounds. Therefore, FNR might be regarded as a secondary NO sensor that nevertheless plays an important role in fine-tuning the cellular response to NO.

FixL – A Sensor Reacting with O_2, CO, and NO The rizobial FixL/FixJ system represents a prototype for O_2-dependent signal transduction in prokaryotes [68]. Rhizobia are nitrogen-fixing soil bacteria that can adapt to a symbiotic lifestyle in leguminous plant roots. Formation of the O_2-sensitive enzyme nitrogenase is controlled by the sensor kinase FixL and its cognate response regulator FixJ. Although FixL proteins from different rhizobial species may have different domain compositions, they all show a common core structure of an N-terminal O_2-sensing PAS domain and a C-terminal histidine kinase domain. The PAS domain coordinates a b-type heme, which is Fe(II) high-spin under anaerobic conditions as present in the root nodules of legumes. In the Fe(II) state, the kinase domain of FixL is active and catalyzes phosphorylation of the transcriptional activator FixJ, which then activates transcription of its target genes. Under aerobic conditions, binding of O_2 to FixL triggers a spin transition of the heme iron. This results in a conformational change that blocks the kinase activity of FixL. The FixL proteins of *Sinorhizobium meliloti* has been shown to react with a number of well-known heme ligands apart from O_2, including CO, NO, CN^-, and imidazole. Like O_2, all of these ligands inhibit the kinase activities of FixL, albeit with different efficiencies [69]. In an assay that measured autophosphorylation of FixL, CO showed fivefold inhibition, while the inhibition by O_2 was 15-fold. In a turnover assay measuring phosphorylation of FixJ, the system was more specific, showing about threefold inhibition of CO compared to more than 100-fold inhibition with O_2. While not yet proven to be of physiological importance, the nonspecificity of FixL may be beneficial for rhizobia since CO has been shown to be a strong inhibitor of nitrogenase.

DosT/DosS/DosR System *Mycobacterium tuberculosis* is the bacterium responsible for human tuberculosis. Typically an infection leads to the establishment of a small population of highly resistant dormant cells that coincides with adaption of the bacteria to reduced oxygen tension. The hypoxic response in *M. tuberculosis* is predominantly controlled by the DosT/DosS/DosR regulatory system [70]. The response regulator DosR is activated from either of the two sensor kinases DosS and DosT by phosphoryl transfer onto an aspartate residue. The paralogs DosS and DosT are heme proteins that both contain C-terminal histidine kinase domains and two N-terminal GAF domains. For DosS, a histidine in the proximal GAF domain has been identified as the most likely ligand of heme iron. Thus DosS is the first example to demonstrate that a GAF domain may bind heme. Under aerobic conditions, autokinase activity of DosT is inhibited by formation of a ferrous DosT-O_2 complex. In the absence of oxygen, DosT also forms complexes with CO and NO. These complexes show enhanced autokinase activity compared to the oxygen-bound form. In contrast to DosT, DosS is oxidized by O_2 to the inactive ferric state. The active ferrous state of DosS can be preserved either by a yet hypothetical reductase or in the presence of NO and/or CO due to formation of nitrosyl and carbonyl complexes, respectively, that posses autokinase activity. Responsiveness of the Dos regulon towards NO and CO has been experimentally verified. It is assumed that endogenous production of CO and NO by the host cells is linked to the formation of dormant *M. tuberculosis* cells *in vivo*.

References

1 Bleecker, A.B. and Kende, H. (2000) Ethylene: a gaseous signal molecule in plants. *Annu. Rev. Cell Dev. Biol.*, **16**, 1–18.

2 Leffler, C.W., Parfenova, H., Jaggar, J.H., and Wang, R. (2006) Carbon monoxide and hydrogen sulfide: gaseous messengers in cerebrovascular circulation. *J. Appl. Physiol.*, **100**, 1065–1076.

3 Lenz, O., Bernhard, M., Buhrke, T., Schwartz, E., and Friedrich, B. (2002) The hydrogen-sensing apparatus in *Ralstonia eutropha*. *J. Mol. Microbiol. Biotechnol.*, **4**, 255–262.

4 Aono, S. (2008) Metal-containing sensor proteins sensing diatomic gas molecules. *Dalton Trans.*, 3137–3146.

5 Tatusov, R.L., Fedorova, N.D., Jackson, J.D., Jacobs, A.R., Kiryutin, B., Koonin, E.V., Krylov, D.M., Mazumder, R., Mekhedov, S.L., Nikolskaya, A.N. et al. (2003) The COG database: an updated version includes eukaryotes. *BMC Bioinformatics*, **4**, 41.

6 Letunic, I., Copley, R.R., Schmidt, S., Ciccarelli, F.D., Doerks, T., Schultz, J., Ponting, C.P., and Bork, P. (2004) SMART 4.0: towards genomic data integration. *Nucleic Acids Res.*, **32**, D142–D144.

7 Finn, R.D., Tate, J., Mistry, J., Coggill, P.C., Sammut, S.J., Hotz, H.R., Ceric, G., Forslund, K., Eddy, S.R., Sonnhammer, E.L. et al. (2008) The Pfam protein families database. *Nucleic Acids Res.*, **36**, D281–D288.

8 Cammack, R. (2001) Hydrogenases and their activities, in *Hydrogen as a Fuel:*

Learning from Nature (eds R. Cammack, M. Frey, and R. Robson), Taylor & Francis, London, pp. 33–56.

9 Vignais, P.M. and Billoud, B. (2007) Occurrence, classification, and biological function of hydrogenases: an overview. *Chem. Rev.*, **107**, 4206–4272.

10 Volbeda, A., Charon, M.H., Piras, C., Hatchikian, E.C., Frey, M., and Fontecilla-Camps, J.C. (1995) Crystal structure of the nickel–iron hydrogenase from *Desulfovibrio gigas*. *Nature*, **373**, 580–587.

11 Fontecilla-Camps, J.C., Volbeda, A., Cavazza, C., and Nicolet, Y. (2007) Structure/function relationships of [NiFe]- and [FeFe]-hydrogenases. *Chem. Rev.*, **107**, 4273–4303.

12 Albracht, S.P.J. (2001) Spectroscopy – the functional puzzle, in *Hydrogen as a Fuel: Learning from Nature* (eds R. Cammack, M. Frey, and R. Robson), Taylor & Francis, London, pp. 110–158.

13 Schwartz, E. and Friedrich, B. (2006) The H_2-metabolizing prokaryotes, in *The Prokaryotes: A Handbook on the Biology of Bacteria*, 3rd edn, vol. 2 (ed. M. Dworkin), Springer, New York, pp. 496–563.

14 Vignais, P.M. and Colbeau, A. (2004) Molecular biology of microbial hydrogenases. *Curr. Issues Mol. Biol.*, **6**, 159–188.

15 Böck, A., King, P.W., Blokesch, M., and Posewitz, M.C. (2006) Maturation of hydrogenases. *Adv. Microb. Physiol.*, **51**, 1–71.

16 Burgdorf, T., Lenz, O., Buhrke, T., van der Linden, E., Jones, A.K., Albracht, S.P., and Friedrich, B. (2005) [NiFe]-hydrogenases of *Ralstonia eutropha* H16: modular enzymes for oxygen-tolerant biological hydrogen oxidation. *J. Mol. Microbiol. Biotechnol.*, **10**, 181–196.

17 Bernhard, M., Buhrke, T., Bleijlevens, B., De Lacey, A.L., Fernandez, V.M., Albracht, S.P., and Friedrich, B. (2001) The H_2 sensor of *Ralstonia eutropha*. Biochemical characteristics, spectroscopic properties, and its interaction with a histidine protein kinase. *J. Biol. Chem.*, **276**, 15592–15597.

18 Lubitz, W., Reijerse, E., and van Gastel, M. (2007) [NiFe] and [FeFe] hydrogenases studied by advanced magnetic resonance techniques. *Chem. Rev.*, **107**, 4331–4365.

19 Winter, G., Buhrke, T., Lenz, O., Jones, A.K., Forgber, M., and Friedrich, B. (2005) A model system for [NiFe] hydrogenase maturation studies: Purification of an active site-containing hydrogenase large subunit without small subunit. *FEBS Lett.*, **579**, 4292–4296.

20 Brecht, M., van Gastel, M., Buhrke, T., Friedrich, B., and Lubitz, W. (2003) Direct detection of a hydrogen ligand in the [NiFe] center of the regulatory H_2-sensing hydrogenase from *Ralstonia eutropha* in its reduced state by HYSCORE and ENDOR spectroscopy. *J. Am. Chem. Soc.*, **125**, 13075–13083.

21 Buhrke, T., Lenz, O., Krauss, N., and Friedrich, B. (2005) Oxygen tolerance of the H_2-sensing [NiFe] hydrogenase from *Ralstonia eutropha* H16 is based on limited access of oxygen to the active site. *J. Biol. Chem.*, **280**, 23791–23796.

22 Duché, O., Elsen, S., Cournac, L., and Colbeau, A. (2005) Enlarging the gas access channel to the active site renders the regulatory hydrogenase HupUV of *Rhodobacter capsulatus* O_2 sensitive without affecting its transductory activity. *FEBS J.*, **272**, 3899–3908.

23 Schwartz, E., Buhrke, T., Gerischer, U., and Friedrich, B. (1999) Positive transcriptional feedback controls hydrogenase expression in *Alcaligenes eutrophus* H16. *J. Bacteriol.*, **181**, 5684–5692.

24 Lenz, O. and Friedrich, B. (1998) A novel multicomponent regulatory system mediates H_2 sensing in *Alcaligenes eutrophus*. *Proc. Natl. Acad. Sci. USA*, **95**, 12474–12479.

25 Buhrke, T., Lenz, O., Porthun, A., and Friedrich, B. (2004) The H_2-sensing complex of *Ralstonia eutropha*: interaction between a regulatory [NiFe] hydrogenase and a histidine protein kinase. *Mol. Microbiol.*, **51**, 1677–1689.

26 Taylor, B.L. and Zhulin, I.B. (1999) PAS domains: internal sensors of oxygen, redox potential, and light. *Microbiol. Mol. Biol. Rev.*, **63**, 479–506.

27 Gebler, A., Burgdorf, T., De Lacey, A.L., Rudiger, O., Martinez-Arias, A., Lenz, O., and Friedrich, B. (2007) Impact of

alterations near the [NiFe] active site on the function of the H$_2$ sensor from *Ralstonia eutropha*. *FEBS J.*, **274**, 74–85.

28 Pryor, W.A., Houk, K.N., Foote, C.S., Fukuto, J.M., Ignarro, L.J., Squadrito, G.L., and Davies, K.J. (2006) Free radical biology and medicine: it's a gas, man! *Am. J. Physiol. Regul. Integr. Comp. Physiol.*, **291**, R491–511.

29 Poulos, T.L. (2006) Soluble guanylate cyclase. *Curr. Opin. Struct. Biol.*, **16**, 736–743.

30 Zumft, W.G. (1997) Cell biology and molecular basis of denitrification. *Microbiol. Mol. Biol. Rev.*, **61**, 533–616.

31 Kampschreur, M.J., Tan, N.C., Picioreanu, C., Jetten, M.S., Schmidt, I., and van Loosdrecht, M.C. (2006) Role of nitrogen oxides in the metabolism of ammonia-oxidizing bacteria. *Biochem. Soc. Trans.*, **34**, 179–181.

32 Corker, H. and Poole, R.K. (2003) Nitric oxide formation by *Escherichia coli*. Dependence on nitrite reductase, the NO-sensing regulator Fnr, and flavohemoglobin Hmp. *J. Biol. Chem.*, **278**, 31584–31592.

33 Fang, F.C. (2004) Antimicrobial reactive oxygen and nitrogen species: concepts and controversies. *Nat. Rev. Microbiol.*, **2**, 820–832.

34 Pohlmann, A., Cramm, R., Schmelz, K., and Friedrich, B. (2000) A novel NO-responding regulator controls the reduction of nitric oxide in *Ralstonia eutropha*. *Mol. Microbiol.*, **38**, 626–638.

35 Rodionov, D.A., Dubchak, I.L., Arkin, A.P., Alm, E.J., and Gelfand, M.S. (2005) Dissimilatory metabolism of nitrogen oxides in bacteria: comparative reconstruction of transcriptional networks. *PLoS Comput. Biol.*, **1**, e55.

36 Saraiva, L.M., Vicente, J.B., and Teixeira, M. (2004) The role of the flavodiiron proteins in microbial nitric oxide deto-xification. *Adv. Microb. Physiol.*, **49**, 77–129.

37 Strube, K., de Vries, S., and Cramm, R. (2007) Formation of a dinitrosyl iron complex by NorA, a nitric oxide-binding di-iron protein from *Ralstonia eutropha* H16. *J. Biol. Chem.*, **282**, 20292–20300.

38 Studholme, D.J. and Dixon, R. (2003) Domain architectures of sigma[54]-dependent transcriptional activators. *J. Bacteriol.*, **185**, 1757–1767.

39 D'Autreaux, B., Tucker, N.P., Dixon, R., and Spiro, S. (2005) A non-haem iron centre in the transcription factor NorR senses nitric oxide. *Nature*, **437**, 769–772.

40 Klink, A., Elsner, B., Strube, K., and Cramm, R. (2007) Characterization of the signaling domain of the NO-responsive regulator NorR from *Ralstonia eutropha* H16 by site-directed mutagenesis. *J. Bacteriol.*, **189**, 2743–2749.

41 Tucker, N.P., D'Autreaux, B., Yousafzai, F.K., Fairhurst, S.A., Spiro, S., and Dixon, R. (2008) Analysis of the nitric oxide-sensing non-heme iron center in the NorR regulatory protein. *J. Biol. Chem.*, **283**, 908–918.

42 Spiro, S. (2007) Regulators of bacterial responses to nitric oxide. *FEMS Microbiol. Rev.*, **31**, 193–211.

43 Yeo, W.S., Lee, J.H., Lee, K.C., and Roe, J.H. (2006) IscR acts as an activator in response to oxidative stress for the *suf* operon encoding Fe–S assembly proteins. *Mol. Microbiol.*, **61**, 206–218.

44 Zumft, W.G. (2002) Nitric oxide signaling and NO dependent transcriptional control in bacterial denitrification by members of the FNR-CRP regulator family. *J. Mol. Microbiol. Biotechnol.*, **4**, 277–286.

45 Harman, J.G. (2001) Allosteric regulation of the cAMP receptor protein. *Biochim. Biophys. Acta*, **1547**, 1–17.

46 Roberts, G.P., Kerby, R.L., Youn, H., and Conrad, M. (2005) CooA, a paradigm for gas sensing regulatory proteins. *J. Inorg. Biochem.*, **99**, 280–292.

47 Unden, G., Achebach, S., Holighaus, G., Tran, H.G., Wackwitz, B., and Zeuner, Y. (2002) Control of FNR function of *Escherichia coli* by O$_2$ and reducing conditions. *J. Mol. Microbiol. Biotechnol.*, **4**, 263–268.

48 Giardina, G., Rinaldo, S., Johnson, K.A., Di Matteo, A., Brunori, M., and Cutruzzola, F. (2008) NO sensing in *Pseudomonas aeruginosa*: structure of the transcriptional regulator DNR. *J. Mol. Biol.*, **378**, 1002–1015.

49 Boczkowski, J., Poderoso, J.J., and Motterlini, R. (2006) CO-metal interaction: vital signaling from a lethal gas. *Trends Biochem. Sci.*, **31**, 614–621.

50 Chung, H.T., Choi, B.M., Kwon, Y.G., and Kim, Y.M. (2008) Interactive relations between nitric oxide (NO) and carbon monoxide (CO): heme oxygenase-1/CO pathway is a key modulator in NO-mediated antiapoptosis and anti-inflammation. *Methods Enzymol.*, **441**, 329–338.

51 Dioum, E.M., Rutter, J., Tuckerman, J.R., Gonzalez, G., Gilles-Gonzalez, M.A., and McKnight, S.L. (2002) NPAS2: a gas-responsive transcription factor. *Science*, **298**, 2385–2387.

52 Oelgeschlager, E. and Rother, M. (2008) Carbon monoxide-dependent energy metabolism in anaerobic bacteria and archaea. *Arch. Microbiol.*, **190**, 257–269.

53 Ragsdale, S.W. (2004) Life with carbon monoxide. *Crit. Rev. Biochem. Mol. Biol.*, **39**, 165–195.

54 Körner, H., Sofia, H.J., and Zumft, W.G. (2003) Phylogeny of the bacterial superfamily of Crp–Fnr transcription regulators: exploiting the metabolic spectrum by controlling alternative gene programs. *FEMS Microbiol. Rev.*, **27**, 559–592.

55 Aono, S., Nakajima, H., Saito, K., and Okada, M. (1996) A novel heme protein that acts as a carbon monoxide-dependent transcriptional activator in *Rhodospirillum rubrum*. *Biochem. Biophys. Res. Commun.*, **228**, 752–756.

56 Marvin, K.A., Kerby, R.L., Youn, H., Roberts, G.P., and Burstyn, J.N. (2008) The transcription regulator RcoM-2 from *Burkholderia xenovorans* is a cysteine-ligated hemoprotein that undergoes a redox-mediated ligand switch. *Biochemistry*, **47**, 9016–9028.

57 Hou, S., Freitas, T., Larsen, R.W., Piatibratov, M., Sivozhelezov, V., Yamamoto, A., Meleshkevitch, E.A., Zimmer, M., Ordal, G.W., and Alam, M. (2001) Globin-coupled sensors: a class of heme-containing sensors in Archaea and Bacteria. *Proc. Natl. Acad. Sci. USA*, **98**, 9353–9358.

58 Cooper, C.E. (1999) Nitric oxide and iron proteins. *Biochim. Biophys. Acta*, **1411**, 290–309.

59 Gaston, B. (1999) Nitric oxide and thiol groups. *Biochim. Biophys. Acta*, **1411**, 323–333.

60 Imlay, J.A. (2008) Cellular defenses against superoxide and hydrogen peroxide. *Annu. Rev. Biochem.*, **77**, 755–776.

61 Sellers, V.M., Johnson, M.K., and Dailey, H.A. (1996) Function of the [2Fe–2S] cluster in mammalian ferrochelatase: a possible role as a nitric oxide sensor. *Biochemistry*, **35**, 2699–2704.

62 Lipinski, P., Starzynski, R.R., Drapier, J.C., Bouton, C., Bartlomiejczyk, T., Sochanowicz, B., Smuda, E., Gajkowska, A., and Kruszewski, M. (2005) Induction of iron regulatory protein 1 RNA-binding activity by nitric oxide is associated with a concomitant increase in the labile iron pool: implications for DNA damage. *Biochem. Biophys. Res. Commun.*, **327**, 349–355.

63 Andrews, S.C., Robinson, A.K., and Rodriguez-Quinones, F. (2003) Bacterial iron homeostasis. *FEMS Microbiol. Rev.*, **27**, 215–237.

64 Hobman, J.L., Wilkie, J., and Brown, N.L. (2005) A design for life: prokaryotic metal-binding MerR family regulators. *Biometals*, **18**, 429–436.

65 Demple, B., Ding, H., and Jorgensen, M. (2002) *Escherichia coli* SoxR protein: sensor/transducer of oxidative stress and nitric oxide. *Methods Enzymol.*, **348**, 355–364.

66 Marshall, H.E., Merchant, K., and Stamler, J.S. (2000) Nitrosation and oxidation in the regulation of gene expression. *FASEB J.*, **14**, 1889–1900.

67 Poole, R.K. and Hughes, M.N. (2000) New functions for the ancient globin family: bacterial responses to nitric oxide and nitrosative stress. *Mol. Microbiol.*, **36**, 775–783.

68 Gilles-Gonzalez, M.A. and Gonzalez, G. (2005) Heme-based sensors: defining characteristics, recent developments, and regulatory hypotheses. *J. Inorg. Biochem.*, **99**, 1–22.

69 Tuckerman, J.R., Gonzalez, G., Dioum, E.M., and Gilles-Gonzalez, M.A. (2002) Ligand and oxidation-state specific regulation of the heme-based oxygen sensor FixL from *Sinorhizobium meliloti*. *Biochemistry*, **41**, 6170–6177.

70 Kumar, A., Toledo, J.C., Patel, R.P., Lancaster, J.R. Jr, and Steyn, A.J. (2007) *Mycobacterium tuberculosis* DosS is a redox sensor and DosT is a hypoxia sensor. *Proc. Natl. Acad. Sci. USA*, **104**, 11568–11573.

18
Signal Transduction by Trigger Enzymes: Bifunctional Enzymes and Transporters Controlling Gene Expression

Fabian M. Commichau and Jörg Stülke

18.1
Introduction

There is only one constant in the life of bacteria – change. To be successful in competition with their fellows, all organisms need to detect these changes and respond to them rapidly and appropriately while preventing the waste of energy and resources. In bacteria, the main level at which these regulatory events occur is transcription. Therefore, several mechanisms allowing a tight control of transcription have evolved.

In each case, regulatory events require the perception of the environmental or internal signal and its transduction to the transcription machinery. Many transcription factors integrate signal perception and the regulatory output (i.e., transcription regulation) in a single molecule, among them the Lac repressor or the cAMP receptor protein from *Escherichia coli*. In many other cases, signal recognition and the regulatory output are embodied in two distinct proteins, subunits, or domains that interact with each other for signal transduction. The paradigm of this type of regulation are the two-component regulatory systems (see also Chapter 8). Similarly, the activity of many sigma factors of the bacterial RNA polymerase is controlled by interactions with anti-sigma factors, which may in turn be controlled by anti-antisigma factors [1]. The separation of the two tasks in signal transduction is very advantageous in evolution as it allows shifts in specificity by gene duplication and specialization. Indeed, transcription factors are still evolving [2].

There are proteins in the cell that are well informed about specific conditions, usually about metabolite availabilities. These proteins are the enzymes that recognize their substrates and that often undergo structural alterations upon interaction with the substrate or during their reaction. This makes the enzymes perfect candidates to share their knowledge with transcription regulators or to regulate gene expression themselves. Moreover, these enzymes are an excellent substrate for the evolution of sensory components of signal transduction systems. Such proteins that exert a second, unrelated function in addition to their primary task are called "moonlighting proteins" [3]. A plethora of enzymes involved in transcription regulation from a variety of biochemical pathways has recently been discovered, suggesting that this regulatory principle is much more common than hitherto acknowledged. These enzymes are collectively designated as trigger enzymes to emphasize their role in signal transduction. Trigger enzymes can exert different functions. They might act as transcription factors by direct binding to either DNA or RNA, or they might modulate the activity of transcription factors either by covalent modification or by protein–protein interactions [4] (see Table 18.1).

Table 18.1 A compilation of trigger enzymes in bacteria.[a]

Organism	Enzyme-catalyzed reaction	Effector/signal	Target
DNA binding (transcription factor)			
E. coli, B. subtilis	BirA, biotin protein ligase	biotin	bir operon
E. coli, S. typhimurium	PutA, bifunctional proline dehydrogenase and Δ^1-pyrroline-5-carboxylate dehydrogenase	proline	promoter region of the put operon
E. coli, S. typhimurium	NadR, nicotinamide mononucleotide adenylyltransferase	NMN/NADH$_2$	promoter region of the NAD$^+$ biosynthetic genes
E. coli	PepA, aminopeptidase	PyrH	carAB operon
RNA binding			
B. subtilis	CitB, aconitase	iron	iron-responsive elements
Covalent modification of a transcription regulator[b]			
E. coli	BglF, β-glucoside permease (EIIB subunit)	salicin	BglG
B. subtilis	LevE, fructose permease (EIIB subunit)	fructose	LevR
B. subtilis	PtsG, glucose permease (EIIB subunit)	glucose	GlcT
Modulation of activity of a DNA-binding protein			
Listeria monocytogenes	GmaR, O-GlcNAc transferase	?	MogR
E. coli	PtsG, glucose permease (EIIB subunit)	glucose	Mlc

Table 18.1 (Continued)

Organism	Enzyme-catalyzed reaction	Effector/signal	Target
E. coli	LysP, lysine permease [35]	lysine	CadC [35]
E. coli	DhaL, DhaK, dihydroxy-acetone kinase (nucleotide and substrate binding subunits)	dihydroxyacetone	DhaR DhaK antagonizes DhaL
E. coli	MalY, βC-S lyase MalK, ATP-binding subunit of the maltose transporter Aes, esterase	maltodextrins, maltotriose	MalT
E. coli	TktA, transketolase	?	MarR
E. coli	PyrH, UMP kinase	pyrimidines	PepA
E. coli	DcuB, fumarate/succinate antiporter	?	DcuS [56]
B. subtilis	CysK, cysteine synthetase A	O-acetylserine, cysteine	CymR [57]
B. subtilis	GlnA, glutamine synthetase	glutamine	TnrA, GlnR
B. subtilis	RocG, glutamate dehydrogenase	glutamate	GltC
Unknown mechanism/regulators that evolved from enzymes			
B. subtilis	PyrR	uridine 5′-monophosphate/5′-phosphoribosyl-α-1′-pyrophosphate	pyr operon
B. subtilis	LipA, lipoic acid synthetase	?	comE operon [58]
T. thermophilus	Mlc	glucose	glc and mal operons
S. pyogenes	LacD.1	dihydroxyacetone phosphate (?)	RopB
B. subtilis	RibR	?	rib operon (mRNA)

a) Detailed references for each trigger enzyme can be found in [4].
b) These trigger enzymes are the best-studied representatives of a large class of sugar permeases that control PRD-containing transcription factors. For a comprehensive compilation, the reader is referred to [24, 55].

18.2
Trigger Enzymes Active as DNA-Binding Transcription Factors

In bacteria, transcription is most commonly controlled by regulator proteins that bind specific DNA sequences in the promoter region. This binding results in either transcription repression or activation. The recognition of a specific sequence requires the presence of a protein domain that is able to recognize and bind a DNA motif. Since DNA recognition motifs are usually not present in enzymes, the acquisition of such a domain is a prerequisite for the evolution of an enzyme to a DNA-binding trigger enzyme.

In proteobacteria, a bifunctional enzyme catalyzes the two-step degradation of proline to glutamate – the two consecutive oxidations of proline and Δ^1-pyrroline-5-carboxylate (see Figure 18.1). In *E. coli* and other enterobacteria this enzyme, PutA, is also involved in transcription regulation of the divergent *putA* and *putP* genes encoding the bifunctional proline utilization protein and the proline transporter, respectively [5]. This regulation is dependent on the availability of proline.

PutA is able to bind directly to five conserved sites in the promoter region of the *putA* and *putP* genes, thereby causing transcriptional repression in the absence of proline [5, 6] (Figure 18.1). On the other hand, the enzymatically active form that links reduction of the cofactor FAD to proline oxidation is a peripheral membrane protein [7] (Figure 18.1). This enzymatically active, membrane-bound form of PutA is unable to bind DNA and to repress transcription [5]. Several studies addressed the

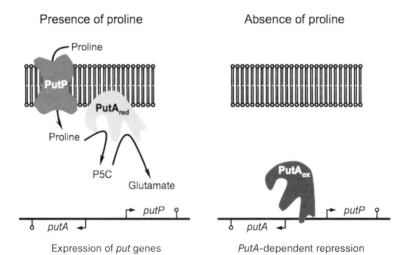

Figure 18.1 Localization of PutA determines its role in proline metabolism. In the presence of proline, the bifunctional enzyme PutA catalyzes the two-step reaction from proline to glutamate. This reduced form of PutA (PutA$_{red}$) is localized to the membrane. The *putP* and *putA* genes, encoding the proline transporter PutP and the enzyme PutA, respectively, are expressed in the presence of proline. In the absence of proline, the oxidized PutA protein (PutA$_{ox}$) binds to the intergenic region of the *putA* and *putP* genes and represses their transcription. P5C, Δ^1-pyrroline-5-carboxylate.

identity of the molecular inducer that causes release of PutA from the DNA, resulting in transcription of the *put* genes. In contrast to many other repressors, two factors contribute to induction – the presence of proline and FAD induces a conformational change in PutA that results in the shuttle of the protein from the DNA to the membrane [8, 9]. Mutation analyses revealed that enzymatic activity is essential for the release of PutA from the DNA and for *put* gene induction [10].

PutA contains two large domains – the N-terminal domain has the proline dehydrogenase (PDH) activity, whereas the C-terminal domain harbors the Δ^1-pyrroline-5-carboxylate dehydrogenase activity [11]. The conformational change that occurs in the presence of proline and FAD was mapped to the PDH domain [9]. The DNA-binding activity of PutA is located in the N-terminal 47 amino acids of the PDH domain that form a ribbon–helix–helix (RHH) motif [12]. The DNA-binding RHH domain is different from the well-known helix–turn–helix motif in that it consists of a β-strand followed by two α-helices. The characteristic of the RHH motif is sequence-specific DNA recognition by residues in the β-sheet formed by two RHH subunits [13].

Similar to PutA, the trigger enzymes BirA and NadR from *E. coli* possess distinct DNA-binding domains (see Table 18.1). Interestingly, BirA seems to act as a transcription factor in many bacteria and even in some Archaea [14].

18.3
Trigger Enzymes Involved in Post-Transcriptional Regulation via Protein–RNA Interaction

RNAs can adopt a variety of structures, and they can bind and interact with virtually any molecule, including metabolites and proteins [15]. Thus, it is not surprising that proteins with very different structures are able to bind RNA, and, indeed, many enzymes interact with RNAs and exert regulatory effects by this interaction (for an excellent review, see [16]).

There is so far only one example of a bacterial trigger enzyme with RNA-binding activity – aconitase. Aconitase catalyzes the reversible conversion of citrate to isocitrate in the tricarboxylic acid (TCA) cycle and requires an iron–sulfur cluster for activity [17]. Under conditions of iron limitation, the TCA cycle cannot operate due to the inactivity of aconitase. However, aconitase can help solve this problem by binding so-called iron-responsive elements (IREs) in the mRNAs of genes involved in iron homeostasis (see Figure 18.2). This was first shown for the human enzyme, but later on also for the aconitases from such diverse bacteria as *E. coli*, *Bacillus subtilis*, and *Mycobacterium tuberculosis* [18–21].

The determination of the structure of IRE-bound aconitase revealed the molecular basis for the two mutually exclusive activities. In the presence of iron, the protein has a compact conformation. In the absence of iron or under conditions of oxidative stress, the iron–sulfur cluster disassembles and the free (apo)aconitase adopts a more open conformation. This opening allows the binding of the IRE (see [17] for review). It is worth noting that only a few bases are conserved in the IREs and that these conserved bases are brought into the right position by secondary structure elements [22].

Sufficient Fe^{2+} Fe^{2+} limitation

Citrate   Isocitrate   5' mRNA binding

Figure 18.2 The RNA-binding function of the aconitase depends on the iron status of the cell. In the presence of iron, the iron–sulfur cluster in the active center of the aconitase is stable and the enzyme catalyzes the formation of isocitrate in the TCA cycle. During iron limitation, the iron–sulfur cluster is absent and the inactive enzyme gains the ability to bind specific mRNAs structures designated as IREs. This binding results in the stabilization or destabilization of the bound mRNAs.

Based on the overall sequence similarity of the eukaryotic and *B. subtilis* aconitases and IREs [20], it seems safe to assume that the *B. subtilis* enzyme follows the same mechanism as outlined above for the mammalian enzyme. *B. subtilis* aconitase binds IREs in the untranslated regions of the *qoxD* and *feuAB* mRNAs. These genes encode the iron-containing protein cytochrome aa_3 oxidase and an iron uptake system, respectively [20].

E. coli possesses two aconitases, of which aconitase B (AcnB) is the main enzyme involved in the TCA cycle. In contrast to the monomeric enzymes of eukaryotes and *B. subtilis*, AcnB has a dimerization domain. An analysis of the requirements of AcnB for RNA binding revealed that it is independent of the iron–sulfur cluster in the active center of the enzyme. Moreover, the arrangement of the RNA in the protein seems to be different from that observed in the mammalian aconitase–IRE complex [23]. This is in good agreement with the observation that the RNA elements recognized by *E. coli* AcnB differ from the classical IREs [19].

There is a large variety of potential RNA structures that might be able to bind diverse ligands and, indeed, many eukaryotic enzymes have been shown to moonlight in RNA interactions. Thus, it would not be surprising if many bacterial enzymes turn out to be trigger enzymes involved in protein–RNA interactions. This might lead to another exciting phase of studying riboregulation after the discovery of riboswitches and small regulatory RNAs.

18.4
Trigger Enzymes Controlling Gene Expression by Signal-Dependent Phosphorylation of Transcription Regulators

In many bacteria, sugars are transported by the phosphoenolpyruvate-dependent phosphotransferase system (PTS) (see also Chapter 19). The PTS permeases are not only involved in sugar transport, they also control the activity of transcription activators and antiterminators by phosphorylating them in response to the availability of the respective substrate. These transcription regulators all share a duplicated so-called PTS regulation domain (PRD) [24]. Phosphorylation of one of these domains by the sugar-specific permease occurs in the absence of the substrate (upon accumulation of phosphorylated permease) and results in the inactivation of the regulators. In the presence of the substrate, the phosphate groups are drained to the sugar, and the regulators become dephosphorylated and, thus, regain activity. This regulatory mechanism has been most intensively studied for the control of β-glucoside transport in *E. coli*, and for glucose and fructose uptake in *B. subtilis* [25–28] (for details, see Chapter 19).

18.5
Trigger Enzymes Controlling the Activity of Transcription Factors by Protein–Protein Interactions

The fourth class of trigger enzymes controls gene expression by modulating the activity of transcription factors – either activators or repressors. The diversity of such interactions reflects the diversity of transcription factors and makes it difficult to predict these trigger enzymes from the primary sequence.

As shown in Table 18.1, there are many regulatory interactions between enzymes and transcription factors, and their detailed presentation would be far beyond the scope of this chapter. Instead, we concentrate on a few interactions for which sufficient information is available.

In *E. coli*, the global transcription regulator Mlc represses the production of several PTS components, including the *ptsG* gene and the *ptsHI crr* operon encoding the glucose permease and the general PTS proteins, respectively [29] (see also Chapters 12 and 19). In the presence of glucose, Mlc is unable to bind its DNA targets, resulting in expression of the controlled genes. However, the mechanism of glucose sensing by Mlc has long remained enigmatic [30, 31]. The *E. coli* glucose permease of the PTS was found to act as a trigger enzyme and to control Mlc activity in addition to its function in glucose transport and phosphorylation. In the presence of glucose, the glucose permease is mainly present in the nonphosphorylated state. This form of the enzyme is capable of binding Mlc, thereby sequestering it and preventing it from DNA binding. In contrast, the phosphorylated glucose permease, the idle form of the enzyme present in the absence of glucose, is unable to bind Mlc and Mlc is free to repress the transcription from its target promoters [32]. The differential ability of the glucose permease to bind Mlc is caused by structural changes in the region of the active site of the IIB domain upon phosphorylation [33]. It is interesting to note that the Mlc repressor of *Thermus thermophilus* also controls gene expression in response to the presence of glucose even though these bacteria do not possess PTS proteins. This Mlc protein binds directly to glucose, which acts as an inducer. This seems to be a remnant of the evolutionary history of Mlc and its relatedness to sugar kinases [34] (see Section 18.6).

Trigger enzyme activity is not restricted to PTS permeases. Recently, the *E. coli* lysine transporter LysP was shown to control the activity of the transcription activator protein CadC in response to the availability of lysine [35]. Interestingly, CadC is a membrane protein with a cytoplasmic DNA-binding domain. It has therefore been hypothesized that CadC itself might sense the lysine concentration. In a series of excellent biochemical and genetic experiments, Tetsch et al. [35] provide compelling evidence that CadC is unable to bind lysine, and that the signal is instead sensed by the lysine transporter LysP and transduced to CadC via interaction of the transmembrane domains of the two proteins. In the absence of lysine, LysP is thought to bind and sequester CadC, thus inhibiting its DNA-binding activity. If lysine is present, then LysP is involved in transporting this amino acid, and CadC becomes free and can now activate the expression of its target operon, cadBA [35].

In *B. subtilis*, nitrogen metabolism is controlled by the global transcription factor TnrA. TnrA activates the expression of genes involved in the utilization of secondary nitrogen sources and in the uptake of ammonium at low concentrations. In contrast, it represses the expression of genes required for ammonium assimilation – the *glnA* gene and the *gltAB* operon encoding glutamine synthetase (GS) and glutamate synthase, respectively [36]. A combined transcriptome and bioinformatic analysis suggested that TnrA controls the expression of about 20 target operons [37]. It is thus the functional equivalent of the global Ntr regulatory system in the enterobacteria [36]. In addition to TnrA, the GlnR and GltC transcription factors specifically control the expression of the *glnRA* and *gltAB* operon encoding glutamine synthetase and glutamate synthase, respectively [36].

The DNA-binding activity of TnrA responds to the nitrogen supply of the cells. It is active if the cells are provided with poor nitrogen sources such as glutamate or nitrate, but inactive in the presence of good nitrogen sources such as glutamine [38]. These findings raised the question of how TnrA senses the nitrogen supply. *In vitro* experiments revealed that the protein is active in transcription activation without cofactors, and that none of the likely metabolite effectors affects DNA binding by TnrA [39]. However, *glnA* mutants defective for GS exhibit constitutive TnrA activity, suggesting that GS exerts negative control over TnrA [38]. Indeed, an interaction between GS and TnrA was detected, and the formation of the complex caused TnrA to lose its DNA-binding ability. The formation of the TnrA–GS complex depends on the presence of glutamine and AMP; that is, it occurs under conditions of feedback inhibition of GS [40]. This provides a direct link between nitrogen supply, enzymatic activity of the GS, and TnrA-controlled gene expression. In the presence of excess glutamine, the GS is feedback-inhibited and forms a complex with TnrA, thus preventing its activity as a transcription factor. In contrast, under nitrogen limitation, GS is enzymatically active in glutamine biosynthesis and free TnrA activates the genes for the utilization of alternative nitrogen sources. This model is strengthened by the observation that mutations of the GS that interfere with feedback inhibition but not with enzymatic activity result in a defect in the interaction and concomitant inactivation of TnrA [41].

Interestingly, the GS does not only control TnrA activity, but also that of GlnR – the repressor protein of the *B. subtilis glnRA* operon. As described for TnrA, the feedback-inhibited GS is able to bind GlnR, but the consequence of this interaction is different – while TnrA is inactivated upon binding of GS, GlnR is activated

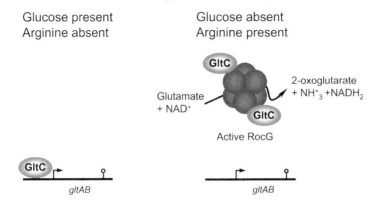

Figure 18.3 Control of glutamate biosynthesis in B. subtilis by the glutamate dehydrogenase RocG. In the presence of the preferred carbon source glucose the transcription factor GltC activates the transcription of the *gltAB* operon. The glutamate synthase is expressed and produces glutamate. In the absence of glucose and in the presence of arginine the catabolic glutamate dehydrogenase RocG is synthesized. This enzyme binds GltC and thereby inhibits the DNA-binding activity of the transcription regulator.

and may now bind its target DNA and repress the expression of the *glnRA* operon. This regulatory mechanism ensures that *glnA*, the structural gene for the GS, is only transcribed in the absence of glutamine [42]. The interaction with GS is required for GlnR activity since the C-terminal domain of GlnR prevents dimerization and thus DNA binding. GS binding is thought to overcome this autoinhibition [43]. A similar mode of GlnR regulation by interaction with the glutamine synthetase does probably operate in other Gram-positive bacteria such as *Streptococcus pneumoniae* [44].

A third example of trigger enzyme activity controlling ammonium assimilation in *B. subtilis* is the regulation of the transcription factor GltC by the glutamate dehydrogenase RocG (see Figure 18.3). RocG participates in arginine degradation and is induced in the presence of arginine. The enzyme catalyzes the last step of the pathway, the conversion of glutamate to 2-oxoglutarate. Thus, arginine utilization results in the formation of glutamate and circumvents the need for glutamate biosynthesis by the glutamate synthase – the product of the *gltAB* operon. This is achieved by a regulatory interaction between RocG and the transcription activator of the *gltAB* operon, GltC [45]. The expression of the *rocG* gene is not only subject to induction by arginine, but also to glucose repression. In consequence, complete repression of *rocG* by glucose does also allow activity of free GltC, and thus expression of the *gltAB* operon and glutamate biosynthesis. This regulatory mechanism couples ammonium assimilation and anabolic reactions to the availability of carbon sources and the flux through glycolysis that results in carbon catabolite repression [46] (see also Chapter 19).

18.6
Evolution of Trigger Enzymes: From Enzymes via Trigger Enzymes to Regulators

Due to their central role in metabolism and their interactions with basically all metabolites in the cell, the enzymes are the "best-informed" proteins in the cell with respect to the metabolic status. This makes them perfect mediators of gene regulation. The trigger enzymes seem to be an evolutionary intermediate as they are related to enzymes as well as to dedicated regulatory proteins. The evolution of these interesting proteins is still continuing, and we can study all stages in this evolution. Two patterns seem to be of special importance: (i) the acquisition of DNA-binding domains by enzymes, and (ii) the functional separation of enzymatic and regulatory functions via duplication events.

An example for the acquisition of a DNA-binding domain is the regulation of proline catabolism. This pathway involves two distinct enzymatic activities that can be performed by two individual proteins, as in the Gram-positive bacteria, or by a single multidomain protein, as in proteobacteria. In many proteobacteria, including *E. coli* and the other enterobacteria, this bifunctional PutA protein has acquired a DNA binding domain (see above, Figure 18.1). In contrast, in *Bradyrhizobium japonicum* and many other α-proteobacteria, PutA lacks a DNA-binding domain and, indeed, these proteins are strictly enzymatic (i.e., they do not act as trigger enzymes) [47].

Another fascinating example for the evolution of an enzyme to a regulator is provided by the so-called ROK (repressor, open reading frame, kinase) family comprising repressors of genes involved in sugar metabolism and sugar kinases [48]. This family includes the glucose kinase that catalyses the first step in glycolysis, and the *B. subtilis* Xyl and the *E. coli* and *T. thermophilus* Mlc repressors. Interestingly, the Xyl repressor binds not only its inducer xylose, but also glucose [49]. Mlc is an excellent example for the ongoing evolution. In *T. thermophilus*, glucose acts as the inducer by binding to the motif conserved in the glucose kinases [34]. The *E. coli* Mlc has lost its ability to bind glucose [34] and Mlc activity is controlled by interaction with the trigger enzyme PtsG (see Section 18.5). This development was probably driven by the invention of the PTS for sugar transport with concomitant phosphorylation of the incoming sugars (see also Chapter 19). Thus, free glucose is not available to control the *E. coli* Mlc protein. Obviously, this family might yield trigger enzymes, although this has not been described so far. Interestingly, the glucose kinase of *Streptomyces coelicolor* is involved in carbon catabolite repression. This protein itself, rather than its enzymatic activity, has the regulatory activity [50]. Thus, this glucose kinase might act as a trigger enzyme.

Similarly to the proteins of the ROK family, transcription repressors of the SorC family may have arisen by evolution from phosphosugar-binding enzymes. This idea results from the similarity of the effector-binding domain of a SorC family member, CggR from *B. subtilis*, with NagB enzymes [51, 52].

There are several cases in which the two functions of trigger enzymes have been separated by gene duplication events. In *Streptococcus pyogenes*, a unique form of carbon catabolite repression is controlled by a specialized trigger enzyme, LacD.1.

This enzyme (as its paralog, LacD.2) has tagatose-1,6-bisphosphate aldolase activity. Under conditions of high glycolytic activity, LacD.1 (but not LacD.2) binds and inactivates the transcriptional activator RopB. If the glycolytic activity is low, LacD.1 is unable to bind RopB, which can then activate transcription of the *speB* gene encoding a protease, thus allowing the utilization of proteins as an alternative carbon source. In contrast to LacD.1, the LacD.2 protein is involved in lactose metabolism rather than in regulation [53, 54]. The two LacD proteins share 71% identical amino acids. Thus, the duplication of an enzyme might even result in the acquisition of regulatory activity for only distantly related functions.

Acknowledgments

The work in our laboratory was supported by grants of the Deutsche Forschungsgemeinschaft and the Fonds der Chemischen Industrie to J.S. and F.M.C.

References

1 Helmann, J.D. (1999) Anti-sigma factors. *Curr. Opin. Microbiol.*, **2**, 135–141.
2 Galvão, T.C. and de Lorenzo, V. (2006) Transcriptional regulators à la carte: engineering new effector specificities in bacterial regulatory proteins. *Curr. Opin. Biotechnol.*, **17**, 34–42.
3 Jeffery, C.J. (1999) Moonlighting proteins. *Trends Biochem. Sci.*, **24**, 8–11.
4 Commichau, F.M. and Stülke, J. (2008) Trigger enzymes: bifunctional proteins active in metabolism and in controlling gene expression. *Mol. Microbiol.*, **67**, 692–702.
5 Ostrovsky de Spicer, P. and Maloy, S. (1993) PutA protein, a membrane-associated flavin dehydrogenase, acts as a redox-dependent transcriptional regulator. *Proc. Natl. Acad. Sci. USA*, **90**, 4295–4298.
6 Zhou, Y., Larson, J.D., Bottoms, C.A., Arturo, E.C., Henzl, M.T., Jenkins, J.L., Nix, J.C., Becker, D.F., and Tanner, J.J. (2008) Structural basis of the transcriptional regulation of the proline utilization regulon by multifunctional PutA. *J. Mol. Biol.*, **381**, 174–188.
7 Brown, E.D. and Wood, J.M. (1993) Conformational change and membrane association of the PutA protein are coincident with reduction of its FAD cofactor by proline. *J. Biol. Chem.*, **268**, 8972–8979.
8 Zhang, W., Zhou, Y., and Becker, D.F. (2004) Regulation of PutA–membrane associations by FAD reduction. *Biochemistry*, **43**, 13165–13174.
9 Zhu, W. and Becker, D.F. (2003) Flavin redox state triggers conformational changes in the PutA protein from *Escherichia coli*. *Biochemistry*, **42**, 5469–5477.
10 Muro-Pastor, A.M. and Maloy, S. (1995) Proline dehydrogenase activity of the transcriptional repressor PutA is required for induction of the *put* operon by proline. *J. Biol. Chem.*, **270**, 9819–9827.
11 Ling, M., Allen, S.W., and Wood, J.M. (1994) Sequence analysis identifies the proline dehydrogenase and Δ^1-pyrroline-5-carboxylate dehydrogenase domains of the multifunctional *Escherichia coli* PutA protein. *J. Mol. Biol.*, **243**, 950–956.
12 Gu, D., Zhou, Y., Kallhoff, V., Baban, B., Tanner, J.J., and Becker, D.F. (2004) Identification and characterization of the DNA-binding domain of the multifunctional PutA flavoenzyme. *J. Biol. Chem.*, **279**, 31171–31176.
13 Schreiter, E.R. and Drennan, C.L. (2007) Ribbon–helix–helix transcription factors: variations on a theme. *Nat. Rev. Microbiol.*, **5**, 710–720.
14 Rodionov, D.A., Mironov, A.A., and Gelfand, M.S. (2002) Conservation of the biotin regulon and the BirA regulatory

signal in eubacteria and archaea. *Genome Res.*, **12**, 1507–1516.

15 Ellington, A.D. and Szostak, J.W. (1990) In vitro selection of RNA molecules that bind specific ligands. *Nature*, **346**, 818–822.

16 Cieśla, J. (2006) Metabolic enzymes that bind RNA: yet another level of cellular regulatory network? *Acta Biochim. Pol.*, **53**, 11–32.

17 Volz, K. (2008) The functional duality of iron regulatory protein 1. *Curr. Opin. Struct. Biol.*, **18**, 106–111.

18 Kaptain, S., Downey, W.E., Tang, C., Philpott, C., Haile, D., Orloff, D.G., Harford, J.B., Rouault, T.A., and Klausner, R.D. (1991) A regulated RNA binding protein also possesses aconitase activity. *Proc. Natl. Acad. Sci. USA*, **88**, 10109–10113.

19 Tang, Y. and Guest, J.R. (1999) Direct evidence for mRNA binding and post-transcriptional regulation by *Escherichia coli* aconitases. *Microbiology*, **145**, 3069–3079.

20 Alén, C. and Sonenshein, A.L. (1999) *Bacillus subtilis* aconitase is an RNA-binding protein. *Proc. Natl. Acad. Sci. USA*, **96**, 10412–10417.

21 Banerjee, S., Nandyala, A.K., Raviprasad, P., Ahmed, N., and Hasnain, S.E. (2007) Iron-dependent RNA-binding activity of *Mycobacterium tuberculosis* aconitase. *J. Bacteriol.*, **189**, 4046–4052.

22 Walden, W.E., Selezneva, A.I., Dupuy, J., Volbeda, A., Fontecilla-Camps, J.C., Theil, E., and Volz, K. (2006) Structure of dual function iron regulatory protein 1 complexed with ferritin IRE-RNA. *Science*, **314**, 1903–1908.

23 Tang, Y., Guest, J.R., Artymiuk, P.J., and Green, J. (2005) Switching aconitase B between catalytic and regulatory modes involves iron-dependent dimer formation. *Mol. Microbiol.*, **56**, 1149–1158.

24 Stülke, J., Arnaud, M., Rapoport, G., and Martin-Verstraete, I. (1998) PRD – a protein domain involved in PTS-dependent induction and carbon catabolite repression of catabolic operons in bacteria. *Mol. Microbiol.*, **28**, 865–874.

25 Amster-Choder, O. and Wright, A. (1990) Regulation of activity of a transcriptional anti-terminator in *E. coli* by phosphorylation in vivo. *Science*, **249**, 540–542.

26 Görke, B. (2003) Regulation of the *Escherichia coli* antiterminator protein BglG by phosphorylation at multiple sites and evidence for transfer of phosphoryl groups between monomers. *J. Biol. Chem.*, **278**, 46219–46229.

27 Martin-Verstraete, I., Charrier, V., Stülke, J., Galinier, A., Erni, B., Rapoport, G., and Deutscher, J. (1998) Antagonistic effects of dual PTS catalyzed phosphorylation on the *Bacillus subtilis* transcriptional activator LevR. *Mol. Microbiol.*, **28**, 293–303.

28 Schmalisch, M., Bachem, S., and Stülke, J. (2003) Control of the *Bacillus subtilis* antiterminator protein GlcT by phosphorylation. Elucidation of the phosphorylation chain leading to inactivation of GlcT. *J. Biol. Chem.*, **278**, 51108–51115.

29 Böhm, A. and Boos, W. (2004) Gene regulation in prokaryotes by subcellular relocalization of transcription factors. *Curr. Opin. Microbiol.*, **7**, 151–156.

30 Plumbridge, J. (1998) Expression of *ptsG*, the gene for the major glucose PTS transporter in *Escherichia coli*, is repressed by Mlc and induced by growth on glucose. *Mol. Microbiol.*, **29**, 1053–1063.

31 Kim, S.-Y., Nam, T.-W., Shin, D., Koo, B.-M., Seok, Y.-J., and Ryu, S. (1999) Purification of Mlc and analysis of its effects on the *pts* expression in *Escherichia coli*. *J. Biol. Chem.*, **274**, 25398–25402.

32 Tanaka, Y., Kimata, K., and Aiba, H. (2000) A novel regulatory role of glucose transporter of *Escherichia coli*: membrane sequestration of a global repressor Mlc. *EMBO J.*, **19**, 5344–5352.

33 Seitz, S., Lee, S.J., Pennetier, C., Boos, W., and Plumbridge, J. (2003) Analysis of the interaction between the global regulator Mlc and EIIB^{Glc} of the glucose-specific phosphotransferase system in *Escherichia coli*. *J. Biol. Chem.*, **278**, 10744–10751.

34 Chevance, F.F.V., Erhardt, M., Lengsfeld, C., Lee, S.J., and Boos, W. (2006) Mlc of *Thermus thermophilus*: a glucose-specific regulator for a glucose/mannose ABC transporter in the absence of the phosphotransferase system. *J. Bacteriol.*, **188**, 6561–6571.

35 Tetsch, L., Koller, C., Haneburger, I., and Jung, K. (2008) The membrane-

36. Fisher, S.H. (1999) Regulation of nitrogen metabolism in *Bacillus subtilis*: vive la différence! *Mol. Microbiol.*, **32**, 223–232.
37. Yoshida, K., Yamaguchi, H., Kinehara, M., Ohki, Y.H., Nakaura, Y., and Fujita, Y. (2003) Identification of additional TnrA-regulated genes of *Bacillus subtilis* associated with a TnrA box. *Mol. Microbiol.*, **49**, 157–165.
38. Wray, L.V. Jr., Ferson, A.E., Rohrer, K., and Fisher, S.H. (1996) TnrA, a transcription factor required for global nitrogen regulation in *Bacillus subtilis*. *Proc. Natl. Acad. Sci. USA*, **93**, 8841–8845.
39. Wray, L.V. Jr., Zalieckas, J.M., and Fisher, S.H. (2000) Purification and *in vitro* activities of the *Bacillus subtilis* TnrA transcription factor. *J. Mol. Biol.*, **300**, 29–40.
40. Wray, L.V. Jr., Zalieckas, J.M., and Fisher, S.H. (2001) *Bacillus subtilis* glutamine synthetase controls gene expression through a protein–protein interaction with transcription factor TnrA. *Cell*, **107**, 427–435.
41. Fisher, S.H. and Wray, L.V. Jr. (2006) Feedback-resistant mutations in *Bacillus subtilis* glutamine synthetase are clustered in the active site. *J. Bacteriol.*, **188**, 5966–5974.
42. Fisher, S.H. and Wray, L.V. Jr. (2008) *Bacillus subtilis* glutamine synthetase regulates its own synthesis by acting as a chaperone to stabilize GlnR–DNA complexes. *Proc. Natl. Acad. Sci. USA*, **105**, 1014–1019.
43. Wray, L.V. and Fisher, S.H. (2008) *Bacillus subtilis* GlnR contains an autoinhibitory C-terminal domain required for the interaction with glutamine synthetase. *Mol. Microbiol.*, **68**, 277–285.
44. Kloosterman, T.G., Hendriksen, W.T., Bijlsma, J.J.E., Bootsma, H.J., van Hijum, S.A.F.T., Kok, J., Hermans, P.W.M., and Kuipers, O.P. (2006) Regulation of glutamine and glutamate metabolism by GlnR and GlnA in *Streptococcus pneumoniae*. *J. Biol. Chem.*, **281**, 25097–25109.
45. Commichau, F.M., Herzberg, C., Tripal, P., Valerius, O., and Stülke, J. (2007) A regulatory protein–protein interaction governs glutamate biosynthesis in *Bacillus subtilis*: the glutamate dehydrogenase RocG moonlights in controlling the transcription factor GltC. *Mol. Microbiol.*, **65**, 642–654.
46. Commichau, F.M., Forchhammer, K., and Stülke, J. (2006) Regulatory links between carbon and nitrogen metabolism. *Curr. Opin. Microbiol.*, **9**, 167–172.
47. Krishnan, N. and Becker, D.F. (2005) Characterization of a bifunctional PutA homologue from *Bradyrhizobium japonicum* and identification of an active site residue that modulates proline reduction of the flavin adenine dinucleotide cofactor. *Biochemistry*, **44**, 9130–9139.
48. Titgemeyer, F., Reizer, J., Reizer, A., and Saier, M.H. Jr. (1994) Evolutionary relationships between sugar kinases and transcriptional repressors in bacteria. *Microbiology*, **140**, 2349–2354.
49. Dahl, M.K., Schmiedel, D., and Hillen, W. (1995) Glucose and glucose-6-phosphate interaction with Xyl repressor proteins from *Bacillus* spp. may contribute to regulation of xylose utilization. *J. Bacteriol.*, **177**, 5467–5472.
50. Kwakman, J.H. and Postma, P.W. (1994) Glucose kinase has a regulatory role in carbon catabolite repression in *Streptomyces coelicolor*. *J. Bacteriol.*, **176**, 2694–2698.
51. Doan, T., Martin, L., Zorrilla, S., Chaix, D., Aymerich, S., Labesse, G., and Declerck, N. (2008) A phospho-sugar binding domain homologous to NagB enzymes regulates the activity of the central glycolytic genes repressor. *Proteins*, **71**, 2038–2050.
52. Řezáčová, P., Kožíšek, M., Moy, S.F., Sieglová, I., Joachimiak, A., Machius, M., and Otwinowski, Z. (2008) Crystal structures of the effector-binding domain of repressor central glycolytic gene regulator from *Bacillus subtilis* reveal ligand-induced structural changes upon binding of several glycolytic intermediates. *Mol. Microbiol.*, **69**, 895–910.

53 Loughman, J.A. and Caparon, M.G. (2006) A novel adaptation of aldolase regulates virulence in *Streptococcus pyogenes*. *EMBO J.*, **25**, 5414–5422.

54 Loughman, J.A. and Caparon, M.G. (2007) Comparative functional analysis of the *lac* operons in *Streptococcus pyogenes*. *Mol. Microbiol.*, **64**, 269–280.

55 Deutscher, J., Francke, C., and Postma, P.W. (2006) How phosphotransferase system-related protein phosphorylation regulates carbohydrate metabolism in bacteria. *Microbiol. Mol. Biol. Rev.*, **70**, 939–1031.

56 Kleefeld, A., Ackermann, B., Bauer, J., Krämer, J., and Unden, G. (2009) The fumarate/succinate antiporter DcuB of *Escherichia coli* is a bifunctional protein with sites for regulation of DcuS-dependent gene expression. *J. Biol. Chem.*, **284**, 265–275.

57 Tanous, C., Soutourina, O., Raynal, B., Hullo, M.F., Mervelet, P., Gilles, A.M., Noirot, P., Danchin, A., England, P., and Martin-Verstraete, I. (2008) The CymR regulator in complex with the enzyme CysK controls cysteine metabolism in *Bacillus subtilis*. *J. Biol. Chem.*, **283**, 35551–35560.

58 Ogura, M. and Tanaka, T. (2009) The *Bacillus subtilis* late competence operon *comE* is transcriptionally regulated by *yutB* and under post-transcription initiation control by *comN* (*yrzD*). *J. Bacteriol.*, **191**, 949–958.

19
Regulation of Carbohydrate Utilization by Phosphotransferase System-Mediated Protein Phosphorylation

Boris Görke and Birte Reichenbach

19.1
Introduction

One of the characteristics of all living organisms is their ability to sense environmental conditions and to respond to them. The information about the carbon source supply is of prime importance for all organisms, including bacteria. This information has several dimensions. (i) If carbon sources are scarce, bacteria can detect gradients of nutrient concentrations and move to places of better nutrient supply. (ii) Most bacteria can utilize a plethora of different carbon sources. However, the functions required for the utilization of a particular carbon source are usually only expressed when this substrate is available in the environment. Substrate-dependent control of gene expression can be achieved by a variety of mechanisms, including control of transcription initiation, attenuation, and translational control. Regulation of the *Escherichia coli lac* operon by the availability of lactose may provide the paradigm for this kind of control [1]. (iii) In their natural habitats bacteria often encounter a mixture of different carbon sources that can potentially be used. In this case, many bacteria are able to choose those carbon sources that permit the most rapid growth and promise the best success in competition with other organisms. For many bacteria glucose is the preferred carbon source [2]. During utilization of glucose, the genes required for the utilization of secondary substrates are usually not expressed and pre-existing enzymes are often inactivated to prevent the waste of resources of the cell. This phenomenon is referred to as carbon catabolite repression (CCR) [3].

Research revealed that the phosphoenolpyruvate (PEP)dependent sugar phosphotransferase system (PTS) represents the central processing unit for regulation of carbohydrate utilization in many bacteria [4]. The PTS is a multiprotein phosphorelay system that couples transport of carbohydrates across the cytoplasmic membrane with their simultaneous phosphorylation (Figure 19.1). The phosphorylation status of the PTS proteins is monitored and used for signaling purposes. Depending on their phosphorylation state, proteins of the PTS phosphorylate or interact with target proteins and control their activities. The PTS links carbohydrate availability and flux through metabolic pathways with the global control of gene expression.

Bacterial Signaling. Edited by Reinhard Krämer and Kirsten Jung
Copyright © 2010 WILEY-VCH Verlag GmbH & Co. KGaA, Weinheim
ISBN: 978-3-527-32365-4

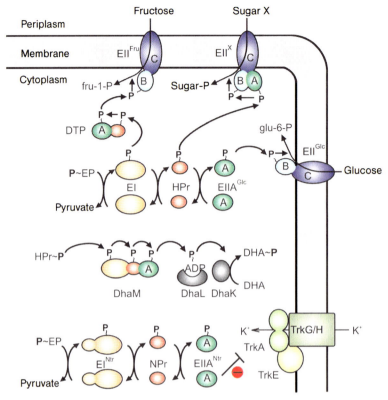

Figure 19.1 PTSs in *Escherichia coli*. The canonical PTS (top) is dedicated to sugar uptake. The two general phosphotransferases EI and HPr transfer phosphoryl groups from PEP to the IIA domains of various carbohydrate-specific EII transporters. The EIIs phosphorylate their substrates during transport across the membrane. As an exception, the fructose-specific EII^{Fru} receives the phosphoryl groups from the diphosphoryl transfer protein (DTP) that contains an HPr-like domain devoted exclusively to transport of fructose. In addition to the canonical PTS, two paralogous PTS are present, which carry out metabolic and regulatory functions, and have no roles in sugar transport. The DhaM protein consists of fused EI-, HPr- and IIA-like domains (middle). DhaM receives phosphoryl groups from HPr and transfers them to the dihyroxyacetone kinase, which consists of the two subunits DhaL and DhaK. The phosphoryl groups are ultimately used to phosphorylate dihydroxyacetone, which is toxic for the cell. The Ntr-PTS consists of EI^{Ntr}, NPr, and EIIA^{Ntr}, which compose a PEP-dependent phosphorylation chain working in parallel to the canonical PTS (bottom). Nonphosphorylated EIIA^{Ntr} binds to subunit TrkA of the Trk K$^+$ transporter and thereby inhibits K$^+$ uptake.

19.2
Unique Features of the Bacterial PTS

The PTS represents the predominant carbohydrate uptake system in many bacteria. The phosphorylation cascade is initiated by Enzyme I (EI), which autophosphorylates with PEP and subsequently transfers the phosphoryl group to the His15 residue within the histidine protein (HPr). EI and HPr serve as general energy couplers for the various membrane-bound carbohydrate-specific Enzyme II (EII) transporters. Most bacteria encode multiple EIIs in their genomes – *Escherichia coli* possesses 21 of these transporters, each dedicated to the uptake of a specific substrate [5]. Usually, the EIIs consist of three domains – HPr delivers the phosphoryl groups to a histidine residue within the IIA domain; subsequently, the phosphoryl group is transferred to a residue in the IIB domain and from there to the sugar during uptake through the membrane-spanning domain IIC (Figure 19.1; for reviews, see [4, 6]).

The PTS is present in many bacteria, but absent in eukaryotes and so far only one archaeal species has been reported to contain a PTS [7]. Phylogenetically, the PTS represents a mosaic system. The EIIs are grouped into seven families, which are not related to each other and have evolved from different sources [6]. The evolution of the PTS also involved extensive horizontal gene transfer events, at least in enteric bacteria such as *E. coli* [7]. Nevertheless, HPr uses the same interaction surface to interact with a large number of unrelated proteins in its host. A signature motif important for carrying out all these interactions has recently been identified for *E. coli* HPr [8].

The PTS was discovered in 1964 as a sugar transport system in *E. coli* [9]. Later, it was recognized to be a global signal transduction system controlling a large number of physiological processes in the cell; notably, it was found to be the key player in CCR [10]. Recent comparative genome analyses suggest that signal transduction might even represent the primary function of the PTS and that its transport functions evolved later [11]. This explains why some bacteria, such as species from the genera *Chlamydia* and *Legionella*, possess the general phosphotransferases EI and HPr, but lack functional EIIs.

What is the molecular basis for signaling by the PTS? It is the phosphorylation status of the PTS proteins that provides the signal(s). Two features of the PTS are crucial in this respect. (i) The high-energy transfer potential of PEP is preserved during phosphoryl transfer between PTS proteins. Therefore, only the transfer of the phosphoryl moiety to the sugar is irreversible. (ii) Autophosphorylation of EI is the rate-limiting step for the phosphorylation of all other PTS proteins [12]. As a result, all proteins of the PTS will preferentially be dephosphorylated when a substrate of the PTS is taken up at high rates and/or when the PEP: pyruvate ratio is low in the cell. In contrast, absence of a PTS substrate and a high PEP: pyruvate ratio favors phosphorylation of the PTS proteins [13]. Two fundamental principles in the mechanism of signal transfer by PTS proteins can be discriminated – either PTS proteins phosphorylate and thereby modulate the activities of target proteins or, alternatively, PTS proteins form complexes with their partners and alter their activities by interaction.

19.3
Phosphorylation of the IIA^{Glc} Subunit of the Glucose Transporter Triggers Global CCR in Enteric Bacteria

In *E. coli* and other enteric bacteria, the soluble IIA^{Glc} subunit of the glucose-specific EII is the central player in sensing, transmitting, and executing CCR (Figure 19.2A;

(A)

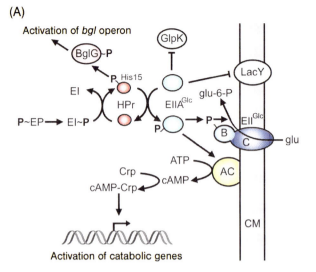

(B)

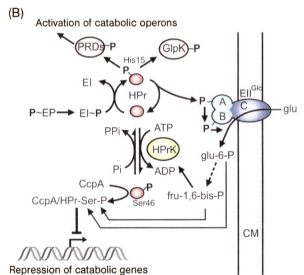

Figure 19.2 Mechanisms of CCR in E. coli (A) and B. subtilis (B). (A) In E. coli the EIIA^{Glc} domain of the glucose transporter is the key player in CCR. In the absence of rapidly metabolizable carbohydrates, EIIA^{Glc} is predominantly phosphorylated. The interaction with EIIA^{Glc}~P stimulates the activity of the adenylate cyclase (AC), resulting in the increased synthesis of cAMP and the formation of cAMP–CRP complexes. Binding of cAMP–CRP to its recognition sites on the DNA activates the expression of secondary catabolic genes. In the presence of preferred carbohydrates, EIIA^{Glc} is preferentially nonphosphorylated and activation of cAMP–CRP-dependent genes cannot occur. In addition, EIIA^{Glc} directly binds and inhibits secondary carbohydrate transporters and metabolic enzymes such as the lactose permease LacY and the glycerol kinase GlpK. HPr contributes to CCR by phosphorylating and activating BglG – the antiterminator protein that controls the β-glucoside (bgl) genes in E. coli. (B) In B. subtilis, CCR is carried out by HPr. Conditions of good carbon source supply increase the intracellular FBP concentration,

for reviews, see [2–4]). IIA^{Glc} binds to the C-terminal domain of adenylate cyclase, which catalyzes synthesis of the global messenger molecule cAMP (see also Chapter 20). In the absence of preferred carbon sources, IIA^{Glc} is predominantly phosphorylated. Exclusively phosphorylated IIA^{Glc} protein is able to activate adenylate cyclase [14]. As a result, cAMP is synthesized and binds to the global transcription regulatory protein cAMP receptor protein (CRP). The cAMP–CRP complex activates expression of numerous catabolic genes and operons. In the presence of preferred carbohydrates, IIA^{Glc} is preferentially unphosphorylated and unable to activate adenylate cyclase. Hence, little cAMP–CRP is formed and systems for the utilization of secondary carbon sources are not expressed. For more details on the role of cAMP in bacterial signaling, see Chapter 20.

19.4
A Second Key Mechanism of CCR: Phosphorylation of IIA^{Glc} Controls Inducer Exclusion in Enteric Bacteria

In addition to its role in the activation of cAMP synthesis, IIA^{Glc} exerts CCR via another pathway – nonphosphorylated IIA^{Glc} directly binds and inhibits various transporters and enzymes involved in the uptake and metabolism of secondary carbon sources (Figure 19.2A). This was demonstrated for the lactose permease LacY, the raffinose permease MelB, the MalK subunit of the maltose transporter, and the metabolic enzyme glycerol kinase GlpK [4]. In addition, galactose utilization is also subject to inducer exclusion. As a result, the intracellular inducers that are necessary for the expression of the genes and operons required for the utilization of these sugars are not formed, and the respective systems are repressed. Therefore, this mechanism has been termed "inducer exclusion." Interestingly, nonphosphorylated IIA^{Glc} binds its target proteins only when the corresponding substrate is available. Binding of the carbohydrate induces structural changes in the transporter that allow the interaction with IIA^{Glc} [15]. This mechanism ensures that IIA^{Glc} only binds to its target when the substrate of the target is present (i.e., useless binding is prevented). In the case of LacY, IIA^{Glc} binds the "outward conformation" of the permease and thereby prevents the structural switch to the "inward conformation" [16].

◀

which activates the kinase activity of HPrK/P. This results in phosphorylation of HPr at its regulatory site, Ser46. HPr(Ser∼P) serves as a cofactor for the global regulatory protein CcpA. The CcpA–HPr(Ser∼P) complex binds to the DNA and represses numerous secondary catabolic genes. Direct binding of FBP and glucose-6-phosphate stimulates this complex formation. Under conditions of poor carbon source supply, the FBP level is low and dephosphorylation of HPr at Ser46 by HPrK/P prevails. Therefore, many catabolic systems are derepressed. However, catabolic systems controlled by PRD-containing regulators require in addition phosphorylation of the respective regulator by HPr(His∼P) for their expression. Similarly, the glycerol kinase GlpK requires HPr(His∼P)-dependent phosphorylation for activity.

The carbohydrate utilization systems that are subject to IIA^{Glc}-exerted inducer exclusion are also regulated by cAMP–CRP at the transcriptional level (see also Chapter 20). CCR of catabolic systems by the simultaneous action of multiple independent mechanisms seems to be a widespread principle. Other catabolic systems are controlled by PTS-mediated phosphorylation of transcriptional regulatory proteins in addition to the cAMP–CRP-mediated CCR (see Section 19.9).

In addition to glucose, which is a substrate of the PTS, many other carbohydrates exert CCR. In *E. coli*, other substrates that are taken up by the PTS at high rates also cause CCR, because their transport leads to a drain and thereby limitation of the phosphoryl groups available within the PTS. Consequently, IIA^{Glc} prevails in its nonphosphorylated form [13]. Moreover, some non-PTS substrates also cause CCR by affecting the PEP: pyruvate ratio in the cell. The metabolism of preferred carbohydrates, which allow the fastest growth, generates a low PEP: pyruvate ratio in the cell, leading to preferential de-phosphorylation of the PTS proteins including IIA^{Glc} [13, 17].

19.5
Phosphorylation of Ser46 of HPr Triggers CCR in Low-GC Gram-Positive Bacteria

CCR also exists in the Gram-positive soil bacterium *Bacillus subtilis* and other Firmicutes. However, the underlying mechanisms are completely different from those operative in *E. coli*. cAMP and IIA^{Glc} have no regulatory role in this case. In contrast, the HPr protein has a key role in CCR (Figure 19.2B). In Firmicutes, HPr can additionally be phosphorylated at a second site, Ser46 [18]. A novel type of a metabolite-controlled ATP-dependent kinase (HPr-kinase/phosphorylase (HPrK/P); see Section 19.6) catalyzes this reaction under conditions of good carbon source supply [19, 20]. Therefore, HPr can exist in four different states: nonphosphorylated, doubly phosphorylated, or phosphorylated at one of both sites. Exclusively Ser46-phosphorylated HPr is active in CCR [21, 22].

HPr(Ser~P) binds to CcpA (*c*atabolite *c*ontrol *p*rotein *A*) – a member of the LacI/GalR repressor protein family [23]. This interaction induces a conformational switch in CcpA allowing binding of the HPr(Ser~P)/CcpA complex to its target sites on the DNA – the *cre* (*c*atabolite *r*esponsive *e*lements) [24]. Binding to these *cre* sites, which are present in the promoter regions of many catabolic genes, usually represses promoter activity. Transcriptome analyses demonstrated that CcpA is a global regulator and controls the expression of 5–10% of the genome [25]. The structures of CcpA, either alone or in complex with HPr(Ser~P), have been solved [26]. The interactions with HPr(Ser~P) induce a rotational movement in the CcpA dimer that brings the DNA-binding domains into a conformation competent in *cre* binding. Both the nonphosphorylated His15 as well as the phosphorylated Ser46 residues in HPr participate in interaction with CcpA. These interactions explain why the other HPr forms are unable to serve as a cofactor for CcpA.

19.6
Phosphorylation of HPr by the Bifunctional Kinase/Phosphorylase Links CCR to the Metabolic State of the Cell in Gram-Positive Bacteria

Phosphorylation and dephosphorylation of HPr at its Ser46 residue are both catalyzed by a single protein – the bifunctional HPrK/P [27]. PrpC, a protein phosphatase of the protein phosphatase 2C family, may also contribute to dephosphorylation of HPr [28]. Although HPrK/P contains a classical Walker A nucleotide-binding motif and uses ATP as phosphoryl group donor, it has no homology to eukaryotic kinases. The mechanism of dephosphorylation of HPr by HPrK/P is rather unusual – unlike classical phosphatases, HPrK/P uses inorganic phosphate as a substrate and produces pyrophosphate. Fructose-1,6-bisphosphate (FBP) has been identified as a key metabolite stimulating the kinase activity. *In vivo*, mutations that prevent the formation of FBP result in a complete relief from CcpA-mediated CCR [29]. *In vitro*, kinase activity of HPrK/P is low below 1 mM FBP and sharply increases at higher concentrations reaching a plateau at 5 mM FBP [30]. FBP has still another important role in CCR – FBP and glucose-6-phosphate directly enhance the interaction between HPr(Ser~P) and CcpA (Figure 19.2B [31]).

In conclusion, the CcpA-mediated CCR pathway in *B. subtilis* seems to be triggered by flux through glycolysis and the metabolic state of the cell [22]. Unlike in *E. coli*, PTS transport activity has no direct effect on the global CCR mechanism, although phosphorylation of HPr plays a key role.

19.7
HPr Controls Inducer Exclusion in Low-GC Gram-Positive Bacteria

Inducer exclusion has also been described in Firmicutes. In these bacteria, HPr carries out this regulation. In *Lactobacillus brevis*, HPr(Ser~P) binds to the galactose permease and thereby inhibits uptake of galactose, when glucose is also present. In *Lactobacillus casei* and *Lactococcus lactis*, HPr(Ser~P) mediates inducer exclusion of the maltose and ribose uptake systems (reviewed in [4]).

In other cases, inducer exclusion involves HPr(His~P)-dependent phosphorylation. In *B. subtilis* and other Firmicutes, the glycerol kinase GlpK requires HPr(His~P)-dependent phosphorylation for activity (for a recent review, see [32]). Therefore, in the presence of glucose or another PTS substrate, HPr(His~P) will preferentially be used for PTS sugar transport and GlpK cannot be activated. This prevents the formation of glycerol-3-phosphate, which is the specific inducer of the glycerol utilization genes. Consequently, utilization of glycerol is repressed by glucose and other PTS sugars. The lactose transport proteins of many *Streptococci* and *Lactobacilli* contain a IIA-like PTS domain although they are no PTS transporters. This IIA domain exclusively serves regulatory functions. Phosphorylation of this domain by HPr(His~P) activates the lactose transporter LacS in *Streptococcus thermophilus* [33]. Therefore, similar to GlpK, LacS is less active in the presence of a preferred PTS carbohydrate.

19.8
Control of Transcription Regulators by EII

In addition to their transport functions, many EII of the PTS directly control the activities of transcription regulators and thereby gene expression. In *E. coli*, the Mlc protein is the master regulator controlling utilization of glucose [34] (see also Chapter 12). In the absence of glucose, Mlc represses the gene encoding the glucose-specific EII. When glucose becomes available, the IIB domain of EII^{Glc} is predominantly nonphosphorylated and able to bind and sequester Mlc at the membrane, resulting in derepression (see also Chapters 18 and 12). It should be emphasized, however, that membrane sequestration of a regulatory protein is not sufficient to prevent its binding to the DNA. For example, it was shown that the Lac repressor retains its capability to repress its target on the chromosome when artificially anchored to the membrane [35]. In fact, upon interaction with the glucose transporter, Mlc looses its structural flexibility that is required for efficient DNA binding [36].

In addition to the control of Mlc by EII^{Glc}, there is a widespread family of transcription regulatory proteins controlled by reversible PTS-mediated phosphorylation. These proteins contain two conserved domains, which are phosphorylated by PTS proteins and have therefore been named PTS regulation domains (phosphotransferase system regulation domain (PRDs) (for reviews, see [4, 32, 37]). The PRDs are homologous to each other and contain two conserved histidines, which are phosphorylated by PTS proteins, respectively (Figure 19.3). There are two groups of PRD-containing regulators. RNA-binding antiterminator proteins of the BglG/SacY family control a riboswitch, which is located in the 5′-leader region of the transcript that encompasses the target genes. The best-studied examples are BglG and LicT, which control the expression of β-glucoside utilization genes in *E. coli* and *B. subtilis*, respectively. In addition, there are DNA-binding transcription activators such as LevR, which controls fructose utilization in *B. subtilis*. The corresponding RNA- or DNA-binding activities are always located at the N-termini followed by the two PRD domains (Figure 19.3). Each PRD regulator positively controls the expression of an assigned EII complex of the PTS, which in turn negatively regulates the PRD regulator. In the absence of substrate, the IIB domain of the transporter phosphorylates the PRD regulator at its histidine residues in PRD1. This prevents the regulators from dimerization, which is a prerequisite for their activity. When substrate becomes available, the phosphoryl groups are transferred back to the EII and the PRD regulator forms dimers, which activate expression of the assigned gene or operon. This elegant induction mechanism was demonstrated for the first time for the *E. coli* BglG protein and has been recapitulated later for many other PRD regulators [37, 38].

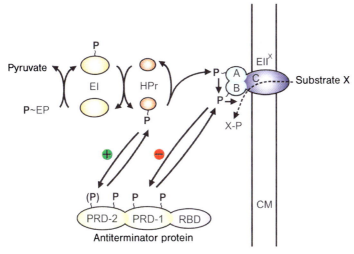

Figure 19.3 Activity control of antiterminator proteins of the BglG/SacY family by antagonistically acting PTS-mediated phosphorylations. Proteins of the BglG/SacY family consist of an RNA binding-domain (RBD) and two PTS regulatory domains, PRD1 and PRD2, which are phosphorylated by PTS proteins. Each antiterminator protein controls the expression of genes dedicated to the uptake and metabolism of a certain PTS sugar. The cognate EII phosphorylates the antiterminator at conserved histidine residues in PRD1 when no substrate is available. These phosphorylations inactivate the antiterminator protein and the corresponding genes are not expressed. When substrate becomes available, the phosphoryl groups are transferred back to the EII, which is a prerequisite for the aniterminators' activity. In addition, many but not all of the antiterminator proteins require HPr(His)-dependent phosphorylation at histidine(s) in PRD2 for activity. This mechanism downregulates the activity of the antiterminator protein when a preferred PTS sugar, such as glucose, becomes available in addition to the cognate substrate.

19.9
Catabolite Control of PRD-Containing Regulators by HPr(His~P)-Mediated Phosphorylation

In addition to negative regulation by their assigned EIIs, many PRD regulators are subject to a second level of regulation: They require HPr-dependent phosphorylation at the conserved histidines in PRD2 for their activity (Figure 19.3), as demonstrated for antiterminators BglG and LicT [39–41]. Hence, these proteins are only active when nonphosphorylated at PRD1 and phosphorylated at PRD2. The HPr-mediated phosphorylation of PRD regulators constitutes a CCR mechanism that is independent of the global CCR [3]. When a preferred PTS sugar like glucose becomes available in addition to the inducer of a PRD-regulator, the phosphoryl groups from HPr are drained towards transport of the preferred substrate. Therefore, the PRD2 cannot be phosphorylated and the PRD regulator is inactive, leading to repression of its target genes. Consequently, this type of regulation allows the hierarchical use of PTS sugars.

Recently, it was reported that the master regulator of virulence in *Bacillus anthracis*, AtxA, is likewise subject to PTS-mediated phosphorylations at histidine residues [42]. The histidine residues are located in a region exhibiting weak homology with two iterative PRDs. A similar region is also present in virulence regulators of other species such as Mga in *Streptococcus pyogenes*. It remains to be shown whether these regulators may constitute a novel family of PTS-regulated transcription factors (reviewed in [32]).

19.10
PTS-Dependent Regulation of Chemotaxis

Both model bacteria, *E. coli* and *B. subtilis*, exhibit chemotaxis towards PTS sugars. It turned out that PTS transport activity is required for the chemotactic response. In both species, the chemotactic response to PTS substrates depends on the phosphorylation state of EI. In *E. coli*, nonphosphorylated EI directly interacts with the CheA histidine kinase that relays the chemotactic signal to the flagellar motor. This interaction inhibits CheA autokinase activity [43], causing less tumbling and therefore movement towards the PTS substrate. In *B. subtilis*, chemotaxis to PTS sugars requires the methyl-accepting receptor protein McpC, which presumably directly interacts with nonphosphorylated EI [44]. For more details, see Chapter 9.

19.11
Regulatory Functions of Paralogous PTSs

Many bacteria possess several HPr and EI homologs, which form phosphoryl transfer chains working in parallel to the canonical PTS. In most cases, these PTS paralogs have regulatory functions rather than a role in sugar transport.

An interesting case is Crh (*c*atabolite *r*epression *H*Pr) – an HPr paralog present in *Bacillus* species. This protein contains the regulatory Ser46 site, but lacks His15, suggesting that it exclusively serves regulatory functions. However, although Crh is phosphorylated by HPrK/P and interacts with CcpA [45], it is dispensable for CCR, whereas HPr is not [22]. This is reflected by the much weaker synthesis level of Crh in comparison to HPr [46]. A recent study suggested that Crh(Ser $\sim$ P) interacts with the central glycolytic enzyme glyceraldehyde-3-phosphate dehydrogenase (GapA) and thereby inhibits its enzymatic activity [47].

E. coli contains four paralogous PTS phosphoryl transfer chains. Recent research gained some insight into the functions of two of them – the dihydroxy-acetone (Dha) PTS and the nitrogen-related (Ntr) PTS (Figure 19.1). During glycolysis, Dha is generated as a byproduct, which is toxic for the cell. Detoxification requires the phosphorylation of Dha by the Dha kinase, which is in turn phosphorylated by DhaM. The DhaM protein consists of a IIA^{Man}-like domain followed by an HPr-like domain and a C-terminal domain with homology to EI (Figure 19.1). DhaM receives the phosphoryl groups from the canonical phosphotransferases EI and HPr. This mechanism couples Dha kinase activity to the phosphorylation state of the canonical PTS, and thereby to sugar transport and flux through glycolysis [48].

The Ntr-PTS consists of EI^{Ntr}, which autophosphorylates with PEP and transfers the phosphoryl groups to the HPr homolog NPr that finally phosphorylates the IIA-like protein IIA^{Ntr} [49, 50]. No final acceptor for the phosphoryl groups delivered to IIA^{Ntr} is known. The Ntr-PTS has been implicated in the regulation of nitrogen metabolism, because it colocalizes with the *rpoN* gene, which encodes an alternative sigma factor that controls the expression of many nitrogen-related genes. However, there is no direct evidence for a role in nitrogen regulation. In contrast, recently it was reported that nonphosphorylated IIA^{Ntr} interacts with subunit TrkA of the low-affinity Trk-K^+ transporter and thereby inhibits K^+ uptake (Figure 19.1, bottom [51]). Interestingly, this regulation is analogous to the mechanism by which IIA^{Glc} brings about inducer exclusion [52]. *In vivo* and *in vitro* experiments demonstrated cross-phosphorylation between proteins of the Ntr-PTS and the canonical PTS [49, 50], suggesting a regulatory connection between carbohydrate metabolism and K^+ transport.

Acknowledgments

Research in the authors' laboratory is supported by grants of the Deutsche Forschungsgemeinschaft to B.G. We are grateful to Jörg Stülke for support. B.R. is a fellow of the "Studienstiftung des Deutschen Volkes."

References

1 Müller-Hill, B. (1996) *The Lac Operon: A Short History of a Genetic Paradigm*, de Gruyter, Berlin.

2 Jahreis, K., Pimentel-Schmitt, E.F., Brückner, R., and Titgemeyer, F. (2008) Ins and outs of glucose transport systems in eubacteria. *FEMS Microbiol. Rev.*, **2**, 891–907.

3 Görke, B. and Stülke, J. (2008) Carbon catabolite repression in bacteria: many ways to make the most out of nutrients. *Nat. Rev. Microbiol.*, **6**, 613–624.

4 Deutscher, J., Francke, C., and Postma, P.W. (2006) How phosphotransferase system-related protein phosphorylation regulates carbohydrate metabolism in bacteria. *Microbiol. Mol. Biol. Rev.*, **70**, 939–1031.

5 Tchieu, J.H., Norris, V., Edwards, J.S., and Saier, M.H. Jr. (2001) The complete phosphotransferase system in *Escherichia coli*. *J. Mol. Microbiol. Biotechnol.*, **3**, 329–346.

6 Lengeler, J.W., Jahreis, K., and Wehmeier, U.F. (1994) Enzymes II of the phosphoenol pyruvate-dependent phosphotransferase systems: their structure and function in carbohydrate transport. *Biochim. Biophys. Acta*, **1188**, 1–28.

7 Comas, I., Gonzalez-Candelas, F., and Zuniga, M. (2008) Unraveling the evolutionary history of the phosphoryl-transfer chain of the phosphoenol-pyruvate: phosphotransferase system through phylogenetic analyses and genome context. *BMC Evol. Biol.*, **8**, 147.

8 Reichenbach, B., Breustedt, D.A., Stülke, J., Rak, B., and Görke, B. (2007) Genetic dissection of specificity determinants in the interaction of HPr with enzymes II of the bacterial phosphoenolpyruvate: sugar phosphotransferase system in *Escherichia coli*. *J. Bacteriol.*, **189**, 4603–4613.

9 Kundig, W., Ghosh, S., and Roseman, S. (1964) Phosphate bound to histidine in a protein as an intermediate in a novel

phospho-transferase system. *Proc. Natl. Acad. Sci. USA*, **52**, 1067–1074.

10. Saier, M.H. and Roseman, S. (1972) Inducer exclusion and repression of enzyme synthesis in mutants of *Salmonella typhimurium* defective in enzyme I of the phosphoenolpyruvate: sugar phosphotransferase system. *J. Biol. Chem.*, **247**, 972–975.

11. Cases, I., Velazquez, F., and de Lorenzo, V. (2007) The ancestral role of the phosphoenolpyruvate–carbohydrate phosphotransferase system (PTS) as exposed by comparative genomics. *Res. Microbiol.*, **158**, 666–670.

12. Weigel, N., Kukuruzinska, M.A., Nakazawa, A., Waygood, E.B., and Roseman, S. (1982) Sugar transport by the bacterial phosphotransferase system. Phosphoryl transfer reactions catalyzed by enzyme I of *Salmonella typhimurium*. *J. Biol. Chem.*, **257**, 14477–14491.

13. Bettenbrock, K., Sauter, T., Jahreis, K., Kremling, A., Lengeler, J.W., and Gilles, E.D. (2007) Correlation between growth rates, EIIA^{Crr} phosphorylation, and intracellular cyclic AMP levels in *Escherichia coli* K-12. *J. Bacteriol.*, **189**, 6891–6900.

14. Park, Y.H., Lee, B.R., Seok, Y.J., and Peterkofsky, A. (2006) In vitro reconstitution of catabolite repression in *Escherichia coli*. *J. Biol. Chem.*, **281**, 6448–6454.

15. Smirnova, I., Kasho, V., Choe, J.Y., Altenbach, C., Hubbell, W.L., and Kaback, H.R. (2007) Sugar binding induces an outward facing conformation of LacY. *Proc. Natl. Acad. Sci. USA*, **104**, 16504–16509.

16. Sondej, M., Weinglass, A.B., Peterkofsky, A., and Kaback, H.R. (2002) Binding of enzyme IIA^{Glc}, a component of the phosphoenolpyruvate: sugar phosphotransferase system, to the *Escherichia coli* lactose permease. *Biochemistry*, **41**, 5556–5565.

17. Hogema, B.M., Arents, J.C., Bader, R., Eijkemans, K., Yoshida, H., Takahashi, H., Aiba, H., and Postma, P.W. (1998) Inducer exclusion in *Escherichia coli* by non-PTS substrates: the role of the PEP to pyruvate ratio in determining the phosphorylation state of enzyme IIA^{Glc}. *Mol. Microbiol.*, **30**, 487–498.

18. Deutscher, J. and Saier, M.H. Jr. (1983) ATP-dependent protein kinase-catalyzed phosphorylation of a seryl residue in HPr, a phosphate carrier protein of the phosphotransferase system in *Streptococcus pyogenes*. *Proc. Natl. Acad. Sci. USA*, **80**, 6790–6794.

19. Galinier, A., Kravanja, M., Engelmann, R., Hengstenberg, W., Kilhoffer, M.C., Deutscher, J., and Haiech, J. (1998) New protein kinase and protein phosphatase families mediate signal transduction in bacterial catabolite repression. *Proc. Natl. Acad. Sci. USA*, **95**, 1823–1828.

20. Reizer, J., Hoischen, C., Titgemeyer, F., Rivolta, C., Rabus, R., Stülke, J., Karamata, D., Saier, M.H. Jr., and Hillen, W. (1998) A novel protein kinase that controls carbon catabolite repression in bacteria. *Mol. Microbiol.*, **27**, 1157–1169.

21. Vadeboncoeur, C., Brochu, D., and Reizer, J. (1991) Quantitative determination of the intracellular concentration of the various forms of HPr, a phosphocarrier protein of the phosphoenolpyruvate: sugar phosphotransferase system in growing cells of oral *streptococci*. *Anal. Biochem.*, **196**, 24–30.

22. Singh, K.D., Schmalisch, M.H., Stülke, J., and Görke, B. (2008) Carbon catabolite repression in *Bacillus subtilis*: a quantitative analysis of repression exerted by different carbon sources. *J. Bacteriol.*, **190**, 7275–7284.

23. Deutscher, J., Kuster, E., Bergstedt, U., Charrier, V., and Hillen, W. (1995) Protein kinase-dependent HPr/CcpA interaction links glycolytic activity to carbon catabolite repression in Gram-positive bacteria. *Mol. Microbiol.*, **15**, 1049–1053.

24. Jones, B.E., Dossonnet, V., Kuster, E., Hillen, W., Deutscher, J., and Klevit, R.E. (1997) Binding of the catabolite repressor protein CcpA to its DNA target is regulated by phosphorylation of its corepressor HPr. *J. Biol. Chem.*, **272**, 26530–26535.

25. Blencke, H.M., Homuth, G., Ludwig, H., Mäder, U., Hecker, M., and Stülke, J. (2003) Transcriptional profiling of gene expression in response to glucose in *Bacillus subtilis*: regulation of the central

metabolic pathways. *Metab. Eng.*, **5**, 133–149.

26 Schumacher, M.A., Allen, G.S., Diel, M., Seidel, G., Hillen, W., and Brennan, R.G. (2004) Structural basis for allosteric control of the transcription regulator CcpA by the phosphoprotein HPr-Ser46-P. *Cell*, **118**, 731–741.

27 Kravanja, M., Engelmann, R., Dossonnet, V., Bluggel, M., Meyer, H.E., Frank, R., Galinier, A., Deutscher, J., Schnell, N., and Hengstenberg, W. (1999) The *hprK* gene of *Enterococcus faecalis* encodes a novel bifunctional enzyme: the HPr kinase/phosphatase. *Mol. Microbiol.*, **31**, 59–66.

28 Singh, K.D., Halbedel, S., Görke, B., and Stülke, J. (2007) Control of the phosphorylation state of the HPr protein of the phosphotransferase system in *Bacillus subtilis*: implication of the protein phosphatase PrpC. *J. Mol. Microbiol. Biotechnol.*, **13**, 165–171.

29 Nihashi, J. and Fujita, Y. (1984) Catabolite repression of inositol dehydrogenase and gluconate kinase syntheses in *Bacillus subtilis*. *Biochim. Biophys. Acta*, **798**, 88–95.

30 Jault, J.M., Fieulaine, S., Nessler, S., Gonzalo, P., Di Pietro, A., Deutscher, J., and Galinier, A. (2000) The HPr kinase from *Bacillus subtilis* is a homo-oligomeric enzyme which exhibits strong positive cooperativity for nucleotide and fructose 1,6-bisphosphate binding. *J. Biol. Chem.*, **275**, 1773–1780.

31 Schumacher, M.A., Seidel, G., Hillen, W., and Brennan, R.G. (2007) Structural mechanism for the fine-tuning of CcpA function by the small molecule effectors glucose 6-phosphate and fructose 1,6-bisphosphate. *J. Mol. Biol.*, **368**, 1042–1050.

32 Görke, B. and Deutscher, J. (2007) The regulatory functions of histidyl-phosphorylated HPr in bacilli, in *Global Regulatory Networks in Bacillus Subtilis* (ed. Y. Fujita), Transworld Research Network, Trivandrum, pp. 1–37.

33 Gunnewijk, M.G. and Poolman, B. (2000) HPr(His~P)-mediated phosphorylation differently affects counterflow and proton motive force-driven uptake via the lactose transport protein of *Streptococcus thermophilus*. *J. Biol. Chem.*, **275**, 34080–34085.

34 Plumbridge, J. (2002) Regulation of gene expression in the PTS in *Escherichia coli*: the role and interactions of Mlc. *Curr. Opin. Microbiol.*, **5**, 187–193.

35 Görke, B., Reinhardt, J., and Rak, B. (2005) Activity of Lac repressor anchored to the *Escherichia coli* inner membrane. *Nucleic Acids Res.*, **33**, 2504–2511.

36 Nam, T.W., Jung, H.I., An, Y.J., Park, Y.H., Lee, S.H., Seok, Y.J., and Cha, S.S. (2008) Analyses of Mlc–IIB^{Glc} interaction and a plausible molecular mechanism of Mlc inactivation by membrane sequestration. *Proc. Natl. Acad. Sci. USA*, **105**, 3751–3756.

37 Stülke, J., Arnaud, M., Rapoport, G., and Martin-Verstraete, I. (1998) PRD – a protein domain involved in PTS-dependent induction and carbon catabolite repression of catabolic operons in bacteria. *Mol. Microbiol.*, **28**, 865–874.

38 Schnetz, K. and Rak, B. (1990) Beta-glucoside permease represses the *bgl* operon of *Escherichia coli* by phosphorylation of the antiterminator protein and also interacts with glucose-specific enzyme III, the key element in catabolite control. *Proc. Natl. Acad. Sci. USA*, **87**, 5074–5078.

39 Görke, B. and Rak, B. (1999) Catabolite control of *Escherichia coli* regulatory protein BglG activity by antagonistically acting phosphorylations. *EMBO J.*, **18**, 3370–3379.

40 Görke, B. (2003) Regulation of the *Escherichia coli* antiterminator protein BglG by phosphorylation at multiple sites and evidence for transfer of phosphoryl groups between monomers. *J. Biol. Chem.*, **278**, 46219–46229.

41 Tortosa, P., Declerck, N., Dutartre, H., Lindner, C., Deutscher, J., and Le Coq, D. (2001) Sites of positive and negative regulation in the *Bacillus subtilis* antiterminators LicT and SacY. *Mol. Microbiol.*, **41**, 1381–1393.

42 Tsvetanova, B., Wilson, A.C., Bongiorni, C., Chiang, C., Hoch, J.A., and Perego, M. (2007) Opposing effects of histidine phosphorylation regulate the AtxA virulence transcription factor in *Bacillus anthracis*. *Mol. Microbiol.*, **63**, 644–655.

43 Lux, R., Jahreis, K., Bettenbrock, K., Parkinson, J.S., and Lengeler, J.W. (1995) Coupling the phosphotransferase system and the methyl-accepting chemotaxis protein-dependent chemotaxis signaling pathways of *Escherichia coli*. *Proc. Natl. Acad. Sci. USA*, **92**, 11583–11587.

44 Kristich, C.J., Glekas, G.D., and Ordal, G.W. (2003) The conserved cytoplasmic module of the transmembrane chemoreceptor McpC mediates carbohydrate chemotaxis in *Bacillus subtilis*. *Mol. Microbiol.*, **47**, 1353–1366.

45 Schumacher, M.A., Seidel, G., Hillen, W., and Brennan, R.G. (2006) Phosphoprotein Crh-Ser46-P displays altered binding to CcpA to effect carbon catabolite regulation. *J. Biol. Chem.*, **281**, 6793–6800.

46 Görke, B., Fraysse, L., and Galinier, A. (2004) Drastic differences in Crh and HPr synthesis levels reflect their different impacts on catabolite repression in *Bacillus subtilis*. *J. Bacteriol.*, **186**, 2992–2995.

47 Pompeo, F., Luciano, J., and Galinier, A. (2007) Interaction of GapA with HPr and its homologue, Crh: novel levels of regulation of a key step of glycolysis in *Bacillus subtilis*? *J. Bacteriol.*, **189**, 1154–1157.

48 Gutknecht, R., Beutler, R., Garcia-Alles, L.F., Baumann, U., and Erni, B. (2001) The dihydroxyacetone kinase of *Escherichia coli* utilizes a phosphoprotein instead of ATP as phosphoryl donor. *EMBO J.*, **20**, 2480–2486.

49 Rabus, R., Reizer, J., Paulsen, I., and Saier, M.H. Jr. (1999) Enzyme I^{Ntr} from *Escherichia coli*. A novel enzyme of the phosphoenolpyruvate-dependent phosphotransferase system exhibiting strict specificity for its phosphoryl acceptor NPr. *J. Biol. Chem.*, **274**, 26185–26191.

50 Zimmer, B., Hillmann, A., and Görke, B. (2008) Requirements for the phosphorylation of the *Escherichia coli* $EIIA^{Ntr}$ protein *in vivo*. *FEMS Microbiol. Lett.*, **286**, 96–102.

51 Lee, C.R., Cho, S.H., Yoon, M.J., Peterkofsky, A., and Seok, Y.J. (2007) *Escherichia coli* enzyme IIA^{Ntr} regulates the K^+ transporter TrkA. *Proc. Natl. Acad. Sci. USA*, **104**, 4124–4129.

52 Ninfa, A.J. (2007) Regulation of carbon and nitrogen metabolism: adding regulation of ion channels and another second messenger to the mix. *Proc. Natl. Acad. Sci. USA*, **104**, 4243–4244.

20
cAMP Signaling in Prokaryotes
Knut Jahreis

20.1
Introduction

Many sensory and regulatory mechanisms that are involved in the control of metabolic fluxes have been identified in prokaryotes during the last few decades, especially in the model organisms *Escherichia coli* K-12 and *Salmonella typhimurium*. On the basis of a variety of systematic approaches such as transcriptomics, proteomics, metabolic flux analysis, and mathematical modeling (or, more generally, by systems biology) we are beginning to get an integrated view on how these regulatory systems generate communication networks and how living cells use these networks to manage the regulation of their metabolic balance under various environmental conditions. The focus of this chapter is on one of the best-studied global regulatory systems in *E. coli* and *S. typhimurium* – the cAMP signaling system, which is the key player in so-called "carbon catabolite repression" (CCR) in enteric bacteria.

20.2
CCR – A Short Historical Account

It has been known for a long time that the presence of a rapidly metabolizable carbon source leads to its immediate utilization and is accompanied by a number of regulatory events that prevent the use of other carbon sources of less energetic value. This was first investigated in detail by Monod [1], who described the phenomenon of diauxic growth behavior. He was able to show that *E. coli* primarily chooses glucose when exposed to a nutrient mixture of glucose and sorbitol. Since then, glucose has been the classical "preferred" carbon source, the effect of which was termed the "glucose effect" or, more generally, CCR. This phenomenon has been studied for decades in order to reveal the molecular mechanisms of carbohydrate transport and regulation. In 1965, cAMP was identified in *E. coli*, and it was demonstrated that its cytoplasmic concentration varies during growth on different carbon sources and at different growth rates [2]. Subsequently, one of the first global regulators, the cAMP receptor protein (CRP), was identified. The description of the phenotypic properties of mutants lacking either the cAMP biosynthetic enzyme, adenylate cyclase (AC; gene *cya*), or CRP (gene *crp*) led to the concept that cAMP represents a bacterial second messenger, which, as a coactivator, modulates the activity of CRP. Further investigations revealed that feast growth conditions (e.g., during growth on glucose) correspond to low cAMP levels and strong catabolite repression, whereas famine conditions lead to high cAMP concentrations and to a relief of repression. Furthermore, it was shown that the cAMP–CRP complex is capable of binding to specific sites within many inducible promoters and activates (or rarely inhibits) transcription by a direct interaction with the RNA polymerase or by bending the DNA. All operons and regulons whose transcription is regulated by this epistatic global control system form the *crp* modulon. Most of these genes are involved in carbohydrate transport and their subsequent utilization, thus *cya* and *crp* mutants cannot use many carbon sources (e.g., lactose, maltose, glycerol) that are utilized as growth substrates by the wild-type strain because synthesis of the catabolic enzymes responsible for the metabolism of these compounds cannot be induced. Further experiments showed that the cAMP–CRP complex is also required for flagellum synthesis and other functions not directly related to carbon catabolism, which further underlines the importance of this system (reviewed in [3]). Finally, the isolation and characterization of mutants with defects in the phosphoenolpyruvate (PEP)-dependent sugar phosphotransferase system (PTS) demonstrated the involvement of this complex carbohydrate transport and signal transduction system in the direct modulation of cAMP levels in a physiologically meaningful manner. A regulatory mechanism was proposed, according to which the phosphorylated form of the EIIA^{Glc} of the PTS stimulates the activity of AC (reviewed in [4, 5]). In contrast, transport of glucose by the PTS leads to a quantitative dephosphorylation of EIIA^{Glc}, and thus to a deactivation of AC and to a decreased cAMP level. This model became generally accepted as the most important mechanism of catabolite repression in enteric bacteria. However, several recent findings led to an extension of this textbook model.

20.3
Regulation of Intracellular cAMP Levels: PTS as a Sensor and Signal Transduction System that Modulates AC Activity

Accurate and fast measurements of intracellular cAMP concentrations are difficult to perform. To obtain reproducible results, the metabolism and the cAMP excretion (see Section 20.5) of cells must be stopped immediately, the cells have to be washed extensively to get rid of the large amounts of extracellular cAMP (up to 95% of total cAMP), and cell lyses have to be minimized. Several groups tried various methods for monitoring intracellular cAMP concentrations, and found that intracellular cAMP concentrations vary between 0.4 µM for cells growing on good substrates (e.g., glucose or glucose-6-phosphate) and 3.3 µM (e.g., on the poor substrate lactate) [6–8]. To avoid difficult extraction methods many groups determined changes in extracellular cAMP or total cAMP amounts under various growth conditions, and found a 30- to 100-fold difference between feast and famine conditions (e.g., [2, 9]). However, it is important to stress that the amounts of extracellular cAMP do not always directly correspond to the intracellular cAMP concentrations, for example, because additional factors seem to modulate cAMP excretion from the cell as well as cAMP degradation (reviewed in [3]). An elegant alternative to determine the active cAMP–CRP concentrations in living cells is the use of constitutively expressed, cAMP–CRP-dependent promoters fused to reporter genes in comparison to equivalent promoter–reporter gene fusions with a cAMP–CRP-independent construct [8, 9].

Transcriptional regulation of the *cya* gene encoding the AC has been studied by using operon and protein *cya–lacZ* reporter gene fusions in both *E. coli* and *S. typhimurium* [10–12]. An approximately two- to fourfold variation in *cya* promoter activity was found in both organisms with the cAMP–CRP complex repressing *cya* transcription. Moreover, Jovanovitch observed a ninefold repression of *cya* promoter activity in cells grown in the presence of high levels of exogenic cAMP. Accordingly, *crp* mutants exhibit maximal activity that is insensitive to fluctuations in response to different carbon sources [13]. These observations mean that in cells actively growing under glucose-repressing conditions, *cya* gene expression is maximal. This effect is in the opposite direction of, and is clearly not responsible for, the cAMP decrease in response to glucose or any other good carbon source. Increased *cya* gene expression in wild-type cells under conditions of carbohydrate excess (AC deactivation) presumably serves to preadapt the bacteria in advance for changes in carbohydrate supply. Thus, when the first carbon source (e.g., glucose) becomes exhausted, the bacteria can instantly produce increasing amounts of cAMP and thereby rapidly adapt to the utilization of a new carbon source, since high cAMP concentrations facilitate transcriptional activation of virtually all carbon catabolic operons concerned with the initiation of exogenous carbon utilization.

From the considerations presented in this section, it appears that the range of variation in *cya* expression is much smaller than the variation in the ability of the cell to synthesize cAMP, which indicates a tight regulation of enzyme activity. The first hint for a direct involvement of the PTS in this process came from the observation that *pts* mutants exhibit very low levels of cAMP synthesis. This finding subsequently led to the model of a post-translational control of Cya activity by one particular component of the PTS – the so-called EIIA^{Glc} [14–17]. In order to understand the function of the PTS in carbohydrate sensing and regulation of catabolite repression, we need to take a careful look at the organization of this important carbohydrate transport and signal transduction system (for a detailed current review, see [18]).

The PTS usually consists of two cytoplasmic energy-coupling proteins, Enzyme I (EI, gene *ptsI*) and histidine-containing protein (HPr, gene *ptsH*), and a range of more than 20 different carbohydrate-specific types of Enzyme II (EIIs), which catalyze concomitant carbohydrate transport and phosphorylation (see also Chapters 18 and 19). The first step in the PTS-typical phosphorylation cascade is catalyzed by EI – a PEP-dependent protein kinase. In the case of the glucose PTS (Figure 20.1), the phosphate group is subsequently transferred to HPr, from HPr∼P to the soluble

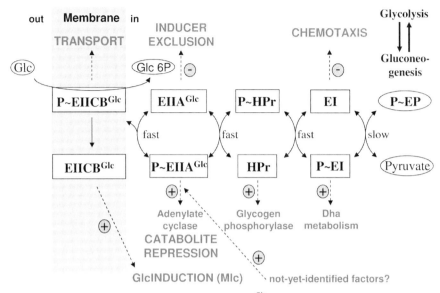

Figure 20.1 The glucose PTS and its targeting subunits and the diversification of information within sensory networks as found within enteric bacteria. The targeting subunits HPr, EIIA^{Glc}, and EIICB^{Glc} carry the information from the CFU "carbon metabolism", as mirrored in the flux of phosphoryl groups through the PTS, to a series of targets in a rapid and transient, although precise way. All functional subunits of the PTS can exist in the phosphorylated or unphosphorylated form, depending on the available carbohydrate and the physiological state of the cell. During fast uptake of a PTS carbohydrate through any EII, EI autophosphorylation appears to be the rate limiting step. Consequently, the amounts of unphosphorylated EI, HPr, EIIA^{Glc}, and EIICB^{Glc} increase transiently. Enteric bacteria use this information to couple the PTS proteins, depending on their degree of phosphorylation, to various key regulator proteins as described in the text. Even HPr has a regulatory function in glycogen metabolism in *E. coli*. Basically, at the onset of the stationary phase when cAMP levels increase, the expression of the *glg* genes involved in glycogen synthesis and degradation are activated by the cAMP–Crp complex, while the activity of the degrading enzyme glycogen phosphorylase (GlgP) is inhibited by HPr~P. In starved cells of the late stationary phase, however, cAMP and HPr~P levels drop, GlgP is activated by binding free HPr, and glycogen degradation begins [90, 91].

EIIA^{Glc} (sometimes also called EIIA^{Crr}, gene *crr*, part of the *ptsHIcrr* operon) and finally from EIIA^{Glc}~P to the glucose-specific membrane protein EIICB^{Glc} (gene *ptsG*), which is responsible for glucose uptake and phosphorylation. All these phosphotransfer reactions between PEP and the EIIB domains of any carbohydrate-specific EII are reversible. Only the final step (i.e., the transfer of the high-energy phosphate group to the substrate) is virtually irreversible. The key to understanding the role of the PTS in all its regulatory functions lies in the complex phosphorylation cycle of its protein kinase EI [19]. Autophosphorylation of EI occurs only in the dimeric state. After phosphorylation, the dimers dissociate and subsequently transfer the phosphate groups to HPr. If the expression of the genes for all proteins of the PTS phosphorylation cascade have been completely induced, this obligatory dimerization of EI monomers prior to autophosphorylation appears to be the rate-limiting step in the EI phosphorylation cycle, since the rate of association/dissociation is very slow in comparison to other protein kinases [19, 20]. These results suggest the following model of the PTS as a sensory system. All functional subunits of the PTS can exist in the phosphorylated or unphosphorylated form, depending on the available carbohydrate and the physiological state of the cell. During fast uptake of a PTS carbohydrate through any EII, EI is dephosphorylated more rapidly by HPr than rephosphorylated at the expense of PEP, in particular because the PEP concentrations are lowered concomitantly. Consequently, the amounts of unphosphorylated EI, HPr, EIIA^{Glc}, and EIICB^{Glc} increase transiently. Enteric bacteria use this information to couple the PTS proteins, depending on their degree of phosphorylation, to various key regulator proteins (Figure 20.1; for a detailed current review on PTS signaling, see [21]).

In the case of cAMP signaling in enteric bacteria, all available evidence indicates that EIIA^{Glc}~P, when bound to Cya, stimulates cAMP synthesis approximately eightfold. As EIIA^{Glc}~P concentrations are high under the conditions of low PTS transport rates and high PEP:pyruvate ratios, conditions normally found in starved cells, starvation triggers an increased cAMP synthesis and, hence, the pleiotropic activation of most members of the *crp* modulon. In contrast, feast conditions cause a dephosphorylation of EIIA^{Glc} and a decrease of the intracellular cAMP concentrations. Except for a few promoters whose activity is lowered by the cAMP–CRP complex, catabolite "repression" thus corresponds to the lack of activation of gene transcription (see Section 20.6.3).

The proposed model of Cya activation by EIIA^{Glc}~P is largely based on mutant analyses. In EIIA^{Glc}, substitutions of the phosphorylable His90 by Ala and the catalytically important His75 by Glu led to a protein that cannot be phosphorylated and, thus, did not activate AC. Moreover, single substitutions of His$_{90}$ by Gln or Glu, which mimics phosphorylation, also abolished the activation of Cya. The replacement of His75 by Glu inhibited the appearance of a steady-state level of phosphorylation of His90 of this mutant protein by HPr~P, yet it was a partial activator of AC [22]. Accordingly, Takahashi *et al.* found that the EIIA^{Glc} His75Glu mutant, which again is barely phosphorylated *in vitro* [22], can also slightly activate Cya [23].

AC seems to consist of at least two functional subunits – an N-terminal catalytic and a C-terminal regulatory domain, which autoinhibits enzyme activity [24, 25]. A single point mutation, which led to a Asp414Asn substitution in the regulatory domain of Cya, gave rise to a defective enzyme that produced cAMP at low rates *in vivo*. Thus, it was proposed that Asp414 of Cya might be involved in EIIA^{Glc}~P binding [26, 27]. This observation led to a simple model of the regulation of AC activity, which suggests that EIIA^{Glc}~P binds to the C-terminal domain of the enzyme and thus prevents the regulatory domain from the inhibition of the catalytic function. However, *in vitro* experiments with purified proteins, performed with the intention of displaying its binding in the activation process, were not conclusive [28, 29]. Instead, several recent findings hint at some, as yet unknown, regulatory component, which is also involved in the activity control of Cya. Using a modified AC, which was tethered to the membrane by fusing it to the transmembrane segment of the serine chemoreceptor Tsr, Park *et al.* [30] showed that *in vitro* this fusion protein is only activated by EIIA^{Glc}~P in the presence of a wild-type *E. coli* extract, which might contain this unidentified regulatory factor. Moreover, a second extension of the simple regulatory model came from the finding that both phosphorylated and unphosphorylated EIIA^{Glc} binds to this Tsr-Cya derivative with similar affinities. However, in accordance with the old model, only the EIIA^{Glc}~P was capable of stimulating cAMP synthesis, meaning that only the phosphorylated EIIA^{Glc} might be able to turn the enzyme into an active state [30]. Recently, Bettenbrock *et al.* [9] demonstrated the existence of a strict correlation between EIIA^{Glc}~P levels and the intracellular cAMP concentrations, but only in cells growing at specific growth rates between 0.3 and 0.7 h^{-1}. Above and below this range the correlation became more and more diffuse, which would again seem to favorably sustain the assertion that as yet unknown regulatory elements exist. Clearly, further experiments are needed to clarify this activation mechanism of Cya.

20.4
Another Extension of the Simple Model: Catabolite Repression by Non-PTS Substrates: The PEP: Pyruvate Ratio is a Key Node in Carbon and Energy Metabolism

As we have seen in the previous section, the reversibility of the phosphoryl group transfer within the PTS allows the bacterial metabolic network to control the phosphorylation state of PTS proteins in various ways (i.e., the PEP: pyruvate ratio is reflected at no additional energy costs in the ratio of each phosphorylated to dephosphorylated PTS component). When structuring complex metabolic and regulatory networks, areas are regularly found in which a large number of enzymatic reactions converge to establish the equilibrium between two metabolic intermediates. Alternatively, information originating from different sources is stored ("integrated") in the configuration or concentration of a signaling molecule such as cAMP. Such "nodes" constitute key parameters, and are the preferred targets for sensing and for global regulatory systems which control these networks [31]. The "pyruvate node" makes up such a central element of carbon catabolism and is accordingly the intracellular target of the PTS as a sensory system. This node couples the upper part of carbon metabolism with its lower part, and, furthermore, connects several catabolic and anabolic pathways. The upper part, mostly responsible for the interconversion of carbohydrates, comprises glycolysis and gluconeogenesis, from which in enteric bacteria emerge the pentose phosphate cycle and polysaccharide synthesis pathways. The lower part comprises the tricarboxylic acid cycle, as well as organic acid and C1-carbon metabolism. The upper and lower part together form the cellular functional unit (CFU) "carbon catabolism/quest for food" [21, 31]. The enzymes clustered around the pyruvate node are regulated in a very sophisticated way according to the needs of the organism and the PEP : pyruvate ratio therefore reflects faithfully the varying flux of their metabolites. Consequently, a PEP-dependent protein kinase EI, rather than the usual ATP-dependent kinases, is optimally suited to permanently measure the intracellular PEP : pyruvate ratio, and to signal this global information to behavioral and regulatory networks.

Accordingly, this mechanism enables the cells to regulate their carbohydrate metabolism by catabolite repression, even by carbohydrates that are not transported by the PTS. Examples of such non-PTS carbohydrates that cause catabolite repression in *E. coli* are gluconate, lactose, and glucose-6-phosphate. When grown on glucose, 95% of the EIIA^{Glc} in *E. coli* is unphosphorylated, which in turn leads to low cAMP concentrations [9]. Surprisingly it was found that EIIA^{Glc} was also partially unphosphorylated in cells grown on several non-PTS carbon sources. For instance, in cells growing on lactose or glucose-6-phosphate, approximately 80% of the EIIA^{Glc} was unphosphorylated. The corresponding PEP: pyruvate ratios were in the range of 0.12–0.21. In contrast, cells growing on the poor substrate succinate, which does not cause any catabolite repression, exhibited approximately 35% unphosphorylated EIIA^{Glc} and a PEP:pyruvate ratio of 1.07 [9]. Two groups [9, 32] found a correlation between the phosphorylation state of EIIA^{Glc} and the intracellular PEP:pyruvate ratio, which comprise the substrate and the product of EI phosphorylation. Thus, a low PEP:pyruvate ratio causes a low EI autophosphorylation activity, which in turn leads to a decreased EIIA^{Glc} phosphorylation and, thus, to a decrease in cAMP concentration. However, the fact that a *crr* mutant which lacks EIIA^{Glc} also exhibits some catabolite repression by glucose-6-phosphate [7, 33] again seems to indicate the involvement of one or more as yet unknown regulatory components of Cya activity.

20.5
cAMP Excretion and Phosphodiesterase Activity

As we have seen in Section 20.3, the intracellular cAMP concentration is predominantly controlled on the level of synthesis. In principle cells could also change the amounts of intracellular cAMP by controlled excretion or degradation by an intracellular cAMP phosphodiesterase, respectively. At least degradation does not seem to play an important role – the *E. coli* cAMP phosphodiesterase (gene *cpd*) has a low affinity for cAMP (K_m approximately between 0.5 and 0.8 mM) [34], which makes it highly unlikely that this enzyme has any influence on the cAMP concentration. Moreover, *cpd* mutants do not differ from wild-type cells in carbon catabolite repression [7] and they still show low cAMP concentrations during growth on glucose [2].

The situation with regard to cAMP excretion is still open for discussion. A putative low-affinity cAMP exporter was biochemically characterized with a K_m in the range of 10 mM [35], the gene encoding this protein, however, has not been identified yet. This low affinity strongly argues against any regulatory function of this transporter. However, the addition of glucose or any other metabolizable carbohydrate to starving *E. coli* or *S. typhimurium* cells, which contain high amounts of cAMP, led to enhanced cAMP excretion into the medium [2, 36]. Moreover, using a rapid sampling method, one group [37] demonstrated that cAMP excretion starts within 3 s after the addition of glucose to starving *E. coli* cells, which might indicate a sophisticated regulatory mechanism. In this context it seems to be important that *crp* mutants not only synthesize more cAMP due to the increase in *cya* expression, but also excrete much higher amounts of cAMP than wild-type cells [38–40], which again might strengthen the case for some kind of regulation of cAMP export.

20.6
Function of the cAMP–CRP Complex

Thus far we have concentrated on the reasonable regulation of cAMP concentrations under various growth conditions. Here, we take a look at the global regulator CRP and the function of the cAMP–CRP complex in gene regulation.

20.6.1
Transcriptional Regulation of the *crp* Gene

The key player in the regulation of *crp* transcription is the cAMP–CRP complex itself, which acts both as a negative and a positive control element for *crp* expression depending on the physiological conditions [41]. Low cAMP concentrations lead to the binding of cAMP–CRP to the *crp* promoter region, which promotes the synthesis of an antisense RNA from a divergent promoter. This antisense RNA is thought to specifically inhibit *crp* transcription by forming a transcription-terminating hairpin loop structure [42, 43]. In contrast, high cAMP concentrations apparently lead to an increase in the cellular levels of CRP, presumably by enhanced *crp* expression [41]. These observations mean that in the presence of glucose (low cAMP), cellular CRP levels decrease, while in the absence of glucose (high cAMP), CRP levels increase. This comprises a response that parallels the changes in intracellular cAMP concentrations. On the one hand, it serves to enhance the intensity of catabolite repression under feast conditions [41]. On the other hand, increased CRP levels under famine conditions enhance the potential of the cell to bring about cAMP–CRP-mediated activation of catabolic operons, if alternative carbon sources are available. The regulation of *crp* expression by factors other than cAMP–CRP adds further complexity to the catabolite repression response. The *crp* transcription is negatively regulated by both the DNA-binding protein Fis [44], the transcription of which is in turn regulated by cAMP–CRP, and by SpoT, the protein that mediates the so-called stringent response [45].

20.6.2
Properties of CRP

The CRP protein has been studied intensively. It was purified and characterized by a variety of biophysical and biochemical techniques, including X-ray analyses of CRP crystals in the presence of cAMP [46] and of the CRP–DNA complex, respectively [47–49]. The protein consists of two identical subunits, each of which can bind one molecule of cAMP. Each single subunit is made up of two distinct domains – the N-terminal domain is responsible for dimerization and forms the cAMP-binding pocket, whereas the C-terminal domain forms the DNA-binding structure. Binding of cAMP leads to massive changes in the tertiary structure and induces sequence-specific DNA recognition (reviewed in [50]). A 22-bp palindromic consensus sequence, 5'-AAaTGTGAtcT/AgaTCACAtTT-3' (the most conserved bases are in capital letters), was identified as a high-affinity site for the cAMP–CRP complex. For example, cAMP–CRP exhibits an approximately 450-fold greater affinity for this consensus sequence in comparison to the naturally occurring *lac* operon CRP-binding site [51]. Thus, variations in this consensus sequence among CRP-binding sites provide one means to determine CRP-dependent promoter strength *in vivo*. Furthermore, regions distal to the CRP site are also known to influence CRP binding [52].

AC-negative mutants, which cannot synthesize cAMP, fail to induce cAMP–CRP-dependent operons and thus cannot utilize many different carbon sources. However, extragenic suppressor mutants can be isolated overnight, which regain the ability to use, for example, lactose, maltose, or glycerol. These suppressor mutations usually map in *crp* and encode so-called CRP* forms of CRP, which contain one or more amino acid substitutions [53]. Several such CRP* derivatives have been characterized in detail [54–56]. These analyses showed that CRP* proteins adopt a conformation that resembles the cAMP-induced conformational change of the wild-type protein, which is capable of binding to CRP-specific binding sites on the DNA. Therefore, CRP* proteins can activate cAMP–CRP-dependent operons in the absence of a functional AC.

20.6.3
cAMP–CRP Complex-Dependent Promoter Activation and Repression

The cAMP–CRP complex activates transcription at several promoters and represses transcription from others (reviewed in [3, 57]). Transcription activation, which was investigated for many different systems (e.g., the operons for lactose, galactose, arabinose, and maltose), is a complex process that depends on the quality and position of the CRP-binding site, as well as on protein–protein interactions between CRP and the RNA polymerase. In general, the cAMP–CRP complex binds upstream of weak RpoD (σ^{70})-dependent promoters, which especially exhibit only a poor similarity to the optimal −35 region consensus sequence (5'-TTGACA-3'). Binding of cAMP–CRP to a region either upstream (class I promoters, e.g., *lacZp*) or at (class II promoters, e.g., *galP1*) the −35 region subsequently leads to enhanced promoter recognition by the RNA polymerase and thus to enhanced transcription. A third class of cAMP–CRP-dependent promoters requires multiple activator molecules for complete transcription activation, that is, two or more CRP molecules, or one or more CRP molecules and one or more operon-specific activator molecules. Examples for these class III promoters include the *araBAD* [58] and the *malK* [59] promoter. As stated, the quality of the CRP-binding site is one important determinant of the promoter strength. Another important factor applies to the position of the CRP-binding site relative to the RNA polymerase-binding site. Experiments with artificial *lacZ* and *melR* promoter derivatives, in which the cAMP–CRP-binding site was moved in respect to the promoter, respectively, revealed that CRP can efficiently activate transcription only when bound to the same face of the DNA helix within a range of helical turns from the RNA polymerase-binding site [60, 61]. Additionally, the extent of CRP-dependent promoter activity decreased with increasing CRP-binding site distance from the start site of transcription [62]. The interaction between the cAMP–CRP complex and the RNA polymerase holoenzyme has been intensively investigated. Depending on the promoter class, CRP makes up to three different activatory contacts with different surfaces of the holo-RNA polymerase. In class II promoters, for example, two discrete contact surfaces of CRP, known as activating region 1 [63] and 2 [64] interact with the C- and the N-terminal domains of the α subunit of the RNA polymerase, respectively, whereas one contact surface of CRP, known as activating region 3, is predicted to interact with the σ^{70} subunit [65]. For detailed reviews on the complex activation process of different promoters by cAMP–CRP, see [66, 67]. One final aspect of transcription activation by cAMP–CRP is the fact that binding of the complex to its specific site leads to DNA-bending effects. For the *lacZ* promoter the bend originates at the center of the CRP-binding site and the magnitude of the bend was estimated at 90–130° [68]. The specific function of CRP-induced bending in the process of promoter activation has yet to be established and, moreover, seems to differ for various promoters [52, 69].

As we have seen, transcription repression by cAMP–CRP takes place both for *cya* and *crp*. Another well-documented example for a direct involvement of this complex in transcription repression is the CytR regulon in *E. coli* that consists of genes involved in pyrimidine metabolism. In the absence of the allosteric effector cytidine, CytR inhibits transcription initiation at CytR regulon promoters by binding to its operator sequence and by making protein–protein interactions with the two CRP dimers, which simultaneously bind to the promoter operator region. The two CRP-binding sites are located immediately upstream and downstream of the CytR-binding site. These CytR–DNA and CytR–CRP interactions block transcription activation by CRP, by sterically blocking the functional activator region 1 of each CRP dimer [70]. Accordingly, this mechanism was also termed "antiactivation" [66].

20.7
cAMP–CRP Modulon and the CFU "Carbohydrate Catabolism/Quest for Food"

Using classical approaches, dozens of carbon catabolic operons were identified that are subject to cAMP–CRP-dependent regulation, most of them being activated under carbohydrate starvation conditions [57]. More recently, several systematic approaches including transcriptome analysis [71], *in vivo* and *in vitro* transcriptional profiling [72], chromatin immunoprecipitation [73], and bioinformatics [74] have been used in an attempt to catalog the entire CRP modulon. Using these approaches, more than 200 promoters were identified that seem to be regulated by cAMP–CRP. As a matter of course, many identified genes that belong to the cAMP–CRP modulon encode enzymes directly involved in carbohydrate metabolism. These genes are part of the CFU "carbohydrate catabolism/quest for food". This genetic unit constitutes, compared to operons and regulons, a higher level of complexity (reviewed in [18, 21, 31]), which can be exemplified by the complex regulatory system of enteric bacterial chemotaxis. The expression of the 40-odd flagellar genes for flagellar structure, assembly, and function is controlled by a three-level regulatory hierarchy. Genes *flhDC* at the highest level are organized in a "master operon", whose products FlhDC are required for the expression of all lower-level flagella genes (see also Chapter 9 and [75] for details). As expression of this master operon, a member of the CRP modulon, is strictly cAMP–CRP-dependent, only starved cells are fully flagellated and swim vigorously [76]. In addition, expression of the master operon is coupled to the cell cycle, which is affected by a number of other global regulatory signals that together control a strict coordination of cell surface rearrangements and cell division (reviewed in [21]). This is just one example of the complexity of specific interactions between different regulatory networks in bacteria.

Apart from these genes, which are more or less directly involved in carbohydrate metabolism, many other cAMP–CRP-regulated genes have been identified that have not yet been characterized in their function and their interconnection to the carbohydrate metabolism. However, global approaches have their limitations and can generate false-positive results. As shown by Hollands *et al.* [77], several genes that were considered to be part of the cAMP–CRP modulon indeed did not show any response to this global regulator. Certainly, follow-up experiments to verify cAMP–CRP involvement in the regulation of genes of interest are absolutely necessary to distinguish between direct and indirect effects.

20.8
Interactions with Other Regulatory Systems

Apparently, there must be some cross-regulation between the carbohydrate metabolism and other global regulatory systems, which control metabolic fluxes and cell maintenance. Indeed, several interactions and overlaps between cAMP–CRP and other regulatory systems have been reported (e.g., between cAMP–CRP and the nitrogen-sensing PII–Ntr system [78] or the osmoregulation system [79]). However, a systematic approach to measure the real effects under various growth conditions is still missing. Therefore, in this section I will focus only on two well-documented examples that also control the carbohydrate metabolism and thus are directly interconnected with cAMP signaling in enteric bacteria.

20.8.1
Inducer Exclusion by Unphosphorylated EIIA^{Glc}

As described before, AC is activated by phosphorylated EIIA^{Glc}. A second immediate control mechanism, involving the PTS as sensing system, rapidly inhibits a number of non-PTS carbohydrate transporters (e.g., lactose permease LacY or MalK of the maltose transport complex) and some key enzymes, in particular those involved in taking up inducing substrates, or in generating the real intracellular inducer, for example, glycerol kinase (GlpK) generating sn-glycerol-phosphate from glycerol. Hence, this mechanism was termed "inducer exclusion" (reviewed in [18], see also Chapter 19). In well-fed cells, for example, during fast growth on glucose, unphosphorylated EIIA^{Glc} accumulates in the cell, binds directly to the mentioned transporters and enzymes, and immediately inhibits their activity. For obvious reasons, this mechanism directly facilitates the regulatory response by the cAMP–CRP system to carbohydrate supply.

20.8.2
Interactions with Other Signaling Systems to Keep the Metabolic Balance: "Anticatabolite Repression" or Glucose Induction by Mlc

E. coli K-12 utilizes different transporters for the uptake of glucose, in particular the glucose PTS (EIICB^{Glc}; gene *ptsG*). The regulation of *ptsG* in *E. coli* is very complex (Fig1ure 20.2). It involves several transcription factors (Mlc, CRP), auxiliary proteins (ArcAB, Fis, MtfA), a small regulatory RNA (SgrS), a small regulatory peptide (SgrT), and three different sigma factors (RpoD, RpoS, RpoH) (reviewed

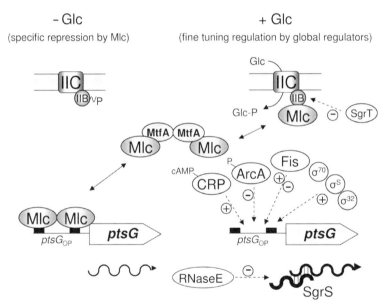

Figure 20.2 Regulation of the major glucose transport system IICB^{Glc} (encoded by *ptsG*) in *E. coli*. In the absence of glucose, *ptsG* expression is repressed by Mlc. In the presence of glucose, Mlc is sequestered to the transporter IICB^{Glc} or to the auxiliary protein MtfA [92]. Fine-tuning of *ptsG* expression under various growth conditions depends on the global-acting transcription factors cAMP–CRP, ArcA∼P, and Fis, and on several sigma factors. The stability of the *ptsG* mRNA is triggered by the small regulatory RNA SgrS. Moreover, a small polypeptide SgrT (also encoded by the *sgrS* gene), downregulates glucose uptake by a direct inhibition of the EIICB^{Glc} transport activity [93]. The expression of *sgrS/T* is regulated by the transcription activator SgrR, which in turn becomes activated during intracellular glucose∼P or fructose∼P stress.

in [80]). A multilateral regulation of EIICB^{Glc} activity seems to be necessary, since downregulation of glucose transport by a factor of 4 leads to a shift in the kinetics of the glucose PTS-dependent phosphorylation cascade in such a way that transport instead of EI autophosphorylation becomes the rate-limiting step ([81] and our unpublished results). This, in turn, would change the regulatory output signal of the PTS. In contrast, uptake of too much glucose caused by an overproduction of EIICB^{Glc} leads to an increase of excretion of overflow metabolites in *E. coli*, which is likewise disadvantageous [82]. The major specific regulator of *ptsG* expression is the repressor Mlc (mnemonic for *makes large colonies* [82], see also Chapters 12, 18 and 19). In contrast to "classical" repressors, the membrane-bound glucose transporter EIICB^{Glc} itself binds and inactivates Mlc, but only when the transport protein is in its dephosphorylated form. Thus, in the absence of glucose, Mlc binds to the *ptsG*p promoter, while in the presence of glucose, the dephosphorylated EIICB^{Glc} sequesters the repressor away from its promoter, allowing induction and enhanced *ptsG* transcription [83–86]. The complex between the membrane-bound EIICB^{Glc} with a cytosolic DNA-binding protein Mlc thus forms an efficient glucose-sensing system. Perhaps not surprisingly, DNA microarray experiments revealed a pleiotropic role of Mlc in the glucose-dependent induction of further carbohydrate transporters and catabolic enzymes (references in [18] and our unpublished results). In this way the EIICB^{Glc}/Mlc system is responsible for the induction of genes in the presence of glucose, whereas the EIIA^{Glc}∼P/cAMP–CRP system is responsible for the induction of genes in the absence of glucose. Several genes have been identified in the meantime, including *ptsG* and the *ptsHIcrr* operon, that are regulated by both systems. At the first glance, this seems to be counterproductive. However, as noted previously, even cells growing on glucose contain residual amounts of cAMP that are sufficient for an effectual level of induction by CRP. In the case of *ptsG* this dual mechanism is effective in restricting the level of transcription in the presence of glucose. Obviously, by using antagonistically acting regulators, this is one way for bacterial cells to keep their metabolism in balance.

20.9
Mathematical and Computer-Assisted Modeling of Catabolite Repression

Metabolism of a cell can be structured into meaningful CFUs, provided all relevant metabolic pathways (i.e., those with a common physiological role) are known. Structuring requires further the knowledge of the true (i.e., measured *in vivo*) kinetics of the various enzymes and the flux of key metabolites through the central nodes of a CFU. This in turn requires detailed knowledge of the regulation of enzyme activity and synthesis, at both the local and the global level, as can be obtained through quantitative transcriptomics, proteomics, and metabolomics. Finally, data on the sensory system(s) that detect(s) the extra- and intracellular stimuli to which a given CFU responds must also be included [31]. The wealth of information on a CFU gathered by this approach can only be handled when assisted by mathematical modeling and computer simulation, as shown by several groups for the PTS and its multiple roles as a global regulator (e.g., in catabolite repression and inducer exclusion [8, 87]). Validation of such models through further experiments becomes imperative whenever the models point to additional parameters, such as diffusion limitations for proteins within the cell [88], or when they indicate major deviations between *in vitro* and *in vivo* data. Modeling of cAMP–CRP-mediated catabolite repression and of inducer exclusion has become a typical example for this approach [89], and will help us to understand how cells are capable of maintaining their metabolic balance under various growth conditions.

20.10
Conclusions

The cAMP–CRP signaling system in enteric bacteria is one of the best-studied global regulatory systems. As we have seen, many details of the signaling pathway and, especially, the role of the PTS in carbohydrate sensing and regulation of carbohydrate fluxes are known. However, there are still many open questions. Are there any as yet unknown regulators of AC activity? Is there a control of cAMP excretion? Which genes encode cAMP transporters? Which genes are definitely part of the CRP modulon? What are the interconnections to other global regulatory systems under various growth conditions? Answers to all these questions will help to improve our knowledge of bacterial physiology.

Acknowledgments

I gratefully acknowledge Joseph Lengeler and Fritz Titgemeyer for helpful discussion, and Lucille Schmieding for help with the manuscript. This work was financially supported by the Deutsche Forschungsgemeinschaft through the "Sonderforschungsbereich" 431 (Teilprojekt P14) and by the Bundesministerium für Bildung und Forschung through the "FORSYS" research program.

References

1 Monod, J. (1942) Recherches sur la croissance des cultures bactériennes. PhD Thesis. University of Paris.
2 Makman, R.S. and Sutherland, E.W. (1965) Adenosine 3′,5′-phosphate in *Escherichia coli*. *J. Biol. Chem.*, **240**, 1309–1314.
3 Botsford, J.L. and Harman, J.G. (1992) Cyclic AMP in prokaryotes. *Microbiol. Rev.*, **56**, 100–122.
4 Postma, P.W. and Lengeler, J.W. (1985) Phosphoenolpyruvate: carbohydrate phosphotransferase systems of bacteria. *Microbiol. Rev.*, **49**, 232–269.
5 Postma, P.W., Lengeler, J.W., and Jacobson, G.R. (1996) Phosphoenolpyruvate: carbohydrate phosphotransferase systems, in *Escherichia coli and Salmonella*, 2nd edn, ASM Press, Washington, DC.
6 Epstein, W., Rothman-Denes, L.B., and Hesse, J. (1975) Adenosine 3′:5′-cyclic monophosphate as mediator of catabolite repression in *Escherichia coli*. *Proc. Natl. Acad. Sci.*, **72**, 2300–2304.

7 Hogema, B.M., Arents, J.C., Inada, T., Aiba, H., van Dam, K., and Postma, P.W. (1997) Catabolite repression by glucose-6-phosphate, gluconate and lactose in *Escherichia coli*. *Mol. Microbiol.*, **24**, 857–867.

8 Wang, J., Gilles, E.D., Lengeler, J.W., and Jahreis, K. (2001) Modeling of inducer exclusion and catabolite repression based on a PTS-dependent sucrose and non-PTS-dependent glycerol transport systems in *Escherichia coli* K-12 and its experimental verification. *J. Biotechnol.*, **92**, 133–158.

9 Bettenbrock, K., Sauter, T., Jahreis, K., Kremling, A., Lengeler, J.W., and Gilles, E.D. (2007) Correlation between growth rates, EIIA^{Crr} phosphorylation, and intracellular cyclic AMP levels in *Escherichia coli* K-12. *J. Bacteriol.*, **189**, 6891–6900.

10 Bankaitis, V.A. and Bassford, P.J. Jr. (1982) Regulation of adenylate cyclase synthesis in *Escherichia coli*: studies with *cya–lac* operon and protein fusion strains. *J. Bacteriol.*, **151**, 1346–1357.

11 Jovanovitch, S.B. (1985) Regulation of *cya–lacZ* fusions by cyclic AMP in *Salmonella typhimurium*. *J. Bacteriol.*, **161**, 641–649.

12 Roy, A., Glaser, P., and Danchin, A. (1988) Aspects of the regulation of adenylate cyclase synthesis in *Escherichia coli* K12. *J. Gen. Microbiol.*, **134**, 359–367.

13 Aiba, H. (1984) Transcription of the *Escherichia coli* adenylate cyclase gene is negatively regulated by cAMP–cAMP receptor protein. *J. Biol. Chem.*, **260**, 3063–3070.

14 Peterkofsky, A. and Gazdar, C. (1975) Interaction of enzyme I of the phosphoenolpyruvate: sugar phosphotransferase system with adenylate cyclase of *Escherichia coli*. *Proc. Natl. Acad. Sci. USA*, **72**, 2920–2924.

15 Saier, M.H.j. and Feucht, B.U. (1975) Coordinate regulation of adenylate cyclase and carbohydrate permeases by the phosphoenolpyruvate: sugar phosphotransferase system in *Salmonella typhimurium*. *J. Biol. Chem.*, **250**, 7078–7080.

16 Harwood, J.P., Gazdar, C., Prasad, C., Peterkofsky, A., Curtis, S.J., and Epstein, W. (1976) Involvement of the glucose enzymes II of the sugar phosphotransferase system in the regulation of adenylate cyclase by glucose in *Escherichia coli*. *J. Biol. Chem.*, **251**, 2462–2468.

17 Feucht, B.U. and Saier, M.H. Jr (1980) Fine control of adenylate cyclase by the phosphoenolpyruvate: sugar phosphotransferase systems in *Escherichia coli* and *Salmonella typhimurium*. *J. Bacteriol.*, **141**, 603–610.

18 Deutscher, J. and Postma, P.W. (2004) How phosphotransferase system-related protein phosphorylation regulates carbohydrate metabolism in bacteria. *Microbiol. Mol. Biol. Rev.*, **70**, 939–1031.

19 Chauvin, F., Brand, L., and Roseman, S. (1996) Enzyme I: the first protein and potential regulator of the bacterial phosphoenolpyruvate: glycose phosphotransferase system. *Res. Microbiol.*, **147**, 471–479.

20 Meadow, N.D., Mattoo, R.L., Savtchenko, R.S., and Roseman, S. (2005) Transient state kinetics of Enzyme I of the phosphoenolpyruvate: glycose phosphotransferase system of *Escherichia coli*: equilibrium and second-order rate constants for the phosphotransfer reactions with phosphoenolpyruvate and HPr. *Biochemistry*, **44**, 12790–12796.

21 Lengeler, J.W. and Jahreis, K. (2009) Bacterial PEP-dependent carbohydrate: phosphotransferase systems couple sensing and global control mechanisms. *Contribut. Microbiol.*, **16**, 65–87.

22 Reddy, P. and Kamireddi, M. (1998) Modulation of *Escherichia coli* adenylyl cyclase activity by catalytic-site mutants of protein IIA^{Glc} of the phosphoenolpyruvate: sugar phosphotransferase system. *J. Bacteriol.*, **180**, 732–736.

23 Takahashi, H., Inada, T., Postma, P., and Aiba, H. (1998) CRP down-regulates adenylate cyclase activity by reducing the level of phosphorylated IIA^{Glc}, the glucose-specific phosphotransferase protein, in *Escherichia coli*. *Mol. Gen. Genet.*, **259**, 317–326.

24 Roy, A., Danchin, A., Joseph, E., and Ullmann, A. (1983) Two functional domains in adenylate cyclase of *Escherichia coli*. *J. Mol. Biol.*, **165**, 197–202.

25 Reddy, P., Hoskins, J., and McKenney, K. (1995) Mapping domains in proteins:

dissection and expression of *Escherichia coli* adenylyl cyclase. *Anal. Biochem.*, **231**, 282–286.

26 Crasnier, M. and Danchin, A. (1990) Characterization of *Escherichia coli* adenylate cyclase mutants with modified regulation. *J. Gen. Microbiol.*, **136**, 1825–1831.

27 Crasnier, M., Dumay, V., and Danchin, A. (1994) The catalytic domain of *Escherichia coli* K-12 adenylate cyclase as revealed by deletion analysis of the *cya* gene. *Mol. Gen. Genet.*, **243**, 409–416.

28 Yang, J.K. and Epstein, W. (1983) Purification and characterization of adenylate cyclase from *Escherichia coli* K12. *J. Biol. Chem.*, **258**, 3750–3758.

29 Reddy, P., Meadow, N., Roseman, S., and Peterkofsky, A. (1985) Reconstitution of regulatory properties of adenylate cyclase in *Escherichia coli* extracts. *Proc. Natl. Acad. Sci. USA*, **82**, 8300–8304.

30 Park, Y.H., Lee, B.R., Seok, Y.J., and Peterkofsky, A. (2006) *In vitro* reconstitution of catabolite repression in *Escherichia coli*. *J. Biol. Chem.*, **281**, 6448–6454.

31 Lengeler, J.W. (2000) Metabolic networks: a signal-oriented approach to cellular models. *Biol. Chem.*, **381**, 911–920.

32 Hogema, B.M., Arents, J.C., Bader, R., Eijkemans, K., Yoshida, H., Takahashi, H., Aiba, H., and Postma, P.W. (1998) Inducer exclusion in *Escherichia coli* by non-PTS substrates: the role of the PEP to pyruvate ratio in determining the phosphorylation state of enzyme IIA^{Glc}. *Mol. Microbiol.*, **30**, 487–498.

33 Dumay, V., Danchin, A., and Crasnier, M. (1996) Regulation of *Escherichia coli* adenylate cyclase activity during hexose phosphate transport. *Microbiology*, **142**, 575–583.

34 Nielsen, L.D. and Rickenberg, H.V. (1974) Cyclic AMP phosphodiesterase of *Escherichia coli*. *Methods Enzymol.*, **38**, 249–256.

35 Goldenbaum, P.E. and Hall, G.A. (1979) Transport of cyclic adenosine 3′,5′-monophosphate across *Escherichia coli* vesicle membranes. *J. Bacteriol.*, **140**, 459–467.

36 Saier, M.H. Jr, Feucht, B.U., and McCaman, M.T. (1975) Regulation of intracellular adenosine cyclic 3′:5′-monophosphate levels in *Escherichia coli* and *Salmonella typhimurium*. Evidence for energy-dependent excretion of the cyclic nucleotide. *J. Biol. Chem.*, **250**, 7593–7601.

37 Schäfer, U. (1999) Automatisierte Probennahme zur Messung intrazellulärer Metabolitendynamik. PhD Thesis. Rheinische Friedrich-Wilhelms University.

38 Botsford, J.L. and Drexler, M. (1978) The cyclic 3′,5′-adenosine monophosphate receptor protein and regulation of cyclic 3′,5′-adenosine monophosphate synthesis in *Escherichia coli*. *Mol. Gen. Genet.*, **165**, 47–56.

39 Fraser, A.D. and Yamazaki, H. (1978) Determination of the rates of synthesis and degradation of adenosine 3′,5′-cyclic monophosphate in *Escherichia coli* CRP$^-$ and CRP$^+$ strains. *Can. J. Biochem.*, **56**, 849–852.

40 Potter, K., Chaloner-Larsson, G., and Yamazaki, H. (1974) Abnormally high rate of cyclic AMP excretion from an *Escherichia coli* mutant deficient in cyclic AMP receptor protein. *Biochem. Biophys. Res. Commun.*, **57**, 379–385.

41 Hanamura, A. and Aiba, H. (1992) A new aspect of transcriptional control of the *Escherichia coli crp* gene: positive autoregulation. *Mol. Microbiol.*, **6**, 2489–2497.

42 Okamoto, K. and Freundlich, M. (1986) Mechanism for the autogenous control of the *crp* operon: transcriptional inhibition by a divergent RNA transcript. *Proc. Natl. Acad. Sci. USA*, **83**, 5000–5004.

43 Okamoto, K., Hara, S., Bhasin, R., and Freundlich, M. (1988) Evidence *in vivo* for autogenous control of the cyclic AMP receptor protein gene (*crp*) in *Escherichia coli* by divergent RNA. *J. Bacteriol.*, **170**, 5076–5079.

44 Gonzalez-Gil, G., Kahmann, R., and Muskhelishvili, G. (1998) Regulation of *crp* transcription by oscillation between distinct nucleoprotein complexes. *EMBO J.*, **17**, 2877–2885.

45 Johansson, J., Balsalobre, C., Wang, S.Y., Urbonaviciene, J., Jin, D.J., Sonden, B., and Uhlin, B.E. (2000) Nucleoid proteins stimulate stringently controlled bacterial

promoters: a link between the cAMP–CRP and the (p)ppGpp regulons in *Escherichia coli*. *Cell*, **102**, 475–485.

46 Weber, I.T. and Steitz, T.A. (1987) Structure of a complex of catabolite gene activator protein and cyclic AMP refined at 2.5 A resolution. *J. Mol. Biol.*, **198**, 311–326.

47 Weber, I.T. and Steitz, T.A. (1984) Model of specific complex between catabolite gene activator protein and B-DNA suggested by electrostatic complementarity. *Proc. Natl. Acad. Sci. USA*, **81**, 3973–3977.

48 Ebright, R.H., Gunasekera, A., Zhang, X.P., Kunkel, T.A., and Krakow, J.S. (1990) Lysine 188 of the catabolite gene activator protein (CAP) plays no role in specificity at base pair 7 of the DNA half site. *Nucleic Acids Res.*, **18**, 1457–1464.

49 Parkinson, G., Wilson, C., Gunasekera, A., Ebright, Y.W., Ebright, R.E., and Berman, H.M. (1996) Structure of the CAP–DNA complex at 2.5 angstroms resolution: a complete picture of the protein–DNA interface. *J. Mol. Biol.*, **260**, 395–408.

50 de Crombrugghe, B., Busby, S., and Buc, H. (1984) Cyclic AMP receptor protein: role in transcription activation. *Science*, **224**, 831–838.

51 Ebright, R.H., Ebright, Y.W., and Gunasekera, A. (1989) Consensus DNA site for the *Escherichia coli* catabolite gene activator protein (CAP): CAP exhibits a 450-fold higher affinity for the consensus DNA site than for the *E. coli lac* DNA site. *Nucleic Acids Res.*, **17**, 10295–10305.

52 Dalma-Weiszhausz, D.D., Gartenberg, M.R., and Crothers, D.M. (1991) Sequence-dependent contribution of distal binding domains to CAP protein–DNA binding affinity. *Nucleic Acids Res.*, **19**, 611–616.

53 Harman, J.G. and Dobrogosz, W.J. (1983) Mechanism of CRP-mediated *cya* suppression in *Escherichia coli*. *J. Bacteriol.*, **153**, 191–199.

54 Blazy, B. and Ullmann, A. (1986) Properties of cyclic AMP-independent catabolite gene activator proteins of *Escherichia coli*. *J. Biol. Chem.*, **261**, 11645–11649.

55 Harman, J.G., McKenney, K., and Peterkofsky, A. (1986) Structure–function analysis of three cAMP-independent forms of the cAMP receptor protein. *J. Biol. Chem.*, **261**, 16332–16339.

56 Harman, J.G., Peterkofsky, A., and McKenney, K. (1988) Arginine substituted for leucine at position 195 produces a cyclic AMP-independent form of the *Escherichia coli* cyclic AMP receptor protein. *J. Biol. Chem.*, **263**, 8072–8077.

57 Kolb, A., Busby, S., Buc, H., Garges, S., and Adhya, S. (1993) Transcriptional regulation by cAMP and its receptor protein. *Annu. Rev. Biochem.*, **62**, 749–795.

58 Zhang, X. and Schleif, R. (1998) Catabolite gene activator protein mutations affecting activity of the *araBAD* promoter. *J. Bacteriol.*, **180**, 195–200.

59 Richet, E., Vidal-Ingigliardi, D., and Raibaud, O. (1991) A new mechanism for coactivation of transcription initiation: repositioning of an activator triggered by the binding of a second activator. *Cell*, **66**, 1185–1195.

60 Ushida, C. and Aiba, H. (1990) Helical phase dependent action of CRP: effect of the distance between the CRP site and the −35 region on promoter activity. *Nucleic Acids Res.*, **18**, 6325–6330.

61 Gaston, K., Bell, A., Kolb, A., Buc, H., and Busby, S. (1990) Stringent spacing requirements for transcription activation by CRP. *Cell*, **62**, 733–743.

62 Valentin-Hansen, P., Holst, B., Sogaard-Andersen, L., Martinussen, J., Nesvera, J., and Douthwaite, S.R. (1991) Design of cAMP–CRP-activated promoters in *Escherichia coli*. *Mol. Microbiol.*, **5**, 433–437.

63 Zhou, Y., Pendergrast, P.S., Bell, A., Williams, R., Busby, S., and Ebright, R.H. (1994) The functional subunit of a dimeric transcription activator protein depends on promoter architecture. *EMBO J.*, **13**, 4549–4557.

64 Williams, R.M., Rhodius, V.A., Bell, A.I., Kolb, A., and Busby, S.J. (1996) Orientation of functional activating regions in the *Escherichia coli* CRP protein during transcription activation at class II promoters. *Nucleic Acids Res.*, **24**, 1112–1118.

65 Rhodius, V.A. and Busby, S.J. (2000) Interactions between activating region 3 of the *Escherichia coli* cyclic AMP receptor

protein and region 4 of the RNA polymerase sigma^{70} subunit: application of suppression genetics. *J. Mol. Biol.*, **299**, 311–324.

66 Busby, S. and Ebright, R.H. (1999) Transcription activation by catabolite activator protein (CAP). *J. Mol. Biol.*, **293**, 199–213.

67 Lawson, C.L., Swigon, D., Murakami, K.S., Darst, S.A., Berman, H.M., and Ebright, R.H. (2004) Catabolite activator protein: DNA binding and transcription activation. *Curr. Opin. Struct. Biol.*, **14**, 10–20.

68 Liu-Johnson, H.N., Gartenberg, M.R., and Crothers, D.M. (1986) The DNA binding domain and bending angle of *E. coli* CAP protein. *Cell*, **47**, 995–1005.

69 Gartenberg, M.R. and Crothers, D.M. (1991) Synthetic DNA bending sequences increase the rate of *in vitro* transcription initiation at the *Escherichia coli lac* promoter. *J. Mol. Biol.*, **219**, 217–230.

70 Valentin-Hansen, P., Sogaard-Andersen, L., and Pedersen, H. (1996) A flexible partnership: the CytR anti-activator and the cAMP–CRP activator protein, comrades in transcription control. *Mol. Microbiol.*, **20**, 461–466.

71 Gosset, G., Zhang, Z., Nayyar, S., Cuevas, W.A., and Saier, M.H. Jr. (2004) Transcriptome analysis of Crp-dependent catabolite control of gene expression in *Escherichia coli. J. Bacteriol.*, **186**, 3516–3524.

72 Zheng, D., Constantinidou, C., Hobman, J.L., and Minchin, S.D. (2004) Identification of the CRP regulon using *in vitro* and *in vivo* transcriptional profiling. *Nucleic Acids Res.*, **32**, 5874–5893.

73 Grainger, D.C., Hurd, D., Harrison, M., Holdstock, J., and Busby, S.J. (2005) Studies of the distribution of *Escherichia coli* cAMP-receptor protein and RNA polymerase along the *E. coli* chromosome. *Proc. Natl. Acad. Sci. USA*, **102**, 17693–17698.

74 Robison, K., McGuire, A.M., and Church, G.M. (1998) A comprehensive library of DNA-binding site matrices for 55 proteins applied to the complete *Escherichia coli* K-12 genome. *J. Mol. Biol.*, **284**, 241–254.

75 Eisenbach, M. (2004) *Bacterial Chemotaxis*, Imperial College Press, London.

76 Vogler, A.P. and Lengeler, J.W. (1987) Indirect role of adenylate cyclase and cyclic AMP in chemotaxis to phosphotransferase system carbohydrates in *Escherichia coli* K-12. *J. Bacteriol.*, **169**, 593–599.

77 Hollands, K., Busby, S.J., and Lloyd, G.S. (2007) New targets for the cyclic AMP receptor protein in the *Escherichia coli* K-12 genome. *FEMS Microbiol. Lett.*, **274**, 89–94.

78 Mao, X.J., Huo, Y.X., Buck, M., Kolb, A., and Wang, Y.P. (2007) Interplay between CRP–cAMP and PII–Ntr systems forms novel regulatory network between carbon metabolism and nitrogen assimilation in *Escherichia coli. Nucleic Acids Res.*, **35**, 1432–1440.

79 Balsalobre, C., Johansson, J., and Uhlin, B.E. (2006) Cyclic AMP-dependent osmoregulation of *crp* gene expression in *Escherichia coli. J. Bacteriol.*, **188**, 5935–5944.

80 Jahreis, K., Pimentel-Schmitt, E.F., Bruckner, R., and Titgemeyer, F. (2008) Ins and outs of glucose transport systems in eubacteria. *FEMS Microbiol. Rev.*, **32**, 891–907.

81 van der Vlag, J., van't Hof, R., van Dam, K., and Postma, P.W. (1995) Control of glucose metabolism by the enzymes of the glucose phosphotransferase system in *Salmonella typhimurium. Eur. J. Biochem.*, **230**, 170–182.

82 Hosono, K., Kakuda, H., and Ichihara, S. (1995) Decreasing accumulation of acetate in a rich medium by *Escherichia coli* on introduction of genes on a multicopy plasmid. *Biosci. Biotechnol. Biochem.*, **59**, 256–261.

83 Lee, S.J., Boos, W., Bouche, J.P., and Plumbridge, J. (2000) Signal transduction between a membrane-bound transporter, PtsG, and a soluble transcription factor, Mlc, of *Escherichia coli. EMBO J.*, **19**, 5353–5361.

84 Nam, T.W., Cho, S.H., Shin, D., Kim, J.H., Jeong, J.Y., Lee, J.H., Roe, J.H., Peterkofsky, A., Kang, S.O., and Ryu, S. et al. (2001) The *Escherichia coli* glucose transporter enzyme IICB^{Glc} recruits the global repressor Mlc. *EMBO J.*, **20**, 491–498.

85 Tanaka, Y., Kimata, K., and Aiba, H. (2000) A novel regulatory role of glucose

86 Zeppenfeld, T., Larisch, C., Lengeler, J.W., and Jahreis, K. (2000) Glucose transporter mutants of *Escherichia coli* K-12 with changes in substrate recognition of IICB^{Glc} and induction behavior of the *ptsG* gene. *J. Bacteriol.*, **182**, 4443–4452.

87 Sauter, T. and Gilles, E.D. (2004) Modeling and experimental validation of the signal transduction via the *Escherichia coli* sucrose phosphotransferase system. *J. Biotechnol.*, **110**, 181–199.

88 Francke, C.P.P., Westerhoff, H.V., Blom, J.G., and Peletier, M.A. (2002) Why the phosphotransferase system of *Escherichia coli* escapes diffusion limitation. *Biophys. J.*, **85**, 612–622.

89 Bettenbrock, K., Fischer, S., Kremling, A., Jahreis, K., Sauter, T., and Gilles, E.D. (2006) A quantitative approach to catabolite repression in *Escherichia coli*. *J. Biol. Chem.*, **281**, 2578–2584.

(continued from previous column) transporter of *Escherichia coli*: membrane sequestration of a global repressor Mlc. *EMBO J.*, **19**, 5344–5352.

90 Koo, B.M. and Seok, Y.-J. (2001) Regulation of glycogen concentration by the histidine-containing phosphocarrier protein HPr in *Escherichia coli*. *J. Microbiol.*, **39**, 24–30.

91 Seok, Y.J., Koo, B.M., Sondej, M., and Peterkofsky, A. (2001) Regulation of *E. coli* glycogen phosphorylase activity by HPr. *J. Mol. Microbiol. Biotechnol.*, **3**, 385–393.

92 Becker, A.K., Zeppenfeld, T., Staab, A., Seitz, S., Boos, W., Morita, T., Aiba, H., Mahr, K., Titgemeyer, F., and Jahreis, K. (2006) YeeI, a novel protein involved in modulation of the activity of the glucose-phosphotransferase system in *Escherichia coli* K-12. *J. Bacteriol.*, **188**, 5439–5449.

93 Wadler, C. and Vanderpool, C.K. (2007) A dual function for a bacterial small RNA: SgrS performs base pairing-dependent regulation and encodes a functional polypeptide. *Proc. Natl. Acad. Sci. USA*, **104**, 20454–20459.

21
c-di-GMP Signaling

Christina Pesavento and Regine Hengge

21.1
Introduction

The first discovery of c-di-GMP in the fruit-degrading bacterium *Acetobacter xylinum* (*Gluconacetobacter xylinus*) dates back to 1987, when Benziman *et al.* identified an allosteric activating factor for membrane-bound cellulose synthase as c-di-GMP and referred to the c-di-GMP-synthesizing and -degrading enzymes as diguanylate cyclases (DGCs) and phosphodiesterases (PDEs), respectively [1]. However, the progression in genome sequencing in the following two decades and the subsequent identification of numerous genes encoding DGCs and PDEs in a large number of bacterial species was required for c-di-GMP to be recognized as a new ubiquitous second messenger in bacteria. Research during the last few years led to the identification of a variety of processes that are subject to c-di-GMP-mediated regulation. As a consequence, a general role of c-di-GMP in promoting functions associated with bacterial biofilm formation and sessility, and as a negative regulator of motility and virulence, has emerged (for recent summaries, see [2–11]).

21.2
Protein Domains Involved in c-di-GMP Signaling

21.2.1
Making and Breaking of c-di-GMP

It was again in *Gluconacetobacter xylinus* that DGC and PDE activities were first linked to a set of proteins containing GGDEF and EAL domains, named after the most prominent conserved sequence motifs found in these domains [12]. Later biochemical studies directly established that the GGDEF domain is able to specifically convert GTP to c-di-GMP and that this activity requires an intact Gly–Gly–Asp–Glu–Phe sequence motif in the active (A)-site of the domain (Figure 21.1) [13]. Many DGCs show a GG**E**EF sequence in the A-site, which is the only deviation from this motif that

Bacterial Signaling. Edited by Reinhard Krämer and Kirsten Jung
Copyright © 2010 WILEY-VCH Verlag GmbH & Co. KGaA, Weinheim
ISBN: 978-3-527-32365-4

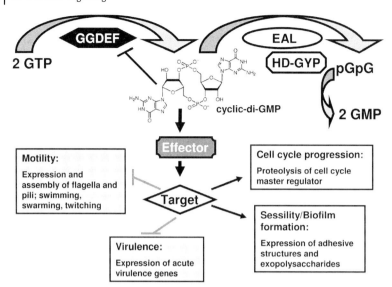

Figure 21.1 Principles of c-di-GMP signaling. Protein domains with DGC and PDE activities, feedback inhibition of DGC activity through c-di-GMP as well as target functions of c-di-GMP signaling are shown. c-di-GMP interferes with cellular functions by binding to effectors (e.g., PilZ domain proteins, GEMM RNAs, etc.), which in turn interact with molecular targets that are part of the c-di-GMP-regulated processes.

still conserves DGC activity [14]. To form a catalytically active enzyme–substrate complex two GTP-carrying GGDEF domains dimerize and the phosphodiester bond is formed by a mechanism closely resembling the two-metal-assisted mechanism suggested for structurally related adenylate cyclases and polymerases [15, 16]. Most active DGCs are subject to allosteric product inhibition. A conserved RXXD motif (inhibitory I-site) in close proximity to the A-site participates in binding of a c-di-GMP dimer that inhibits DGC activity [15, 17]. Several mechanisms of product inhibition have been suggested, including direct communication between the I- and A-site upon binding of c-di-GMP [17], and c-di-GMP-mediated cross-linking of GGDEF domains within a dimer or to additional protein domains in a manner that coordinates the formation of a catalytically inactive enzyme conformation [15, 16, 18]. Feedback control of DGC activity probably prevents excessive GTP consumption, establishes an upper limit for the cellular c-di-GMP concentration, and increases stability of c-di-GMP signaling towards stochastic perturbations [17].

Degradation of c-di-GMP is mediated by the EAL domain, which specifically cleaves c-di-GMP into its linear form 5′pGpG (assigned as PDE-A activity) that is then further degraded by an as-yet uncharacterized, comparably slow, and most likely unspecific reaction to GMP (termed PDE-B activity) (Figure 21.1) [19, 20]. The residues required for activity exceed those found in the EAL (Glu–Ala–Leu) site and hydrolysis of c-di-GMP is thought to be mediated by a general base-catalyzed mechanism involving Mg^{2+} (or Mn^{2+}) [21]. In addition to a strong dependence on Mg^{2+} or Mn^{2+}, PDE activity is generally strongly inhibited by Ca^{2+} or Zn^{2+}

ions [19, 20, 22]. Dimerization of EAL domains does not seem to be required for the catalytic mechanism of c-di-GMP hydrolysis [20, 21], but several studies suggested that EAL domain proteins as well as single EAL domains can occur in di- or oligomeric conformations [21, 23, 24], the role of which has yet to be determined.

Another less-frequent domain involved in the turnover of c-di-GMP is the HD-GYP domain (named after the characteristic sequence signatures), which constitutes a subgroup of the HD superfamily of metal-dependent phosphohydrolases [25]. PDE activity of the HD-GYP domain converts c-di-GMP to GMP via pGpG, indicating that the HD-GYP domain might differ from the EAL domain in its activity against pGpG (Figure 21.1) [26]. An isolated HD-GYP domain was found to be able to interact with several GGDEF domains [27], suggesting a potential role for protein–protein interactions in signaling mediated by this domain.

21.2.2
Composite GGDEF, EAL, and HD-GYP Proteins

One common aspect that had made the assignment of PDE and DGC activities to GGDEF, EAL, and HD-GYP domains difficult is the frequent combination of GGDEF domains with EAL or HD-GYP domains in single proteins. The majority of such composite proteins studied *in vitro* so far only show either PDE or DGC activity [19, 20, 28, 29]. However, the example of a GGDEF and EAL domain-containing protein from *Rhodobacter sphaeroides* suggests that bifunctional proteins may exist. The full-length version of this protein shows PDE activity, while DGC activity of the GGDEF domain could be demonstrated upon removal of the EAL domain [24]. Whether both activities play a role *in vivo* and how they might be coordinated has yet to be shown. In many composite proteins, however, one domain shows deviations from the conserved residues required for activity, thus accounting for the enzymatic inactivity of this domain. In a GGDEF/EAL protein from *Caulobacter crescentus*, a catalytically inactive GGDEF domain plays a role in allosteric regulation of the neighboring EAL domain by GTP. The GEDEF sequence in the A-site of the enzyme is still able to bind GTP and in turn exerts a stimulatory effect on the PDE activity of the EAL domain [19].

21.2.3
Recruitment of GGDEF and EAL Domains for c-di-GMP-Unrelated Functions

Finally, in addition to GGDEF and EAL domains involved in the "making and breaking" of c-di-GMP and to the recruitment of catalytically inactive GGDEF domains as sensory domains in composite proteins, an increasing number of reports indicate that some proteins with GGDEF and EAL domains do not show any c-di-GMP-associated activity at all, but instead have adopted c-di-GMP-unrelated functions. In these cases the components of the c-di-GMP signaling system have lost their original activities, and often show highly degenerate GGDEF and EAL motifs [30, 31]. These "non-c-di-GMP-associated" GGDEF and EAL proteins have been shown to control such diverse processes as degradation of small regulatory RNAs

[30], biofilm gene expression [32], and transcriptional regulation in response to blue light [31]. The latter case involves a BLUF (blue light using flavin)/EAL domain protein, which acts as an antirepressor that directly interacts with a repressor and releases it from its operator in response to blue-light irradiation [31]. These diverse roles GGDEF and EAL domains are able to adopt strongly indicate the need for a thorough biochemical and physiological analysis before unambiguous functions can be assigned to distinct GGDEF and EAL proteins.

21.2.4
Regulation of DGC and PDE Activity and Expression

The majority of GGDEF, EAL, and HD-GYP domains are joined to N-terminal sensory input domains comprising transmembrane sensor domains as well as phosphoacceptor and other cytoplasmic sensor domains [33, 34]. The great variety of these sensory input domains suggests that the c-di-GMP signaling system is able to integrate a large variety of signals. However, both the nature of the signals and the mechanism of transduction into a specific output of the protein are still unknown in most cases. Only in a few examples could the transformation of a specific signal sensed by a sensory domain into a specific output of a GGDEF or EAL domain protein be clarified in more detail. These include the PDE activities of PDE A1 from *Acetobacter xylinum* [35] and Dos (direct oxygen sensor) from *Escherichia coli*, [29, 36–38], which are both regulated in response to the binding of oxygen and other gases, and the blue-light-dependent transcriptional regulation involving the EAL protein YcgF and the repressor protein YcgE in *E. coli* [31, 39, 40]. In *C. crescentus* phosphorylation of the response regulator and GGDEF protein PleD at its receiver domain induces dimerization and DGC activity as well as polar localization [13, 16, 41]. Phosphorylation of the response regulator and PDE VieA by its cognate sensor kinase VieS also plays a role in c-di-GMP-mediated regulation of biofilm formation and motility in *Vibrio cholerae*. However, in this case phosphorylation of VieA does not directly stimulate PDE activity, but has an indirect effect by enhancing VieA transcriptional autoregulation and thus increasing VieA levels in the cell [42]. Specific and differential control of expression of GGDEF and EAL genes represents another level of regulation employed to coordinate DGC and PDE activities in the cell [28, 43].

21.2.5
c-di-GMP-Binding Effectors

While details about c-di-GMP synthesis and degradation were elucidated relatively rapidly, the molecular mechanisms by which c-di-GMP acts on its targets remained obscure for a long time despite significant efforts to identify effector proteins that bind c-di-GMP and mediate specific phenotypic outputs. A breakthrough in this field came from a bioinformatics analysis that identified the so called PilZ domain, named after its presence in a *Pseudomonas aeruginosa* protein involved in twitching motility. A similar phylogenetic distribution as those of the GGDEF, EAL, and HD-GYP domains, and the presence of the PilZ domain in proteins involved in processes known to be regulated by c-di-GMP, strongly indicated a role in c-di-GMP signaling [44]. Several subsequent biochemical and phenotypic analyses established the PilZ domain as one of the long-sought c-di-GMP-binding proteins, demonstrating that c-di-GMP binding to PilZ proteins is an essential step in c-di-GMP-mediated regulation of motility, extracellular polysaccharide synthesis, biofilm formation, and virulence in many organisms [45–48]. These studies also revealed a conserved RxxxR motif at the N-terminal end of the PilZ domain that is involved in c-di-GMP binding. Analysis of the structure of a PilZ domain protein suggests that c-di-GMP binding to the region carrying this motif induces conformational changes that activate the effector protein [49]. However, the facts that PilZ domains exist in both single and multidomain proteins, and that regions outside the PilZ domain can be involved in c-di-GMP binding [49], indicate that the mechanisms of c-di-GMP- induced effector activation involving the PilZ domain might be more diverse.

Furthermore, additional c-di-GMP effectors that do not carry PilZ domains have been identified. In *P. aeruginosa* the transcription factor FleQ as well as the putative inner membrane protein PelD, both involved in the expression of extracellular polysaccharides, specifically bind c-di-GMP – the latter protein employing a c-di-GMP-binding site very similar to the I-site shown to be involved in feedback inhibition of DGCs (see Section 21.2.1) [50, 51]. Interestingly, the *C. crescentus* GGDEF protein PopA, which features a degenerate A-site, showing no DGC activity, but an intact I-site able to bind c-di-GMP, functions as a c-di-GMP effector. It mediates recruitment of the cell cycle regulator CtrA to the differentiating cell pole where it is degraded [52].

An entirely different type of c-di-GMP effector is represented by a class of riboswitches recently discovered in the upstream region of various genes found in several bacterial species. These mRNA motifs, termed GEMM (genes for the environment, membranes and motility) RNAs after the physiological context of the genes possibly regulated by this motif [53], undergo structural modulations upon binding to c-di-GMP and in consequence alter the expression of downstream genes [54]. This demonstrates that c-di-GMP binding is not restricted to proteins and suggests that our current knowledge of effector components as well as of mechanisms by which c-di-GMP affects its targets is far from being complete.

c-di-GMP bound to these known and unknown effector molecules then controls various target processes (Figure 21.1), including such diverse biological processes as transcriptional regulation [28, 51], flagellar motor function [45, 55], enzymatic activity [1], organelle assembly [56, 57], and localized proteolysis in cell cycle control [52].

21.3
Signaling Specificity

Comparative analysis of complete microbial genomes revealed the presence of multiple copies of genes encoding GGDEF, EAL, and HD-GYP domains in the genomes of most bacterial species, while they are absent from the sequenced genomes of the Archaea and eukaryotes [33, 34]. The numbers of GGDEF, EAL, and HD-GYP encoding genes vary vastly between the species, with generally higher numbers in Gram-negative bacteria (e.g., *E. coli* encoding 19 GGDEF and 17 EAL domains) than in Gram-positive bacteria (e.g., *B. subtilis* encoding four GGDEF and three EAL domains), and their high abundance in free-living bacteria mirrors their importance in signal transduction and adaptational processes in complex environments [33]. Considering this abundance of proteins with redundant enzymatic activities within one organism, one of the most puzzling aspects of c-di-GMP-mediated signaling remains the question of how signaling specificity can be achieved in these systems. Although the molecular details of how single GGDEF, EAL, and HD-GYP domain proteins within such complex networks are able to specifically affect certain c-di-GMP-dependent target processes remain mostly elusive, sequestration of c-di-GMP control systems has been suggested as a solution to this problem [4, 26, 28, 58, 59]. Different modes of sequestration can be envisaged. Through temporal regulatory sequestration only a subgroup of the DGCs and PDEs encoded in the genome could be present and active at any given point in the cell cycle or under any growth condition. This could be realized through specific regulatory expression patterns of the corresponding genes combined with tight control of enzymatic activities through the N-terminal sensory input domains. Alternatively, certain DGCs and PDEs might form complexes with their respective effectors and target molecules, yielding microcompartments in which the components of one c-di-GMP signaling module are functionally sequestered. Similarly, spatial restriction of components of the c-di-GMP signaling network to specific locations within the cell (e.g., to the cell poles) could confine specific c-di-GMP signaling to the immediate vicinity of colocalized DGCs, PDEs, effector, and target molecules. Evidence for all modes of sequestration can be found in the following sections, in which the physiology of c-di-GMP signaling will be illustrated using three different model organisms.

21.4
c-di-GMP Signaling in *E. coli*

During transition from the postexponential to the stationary phase (in complex medium), *E. coli* exhibits an easy to study lifestyle switch, in which c-di-GMP regulation plays a crucial role (Figure 21.2). In the course of this transition, *E. coli* cells can switch from a single-cell planktonic state, in which they express flagella and thus are motile, to a sedentary state, in which the cells adhere to each other and to surfaces with the help of adhesive surface appendices and an excreted extracellular matrix, thus forming biofilms. These two states reflect two fundamentally different lifestyles, which many bacteria can adopt depending on the environmental conditions.

E. coli features a total of 29 GGDEF/EAL genes, coding for 12 GGDEF domain proteins, 10 EAL domain proteins, and seven proteins harboring both domains.

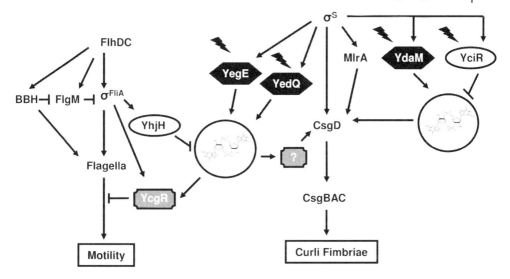

Figure 21.2 C-di-GMP-mediated coordination of motility and curli expression in E. coli. The regulatory cascades governing motility and curli expression are shown. DGCs, PDEs, and effector molecules are indicated by the same symbols as used in Figure 21.1, and potential signaling input regulating enzymatic activities via N-terminal sensory domains is indicated by black bolts. BBH, basal body–hook. (Modified from [11]).

Interestingly, the majority of the GGDEF and EAL genes, for which expression could be shown, are under the control of the general stress and stationary-phase sigma factor σ^S, indicating an important role for c-di-GMP signaling in the stationary phase of growth [43]. Consistently, several GGDEF and EAL proteins have been implicated in the regulation of curli synthesis. Curli fimbriae are adhesive structures that mediate cell–cell as well as cell–surface contacts and curli synthesis is induced as part of the general stress response when E. coli or Salmonella enter the stationary phase [28, 59–62]. In both species, expression of curli fimbriae is generally restricted to temperatures below 30 °C [60, 62], responds to a diverse set of environmental stimuli [63], and is regulated by a complex transcription factor cascade under multiple feedforward control by the master regulator σ^S. σ^S regulates the expression of the MerR-like regulator MlrA, which, together with σ^S, activates the transcription of the essential curli regulator CsgD. CsgD in turn induces the expression of the curli structural operon *csgBAC* [61, 64–67]. In E. coli σ^S also controls the expression of the GGDEF domain protein and DGC YdaM, and the composite GGDEF/EAL protein and PDE YciR. These two proteins antagonistically control curli expression at the level of *csgD* transcription [28]. c-di-GMP synthesized by YdaM is absolutely required for *csgD* transcription, while YciR counteracts YdaM activity. CsgD expression seems the only target process affected by YdaM and YciR, indicating that distinct GGDEF and EAL proteins are able to specifically and very precisely affect distinct target processes in the cell. In contrast, the σ^S-dependent composite GGDEF/EAL protein YegE not only positively influences curli synthesis, but also has a negative effect on motility [59].

c-di-GMP has been shown to inhibit motility in *E. coli* and *Salmonella* by binding to the c-di-GMP receptor protein YcgR, which in turn interferes with flagellar motor function [45, 59]. YegE as well as another σ^S-dependent GGDEF protein, YedQ, are essential for the generation of the motility-inhibiting c-di-GMP [55, 59]. The c-di-GMP that is positively controlled by YegE and YedQ is subject to degradation by the PDE YhjH – an EAL protein and member of the flagellar regulon. YhjH keeps c-di-GMP levels low for as long as flagellar gene expression proceeds, thus counteracting YcgR-mediated inhibition of motility [45, 55, 59]. However, when cells enter the stationary phase, flagellar gene expression and, in consequence, YhjH expression ceases [59, 68, 69]. At the same time σ^S-mediated induction of YegE (and to a lower extent induction of YedQ) leads to increased c-di-GMP synthesis, finally overcoming the negative effect of YhjH. The resulting elevated c-di-GMP level not only inhibits motility via YcgR, but also contributes to stimulation of CsgD expression [59]. Thus, YegE/YhjH-controlled c-di-GMP connects inhibition of motility to activation of curli synthesis, demonstrating a broader target spectrum for this c-di-GMP control module than for the YdaM/YciR module, which is active in parallel. This suggests that the two sets of proteins might affect distinct pools of c-di-GMP [59]. In addition, the fact that a different protein, YeaJ, takes over the roles of YegE and YedQ in motility inhibition at 37 °C and that this role coincides with a preferential expression of the *yeaJ* gene at 37 °C, indicates that specific expression patterns of GGDEF and EAL genes might play a role in establishing distinct functions of certain GGDEF and EAL proteins [43, 59].

Interestingly, YegE/YhjH controlled c-di-GMP influences CsgD expression via an effector distinct from YcgR [59]. Although the identity of this effector as well as the direct target mechanism remain to be shown, the use of different effector molecules, controlled by the same c-di-GMP pool, but dedicated to different target processes, also raises further possibilities for establishing signaling specificity (e.g., by making use of different effector affinities for c-di-GMP). Accessibility of the c-di-GMP by different effectors also indicates that YegE/YhjH controls a pool of freely diffusible c-di-GMP.

At least one more GGDEF/EAL protein, YhdA, is involved in curli synthesis by affecting the expression of the essential curli activator gene *ydaM* as well as of *csgD* [43]. YhdA, however, has degenerate GGDEF and EAL motifs, and was shown to affect the turnover of small regulatory RNAs in the Csr system [30], suggesting that small RNAs are also involved in curli control. This indicates the presence of a fine-tuned network consisting of GGDEF and EAL proteins that are potentially able to integrate diverse environmental or intracellular stimuli into a tight control of curli expression at all possible levels. Moreover, c-di-GMP regulates the synthesis of cellulose, another component of the extracellular matrix contributing to the multi-cellular morphotype in many *E. coli* and *Salmonella* strains [70]. CsgD, the regulator of curli synthesis, also induces expression of YaiC (or the *Salmonella* homolog AdrA) – a GGDEF protein whose DGC activity is required for cellulose synthase activity [70–72]. Finally, some pathogenic *E. coli* employ the c-di-GMP system for the regulation of certain virulence factors. In adherent-invasive *E. coli*, it was shown that c-di-GMP mediates coregulation of motility and type 1 pili, which are required for adhesion to and invasion of host cells. In this strain the PDE YhjH degrades c-di-GMP that otherwise inhibits the synthesis of type 1 pili, thus linking flagellar and type 1 pili synthesis [73].

Intriguingly, while the overall architecture of the control cascades governing curli expression in *E. coli* and *Salmonella* are the same, the two species show significant differences concerning the c-di-GMP signaling input into curli regulation. The closely related species do not only differ in the number of GGDEF and EAL domain proteins encoded in their genomes (29 in *E. coli* versus 19 in *Salmonella*), but also show a surprisingly large divergence in the identity of the proteins present with a limited overlap of homologous proteins. Consistently, contributions of single GGDEF and EAL proteins, and also of those proteins present as homologs in both species, to the regulation of certain phenotypes differ between these organisms. Thus, *Salmonella* does not have YdaM, but uses another GGDEF protein (not present in *E. coli*) for activating curli synthesis, and not YhjH (as in *E. coli*), but YjcC seems to be a major PDE in the postexponential phase [43, 58, 74]. The physiological significance of these differences as well as the cause of the divergence in the closely related organisms remain elusive. However, one possible explanation for divergent functions of homologous proteins in *Salmonella* and *E. coli* could be based on simple variations in the expression patterns of the corresponding genes due to promoter mutations that alter the expression pattern of a GGDEF or EAL gene. Such alterations in the pattern of temporal sequestration might expose the c-di-GMP processed by certain PDEs and DGCs to a different set of effectors and targets not present during expression of the homologous proteins in the other species. The observation that higher expression levels of the GGDEF protein YedQ, as observed in certain *E. coli* strains, endow the cell with a way to bypass CsgD/AdrA-dependent induction of cellulose synthesis [75] further supports the idea that alterations in the expression patterns of DGCs and PDEs might easily rewire these signaling enzymes to new effector/target systems, thus accounting for the great flexibility of the c-di-GMP signaling system.

21.5
c-di-GMP signaling in *V. cholerae*

The Gram-negative γ-proteobacterium *V. cholerae* is the causative agent of the severe diarrheal disease cholera and represents a paradigm organism for the role of c-di-GMP signaling in virulence gene regulation. It harbors an even more impressive set of genes potentially involved in c-di-GMP signaling than *E. coli*, comprising genes coding for 31 GGDEF proteins, 12 EAL proteins, 10 proteins with both domains, and nine HD-GYP domain proteins [33]. c-di-GMP-mediated signaling has been shown to play an important role in the regulation of biofilm formation as well as in the regulation of virulence factors, thus contributing to both the persistence of the pathogen in aquatic environments and the infection of the host. Virulence gene expression is controlled by a complex network comprising numerous transcriptional regulators as well as several quorum sensing (QS) systems and is linked to regulation of biofilm formation via common regulatory components (reviewed in [76]). In the classical biotype of *V. cholerae*, the PDE VieA negatively regulates biofilm formation by influencing *vps* gene expression [77] – a gene cluster required for the production of an exopolysaccharide essential for *V. cholerae* biofilm formation [78, 79]. On the other hand, low c-di-GMP levels established by VieA induce the expression of virulence genes [80], thus indicating an inverse regulation of virulence and biofilm genes by c-di-GMP. This suggests a role for c-di-GMP-mediated signaling in the transition between the host and the environment. Further support for this role comes from the observation that three genes encoding GGDEF proteins are induced late during infection and enhance the environmental fitness of *V. cholerae*, thereby preparing the bacterium for the exit from the host [81]. Variations in c-di-GMP levels at different stages in the *V. cholerae* life cycle are also influenced by a link to the QS systems operating in *V. cholerae*. HapR, the central regulator affected by multiple QS systems, was shown to control the expression of 14 GGDEF and EAL genes, ultimately resulting in a reduction of cellular c-di-GMP levels at high cell density as reached *in vivo* and, in consequence, a repression of biofilm genes [82]. The link between c-di-GMP signaling and QS in *V. cholerae*, which results in an integration of information about both the surrounding bacterial population and the local environment into biofilm formation [82], is extended by several levels of mutual control between the two signaling systems. These comprise the expression control of further GGDEF and EAL genes mediated by components of the QS system [83] as well as regulation of the latter components through GGDEF and EAL genes [84], and show that QS and c-di-GMP signaling are closely inter-related in *V. cholerae*.

Another interesting regulatory link connects the cAMP–cAMP receptor protein (CRP) regulatory network with c-di-GMP regulation of biofilm formation in *V. cholerae*. A set of genes encoding GGDEF, EAL, and HD-GYP proteins is differentially regulated by cAMP–CRP and increased expression of the negatively cAMP–CRP-controlled GGDEF gene *cdgA* in the *crp* mutant was shown to be, in part, responsible for the increased biofilm formation exhibited by this mutant [85].

Several studies established a direct involvement of c-di-GMP signaling in the regulation of the different phase variants of *V. cholerae*. The rugose variant produces more exopolysaccharide, mediating increased biofilm formation and higher resistance to several stresses compared to the other, smooth variant [79]. The rugose variant was demonstrated to have higher intracellular c-di-GMP levels than the smooth variant [86], and several GGDEF and EAL genes, which are differentially expressed between the two phase variants [87] are involved in the regulation of motility, biofilm formation, and the switch frequency between smooth and rugose variants [84, 86, 88–90]. The GGDEF and EAL proteins involved in this regulation show differences in the timing of the regulation they exert as well as in the mode of action by which they influence certain phenotypes [84, 86], supporting the concept of functional differentiation in the complex GGDEF/EAL network employed by this organism.

Comparison of the contribution of single GGDEF and EAL proteins to virulence gene regulation in different biotypes of *V. cholerae* reveals specific differences. While the PDE VieA regulates virulence gene expression in the classical *V. cholerae* biotype, it does not do so in the El Tor biotype [80, 91]. Consistent with this, the expression of the *vieSAB* operon under virulence gene-inducing conditions is higher in the classical biotype than in the El Tor biotype with VieA controlling the expression of a large subset of genes in the classical biotype and hardly any genes in the El Tor biotype [91]. Again, this shows that functional divergence between the components of the c-di-GMP signaling system, seen even between different biotypes of one species, can be accompanied by and might be due to differences in the expression of GGDEF and EAL genes. This further highlights the great flexibility of the c-di-GMP system. Additional complexity is provided by phase variant-specific differences. For instance, a presumable PDE mutant, which shows slightly elevated c-di-GMP levels in a rugose El Tor biotype variant, displays induced virulence gene expression, while the same mutation in the smooth variant reduces virulence gene expression [84].

21.6
c-di-GMP Signaling in *C. crescentus*

Cyclic di-GMP signaling in the aquatic α-proteobacterium *C. crescentus* represents a paradigm of spatial sequestration of DGC and PDE activities and c-di-GMP-mediated control. The cell cycle of *C. crescentus* comprises an asymmetric cell division, yielding a sessile stalked and a motile swarmer daughter cell. The surface-adherent, stalked cell immediately undergoes the next round of replication, while cell cycle progression is blocked in the swarmer cell until it differentiates into a stalked cell. The two cell types feature different polar structures – the stalk with an adhesive holdfast in the sessile cell, and a flagellum and pili in the swarmer cell. The transition from swarmer to stalked cells involves remodeling of the pole – a loss of pili and flagellum and synthesis of the holdfast and stalk. The control of these morphological changes is integrated with cell cycle control by a complex regulatory network comprising several master regulators and their control through two-component signal transduction systems (reviewed in [92, 93]). Several GGDEF and EAL domain proteins have been implicated in the control of polar development, cell cycle progression, and cell polarity. Consistent with their functional role these proteins show distinct positioning within the cell at certain times in the cell cycle. The response regulator and DGC PleD is evenly distributed in the swarmer cell, but concentrates at the emerging stalked pole during swarmer-to-stalked cell transition [13]. Targeting to the stalked pole is mediated by phosphorylation of PleD which induces dimerization, a prerequisite for both DGC activity of the PleD GGDEF domain, and polar localization of the protein [13, 41]. PleD activity is essential for loss of motility, flagellum ejection, and stalk biogenesis (i.e., essential steps in swarmer-to-stalked cell transition) [94–96]. Simultaneous sequestration and activation of DGC activity strongly indicates local production of c-di-GMP at the stalked pole, resulting in inhibition of motility and induction of adhesive structures. The phosphorylation status of PleD is controlled through the activities of the histidine kinases PleC and DivJ, which are differentially localized to opposite poles as a function of the cell cycle [13, 96, 97], thus leading to a coordinate regulation of polar differentiation and cell cycle progression through PleD.

This role for c-di-GMP mediated signaling in *C. crescentus* is extended by the participation of the EAL protein TipF in the establishment of cell polarity during cell division. TipF is required for flagellum synthesis and, together with the protein TipN, mediates correct positioning of the flagellum at the nascent cell pole. During cell division TipF and TipN, which are also required for normal cell division, dynamically localize to the division septum and subsequently to the newborn pole, specifying the future flagellum assembly site [98]. The function of TipF relies on its EAL domain, again suggesting a locally restricted turnover of c-di-GMP, possibly resulting in decreased levels of the second messenger at the site of flagellum biogenesis.

In addition, c-di-GMP also contributes to the control of the cell cycle itself. The GGDEF protein PopA functions as a c-di-GMP effector protein by binding of c-di-GMP to its conserved I-site motif, while the A-site is degenerate and inactive (see Section 21.2.5). During swarmer-to-stalked cell transition, PopA targets CtrA, one of the master regulators of the cell cycle, to the incipient stalked pole [52]. Sequestration of CtrA to this pole at this specific point in the cell cycle results in its degradation by the colocalized ClpXP protease, which is targeted to the pole by a converging pathway [52, 99, 100]. Thereby, CtrA-mediated inhibition of replication is relieved, thus allowing for the cell cycle to proceed. PopA sequestration to the incipient stalked pole requires c-di-GMP binding, and CtrA sequestration is realized by an interaction between PopA and the mediator protein RcdA, which in turn binds CtrA [52]. As CtrA localization to the pole, and therefore its degradation, is ultimately regulated by c-di-GMP, c-di-GMP signaling is involved in controlling the timing of CtrA proteolysis – an essential step in the progression of the cell cycle [52].

21.7
Conclusions and Outlook

Extensive research over the last few years, inspired by the recognition of the ubiquity, multiplicity, and diversity of components of the c-di-GMP signaling system in bacteria, has established c-di-GMP as an important factor in the regulation of important bacterial phenotypes including biofilm formation, motility, and pathogenicity. C-di-GMP signaling generally plays a role in the regulation of bacterial lifestyle transitions, such as the transition between motile and sedentary cells, as described above for *E. coli* and *C. crescentus*, or in the transition between the host and the environment in pathogenic bacteria such as *V. cholerae*.

Many unresolved questions remain to be answered in order to gain a comprehensive view of the detailed mechanisms of c-di-GMP signaling and its importance in bacteria. These concern, for example, the generation of signaling specificity in the complex c-di-GMP signaling systems found in many species, the great adaptive and evolutionary flexibility of the c-di-GMP system, and the role of the many degenerate GGDEF and EAL proteins with c-di-GMP-independent regulatory functions. For a comprehensive understanding it will also be essential to elucidate how c-di-GMP signaling is integrated with other regulatory systems. As discussed in Sections 21.4 to 21.6, some regulatory links have already been identified connecting, for instance, the c-di-GMP signaling system with the σ^S-controlled general stress response network in *E. coli*, with cAMP signaling and QS in *V. cholerae,* and with cell cycle progression in *C. crescentus*, but our knowledge of these regulatory integration is far from being complete. It will be particularly interesting to elucidate how such regulatory links are used by different species for c-di-GMP signaling in regulatory contexts specific to the lifestyle and ecological niches of the respective species.

Acknowledgments

Research performed in the laboratory of R.H. that is mentioned in this chapter has been funded by the Deutsche Forschungsgemeinschaft, the Dr. Hans-Messner-Stiftung, and the Fonds der Chemischen Industrie.

References

1. Ross, P., Weinhouse, H., Aloni, Y., Michaeli, D., Weinberger-Ohana, P., Mayer, R., Braun, S., de Vroom, E., van der Marel, G.A., van Boom, J.H. *et al.* (1987) Regulation of cellulose synthesis in *Acetobacter xylinum* by cyclic diguanylic acid. *Nature*, **325**, 279–281.
2. Romling, U., Gomelsky, M., and Galperin, M.Y. (2005) C-di-GMP: the dawning of a novel bacterial signalling system. *Mol. Microbiol.*, **57**, 629–639.
3. Romling, U. and Amikam, D. (2006) Cyclic di-GMP as a second messenger. *Curr. Opin. Microbiol.*, **9**, 218–228.
4. Jenal, U. and Malone, J. (2006) Mechanisms of cyclic-di-GMP signaling in bacteria. *Annu. Rev. Genet.*, **40**, 385–407.
5. Ryan, R.P., Fouhy, Y., Lucey, J.F., and Dow, J.M. (2006) Cyclic di-GMP signaling in bacteria: recent advances and new puzzles. *J. Bacteriol.*, **188**, 8327–8334.
6. Dow, J.M., Fouhy, Y., Lucey, J.F., and Ryan, R.P. (2006) The HD-GYP domain, cyclic di-GMP signaling, and bacterial virulence to plants. *Mol. Plant-Microbe Interact.*, **19**, 1378–1384.
7. Tamayo, R., Pratt, J.T., and Camilli, A. (2007) Roles of cyclic diguanylate in the regulation of bacterial pathogenesis. *Annu. Rev. Microbiol.*, **61**, 131–148.
8. Cotter, P.A. and Stibitz, S. (2007) c-di-GMP-mediated regulation of virulence and biofilm formation. *Curr. Opin. Microbiol.*, **10**, 17–23.
9. Wolfe, A.J. and Visick, K.L. (2008) Get the message out: cyclic-di-GMP regulates multiple levels of flagellum-based motility. *J. Bacteriol.*, **190**, 463–475.
10. Hengge, R. (2009) Principles of cyclic-di-GMP signalling in bacteria. *Nat. Rev.*, **7**, 263–273.
11. Pesavento, C. and Hengge, R. (2009) Bacterial nucleotide-based second messengers. *Curr. Opin. Microbiol.*, **12**, 170–176.
12. Tal, R., Wong, H.C., Calhoon, R., Gelfand, D., Fear, A.L., Volman, G., Mayer, R., Ross, P., Amikam, D., Weinhouse, H. *et al.* (1998) Three *cdg* operons control cellular turnover of cyclic di-GMP in *Acetobacter xylinum*: genetic organization and occurrence of conserved domains in isoenzymes. *J. Bacteriol.*, **180**, 4416–4425.
13. Paul, R., Weiser, S., Amiot, N.C., Chan, C., Schirmer, T., Giese, B., and Jenal, U. (2004) Cell cycle-dependent dynamic

localization of a bacterial response regulator with a novel di-guanylate cyclase output domain. *Genes Dev.*, **18**, 715–727.

14 Malone, J.G., Williams, R., Christen, M., Jenal, U., Spiers, A.J., and Rainey, P.B. (2007) The structure–function relationship of WspR, a *Pseudomonas fluorescens* response regulator with a GGDEF output domain. *Microbiology*, **153**, 980–994.

15 Chan, C., Paul, R., Samoray, D., Amiot, N.C., Giese, B., Jenal, U., and Schirmer, T. (2004) Structural basis of activity and allosteric control of diguanylate cyclase. *Proc. Natl. Acad. Sci. USA*, **101**, 17084–17089.

16 Wassmann, P., Chan, C., Paul, R., Beck, A., Heerklotz, H., Jenal, U., and Schirmer, T. (2007) Structure of BeF_3^--modified response regulator PleD: implications for diguanylate cyclase activation, catalysis, and feedback inhibition. *Structure*, **15**, 915–927.

17 Christen, B., Christen, M., Paul, R., Schmid, F., Folcher, M., Jenoe, P., Meuwly, M., and Jenal, U. (2006) Allosteric control of cyclic di-GMP signaling. *J. Biol. Chem.*, **281**, 32015–32024.

18 De, N., Pirruccello, M., Krasteva, P.V., Bae, N., Raghavan, R.V., and Sondermann, H. (2008) Phosphorylation-independent regulation of the diguanylate cyclase WspR. *PLoS Biol.*, **6**, e67.

19 Christen, M., Christen, B., Folcher, M., Schauerte, A., and Jenal, U. (2005) Identification and characterization of a cyclic di-GMP-specific phosphodiesterase and its allosteric control by GTP. *J. Biol. Chem.*, **280**, 30829–30837.

20 Schmidt, A.J., Ryjenkov, D.A., and Gomelsky, M. (2005) The ubiquitous protein domain EAL is a cyclic diguanylate-specific phosphodiesterase: enzymatically active and inactive EAL domains. *J. Bacteriol.*, **187**, 4774–4781.

21 Rao, F., Yang, Y., Qi, Y., and Liang, Z.X. (2008) Catalytic mechanism of cyclic di-GMP-specific phosphodiesterase: a study of the EAL domain-containing RocR from *Pseudomonas aeruginosa*. *J. Bacteriol.*, **190**, 3622–3631.

22 Tamayo, R., Tischler, A.D., and Camilli, A. (2005) The EAL domain protein VieA is a cyclic diguanylate phosphodiesterase. *J. Biol. Chem.*, **280**, 33324–33330.

23 Bobrov, A.G., Kirillina, O., and Perry, R.D. (2005) The phosphodiesterase activity of the HmsP EAL domain is required for negative regulation of biofilm formation in *Yersinia pestis*. *FEMS Microbiol. Lett.*, **247**, 123–130.

24 Tarutina, M., Ryjenkov, D.A., and Gomelsky, M. (2006) An unorthodox bacteriophytochrome from *Rhodobacter sphaeroides* involved in turnover of the second messenger c-di-GMP. *J. Biol. Chem.*, **281**, 34751–34758.

25 Galperin, M.Y., Natale, D.A., Aravind, L., and Koonin, E.V. (1999) A specialized version of the HD hydrolase domain implicated in signal transduction. *J. Mol. Microbiol. Biotechnol.*, **1**, 303–305.

26 Ryan, R.P., Fouhy, Y., Lucey, J.F., Crossman, L.C., Spiro, S., He, Y.W., Zhang, L.H., Heeb, S., Camara, M., Williams, P. *et al.* (2006) Cell–cell signaling in *Xanthomonas campestris* involves an HD-GYP domain protein that functions in cyclic di-GMP turnover. *Proc. Natl. Acad. Sci. USA*, **103**, 6712–6717.

27 Andrade, M.O., Alegria, M.C., Guzzo, C.R., Docena, C., Rosa, M.C., Ramos, C.H., and Farah, C.S. (2006) The HD-GYP domain of RpfG mediates a direct linkage between the Rpf quorum-sensing pathway and a subset of diguanylate cyclase proteins in the phytopathogen *Xanthomonas axonopodis* pv *citri*. *Mol. Microbiol.*, **62**, 537–551.

28 Weber, H., Pesavento, C., Possling, A., Tischendorf, G., and Hengge, R. (2006) Cyclic-di-GMP-mediated signalling within the sigma network of *Escherichia coli*. *Mol. Microbiol.*, **62**, 1014–1034.

29 Takahashi, H. and Shimizu, T. (2006) Phosphodiesterase activity of Ec DOS, a heme-regulated enzyme from *Escherichia coli*, toward 3′,5′-cyclic diguanylic acid is obviously enhanced by O_2 and CO binding. *Chem. Lett.*, **35**, 970–971.

30 Suzuki, K., Babitzke, P., Kushner, S.R., and Romeo, T. (2006) Identification of a novel regulatory protein (CsrD) that targets the global regulatory RNAs CsrB and CsrC for degradation by RNase E. *Genes Dev.*, **20**, 2605–2617.

31 Tschowri, N., Busse, S., and Hengge, R. (2008) The BLUF-EAL protein YcgF acts as a direct anti-repressor in a blue light stress response of *Escherichia coli*. *Genes Dev.*, **23**, 522–534.

32 Holland, L.M., O'Donnell, S.T., Ryjenkov, D.A., Gomelsky, L., Slater, S.R., Fey, P.D., Gomelsky, M., and O'Gara, J.P. (2008) A staphylococcal GGDEF domain protein regulates biofilm formation independently of cyclic dimeric GMP. *J. Bacteriol.*, **190**, 5178–5189.

33 Galperin, M.Y., Nikolskaya, A.N., and Koonin, E.V. (2001) Novel domains of the prokaryotic two-component signal transduction systems. *FEMS Microbiol. Lett.*, **203**, 11–21.

34 Galperin, M.Y. (2004) Bacterial signal transduction network in a genomic perspective. *Environ. Microbiol.*, **6**, 552–567.

35 Chang, A.L., Tuckerman, J.R., Gonzalez, G., Mayer, R., Weinhouse, H., Volman, G., Amikam, D., Benziman, M., and Gilles-Gonzalez, M.A. (2001) Phosphodiesterase A1, a regulator of cellulose synthesis in *Acetobacter xylinum*, is a heme-based sensor. *Biochemistry*, **40**, 3420–3426.

36 Tanaka, A., Takahashi, H., and Shimizu, T. (2007) Critical role of the heme axial ligand, Met95, in locking catalysis of the phosphodiesterase from *Escherichia coli* (Ec DOS) toward Cyclic diGMP. *J. Biol. Chem.*, **282**, 21301–21307.

37 El-Mashtoly, S.F., Takahashi, H., Shimizu, T., and Kitagawa, T. (2007) Ultraviolet resonance Raman evidence for utilization of the heme 6-propionate hydrogen-bond network in signal transmission from heme to protein in Ec DOS protein. *J. Am. Chem. Soc.*, **129**, 3556–3563.

38 El-Mashtoly, S.F., Nakashima, S., Tanaka, A., Shimizu, T., and Kitagawa, T. (2008) Roles of Arg-97 and Phe-113 in regulation of distal ligand binding to heme in the sensor domain of Ec DOS protein. Resonance Raman and mutation study. *J. Biol. Chem.*, **283**, 19000–19010.

39 Rajagopal, S., Key, J.M., Purcell, E.B., Boerema, D.J., and Moffat, K. (2004) Purification and initial characterization of a putative blue light-regulated phosphodiesterase from *Escherichia coli*. *Photochem. Photobiol.*, **80**, 542–547.

40 Hasegawa, K., Masuda, S., and Ono, T.A. (2006) Light induced structural changes of a full-length protein and its BLUF domain in YcgF (Blrp), a blue-light sensing protein that uses FAD (BLUF). *Biochemistry*, **45**, 3785–3793.

41 Paul, R., Abel, S., Wassmann, P., Beck, A., Heerklotz, H., and Jenal, U. (2007) Activation of the diguanylate cyclase PleD by phosphorylation-mediated dimerization. *J. Biol. Chem.*, **282**, 29170–29177.

42 Martinez-Wilson, H.F., Tamayo, R., Tischler, A.D., Lazinski, D.W., and Camilli, A. (2008) The *Vibrio cholerae* hybrid sensor kinase VieS contributes to motility and biofilm regulation by altering cyclic diguanylate level. *J. Bacteriol.*, **190**, 6439–6447.

43 Sommerfeldt, N., Possling, A., Becker, G., Pesavento, C., Tschowri, N., and Hengge, R. (2008) Expression patterns and differential input into curli fimbriae regulation of all GGDEF/EAL genes in *Escherichia coli*. *Microbiology*, **155**, 1318–1331.

44 Amikam, D. and Galperin, M.Y. (2006) PilZ domain is part of the bacterial c-di-GMP binding protein. *Bioinformatics*, **22**, 3–6.

45 Ryjenkov, D.A., Simm, R., Romling, U., and Gomelsky, M. (2006) The PilZ domain is a receptor for the second messenger c-di-GMP: the PilZ domain protein YcgR controls motility in enterobacteria. *J. Biol. Chem.*, **281**, 30310–30314.

46 Christen, M., Christen, B., Allan, M.G., Folcher, M., Jeno, P., Grzesiek, S., and Jenal, U. (2007) DgrA is a member of a new family of cyclic diguanosine monophosphate receptors and controls flagellar motor function in *Caulobacter crescentus*. *Proc. Natl. Acad. Sci. USA*, **104**, 4112–4117.

47 Pratt, J.T., Tamayo, R., Tischler, A.D., and Camilli, A. (2007) PilZ domain proteins bind cyclic diguanylate and regulate diverse processes in *Vibrio cholerae*. *J. Biol. Chem.*, **282**, 12860–12870.

48 Merighi, M., Lee, V.T., Hyodo, M., Hayakawa, Y., and Lory, S. (2007) The second messenger bis-(3′–5′)-cyclic-GMP and its PilZ domain-containing receptor Alg44 are required for alginate biosynthesis in *Pseudomonas aeruginosa*. *Mol. Microbiol.*, **65**, 876–895.

49 Benach, J., Swaminathan, S.S., Tamayo, R., Handelman, S.K., Folta-Stogniew, E., Ramos, J.E., Forouhar, F., Neely, H., Seetharaman, J., Camilli, A. et al. (2007) The structural basis of cyclic diguanylate signal transduction by PilZ domains. *EMBO J.*, **26**, 5153–5166.

50 Lee, V.T., Matewish, J.M., Kessler, J.L., Hyodo, M., Hayakawa, Y., and Lory, S. (2007) A cyclic-di-GMP receptor required for bacterial exopolysaccharide production. *Mol. Microbiol.*, **65**, 1474–1484.

51 Hickman, J.W. and Harwood, C.S. (2008) Identification of FleQ from *Pseudomonas aeruginosa* as a c-di-GMP-responsive transcription factor. *Mol. Microbiol.*, **69**, 376–389.

52 Duerig, A., Abel, S., Folcher, M., Nicollier, M., Schwede, T., Amiot, N., Giese, B., and Jenal, U. (2009) Second messenger mediated spatiotemporal control of protein degradation regulates bacterial cell cycle progression. *Genes Dev.*, **23**, 93–104.

53 Weinberg, Z., Barrick, J.E., Yao, Z., Roth, A., Kim, J.N., Gore, J., Wang, J.X., Lee, E.R., Block, K.F., Sudarsan, N. et al. (2007) Identification of 22 candidate structured RNAs in bacteria using the CMfinder comparative genomics pipeline. *Nucleic Acids Res.*, **35**, 4809–4819.

54 Sudarsan, N., Lee, E.R., Weinberg, Z., Moy, R.H., Kim, J.N., Link, K.H., and Breaker, R.R. (2008) Riboswitches in eubacteria sense the second messenger cyclic di-GMP. *Science*, **321**, 411–413.

55 Girgis, H.S., Liu, Y., Ryu, W.S., and Tavazoie, S. (2007) A comprehensive genetic characterization of bacterial motility. *PLoS Genet.*, **3**, 1644–1660.

56 Huang, B., Whitchurch, C.B., and Mattick, J.S. (2003) FimX, a multidomain protein connecting environmental signals to twitching motility in *Pseudomonas aeruginosa*. *J. Bacteriol.*, **185**, 7068–7076.

57 Kazmierczak, B.I., Lebron, M.B., and Murray, T.S. (2006) Analysis of FimX, a phosphodiesterase that governs twitching motility in *Pseudomonas aeruginosa*. *Mol. Microbiol.*, **60**, 1026–1043.

58 Kader, A., Simm, R., Gerstel, U., Morr, M., and Romling, U. (2006) Hierarchical involvement of various GGDEF domain proteins in rdar morphotype development of *Salmonella enterica* serovar Typhimurium. *Mol. Microbiol.*, **60**, 602–616.

59 Pesavento, C., Becker, G., Sommerfeldt, N., Possling, A., Tschowri, N., Mehlis, A., and Hengge, R. (2008) Inverse regulatory coordination of motility and curli-mediated adhesion in *Escherichia coli*. *Genes Dev.*, **22**, 2434–2446.

60 Arnqvist, A., Olsen, A., Pfeifer, J., Russell, D.G., and Normark, S. (1992) The Crl protein activates cryptic genes for curli formation and fibronectin binding in *Escherichia coli* HB101. *Mol. Microbiol.*, **6**, 2443–2452.

61 Arnqvist, A., Olsen, A., and Normark, S. (1994) Sigma S-dependent growth-phase induction of the *csgBA* promoter in *Escherichia coli* can be achieved *in vivo* by sigma 70 in the absence of the nucleoid-associated protein H-NS. *Mol. Microbiol.*, **13**, 1021–1032.

62 Olsen, A., Jonsson, A., and Normark, S. (1989) Fibronectin binding mediated by a novel class of surface organelles on *Escherichia coli*. *Nature*, **338**, 652–655.

63 Gerstel, U. and Romling, U. (2003) The *csgD* promoter, a control unit for biofilm formation in *Salmonella typhimurium*. *Res. Microbiol.*, **154**, 659–667.

64 Brown, P.K., Dozois, C.M., Nickerson, C.A., Zuppardo, A., Terlonge, J., and Curtiss, R. 3rd. (2001) MlrA, a novel regulator of curli (AgF) and extracellular matrix synthesis by *Escherichia coli* and *Salmonella enterica* serovar Typhimurium. *Mol. Microbiol.*, **41**, 349–363.

65 Hammar, M., Arnqvist, A., Bian, Z., Olsen, A., and Normark, S. (1995) Expression of two *csg* operons is required for production of fibronectin- and Congo

Red-binding curli polymers in *Escherichia coli* K-12. *Mol. Microbiol.*, **18**, 661–670.

66 Olsen, A., Arnqvist, A., Hammar, M., Sukupolvi, S., and Normark, S. (1993) The RpoS sigma factor relieves H-NS-mediated transcriptional repression of *csgA*, the subunit gene of fibronectin-binding curli in *Escherichia coli*. *Mol. Microbiol.*, **7**, 523–536.

67 Romling, U., Bian, Z., Hammar, M., Sierralta, W.D., and Normark, S. (1998) Curli fibers are highly conserved between *Salmonella typhimurium* and *Escherichia coli* with respect to operon structure and regulation. *J. Bacteriol.*, **180**, 722–731.

68 Adler, J. and Templeton, B. (1967) The effect of environmental conditions on the motility of *Escherichia coli*. *J. Gen. Microbiol.*, **46**, 175–184.

69 Amsler, C.D., Cho, M., and Matsumura, P. (1993) Multiple factors underlying the maximum motility of *Escherichia coli* as cultures enter post-exponential growth. *J. Bacteriol.*, **175**, 6238–6244.

70 Zogaj, X., Nimtz, M., Rohde, M., Bokranz, W., and Romling, U. (2001) The multicellular morphotypes of *Salmonella typhimurium* and *Escherichia coli* produce cellulose as the second component of the extracellular matrix. *Mol. Microbiol.*, **39**, 1452–1463.

71 Romling, U., Rohde, M., Olsen, A., Normark, S., and Reinkoster, J. (2000) AgfD, the checkpoint of multicellular and aggregative behaviour in *Salmonella typhimurium* regulates at least two independent pathways. *Mol. Microbiol.*, **36**, 10–23.

72 Simm, R., Morr, M., Kader, A., Nimtz, M., and Romling, U. (2004) GGDEF and EAL domains inversely regulate cyclic di-GMP levels and transition from sessility to motility. *Mol. Microbiol.*, **53**, 1123–1134.

73 Claret, L., Miquel, S., Vieille, N., Ryjenkov, D.A., Gomelsky, M., and Darfeuille-Michaud, A. (2007) The flagellar sigma factor FliA regulates adhesion and invasion of Crohn disease-associated *Escherichia coli* via a cyclic dimeric GMP-dependent pathway. *J. Biol. Chem.*, **282**, 33275–33283.

74 Simm, R., Lusch, A., Kader, A., Andersson, M., and Romling, U. (2007) Role of EAL-containing proteins in multicellular behavior of *Salmonella enterica* serovar Typhimurium. *J. Bacteriol.*, **189**, 3613–3623.

75 Da Re, S. and Ghigo, J.M. (2006) A CsgD-independent pathway for cellulose production and biofilm formation in *Escherichia coli*. *J. Bacteriol.*, **188**, 3073–3087.

76 Matson, J.S., Withey, J.H., and DiRita, V.J. (2007) Regulatory networks controlling *Vibrio cholerae* virulence gene expression. *Infect. Immun.*, **75**, 5542–5549.

77 Tischler, A.D. and Camilli, A. (2004) Cyclic diguanylate (c-di-GMP) regulates *Vibrio cholerae* biofilm formation. *Mol. Microbiol.*, **53**, 857–869.

78 Ali, A., Mahmud, Z.H., Morris, J.G. Jr., Sozhamannan, S., and Johnson, J.A. (2000) Sequence analysis of TnphoA insertion sites in *Vibrio cholerae* mutants defective in rugose polysaccharide production. *Infect. Immun.*, **68**, 6857–6864.

79 Yildiz, F.H. and Schoolnik, G.K. (1999) *Vibrio cholerae* O1 El Tor: identification of a gene cluster required for the rugose colony type, exopolysaccharide production, chlorine resistance, and biofilm formation. *Proc. Natl. Acad. Sci. USA*, **96**, 4028–4033.

80 Tischler, A.D. and Camilli, A. (2005) Cyclic diguanylate regulates *Vibrio cholerae* virulence gene expression. *Infect. Immun.*, **73**, 5873–5882.

81 Schild, S., Tamayo, R., Nelson, E.J., Qadri, F., Calderwood, S.B., and Camilli, A. (2007) Genes induced late in infection increase fitness of *Vibrio cholerae* after release into the environment. *Cell Host Microbe*, **2**, 264–277.

82 Waters, C.M., Lu, W., Rabinowitz, J.D., and Bassler, B.L. (2008) Quorum sensing controls biofilm formation in *Vibrio cholerae* through modulation of cyclic di-GMP levels and repression of vpsT. *J. Bacteriol.*, **190**, 2527–2536.

83 Kovacikova, G., Lin, W., and Skorupski, K. (2005) Dual regulation of genes involved in acetoin biosynthesis and motility/biofilm formation by the virulence activator AphA and the acetate-responsive LysR-type regulator AlsR in

Vibrio cholerae. Mol. Microbiol., **57**, 420–433.

84 Lim, B., Beyhan, S., and Yildiz, F.H. (2007) Regulation of *Vibrio* polysaccharide synthesis and virulence factor production by CdgC, a GGDEF/EAL domain protein, in *Vibrio cholerae*. *J. Bacteriol.*, **189**, 717–729.

85 Fong, J.C. and Yildiz, F.H. (2008) Interplay between cAMP–CRP and c-di-GMP signaling in *Vibrio cholerae* biofilm formation. *J. Bacteriol.*, **190**, 6646–6659.

86 Lim, B., Beyhan, S., Meir, J., and Yildiz, F.H. (2006) Cyclic-diGMP signal transduction systems in *Vibrio cholerae*: modulation of rugosity and biofilm formation. *Mol. Microbiol.*, **60**, 331–348.

87 Yildiz, F.H., Liu, X.S., Heydorn, A., and Schoolnik, G.K. (2004) Molecular analysis of rugosity in a *Vibrio cholerae* O1 El Tor phase variant. *Mol. Microbiol.*, **53**, 497–515.

88 Rashid, M.H., Rajanna, C., Ali, A., and Karaolis, D.K. (2003) Identification of genes involved in the switch between the smooth and rugose phenotypes of *Vibrio cholerae*. *FEMS Microbiol. Lett.*, **227**, 113–119.

89 Bomchil, N., Watnick, P., and Kolter, R. (2003) Identification and characterization of a *Vibrio cholerae* gene, *mbaA*, involved in maintenance of biofilm architecture. *J. Bacteriol.*, **185**, 1384–1390.

90 Beyhan, S. and Yildiz, F.H. (2007) Smooth to rugose phase variation in *Vibrio cholerae* can be mediated by a single nucleotide change that targets c-di-GMP signalling pathway. *Mol. Microbiol.*, **63**, 995–1007.

91 Beyhan, S., Tischler, A.D., Camilli, A., and Yildiz, F.H. (2006) Differences in gene expression between the classical and El Tor biotypes of *Vibrio cholerae* O1. *Infect. Immun.*, **74**, 3633–3642.

92 Laub, M.T., Shapiro, L., and McAdams, H.H. (2007) Systems biology of *Caulobacter. Annu. Rev. Genet.*, **41**, 429–441.

93 Lawler, M.L. and Brun, Y.V. (2007) Advantages and mechanisms of polarity and cell shape determination in *Caulobacter crescentus. Curr. Opin. Microbiol.*, **10**, 630–637.

94 Aldridge, P. and Jenal, U. (1999) Cell cycle-dependent degradation of a flagellar motor component requires a novel-type response regulator. *Mol. Microbiol.*, **32**, 379–391.

95 Hecht, G.B. and Newton, A. (1995) Identification of a novel response regulator required for the swarmer-to-stalked-cell transition in *Caulobacter crescentus. J. Bacteriol.*, **177**, 6223–6229.

96 Aldridge, P., Paul, R., Goymer, P., Rainey, P., and Jenal, U. (2003) Role of the GGDEF regulator PleD in polar development of *Caulobacter crescentus. Mol. Microbiol.*, **47**, 1695–1708.

97 Wheeler, R.T. and Shapiro, L. (1999) Differential localization of two histidine kinases controlling bacterial cell differentiation. *Mol. Cell*, **4**, 683–694.

98 Huitema, E., Pritchard, S., Matteson, D., Radhakrishnan, S.K., and Viollier, P.H. (2006) Bacterial birth scar proteins mark future flagellum assembly site. *Cell*, **124**, 1025–1037.

99 Iniesta, A.A., McGrath, P.T., Reisenauer, A., McAdams, H.H., and Shapiro, L. (2006) A phospho-signaling pathway controls the localization and activity of a protease complex critical for bacterial cell cycle progression. *Proc. Natl. Acad. Sci. USA*, **103**, 10935–10940.

100 McGrath, P.T., Iniesta, A.A., Ryan, K.R., Shapiro, L., and McAdams, H.H. (2006) A dynamically localized protease complex and a polar specificity factor control a cell cycle master regulator. *Cell*, **124**, 535–547.

22
ppGpp Signaling
Rolf Wagner

22.1
Introduction

ppGpp (guanosine 5′,3′-(bis)diphosphate) and pppGpp (guanosine 5′-triphosphate, 3′-diphosphate) are known as hormone-like effector molecules, often termed alarmones, responsible for a global bacterial response to nutritional deprivation and stress. The accumulating nucleotide derivatives were first observed upon amino acid starvation and termed "magic spots" due to their rapid appearance on thin-layer chromatograms of extracts from starved cells. Concomitant with the increase of (p)ppGpp it was found that cells undergo a dramatic change in physiology with an instantaneous decline in stable RNA (rRNA and tRNA) synthesis as the most outstanding reaction. This phenomenon, termed "stringent response", had been studied for more than 40 years and had led to many controversial results in the past. Although a more consistent picture has emerged today, we do not understand all the mechanistic details behind this central control system of bacterial physiology. This chapter focuses on the role of (p)ppGpp as global signaling molecule; because it cannot cover all the aspects of this complex phenomenon, the reader is referred to several comprehensive recent reviews [1–7].

Although a decline in stable RNA synthesis is the most striking effect triggered by elevated (p)ppGpp levels, the physiological changes associated with the concentration of these alarmones are effectively global and not restricted to bacteria undergoing amino acid starvation. The concentration of the effector molecule (p)ppGpp has been recognized as a central response, linking different kinds of stress reactions crucial for the adaptation and adjustment of cell growth in response to altered conditions. As such, (p)ppGpp levels are also involved in controlling steady-state growth. Compelling evidence has demonstrated an almost linear inverse correlation between bacterial growth rates and the basal cellular ppGpp concentration, which in contrast to the stringent control under amino acid deprivation varies in the micromolar range [8]. The stringent control phenomenon more or less denotes the situation of very high (p)ppGpp concentrations and this response affects all levels of gene expression, inducing changes in transcription, translation, and replication. Moreover, it is important for bacterial differentiation, virulence induction, and long-term

Bacterial Signaling. Edited by Reinhard Krämer and Kirsten Jung
Copyright © 2010 WILEY-VCH Verlag GmbH & Co. KGaA, Weinheim
ISBN: 978-3-527-32365-4

persistence under stress. Hence, (p)ppGpp signaling generally promotes survival under unfavorable conditions. One essential feature of ppGpp signaling is the speed at which it reprograms cell physiology and, due to the very short half-life of the mediator(s) (p)ppGpp, allows the cell to resume normal growth upon changing conditions. Apart from being almost universal in the bacterial kingdom, recent studies have demonstrated that (p)ppGpp signaling is also found in plants, where it contributes to the defense systems and is possibly involved in coping with various forms of environmental stresses plants have to withstand.

22.2
Induction of the Effector (p)ppGpp Through Synthesis and Degradation

After induction, the accumulation of (p)ppGpp to steady-state levels is accomplished by enzymes that synthesize and degrade the effector molecules in response to metabolic or environmental changes (Figure 22.1). The enzymes responsible for the synthesis and turnover of (p)ppGpp in all organisms today are termed RSHs (Rel Spo homologs), based on the RelA (synthetase I) and SpoT (synthetase II) enzymes, which are responsible for this reaction in *E. coli* and many other Gram-negative bacteria from the β and γ subdivision of proteobacteria. RelA is known as a ribosome-associated protein, which synthesizes (p)ppGpp from GTP or GDP by transfer of pyrophosphate from ATP to the respective ribose 3′-OH position. The reaction is induced upon the appearance of deacylated transfer RNAs, which enter the ribosomal active site due to shortage of amino acids or defects in aminoacyl synthetases. The pentaphosphate pppGpp normally emerges as the first product, which is then rapidly converted to ppGpp by a 5′-phosphohydrolase. Accumulation of the two effector molecules is rapid, reaching millimolar concentrations in the cell within minutes. A second route for (p)ppGpp synthesis (in Gram-negative bacteria) is catalyzed by the bifunctional enzyme SpoT. This enzyme has a major function in (p)ppGpp degradation, but exhibits an independent synthesizing activity, which is induced upon deprivation of carbon sources, iron, phosphate, fatty acids, or general stress conditions. By shifting the two opposing hydrolytic and synthesizing activities, SpoT is responsible for preserving and adjusting the basal level of (p)ppGpp (in the micromolar range) responsible for maintenance or changes in cell growth. The mechanisms how this balance is regulated and how SpoT senses the need to adjust the opposing synthetic or hydrolytic activities are largely unknown, except for some specific cases. A recent study has shed light on the question how SpoT activity can be modulated by the acyl carrier protein (ACP) – an essential cofactor for fatty acid biosynthesis that had been shown to interact directly with SpoT [9]. The 9-kDa ACP protein must be modified post-translationally by a 4′-phosphopantethein prosthetic group, enabling the binding of fatty acid intermediates. Depending on the status of fatty acid metabolism, the fraction of the functional form of ACP is altered and the conformation of the bound ACP changes. This altered conformation results in an allosteric transition in SpoT. As a consequence, the balance of the two opposing activities of SpoT is turned, leading to the accumulation of (p)ppGpp and thus linking

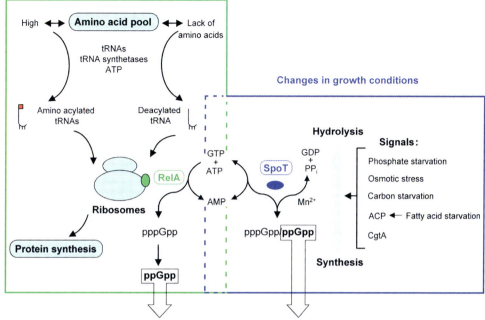

Figure 22.1 Scheme for the induction, synthesis, and turnover of ppGpp in *E. coli*. In *E. coli* two pathways control the accumulation of ppGpp. The left panel shows RelA-dependent (p)ppGpp synthesis from GTP and ATP, according to the stringent response, which is induced when deacylated tRNAs bind to the ribosomal active site. Hence, this reaction depends on the ratio of acylated versus deacylated tRNAs. The right panel shows the SpoT-dependent synthesis and hydrolysis of ppGpp as a function of changing growth conditions. The balance of this reaction is affected by the signals indicated. See text for details.

fatty acid starvation to the SpoT-dependent network. The same mechanism might also account for the SpoT-mediated synthesis of (p)ppGpp upon carbon source starvation, because carbon deprivation ultimately also leads to fatty acid starvation. Moreover, the activity of SpoT might generally be modulated by a number of additional cellular factors, such as small G-proteins, and thereby linked to diverse inputs. One example is the conserved bacterial G-protein CgtA (Obg), which had been shown to bind to SpoT, thereby causing reduced ppGpp levels in a nutrient-rich environment [10, 11].

Bacillus subtilis and many other Gram-positive bacteria lack SpoT homologs and a single bifunctional RSH enzyme appears to be responsible for (p)ppGpp synthesis. Recently, the existence of two small enzymes with synthesis activity, but lacking the hydrolase, have been characterized in *Streptococcus mutans*, *Bacillus subtilis*, and bacteria belonging to the phylum Firmicutes [12, 13]. These enzymes coexist with the bifunctional RSH enzymes, yet their contribution to the regulation of (p)ppGpp synthesis is presently not known.

22.3
ppGpp – A *Bona Fide* Global Regulator

According to classical terms, (p)ppGpp is often understood as the elicitor of the stringent control, which is characterized by a rapid decline in stable RNA synthesis and growth arrest. It must be emphasized, however, that ppGpp signaling is much more general and actually constitutes a real global control system, interfering with all major biosynthetic reactions, such as replication, transcription, and translation, which are basic for gene expression, cell growth, or maintenance (Figure 22.2). Moreover, central pathways for maintenance and survival, such as cell cycle signals, adaptation to physical stress, induction of virulence, altered mutation frequency, bacterial programmed cell death, phage development, or quorum sensing (QS), are all integrated in the ppGpp signaling network.

22.3.1
Transcriptional Profiling in Different Bacterial Strains

Recent genome-wide studies in different bacterial strains, taking advantage of the "omics" methodology, have made comprehensive contributions to our present knowledge of the many genes affected by (p)ppGpp. Next to *E. coli*, the list of strains that have been analyzed by transcriptional profiling includes *Streptococcus mutans*, *Streptomyces coelicolor*, or the Gram-positive organisms *Corynebacterium glutamicum*, *Bacillus subtilis*, and *Vibrio cholerae*. Microarray studies under conditions of the

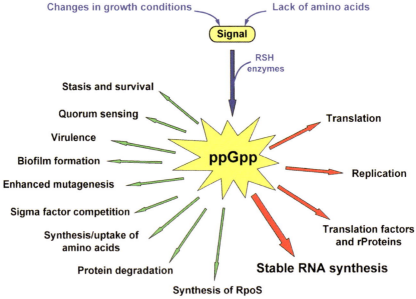

Figure 22.2 Global effects of ppGpp on cell physiology. The cartoon depicts the global position of ppGpp as a signaling molecule for bacterial physiology.

stringent response or under growth arrest have been performed, comparing bacteria that can or cannot synthesize ppGpp due to mutations in *rsh* genes. Some studies were complemented by a proteome or time-resolved approach to facilitate the distinction between primary and secondary effects. Only some of the major conclusions from the long list of detailed data are summarized here [10, 14–21].

Generally, genes constituting the transcription and translation apparatus were shown to be downregulated, as were DNA synthesis and cell growth, while genes involved in the stationary-phase stress response or genes for the nucleoid-associated proteins, which function as regulators during environmental changes, were induced. Moreover, addiction module toxin antitoxin genes involved in bacterial programmed cell death or virulence genes were also affected. Among the upregulated genes, stress response sigma factors *rpoE*, *rpoH*, and *rpoS* were frequently found. In *E. coli*, induction of amino acid biosynthetic genes was found restricted to attenuator-regulated operons.

22.3.2
Lack of (p)ppGpp Signaling in Obligate Intracellular Bacterial Pathogens and Archaea

While all the above studies show that the response of (p)ppGpp is very similar in different bacteria living in free environments, the situation becomes totally different for species that have lost their capacity to exist as free-living organisms. Most notably, host-adapted obligate intracellular bacterial pathogens, such as *Chlamydia*, have lost the genes for stringent control during reductive evolution [22]. Likewise, it had been found in different strains of the eubacterium *Helicobacter pylori* that no stringent response is elicited although the enzymes responsible for (p)ppGpp production and turnover are present. This observation suggests that adaptation of cell physiology mediated by the stringent control is less important for the fitness of prokaryotes growing in protected niches.

Archaea have also been analyzed as to whether they show stringent control of stable RNA synthesis and respond to amino acid deprivation as eubacteriae. The studies show that Archaea can have either stringent or relaxed phenotypes with respect to stable RNA accumulation upon inhibition of tRNA aminoacylation. Although the arrest of rRNA synthesis is widespread in Archea, the alarmones (p)ppGpp are completely absent during normal growth or after induction of the stringent reaction in these organisms [23–25]. This finding points to RNA polymerase as an important component of the ppGpp-mediated mechanism of the stringent control because archaeal RNA polymerases are distinctly different from their eubacterial counterparts.

22.3.3
ppGpp in Plants

A number of studies have shown that eubacterial stringent control mediated by ppGpp has been conserved during evolution of the chloroplast from the photosynthetic bacterial symbiont. RSH enzymes had been identified in chloroplasts of higher plants [26, 27] and their activity to synthesize (p)ppGpp in a ribosome-associated reaction has been demonstrated, for instance, in extracts from pea [28–30]. Moreover, RSH-like enzymes have been identified as Ca^{2+}-activated (p)ppGpp synthetases in the chloroplasts of monocotyledonous and dicotyledonous land plants, where they regulate the genetic systems of these organelles [31]. Apparently, ppGpp signaling in plants does not only serve as a system to sense nutritional changes, but is integrated in the plant signaling cascade to cope with stresses, such as wounding, the action of hormones (e.g., jasmonic acid), or physical stress. Moreover, it has also been documented that ppGpp controls fertilization in plants. A recent mutational study in *Arabidopsis* indicated that plant reproduction is under the control of chloroplast function through ppGpp signaling [32].

22.3.4
(p)ppGpp as a Mediator of Bacterial Social Behavior and Cell–Cell Signaling Mechanisms

Bacterial cells have evolved systems to communicate with each other or even across species, enabling a colony, for instance, to determine the population density – a phenomenon termed QS. QS induces an altered pattern of gene expression, giving rise to important phenomena, such as biofilm formation or invasion of host cells by bacterial pathogens [33]. High population density, as found in biofilms, is consistent with limited availability of nutrients and thus might be comparable with a situation of starvation. This view is supported by recent studies demonstrating a more central role of QS in bacterial physiology, such as response to stress or entry into stationary growth [34]. A connection between QS and the stress- and starvation-sensing (p)ppGpp mechanism is thus obvious. In fact, QS and biofilm formation have been shown to be coupled to the ppGpp signaling pathway [35–37].

22.3.5
(p)ppGpp Signaling in Virulence and Pathogen–Host Interaction

It is not surprising that ppGpp signaling is also involved in pathogenicity, where bacteria–host interactions are important and ppGpp signaling has unequivocally been demonstrated to be involved in bacterial virulence gene expression [38]. For instance, the expression of *hilA* and *invF*, two major transcriptional activators for *Salmonella typhimurium* pathogenicity island-1 genes, is severely reduced in (p)ppGpp-free strains, indicating that (p)ppGpp plays a central role as regulator for virulence gene expression in *S. typhimurium* [39]. Moreover, it could be demonstrated that mutants defective in the synthesis of the stringent signal molecules have virtually lost their virulence. Such avirulent mutants have been used successfully as live vaccine for the immunization of mice [40, 41]. Other examples show that the adherence capacity of enterohemorrhagic *E. coli* is dependent on an increase of the cellular ppGpp concentration (and also depends on the expression of DksA – the accessory protein for stringent regulation), suggesting that the regulation of virulence genes in enterohemorrhagic *E. coli* is also integrated in the stringent response system of *E. coli* [42].

RelA and SpoT homologs are known to be implicated in a differential way in bacterial virulence, but the mechanisms how (p)ppGpp controls virulence gene expression are not completely understood. It seems clear, however, that in addition to the activity of RNA polymerase, indirect effects such as (p)ppGpp-dependent sigma factor competition or induction of other stress factors play a major role. One has to conclude, therefore, that due to the complexity of the regulatory pathways the relationship between the stringent response and virulence might be unique for each pathogen and the virulence genes it controls [43].

22.3.6
(p)ppGpp as a Regulator for Toxin–Antitoxin Systems in Bacterial Programmed Cell Death

The suicide function of toxin–antitoxins was originally described in prokaryotes as an extrachromosomal system responsible for postsegregational killing. Many bacteria contain chromosomal addiction modules, whilst also encoding stable toxins and labile antitoxins involved in programmed cell death or bacteriostasis. Programmed cell death is induced when the antitoxins are degraded. Typical well-studied examples are the *mazEF(G)* or the *relBE* systems [44, 45]. There has been some debate about the advantage of programmed cell death for unicellular systems such as bacteria. It is clear, however, that the altruistic suicide of an individual cell may be beneficial for the survival of the whole cell population. Programmed cell death, for instance, can arrest the spread of infectious phages [46] or it may otherwise act to maintain the genomic stability of a population by eliminating cells that carry extensive genomic defects. It has also been discussed that under severe nutritional stress the death of part of the population may provide nutrients for the surviving cells – a situation that can be described as bacterial cannibalism. Interestingly, ppGpp signaling is considered to play a role in stress-induced suicide modules. For instance, the expression of the *E. coli mazEFG* system, encoded downstream of the *relA* gene, depends on the cellular level of (p)ppGpp. In addition, MazG has been shown to cause depletion of the (p)ppGpp level synthesized by RelA, also suggesting a role of ppGpp signaling in programmed cell death [45, 47].

22.3.7
Persister Cells and Enhanced Mutation Frequency

Some rare phenomena are known in bacterial adaptation for which ppGpp has been suggested as an inducing agent. One such phenomenon, termed persistence, is related to the ability of all bacteria to survive long-term exposure to antibiotics at a high frequency. Persister cells can be characterized by a noninherited phenotypic switch to a slow-growing state. Such dormant (nondividing) cells are important in biofilm formation and in the development of multidrug tolerance. Recent studies with *relA* and *spoT* mutants clearly suggest that the development of high persistence depends on the synthesis of ppGpp [48–51].

A different phenomenon has been described for which ppGpp might be responsible – stress-directed adaptive mutation. Under such conditions an increased mutation rate could be correlated to transcription events that, due to ppGpp-dependent arrest or pausing, result in localized supercoiling or prolonged unpaired DNA regions. Such single-stranded or conformationally deformed DNA regions are especially vulnerable to base modifications, leading to enhanced mutation frequencies [52].

22.4
Effects on Macromolecular Synthesis

As evidenced from genome-wide transcriptome and proteome analyses, and from direct studies of (p)ppGpp effects on replication, there is no doubt today that all three macromolecular synthesis reactions of the cell – replication, transcription, and translation – are directly affected by this global effector.

22.4.1
Role of ppGpp as an Inhibitor of Replication

Effects of (p)ppGpp on replication have been known for a long time, although our understanding of the mechanism has been changed [53]. Originally, in *E. coli*, inhibition of replication was believed to occur at the stage of initiation involving the initiator protein DnaA. In contrast, in *B. subtilis*, the stringent response inhibits replication elongation. It was proposed that the termination protein Rtp causes replication arrest at a site 100–200 kb to the left and right of *oriC* [54–56]. A recent study of the nutritional control of DNA replication in *B. subtilis* revealed, however, that after induction of the stringent response replication was immediately inhibited, regardless of where the replication forks were located on the chromosome. Moreover, it was shown that increased (p)ppGpp levels and not decreased GTP concentrations were responsible for the regulation, and that primase (DnaG) was the target for (p)ppGpp-dependent inhibition. The study also showed that RecA was not recruited to the arrested replication fork, which would activate a recombination or repair pathway. Hence, no disruption of the replication fork occurs during arrest [57]. This means that inhibition is reversible and replication can be resumed, once nutrients become available again, without the need for reinitiation of the replication cycle. In addition, the genomic stability of cells could be maintained during starving conditions.

Of particular interest is the fact that the primary target for (p)ppGpp-dependent inhibition of replication is primase, which is actually a DNA-dependent RNA synthesizing enzyme, analogous to RNA polymerase. As such, primase has a similar active center with a conserved divalent metal-binding site and conserved structural features for the template-directed binding of NTPs. Hence, it is plausible to speculate that (p)ppGpp might bind to and inhibit primase in a very similar manner as proposed for *E. coli* RNA polymerase.

22.4.2
Inhibition of Translation: Effect on Initiation Factor 2

Regulation of translation by (p)ppGpp is ultimately a consequence of the downregulation of the components that constitute the translation machinery – rRNAs, ribosomal proteins but also transfer RNAs, as well as the translation elongation factors EF-Tu and EF-G. The question whether the translation cycle itself is directly affected by the stringent response mediator(s) (p)ppGpp is less clear. The instantaneous decline in protein synthesis rates, however, argues for the assumption that the pre-existing protein synthesis apparatus must also be affected directly, because the observed response usually is too rapid to be explained by decay or altered *de novo* synthesis of ribosomes. Recently, direct binding of ppGpp to the translation initiation factor 2 (IF2) has been demonstrated, providing strong support for a direct inhibition of translation when ppGpp levels increase in the cell. The effector molecule binds with comparable affinity to the same nucleotide-binding site of IF2 as GTP and thereby inhibits IF2-dependent 30S initiation complex formation as well as dipeptide formation. As under conditions of the stringent control the ppGpp level reaches, or even exceeds, those of GTP, which itself simultaneously declines, the two nucleotides can replace each other effectively. IF2 has therefore been suggested as a metabolic sensor, which, depending on the nutritional situation, oscillates between the active GTP-bound and the inactive ppGpp-bound form, thus linking translation directly to ppGpp signaling [58].

22.5
Regulation of Transcription: RNA Polymerase is the Target

Although inhibition of stable RNA transcription is the most prominent effect of (p)ppGpp-dependent regulation, which has been studied most extensively over the years, a complete understanding of the detailed mechanism behind this phenomenon is still pending. There is no doubt today that RNA polymerase is the target of the mediator molecules. This has been established genetically, by studies with RNA polymerase mutants [59, 60], biochemically, using binding experiments [59, 61], and finally by high-resolution X-ray analysis of ppGpp–RNA polymerase cocrystals [62]. The presentation of the 2.7-Å structure of RNA polymerase in complex with ppGpp provided a major breakthrough in studying ppGpp-dependent transcriptional control and opened the path for many testable new experiments to understand the mechanism of the stringent control. The structure obtained from *Thermus thermophilus* RNA polymerase holoenzyme in complex with ppGpp revealed that the effector molecule binds to a single site, but in two alternate orientations near the active center of the enzyme. Consistent with the structure it was suggested that ppGpp forms base pairs with cytosines of the nontemplate DNA strand and may affect the coordination of the conserved Mg^{2+} ions located in the active site. The structure also suggests that ppGpp might affect the binding of incoming NTP substrates by modulating the active site configuration. Several of the conclusions derived from the structural investigation have recently been challenged, however [63]. RNA polymerase mutants with amino acid substitutions in the vicinity of the ppGpp-binding site had been constructed in *E. coli* without significantly reducing the ppGpp effects on transcription. This finding leads the authors to suggest that the ppGpp-binding pocket identified in the *T. thermophilus* structure is likely not the one responsible for regulation in *E. coli* RNA polymerase. Strangely, in the original study one of the substitutions was shown to reduce ppGpp sensitivity in *E. coli* [62]. Whether differences in the presence of the ω subunit in the two RNA polymerase preparations was responsible for the observed discrepancy, as the authors suggest, must await further studies.

22.5.1
Role of RNA Polymerase ω Subunit

A role for (p)ppGpp signaling has been described with a long controversial history for the smallest RNA polymerase subunit ω [64]. Initially, ω was shown to support RNA polymerase subunit assembly and stability [65]. The fact that the gene for ω is cotranscribed in a single operon together with *spoT* has given rise to the idea that ω might be involved in ppGpp signaling [66]. This assumption had been substantiated experimentally [67], but was later challenged by the finding that the stringent response was still observed in a strain where the gene for ω had been deleted [68]. These apparently conflicting reports have recently been solved by studies employing reconstituted RNA polymerase preparations together with the stringent control cofactor DksA [69]. The authors demonstrated that DksA eliminates the ω requirement for the response of RNA polymerase to ppGpp. They could show that DksA is able to rescue the ppGpp unresponsiveness of RNA polymerase preparations lacking the ω subunit, indicating that it likely participates in the stringent control mechanism.

22.5.2
Promoter-Specific Effects of (p)ppGpp

Most studies on the role of ppGpp for transcription have been focused on initiation, which generally is considered to be the rate-limiting step. Consequently, many open questions are concerned with the particular DNA structures that might define ppGpp-sensitive promoters. A GC-rich sequence element downstream of the −10 recognition sequence was first recognized as a common denominator, important for promoters under negative stringent control in *E. coli* [70]. This structure has been termed the "stringent discriminator" and it was subsequently shown to be a necessary, but not sufficient, structural element causing negative (p)ppGpp-dependent control, for both stringent and growth rate regulation [71–73]. Promoters under positive stringent control were reported to have AT-rich discriminator sequences instead, such as the *his* operon promoter [74]. A mechanistic explanation for the role of the discriminator sequence in regulation of transcription has become conceivable according to the recent finding that σ^{70} subunit region 1.2 (Met102) makes direct contact with nucleotide −7 of the nontemplate strand, which is part of the discriminator. It was shown that these contacts are suboptimal for rRNA promoters, containing a perfect discriminator sequence. Such nonoptimal contacts are considered as a thermodynamic basis for the short-lived complexes, which are supposed to be essential for the mechanism of ppGpp-directed inhibition (see Section 22.5.3) [75, 76].

An additional criterion characteristic for σ^{70} promoters under negative stringent control is a deviation from the consensus 17-base-pair spacer distance. Based on the altered spacer length, stringent promoters can generally be predicted to be twist-sensitive and thus also to depend on the superhelicity [77].

Although the major impact on regulation occurs during the initiation phase of transcription, it should not be neglected that regulation has also been demonstrated during the elongation phase and in special cases could be a decisive step in regulation. Often RNA polymerase pausing is responsible for regulation during the elongation phase. Changes in the transcription efficiency of specific genes due to ppGpp-dependent RNA polymerase pausing have been demonstrated in different studies, explaining both direct or indirect inhibition as a consequence of RNA polymerase redistribution [78–80].

22.5.3
Rate-Limiting Step in ppGpp-Dependent Transcription Initiation

The functional analysis of promoters negatively affected by ppGpp has shown that they do not form stable open complexes with RNA polymerase. In contrast, RNA polymerase complexes with stringently regulated promoters share in common short half-lives and tend to dissociate readily. Productive transcription of promoters under negative stringent control apparently depends on prior binding of the initiating NTP substrate, which is a prerequisite for the formation of a stable ternary complex – the essential intermediate for the forward reaction of the transcription cycle. In numerous concurrent studies ppGpp has been reported to reduce the half-life of open RNA polymerase–promoter complexes, irrespective of the nature of the promoter [81–83]. Only sensitive promoters are inhibited, however, since only short-lived open complexes are effectively dissociated in the presence of ppGpp, while stable open promoter complexes have time enough to undergo irreversible steps of transcription initiation, even in the presence of ppGpp. Hence, within the initiation pathway, open complex formation has been proposed as the crucial point for ppGpp action, consistent with the assumption that this step is rate-limiting for ppGpp-sensitive promoters. It should be noted, however, that intrinsically unstable open complexes are not sufficient to explain ppGpp-sensitivity in all cases. The λpR promoter, for instance, forms very stable open complexes, but is under negative stringent control [84]. Moreover, the observation that ppGpp-dependent inhibition can be reversed by high concentrations of the respective initiating NTP has lead to an alternative explanation of the inhibitory step during the initiation cycle [7]. According to the observation that ppGpp-dependent inhibition can be reversed by competition with initiating NTPs the crucial step in regulation might not be the destabilization of the open promoter complex, but could rather be triggered by a change in the apparent K_M value for binding of the first NTP and formation of the ternary complex, which actually could be the rate-limiting step for productive transcription [85]. The importance of the starting NTP concentration for transcription initiation from different stringent promoters has been documented, supporting such a mechanism [86, 87]. According to the high-resolution structure of RNA polymerase in complex with ppGpp the change in affinity of the starting NTP might be brought about by restructuring the functionally important conserved Mg^{2+} ion in the catalytic site of RNA polymerase through coordination with the phosphate groups of bound ppGpp [62].

22.5.4
Different Mechanism of rRNA Regulation Between *E. coli* and *B. subtilis*

The role of the starting nucleotide in ppGpp-dependent transcriptional regulation has received additional importance, comparing its regulatory effects in *E. coli* and *B. subtilis*. In contrast to *E. coli*, where ATP is the predominant nucleotide at the start site for stable RNA genes the Gram-positive bacteria start with GTP. The promoter strength for stable RNA genes in such strains correlates with the intracellular GTP concentration. The concentration of the GTP pool, however, is reduced as a function of increasing ppGpp concentrations. According to recent studies in *B. subtilis* ppGpp does not inhibit RNA polymerase directly, but appears to inactivate IMP dehydrogenase activity, which causes a reduction of the GTP pool size [88, 89]. The importance of the starting nucleotide ATP or GTP upon the stringent response in *B. subtilis* and *E. coli* has been verified by the corresponding promoter mutations [90] (Kolmsee *et al.*, unpublished). Unlike the mechanism proposed for *E. coli*, regulation of rRNA synthesis in *T. thermophilus* has been described to be downregulated in a similar way as shown for *B. subtilis* [89].

22.5.5
Involvement of Coregulators: RNA Polymerase Secondary Channel-Binding Proteins

The protein DksA, which was originally identified as suppressor of a *dnaK* mutant, has been shown to potentiate the function of ppGpp *in vitro* and *in vivo* [83]. The high-resolution structure of DksA provides a key to understand how it interacts with RNA polymerase and how it might affect the stringent control. The protein shows strong structural homology with the transcription elongation factors GreA and GreB, consisting of a long coiled coil domain that fits exactly into the RNA polymerase secondary channel, where it positions two conserved acidic residues (aspartic acid) close to the catalytic site of the enzyme. This arrangement is believed to affect the coordination of a Mg^{2+} ion in the active site and to stabilize ppGpp binding to RNA polymerase [91]. DksA does not only augment the ppGpp effect at promoters under negative stringent control (e.g., rRNA promoters), but also at promoters that are activated, such as amino acid biosynthetic promoters [92, 93]. Although DksA is considered as a potentiating cofactor for ppGpp, it is clear that it has a number of functions on its own, such as DNA repair, protein folding, virulence, or bacterial motility [94, 95].

A Gre factor homolog (Gfh1) is found in *Thermus aquaticus* [96], while no *dksA* homologs are identified in the genomes of *T. thermophilus* and *T. aquaticus* [96]. Recently, the F plasmid-encoded TraR protein was identified as a DksA homolog that mimics ppGpp/DksA effects, even in the absence of ppGpp [97].

Based on sequence comparison, a recent study has identified a new family of protein genes, named *rnk*, which according to their structural similarity with GreA/B or DksA are suggested to act as secondary channel binders. Rnk from *E. coli* has a much shorter C-terminal coiled-coil domain and although it competes with RNA polymerase secondary channel-binding proteins *in vitro*, it has neither transcript cleavage properties nor does it augment the effect of ppGpp and seems dispensable for stable RNA regulation [98].

22.5.6
Positive Stringent Control

The mechanism of positive stringent control had been a matter of debate for a long time. Direct activation of transcription for σ^{70}-dependent promoters could not be shown by pure *in vitro* transcription systems and required coupled transcription–translation protocols [99]. As a notable exception, direct positive regulation was published for the λpaQ promoter [100]. Although with this promoter the half-life of open complex formation was similarly reduced in the presence of ppGpp, as noted for other promoters, activation was proposed to result from increased rates of productive open complex formation. Generally, positive regulation was considered to be passive and promoters under positive stringent control (e.g., promoters for amino acid biosynthesis operons) were believed to be more sensitive to competition for RNA polymerase *in vivo*. Hence, it was suggested that due to the inhibition of stable RNA promoters under high ppGpp conditions enough RNA polymerase is liberated and redistributed to stimulate transcription from positively regulated promoters [101].

The situation is different for promoters directed by alternative sigma factors or *in vitro* transcription reactions in the presence of DksA. Promoters dependent on the alternative sigma factor σ^S require ppGpp for induction during the stationary phase [102]. This requirement can also be demonstrated in a purified *in vitro* transcription system (Reckendrees and Wagner, unpublished). Moreover, in the presence of the ppGpp-potentiating protein DksA, the direct activation of amino acid biosynthesis promoters has been reported in an *in vitro* system [92]. Clearly, the mechanism responsible for activation must be different from inhibition and probably involves restructuring of the RNA polymerase active site conformation. Whatever the mechanism is for positive control, we do not know a precise answer yet.

22.5.7
Passive Regulation by Sigma Factor Competition – Direct Versus Indirect Effects

Regulation by (p)ppGpp is not restricted to σ^{70}-dependent transcription, but rather occurs also with core RNA polymerase associated with one of the alternative sigma factors σ^S, σ^N, σ^H, or σ^E [103–106]. In the case of σ^S, ppGpp increases both the expression and activity in response to several starvation conditions, while for σ^N only the activity seems to be affected. It is believed that activation of σ^S or σ^N is indirect because the effect could initially not be shown *in vitro* with isolated components [105]. Therefore, it was assumed that ppGpp-dependent activation of alternative sigma factors is brought about in a passive way by altering the competition among sigma factors for core RNA polymerase. This passive regulation was proposed to be a consequence of downregulation of intensely transcribed stable RNA genes, which sequester most of the available RNA polymerase holoenzymes, such that the limiting pool of core enzymes increases. The relative affinity of sigma factors to core RNA polymerase at low ppGpp concentration is in favor of σ^{70}, while at elevated effector concentrations alternative sigma factors are able to compete efficiently with the housekeeping sigma factor. Competition in favor of the alternative sigma factors is further advanced through the action of anti-sigma factor Rsd and increased expression or altered degradation of some of the alternative factors. All of these reactions are additionally controlled by ppGpp [107–110]. In line with this assumption it was shown *in vivo* that the alternative sigma factors σ^S, σ^N, or σ^H compete significantly better against σ^{70} in the presence of ppGpp [103, 104]. Moreover, the fraction of σ^{70} bound to the core is lower in the presence of ppGpp [111]. Obviously, the presence of ppGpp alters the competitiveness of sigma factors, thereby contributing to the adaptation to physiological stress. Recent *in vitro* experiments demonstrated, however, that ppGpp can also directly activate σ^E- or σ^S-dependent transcription [112] (Reckendrees and Wagner, unpublished). Apparently, more than one mechanism might be involved in the ppGpp-dependent change of sigma factor activity and/or competition.

References

1 Potrykus, K. and Cashel, M. (2008) (p)ppGpp: still magical? *Annu. Rev. Microbiol.*, **62**, 35–51.
2 Srivatsan, A. and Wang, J.D. (2008) Control of bacterial transcription, translation and replication by (p)ppGpp. *Curr. Opin. Microbiol.*, **11**, 100–105.
3 Szalewska-Palasz, A., Wegrzyn, G., and Wegrzyn, A. (2007) Mechanisms of physiological regulation of RNA synthesis in bacteria: new discoveries breaking old schemes. *J. Appl. Genet.*, **48**, 281–294.
4 Jain, V., Kumar, M., and Chatterji, D. (2006) ppGpp: stringent response and survival. *J. Microbiol.*, **44**, 1–10.
5 Braeken, K., Moris, M., Daniels, R., Vanderleyden, J., and Michiels, J. (2006) New horizons for (p)ppGpp in bacterial and plant physiology. *Trends Microbiol.*, **14**, 45–54.
6 Magnusson, L.U., Farewell, A., and Nystrom, T. (2005) ppGpp: a global regulator in *Escherichia coli*. *Trends Microbiol.*, **13**, 236–242.
7 Wagner, R. (2002) Regulation of ribosomal RNA synthesis in *E. coli*: effects of the global regulator guanosine tetraphosphate (ppGpp). *J. Mol. Microbiol. Biotechnol.*, **4**, 331–340.
8 Ryals, J., Little, R., and Bremer, H. (1982) Temperature dependence of RNA synthesis parameters in *Escherichia coli*. *J. Bacteriol.*, **151**, 879–887.
9 Battesti, A. and Bouveret, E. (2006) Acyl carrier protein/SpoT interaction, the switch linking SpoT-dependent stress response to fatty acid metabolism. *Mol. Microbiol.*, **62**, 1048–1063.
10 Raskin, D.M., Judson, N., and Mekalanos, J.J. (2007) Regulation of the

stringent response is the essential function of the conserved bacterial G protein CgtA in *Vibrio cholerae*. *Proc. Natl. Acad. Sci. USA*, **104**, 4636–4641.

11 Jiang, M., Sullivan, S.M., Wout, P.K., and Maddock, J.R. (2007) G-protein control of the ribosome-associated stress response protein SpoT. *J. Bacteriol.*, **189**, 6140–6147.

12 Lemos, J.A., Lin, V.K., Nascimento, M.M., Abranches, J., and Burne, R.A. (2007) Three gene products govern (p)ppGpp production by *Streptococcus mutans*. *Mol. Microbiol.*, **65**, 1568–1581.

13 Nanamiya, H., Kasai, K., Nozawa, A., Yun, C.S., Narisawa, T., Murakami, K., Natori, Y., Kawamura, F., and Tozawa, Y. (2008) Identification and functional analysis of novel (p)ppGpp synthetase genes in *Bacillus subtilis*. *Mol. Microbiol.*, **67**, 291–304.

14 Chang, D.-E., Smalley, D.J., and Conway, T. (2002) Gene expression profiling of *Escherichia coli* growth transitions: an expanded stringent control model. *Mol. Microbiol.*, **45**, 289–306.

15 Traxler, M.F., Summers, S.M., Nguyen, H.T., Zacharia, V.M., Hightower, G.A., Smith, J.T., and Conway, T. (2008) The global, ppGpp-mediated stringent response to amino acid starvation in *Escherichia coli*. *Mol. Microbiol.*, **68**, 1128–1148.

16 Durfee, T., Hansen, A.M., Zhi, H., Blattner, F.R., and Jin, D.J. (2008) Transcription profiling of the stringent response in *Escherichia coli*. *J. Bacteriol.*, **190**, 1084–1096.

17 Eymann, C., Homuth, G., Scharf, C., and Hecker, M. (2002) *Bacillus subtilis* functional genomics: global characterization of the stringent response by proteome and transcriptome analysis. *J. Bacteriol.*, **184**, 2500–2520.

18 Brockmann-Gretza, O. and Kalinowski, J. (2006) Global gene expression during stringent response in *Corynebacterium glutamicum* in presence and absence of the *rel* gene encoding (p)ppGpp synthase. *BMC Genomics*, **7**, 230.

19 Hesketh, A., Chen, W.J., Ryding, J., Chang, S., and Bibb, M. (2007) The global role of ppGpp synthesis in morphological differentiation and antibiotic production in *Streptomyces coelicolor* A3(2). *Genome Biol.*, **8**, R161.

20 Nascimento, M.M., Lemos, J.A., Abranches, J., Lin, V.K., and Burne, R.A. (2008) Role of RelA of *Streptococcus mutans* in global control of gene expression. *J. Bacteriol.*, **190**, 28–36.

21 Traxler, M.F., Chang, D.E., and Conway, T. (2006) Guanosine 3′,5′-bispyrophosphate coordinates global gene expression during glucose–lactose diauxie in *Escherichia coli*. *Proc. Natl. Acad. Sci. USA*, **103**, 2374–2379.

22 Ouellette, S.P., Hatch, T.P., AbdelRahman, Y.M., Rose, L.A., Belland, R.J., and Byrne, G.I. (2006) Global transcriptional upregulation in the absence of increased translation in *Chlamydia* during IFN-gamma-mediated host cell tryptophan starvation. *Mol. Microbiol.*, **62**, 1387–1401.

23 Cellini, A., Scoarughi, G.L., Poggiali, P., Santino, I., Sessa, R., Donini, P., and Cimmino, C. (2004) Stringent control in the archaeal genus *Sulfolobus*. *Res. Microbiol.*, **155**, 98–104.

24 Cimmino, C., Scoarughi, G.L., and Donini, P. (1993) Stringency and relaxation among the halobacteria. *J. Bacteriol.*, **175**, 6659–6662.

25 Scoarughi, G.L., Cimmino, C., and Donini, P. (1995) Lack of production of (p)ppGpp in *Halobacterium volcanii* under conditions that are effective in the eubacteria. *J. Bacteriol.*, **177**, 82–85.

26 van der Biezen, E.A., Sun, J., Coleman, M.J., Bibb, M.J., and Jones, J.D. (2000) *Arabidopsis* RelA/SpoT homologs implicate (p)ppGpp in plant signalling. *Proc. Natl. Acad. Sci. USA*, **97**, 3747–3752.

27 Givens, R.M., Lin, M.H., Taylor, D.J., Mechold, U., Berry, J.O., and Hernandez, V.J. (2004) Inducible expression, enzymatic activity, and origin of higher plant homologues of bacterial RelA/SpoT stress proteins in *Nicotiana tabacum*. *J. Biol. Chem.*, **279**, 7495–7504.

28 Takahashi, K., Kasai, K., and Ochi, K. (2004) Identification of the bacterial alarmone guanosine 5′-diphosphate

3′-diphosphate (ppGpp) in plants. *Proc. Natl. Acad. Sci. USA*, **101**, 4320–4324.

29 Kasai, K., Kanno, T., Endo, Y., Wakasa, K., and Tozawa, Y. (2004) Guanosine tetra- and pentaphosphate synthase activity in chloroplasts of a higher plant: association with 70S ribosomes and inhibition by tetracycline. *Nucleic Acids Res.*, **32**, 5732–5741.

30 Kasai, K., Usami, S., Yamada, T., Endo, Y., Ochi, K., and Tozawa, Y. (2002) A RelA–SpoT homolog (Cr-RSH) identified in *Chlamydomonas reinhardtii* generates stringent factor *in vivo* and localizes to chloroplasts *in vitro*. *Nucleic Acids Res.*, **30**, 4985–4992.

31 Tozawa, Y., Nozawa, A., Kanno, T., Narisawa, T., Masuda, S., Kasai, K., and Nanamiya, H. (2007) Calcium-activated (p)ppGpp synthetase in chloroplasts of land plants. *J. Biol. Chem.*, **282**, 35536–35545.

32 Masuda, S., Mizusawa, K., Narisawa, T., Tozawa, Y., Ohta, H., and Takamiya, K. (2008) The bacterial stringent response, conserved in chloroplasts, controls plant fertilization. *Plant Cell Physiol.*, **49**, 135–141.

33 Bassler, B.L. (1999) How bacteria talk to each other: regulation of gene expression by quorum sensing. *Curr. Opin. Microbiol.*, **2**, 582–587.

34 Lazazzera, B.A. (2000) Quorum sensing and starvation: signals for entry into stationary phase. *Curr. Opin. Microbiol.*, **3**, 177–182.

35 van Delden, C., Comte, R., and Bally, A.M. (2001) Stringent response activates quorum sensing and modulates cell density-dependent gene expression in *Pseudomonas aeruginosa*. *J. Bacteriol.*, **183**, 5376–5384.

36 Balzer, G.J. and McLean, R.J. (2002) The stringent response genes *relA* and *spoT* are important for *Escherichia coli* biofilms under slow-growth conditions. *Can. J. Microbiol.*, **48**, 675–680.

37 Moris, M., Braeken, K., Schoeters, E., Verreth, C., Beullens, S., Vanderleyden, J., and Michiels, J. (2005) Effective symbiosis between *Rhizobium etli* and *Phaseolus vulgaris* requires the alarmone ppGpp. *J. Bacteriol.*, **187**, 5460–5469.

38 Thompson, A., Rolfe, M.D., Lucchini, S., Schwerk, P., Hinton, J.C., and Tedin, K. (2006) The bacterial signal molecule, ppGpp, mediates the environmental regulation of both the invasion and intracellular virulence gene programs of *Salmonella*. *J. Biol. Chem.*, **281**, 30112–30121.

39 Pizarro-Cerda, J. and Tedin, K. (2004) The bacterial signal molecule, ppGpp, regulates *Salmonella* virulence gene expression. *Mol. Microbiol.*, **52**, 1827–1844.

40 Na, H.S., Kim, H.J., Lee, H.C., Hong, Y., Rhee, J.H., and Choy, H.E. (2006) Immune response induced by *Salmonella typhimurium* defective in ppGpp synthesis. *Vaccine*, **24**, 2027–2034.

41 Song, M., Kim, H.J., Kim, E.Y., Shin, M., Lee, H.C., Hong, Y., Rhee, J.H., Yoon, H., Ryu, S., Lim, S. *et al.* (2004) ppGpp-dependent stationary phase induction of genes on *Salmonella* pathogenicity island 1. *J. Biol. Chem.*, **279**, 34183–34190.

42 Nakanishi, N., Abe, H., Ogura, Y., Hayashi, T., Tashiro, K., Kuhara, S., Sugimoto, N., and Tobe, T. (2006) ppGpp with DksA controls gene expression in the locus of enterocyte effacement (LEE) pathogenicity island of enterohaemorrhagic *Escherichia coli* through activation of two virulence regulatory genes. *Mol. Microbiol.*, **61**, 194–205.

43 Godfrey, H.P., Bugryshev, J.V., and Cabello, F.C. (2002) The role of the stringent response in the pathogenesis of bacterial infections. *Trends Microbiol.*, **10**, 349–351.

44 Engelberg-Kulka, H. and Glaser, G. (1999) Addiction modules and programmed cell death and antideath in bacterial cultures. *Annu. Rev. Microbiol.*, **53**, 43–70.

45 Aizenman, E., Engelberg-Kulka, H., and Glaser, G. (1996) An *Escherichia coli* chromosomal "addiction module" regulated by guanosine [corrected] 3′,5′-bispyrophosphate: a model for programmed bacterial cell death. *Proc. Natl. Acad. Sci. USA*, **93**, 6059–6063.

46 Hazan, R. and Engelberg-Kulka, H. (2004) *Escherichia coli* mazEF-mediated cell death as a defense mechanism that

inhibits the spread of phage P1. *Mol. Genet. Genomics*, **272**, 227–234.

47 Gross, M., Marianovsky, I., and Glaser, G. (2006) MazG – a regulator of programmed cell death in *Escherichia coli*. *Mol. Microbiol.*, **59**, 590–601.

48 Hansen, S., Lewis, K., and Vulic, M. (2008) The role of global regulators and nucleotide metabolism in antibiotic tolerance in *Escherichia coli*. *Antimicrob. Agents Chemother.*, **52**, 2718–2726.

49 Lewis, K. (2007) Persister cells, dormancy and infectious disease. *Nat. Rev. Microbiol.*, **5**, 48–56.

50 Gardner, A., West, S.A., and Griffin, A.S. (2007) Is bacterial persistence a social trait? *PLoS ONE*, **2**, e752.

51 Korch, S.B., Henderson, T.A., and Hill, T.M. (2003) Characterization of the *hipA7* allele of *Escherichia coli* and evidence that high persistence is governed by (p)ppGpp synthesis. *Mol. Microbiol.*, **50**, 1199–1213.

52 Wright, B.E. (2004) Stress-directed adaptive mutations and evolution. *Mol. Microbiol.*, **52**, 643–650.

53 Zyskind, J.W. and Smith, D.W. (1992) DNA replication, the bacterial cell cycle, and cell growth. *Cell*, **69**, 5–8.

54 Autret, S., Levine, A., Vannier, F., Fujita, Y., and Seror, S.J. (1999) The replication checkpoint control in *Bacillus subtilis*: identification of a novel RTP-binding sequence essential for the replication fork arrest after induction of the stringent response. *Mol. Microbiol.*, **31**, 1665–1679.

55 Levine, A., Autret, S., and Seror, S.J. (1995) A checkpoint involving RTP, the replication terminator protein, arrests replication downstream of the origin during the Stringent Response in *Bacillus subtilis*. *Mol. Microbiol.*, **15**, 287–295.

56 Levine, A., Vannier, F., Dehbi, M., Henckes, G., and Seror, S.J. (1991) The stringent response blocks DNA replication outside the *ori* region in *Bacillus subtilis* and at the origin in *Escherichia coli*. *J. Mol. Biol.*, **219**, 605–613.

57 Wang, J.D., Sanders, G.M., and Grossman, A.D. (2007) Nutritional control of elongation of DNA replication by (p)ppGpp. *Cell*, **128**, 865–875.

58 Milon, P., Tischenko, E., Tomsic, J., Caserta, E., Folkers, G., La Teana, A., Rodnina, M.V., Pon, C.L., Boelens, R., and Gualerzi, C.O. (2006) The nucleotide-binding site of bacterial translation initiation factor 2 (IF2) as a metabolic sensor. *Proc. Natl. Acad. Sci. USA*, **103**, 13962–13967.

59 Chatterji, D., Fujita, N., and Ishihama, A. (1998) The mediator for stringent control, ppGpp, binds to the beta-subunit of *Escherichia coli* RNA polymerase. *Genes Cells*, **3**, 279–287.

60 Toulokhonov, I., Artsimovitch, I., and Landick, R. (2001) Allosteric control of RNA polymerase by a site that contacts nascent RNA hairpins. *Science*, **292**, 730–733.

61 Reddy, P.S., Raghavan, A., and Chatterji, D. (1995) Evidence for a ppGpp-binding site on *Escherichia coli* RNA polymerase: proximity relationship with the rifampicin-binding domain. *Mol. Microbiol.*, **15**, 255–265.

62 Artsimovitch, I., Patlan, V., Sekine, S., Vassylyeva, M.N., Hosaka, T., Ochi, K., Yokoyama, S., and Vassylyev, D.G. (2004) Structural basis for transcription regulation by alarmone ppGpp. *Cell*, **117**, 299–310.

63 Vrentas, C.E., Gaal, T., Berkmen, M.B., Rutherford, S.T., Haugen, S.P., Ross, W., and Gourse, R.L. (2008) Still looking for the magic spot: the crystallographically defined binding site for ppGpp on RNA polymerase is unlikely to be responsible for rRNA transcription regulation. *J. Mol. Biol.*, **377**, 551–564.

64 Mathew, R. and Chatterji, D. (2006) The evolving story of the omega subunit of bacterial RNA polymerase. *Trends Microbiol.*, **14**, 450–455.

65 Ghosh, P., Ishihama, A., and Chatterji, D. (2001) *Escherichia coli* RNA polymerase subunit omega and its N-terminal domain bind full-length beta' to facilitate incorporation into the alpha2beta subassembly. *Eur. J. Biochem.*, **268**, 4621–4627.

66 Gentry, D.R. and Burgess, R.R. (1989) *rpoZ*, encoding the omega subunit of *Escherichia coli* RNA polymerase, is in the same operon as *spoT*. *J. Bacteriol.*, **171**, 1271–1277.

67 Igarashi, K., Fujita, N., and Ishihama, A. (1989) Promoter selectivity of *Escherichia coli* RNA polymerase: omega factor is responsible for the ppGpp sensitivity. *Nucleic Acids Res.*, **17**, 8755–8765.

68 Gentry, D., Xiao, H., Burgess, R., and Cashel, M. (1991) The omega subunit of *Escherichia coli* K-12 RNA polymerase is not required for stringent RNA control *in vivo*. *J. Bacteriol.*, **173**, 3901–3903.

69 Vrentas, C.E., Gaal, T., Ross, W., Ebright, R.H., and Gourse, R.L. (2005) Response of RNA polymerase to ppGpp: requirement for the ω subunit and relief of this requirement by DksA. *Genes Dev.*, **19**, 2378–2387.

70 Travers, A.A. (1980) Promoter sequence for stringent control of bacterial RNA synthesis. *J. Bacteriol.*, **141**, 973–976.

71 Mizushima-Sugano, J. and Kaziro, Y. (1985) Regulation of the expression of the *tufB* operon: DNA sequences directly involved in the stringent control. *EMBO J.*, **4**, 1053–1058.

72 Zacharias, M., Göringer, H.U., and Wagner, R. (1989) Influence of the GCGC discriminator motif introduced into the ribosomal RNA P2- and *tac* promoter on growth rate control and stringent sensitivity. *EMBO J.*, **11**, 3357–3363.

73 Davies, I.J. and Drabble, W.T. (1996) Stringent and growth-rate-dependent control of the *gua* operon of *Escherichia coli* K-12. *Microbiology*, **142**, 2429–2437.

74 Riggs, D.L., Müller, R.D., Kwan, H.-S., and Artz, S.W. (1986) Promoter domain mediates guanosine tetraphosphate activation of the histidine operon. *Proc. Natl. Acad Sci. USA*, **83**, 9333–9337.

75 Haugen, S.P., Berkmen, M.B., Ross, W., Gaal, T., Ward, C., and Gourse, R.L. (2006) rRNA promoter regulation by nonoptimal binding of sigma region 1.2: an additional recognition element for RNA polymerase. *Cell*, **125**, 1069–1082.

76 Haugen, S.P., Ross, W., Manrique, M., and Gourse, R.L. (2008) Fine structure of the promoter–σ region 1.2 interaction. *Proc. Natl. Acad. Sci. USA*, **105**, 3292–3297.

77 Figueroa-Bossi, N., Guérin, M., Rahmouni, R., Leng, M., and Bossi, L. (1998) The supercoiling sensitivity of a bacterial tRNA promoter parallels its responsiveness to stringent control. *EMBO J.*, **17**, 2359–2367.

78 Krohn, M. and Wagner, R. (1996) Transcriptional pausing of RNA polymerase in the presence of guanosine tetraphosphate depends on the promoter and gene sequence. *J. Biol. Chem.*, **271**, 23884–23894.

79 Vogel, U. and Jensen, K.F. (1994) Effects of guanosine 3′,5′-bisdiphosphate (ppGpp) on rate of transcription elongation in isoleucine-starved *Escherichia coli*. *J. Biol. Chem.*, **269**, 16236–16241.

80 Vogel, U. and Jensen, K.F. (1995) Effects of the antiterminator boxA on transcription elongation kinetics and ppGpp inhibition of transcription elongation in *Escherichia coli*. *J. Biol. Chem.*, **270**, 18335–18340.

81 Raghavan, A. and Chatterji, D. (1998) Guanosine tetraphosphate-induced dissociation of open complexes at the *Escherichia coli* ribosomal protein promoters *rplJ* and *rpsA* P1: nanosecond depolarization spectroscopic studies. *Biophys. Chem.*, **75**, 21–32.

82 Barker, M.M., Gaal, T., Josaitis, C.A., and Gourse, R.L. (2001) Mechanism of regulation of transcription initiation by ppGpp. I. Effects of ppGpp on transcription initiation *in vivo* and *in vitro*. *J. Mol. Biol.*, **305**, 673–688.

83 Paul, B.J., Barker, M.M., Ross, W., Schneider, D.A., Webb, C., Foster, J.W., and Gourse, R.L. (2004) DksA: a critical component of the transcription initiation machinery that potentiates the regulation of rRNA promoters by ppGpp and the initiating NTP. *Cell*, **118**, 311–322.

84 Potrykus, K., Wegrzyn, G., and Hernandez, V.J. (2002) Multiple mechanisms of transcription inhibition by ppGpp at the lambda PR promoter. *J. Biol. Chem.*, **277**, 43785–43791.

85 Jöres, L. and Wagner, R. (2003) Essential steps in the ppGpp-dependent regulation of bacterial ribosomal RNA promoters can be explained by substrate competition. *J. Biol. Chem.*, **278**, 16834–16843.

86 Gaal, T., Bartlett, M.S., Ross, W., Turnbough, C., and Gourse, R.L. (1997)

Transcription regulation by initiating NTP concentration: rRNA synthesis in bacteria. *Science*, **278**, 2092–2097.

87 Schneider, D.A. and Gourse, R.L. (2003) Changes in *Escherichia coli* rRNA promoter activity correlate with changes in initiating nucleoside triphosphate and guanosine 5′-diphosphate 3′-diphosphate concentrations after induction of feedback control of ribosome synthesis. *J. Bacteriol.*, **185**, 6185–6191.

88 Krasny, L. and Gourse, R.L. (2004) An alternative strategy for bacterial ribosome synthesis: *Bacillus subtilis* rRNA transcription regulation. *EMBO J*, **23**, 4473–4483.

89 Kasai, K., Nishizawa, T., Takahashi, K., Hosaka, T., Aoki, H., and Ochi, K. (2006) Physiological analysis of the stringent response elicited in an extreme thermophilic bacterium, *Thermus thermophilus*. *J. Bacteriol.*, **188**, 7111–7122.

90 Krasny, L., Tiserova, H., Jonak, J., Rejman, D., and Sanderova, H. (2008) The identity of the transcription +1 position is crucial for changes in gene expression in response to amino acid starvation in *Bacillus subtilis*. *Mol. Microbiol.*, **69**, 42–54.

91 Perederina, A., Svetlov, V., Vassylyiva, M.N., Tahirov, T.H., Ykoyama, S., Artsimovitch, I., and Vassylyev, D.G. (2004) Regulation through the secondary channel – structural framework for ppGpp–DksA synergism during transcription. *Cell*, **118**, 297–309.

92 Paul, B.J., Berkmen, M.B., and Gourse, R.L. (2005) DksA potentiates direct activation of amino acid promoters by ppGpp. *Proc. Natl. Acad. Sci. USA*, **102**, 7823–7828.

93 Perron, K., Comte, R., and van Delden, C. (2005) DksA represses ribosomal gene transcription in *Pseudomonas aeruginosa* by interacting with RNA polymerase on ribosomal promoters. *Mol. Microbiol.*, **56**, 1087–1102.

94 Magnusson, L.U., Gummesson, B., Joksimovic, P., Farewell, A., and Nystrom, T. (2007) Identical, independent, and opposing roles of ppGpp and DksA in *Escherichia coli*. *J. Bacteriol.*, **189**, 5193–5202.

95 Potrykus, K., Vinella, D., Murphy, H., Szalewska-Palasz, A., D'Ari, R., and Cashel, M. (2006) Antagonistic regulation of *Escherichia coli* ribosomal RNA rrnB P1 promoter activity by GreA and DksA. *J. Biol. Chem.*, **281**, 15238–15248.

96 Lamour, V., Hogan, B.P., Erie, D.A., and Darst, S.A. (2006) Crystal structure of *Thermus aquaticus* Gfh1, a Gre-factor paralog that inhibits rather than stimulates transcript cleavage. *J. Mol. Biol.*, **356**, 179–188.

97 Blankschien, M.D., Potrykus, K., Grace, E., Choudhary, A., Vinella, D., Cashel, M., and Herman, C. (2009) TraR, a homolog of a RNAP secondary channel interactor, modulates transcription. *PLoS Genet.*, **5**, e1000345.

98 Lamour, V., Rutherford, S.T., Kuznedelov, K., Ramagopal, U.A., Gourse, R.L., Severinov, K., and Darst, S.A. (2008) Crystal structure of *Escherichia coli* Rnk, a new RNA polymerase-interacting protein. *J. Mol. Biol.*, **383**, 367–379.

99 Choy, H.E. (2000) The study of guanosine 5′-diphosphate 3′-diphosphate-mediated transcription regulation *in vitro* using a coupled transcription- translation system. *J. Biol. Chem.*, **275**, 6783–6789.

100 Potrykus, K., Wegrzyn, G., and Hernandez, V.J. (2004) Direct stimulation of the lambda paQ promoter by the transcription effector guanosine-3′,5′-(bis)pyrophosphate in a defined *in vitro* system. *J. Biol. Chem.*, **279**, 19860–19866.

101 Barker, M.M., Gaal, T., and Gourse, R.L. (2001) Mechanism of regulation of transcription initiation by ppGpp. II. Models for positive control based on properties of RNAP mutants and competition for RNAP. *J. Mol. Biol.*, **305**, 689–702.

102 Kvint, K., Farewell, A., and Nystrom, T. (2000) RpoS-dependent promoters require guanosine tetraphosphate for induction even in the presence of high levels of sigma^S. *J. Biol. Chem.*, **275**, 14795–14798.

103 Jishage, M., Kvint, K., Shingler, V., and Nystrom, T. (2002) Regulation of sigma factor competition by the alarmone ppGpp. *Genes Dev.*, **16**, 1260–1270.

104 Laurie, A.D., Bernardo, L.M., Sze, C.C., Skarfstad, E., Szalewska-Palasz, A., Nystrom, T., and Shingler, V. (2003) The role of the alarmone (p)ppGpp in sigma N competition for core RNA polymerase. *J. Biol. Chem.*, **278**, 1494–1503.

105 Costanzo, A. and Ades, S.E. (2006) Growth phase-dependent regulation of the extracytoplasmic stress factor, σ^E, by guanosine 3′,5′-bispyrophosphate (ppGpp). *J. Bacteriol.*, **188**, 4627–4634.

106 Bernardo, L.M., Johansson, L.U., Solera, D., Skarfstad, E., and Shingler, V. (2006) The guanosine tetraphosphate (ppGpp) alarmone, DksA and promoter affinity for RNA polymerase in regulation of sigma-dependent transcription. *Mol. Microbiol.*, **60**, 749–764.

107 Jishage, M. and Ishihama, A. (1998) A stationary phase protein in *Escherichia coli* with binding activity to the major σ subunit of RNA polymerase. *Proc. Natl. Acad. Sci. USA*, **95**, 4953–4958.

108 Brown, L., Gentry, D., Elliot, T., and Cashel, M. (2002) DksA affects ppGpp induction of RpoS at a translational level. *J. Bacteriol.*, **184**, 4455–4465.

109 Hirsch, M. and Elliott, T. (2002) Role of ppGpp in *rpoS* stationary-phase regulation in *Escherichia coli*. *J. Bacteriol.*, **184**, 5077–5087.

110 Bougdour, A. and Gottesman, S. (2007) ppGpp regulation of RpoS degradation via anti-adaptor protein IraP. *Proc. Natl. Acad. Sci. USA*, **104**, 12896–12901.

111 Hernandez, V.J. and Cashel, M. (1995) Changes in conserved region 3 of *Escherichia coli* σ^{70} mediate ppGpp-dependent functions *in vivo*. *J. Mol. Biol.*, **252**, 536–549.

112 Costanzo, A., Nicoloff, H., Barchinger, S.E., Banta, A.B., Gourse, R.L., and Ades, S.E. (2008) ppGpp and DksA likely regulate the activity of the extracytoplasmic stress factor sigma^E in *Escherichia coli* by both direct and indirect mechanisms. *Mol. Microbiol.*, **67**, 619–632.

23
Sensory RNAs
Franz Narberhaus

23.1
Introduction

Sensing of environmental parameters is particularly important in unicellular organisms as they live in an ever-changing microenvironment. Bacteria have evolved a number of protective strategies to cope with changes in nutrient supply, osmolarity, pH, and temperature. To be aware of the actual conditions, accurate measurement of a number of physical and chemical parameters is required. The input signals must then be transduced in order to mount an appropriate and rapid adaptive response.

We all appreciate the power of proteins as sensory and regulatory molecules. In fact, many chapters of this book are devoted to protein-mediated control mechanisms. Proteins certainly play an imminent role in controlling bacterial stress responses – by protein kinases or alternative sigma factors, to name the most prominent examples. Proteins are usually considered to be more versatile than RNA, as they are composed of 20 amino acids that can be arranged in a modular fashion to achieve specific functions in signal recognition and transduction.

Although limited in its number of building blocks and functional groups, RNA turns out to be remarkably versatile in its sensory and regulatory potential. There is a rapidly growing body of evidence that gene regulation can very well occur without the initial participation of proteins [1]. Numerous riboswitches control gene expression through changes in their conformation in response to small metabolites [2]. Up to 100 different regulatory RNAs per microorganism control gene expression by RNA–RNA or RNA–protein interactions [3]. Although proteins, such as the RNA chaperone Hfq [4], might play a supportive role in these processes, their precise contribution is not yet very clear.

23.2
RNA as a Regulatory Molecule

Why can only four nucleotides in the RNA be sufficient to control diverse cellular processes? Usually this type of control relies on the regulated formation of RNA structures. Often not full complementary but only limited, incomplete base pairing and weak noncanonical base pairs are the key to selective target recognition. It is important to recognize that it is not just Watson–Crick-like base pairing that is involved. There are many other ways in which RNA nucleotides can interact with each other to fold into complex structures [5]. Long-range interactions, including loop–loop interactions and pseudoknots, extend the versatility of RNA as a regulatory molecule.

We distinguish two types of regulatory RNAs. Trans-encoded small regulatory RNAs (sRNAs) are expressed from somewhere on the chromosome and usually control gene expression on some other location. sRNAs play important roles in stress management, in particular during outer membrane stress and in bacterial virulence [6, 7]. Often, expression of sRNAs is environmentally controlled. Hence, it is not the regulatory RNAs themselves, but the factors that control their expression that are responsive to external signals.

This is different in the second class of regulatory RNA molecules that are encoded within a transcript and referred to as cis-encoded RNAs. Riboswitches and RNA thermometers belong to this category. They are the focus of this chapter as they directly perceive an environmental signal and use it to control the fate of the downstream message.

23.3
Riboswitches

A remarkable advance in our understanding of the sensory and regulatory potential of RNA has come from the discovery of riboswitches. The term "riboswitch" initially appeared in the literature when a thiamine pyrophosphate (TPP)-responsive RNA was described in 2002 [8]. In 2008, the query "riboswitch" retrieved almost 200 entries from the PubMed library, indicating how rapidly this research field is developing. Riboswitches are complex RNA structures that are located primarily in the 5′-untranslated region (5′-UTR) of certain protein-encoding mRNAs. They are composed of at least two separate modules – a ligand-binding region commonly referred to as the aptamer domain and a so-called expression platform [9]. The aptamer is a highly structured element that forms a selective binding pocket for the target molecule. It must specifically recognize its ligand and discriminate against similar molecules in a crowded intracellular environment. Strict binding specificity correlates with restrictions in sequence and structure variability, even in distantly related organisms. This conservation has been exploited for the bioinformatic identification and classification of riboswitches [10]. More than a dozen classes of riboswitches have been reported so far and there is no reason to believe that this might be the end of it. Many, but not all, ligands carry phosphate or nucleotide-like moieties. Target molecules range from the Mg^{2+} ion [11] to the amino acids glycine or lysine [12, 13], the purine nucleotides adenine and guanine [14, 15], the second messenger c-di-GMP [16], the peptidoglycan building block glucosamine-6-phosphate (GlcN6P) [17] all the way to larger and bulkier compounds like the vitamin B_{12}

precursor adenosyl-cobalamin (AdoCbl) [18], the vitamin B_1 derivative TPP [8], flavin mononucleotide [19, 20], S-adenosylmethionine (SAM) [21, 22], S-adenosylhomocysteine [23], or the molybdenum-cofactor Moco [24]. Each riboswitch class is defined by a characteristic sequence and unique structural features. Interestingly, there are several ways to detect SAM because four distinct riboswitch classes (called SAM-I through SAM-IV) have been reported [25].

How is binding of a target molecule to the aptamer domain converted into genetic control of fundamental metabolic pathways? As the name indicates, riboswitches are thought to exist in two alternative conformations – a ligand-free and a ligand-bound state. Association of the appropriate metabolite with the aptamer domain causes an allosteric rearrangement in the adjoining expression platform that determines whether the gene will be expressed. There are several ways how such conformational changes can impact gene expression (Figure 23.1). Most riboswitches operate as OFF switches that control transcription or translation of genes coding for the biosynthetic enzymes or the import system of its ligand. In the absence or at low intercellular concentrations of an important metabolite, expression of genes responsible for its production or uptake should be ON. However, if this compound is abundant these genes should be shut OFF to avoid the waste of cellular resources. This can be accomplished in two ways, which can be exemplified by known TPP riboswitches. Thiamine biosynthesis in *E. coli* is controlled by a riboswitch in the *thiM* gene [8]. The ribosome-binding site is accessible in the ligand-free mRNA structure (Figure 23.1A). When TPP is bound, translation is shut off by trapping the Shine–Dalgarno (SD) sequence in a helical structure. *Bacillus subtilis* uses a different strategy, in which thiamine biosynthesis is regulated by a transcriptional control element [19]. In the absence of the ligand, the 5′-UTR folds into a structure that does not interfere with transcription elongation (Figure 23.1B). Metabolite binding introduces a mutually exclusive structure containing a typical Rho-independent transcription terminator – a GC-rich stem–loop followed by a polyU tract. The RNA polymerase responds to formation of this structure and terminates transcription prematurely.

Species- and gene-dependent association of the conserved TPP aptamer with different expression platforms underscores the modularity and versatility of riboswitches. A principal difference between the two types of control depicted in Figure 23.1(A and B) is that riboswitch-induced transcription termination influences expression of an entire operon, whereas translational control is restricted to a single gene. For unknown reasons, there is a strong bias towards transcriptional control in Gram-positive bacteria, whereas Gram-negatives tend to use riboswitches as translational control elements [26].

Although most metabolite-loaded riboswitches inhibit gene expression, there also are examples of positive regulation. Here, the default state is OFF because a transcription terminator structure is formed (Figure 23.1C). Binding of a metabolite causes a structural rearrangement in the 5′-UTR that disrupts the terminator and favors an antiterminator conformation that allows the RNA polymerase to proceed with transcription. Such a genetic ON switch occurs in the *B. subtilis ydhL* transcript that binds adenine and is predicted to encode a purine efflux pump [14].

Riboswitches

(A) **Translation initation** (OFF switch)

(B) **Transcription termination** (OFF switch)

(C) **Transcription termination** (ON switch)

(D) **Ribozyme activity** (OFF switch)

RNA thermometers

(E) **Translation initiation**

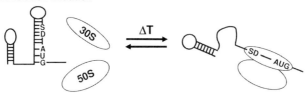

Figure 23.1 Schematic illustration of regulatory principles of riboswitches (A–D) and RNA thermometers (E). In each case, the genetic processes that are controlled are indicated. Stars represent metabolites. Abbreviations (not already defined in text): 30S and 50S, ribosomal subunits; RNAP, RNA polymerase; AUG and UUUUUU, ribonucleotide sequences; ΔT, temperature change.

The aptamer domain of the adenine riboswitch is remarkably similar in sequence and structure to the guanine aptamer upstream of the *B. subtilis xpt–pbuX* operon. This element functions as a transcriptional OFF switch controlling purine biosynthesis [15]. The purine riboswitches provided valuable insights into the exquisite substrate specificity of riboswitches. Adenine and guanine aptamer modules strictly discriminate between these purines and their analogs [14, 15]. Specificity is mediated by the formation of a Watson–Crick base pair in the core of the purine-binding pocket. Adenine binding depends on a critical U residue, which is exchanged for a C in the guanine aptamer [27, 28].

All riboswitches described above bind a single metabolite and function as a simple genetic switch. In rare cases, two aptamers are combined in tandem arrangement. A composite glycine-dependent riboswitch controls expression of a glycine cleavage system by integrating cooperative binding of two glycine molecules to independent binding sites [13]. Only occupation of both sites by glycine efficiently disrupts a transcription terminator. This genetic ON switch ensures that adequate amounts of glycine are maintained for protein biosynthesis and only excess glycine is fed into catabolic pathways.

Combination of two different aptamer modules allows integration of separate signals. The 5′-UTR of the *metE* mRNA from *Bacillus clausii* contains a tandem SAM and AdoCbl riboswitch [29]. Since there is a more efficient AdoCbl-dependent pathway for methionine biosynthesis in this organism, the MetE pathway should be repressed when AdoCbl and SAM (a methionine-containing compound) are abundant. Each aptamer overlaps with a transcriptional control element that shifts towards the terminator structure in the ligand-bound state. This architecture implies that the two ligands can independently repress *metE* expression, which was shown to be the case.

Yet another riboswitch mechanism is exemplified by the *B. subtilis glmS* riboswitch (Figure 23.1D). Binding of GlcN6P induces ribozyme activity that cleaves the *glmS* transcript near the 5′-end [17]. Self-cleavage specifically targets the mRNA for degradation by the RNase J1 enzyme [30]. Like other riboswitches, the *glmS* aptamer discriminates against even closely related metabolites. This selectivity precisely couples the GlcN6P amount to sugar metabolism and cell wall biosynthesis.

Riboswitches are very widespread and versatile gene control elements in bacteria. In some species they are very abundant [10]. In *B. subtilis*, 2% of all genes are predicted to be under riboswitch control [15]. Genome searches identified potential riboswitches in all three domains of life [10]. However, so far only the TPP element has been found in Archaea, fungi, and plants [10, 31, 32]. It simply might a matter of time before more riboswitches are discovered in eukaryotes. Riboswitch-mediated control mechanisms in eukaryotes still are largely unexplored. It is interesting that TPP riboswitches have been predicted in various places, namely in 5-UTRs, in introns, and in 3′-UTRs. In the filamentous fungi *Aspergillus oryzae* and *Neurospora crassa* they can activate or repress gene expression by controlling alternative splicing [31, 33]. In plants, TPP-mediated alternative folding of the riboswitch controls splicing and alternative 3′-end mRNA processing [34].

Recently, the three-dimensional structure of several riboswitches has been determined at atomic resolution. Several riboswitches have a fork-like architecture, in which two helical segments almost completely envelope the metabolite [35–38]. Nucleotides in junctional regions, bulges, and terminal loops are engaged in long-range interactions that stabilize the tertiary structure. The overall architecture of the TPP riboswitch from *E. coli* and *Arabidopsis* is almost identical, suggesting a common phylogenetic origin of this element [37, 38]. An independent evolutionary solution has been established in the *glmS* ribozyme. Its structure differs from other riboswitches in several ways. It adopts a rigid fold that does not undergo substantial conformational changes upon metabolite binding [39]. The binding pocket is solvent accessible and the ligand itself is thought to be involved in catalysis.

The modular nature and regulatory potential of riboswitches can be exploited for applied purposes [40, 41]. Artificial RNA regulators have been developed for conditional gene expression in various systems. Ogawa and Maeda developed an artificial ribozyme-based ON switch in *E. coli* that cleaves and releases the anti-SD sequence in the presence of the cofactor theophylline [42]. Engineered tetracycline- or neomycin-binding riboswitches are able to modulate gene expression in yeast either by controlling translation initiation [43, 44] or pre-mRNA splicing [45]. Finally, engineered riboswitches have been shown to autonomously control gene expression by alternative splicing in plants [46]. These arbitrarily selected examples illustrate the enormous potential of riboswitches to control gene expression in prokaryotes and eukaryotes. Another promising application is the integration of riboswitches into high-precision biosensors that respond specifically to the presence of their corresponding ligand [47]. Some riboswitches may even become useful drug targets [48]. If a metabolite analog were designed that targets the aptamer domain of an OFF riboswitch, this might have a lethal effect to the bacterium. Most promising are riboswitches with an accessible binding pocket like the *glmS* riboswitch [39].

23.4
RNA Thermometers

All RNA thermometers known so far are translational control elements. In contrast to riboswitches they register a physical input – changes in temperature – rather than chemical signals. Most RNA thermometers make use of an elegant and conceptually simple mechanism. The SD sequence is trapped in a hairpin structure, which is stable only at low temperatures (Figure 23.1E). Increasing temperature destabilizes the structure such that the ribosome-binding site becomes accessible, allowing translation to be initiated [49]. RNA thermometers operating by this principle are found in the 5′-UTR of bacterial heat shock and virulence genes.

The first RNA thermometer acting by the melting mechanism was described in the *E. coli rpoH* gene, which codes for the alternative sigma factor σ^{32} or RpoH [50, 51]. It is a somewhat atypical example as most of the regulatory sequences reside in the coding region. Two segments (A and B) up to 220 nucleotides into the open reading frame of *rpoH* contribute to an extensive RNA structure that blocks entry of the ribosome to the SD sequence. Disruption of the structure at heat shock temperatures liberates the ribosome-binding site and enhances translation of the sigma factor, resulting in the rapid induction of the heat shock response.

The most abundant bacterial RNA thermometer is the repression of heat shock gene expression (ROSE) element. It was discovered in rhizobia [52], and is widespread in numerous α- and γ-proteobacteria, including *E. coli* and *Salmonella*, where it controls the expression of small heat shock genes [53]. The 60- to 100-nucleotide-long ROSE element is located in the 5′-UTR of these genes. It acquires a complex structure comprised of up to four stem–loops. The final hairpin contains the SD sequence and in some cases also the AUG start codon. The ROSE element is characterized by only a few conserved nucleotides (UYGCU, in which Y is a pyrimidine) that pair with the SD sequence [49, 53]. Short internal loops and bulges in the computer-predicted final structure are thought to create a thermolabile structure that melts as the temperature increases. Introduction of mismatches into this structure relieves repression at low temperature, whereas mutations that stabilize the structure abolish induction at heat stress temperatures [52, 54]. The high-resolution nuclear magnetic resonance structure of the temperature-responsive ROSE hairpin revealed the molecular basis for temperature sensing [55]. The actual structure differs from the computer-predicted one and contains several noncanonical base pairs, including a GG pair and a base triple that renders the structure vulnerable to high temperatures.

Another RNA thermometer that might turn out to be widespread is the fourU element. It was initially described upstream of the small heat shock gene *agsA* in *Salmonella* [56]. At 60 nucleotides, it is among the shortest natural RNA thermometers. The predicted structure contains two hairpins and, as the name indicates, four uridine residues are used to interact with the SD sequence. Structure-probing experiments revealed temperature-controlled opening of the second hairpin and ribosome binding was demonstrated to occur only at heat shock temperatures.

Four uridines are well suited as a building block to loosely pair with AGGA of the SD sequence. Therefore, it is not surprising that additional fourU thermometers were predicted upstream of several bacterial heat shock and virulence genes [56]. An interesting candidate is a fourU-like structure in the 5′-UTR of the *Yersinia lcrF* (*virF*) gene, which encodes a regulator of the virulence response. Translation of this gene is inefficient at 26 °C, but is induced at 37 °C. The predicted RNA structure upstream of the gene led to the formulation of a RNA thermometer-like model already in 1993 [57].

A temperature of 37 °C is an important signal for many mammalian pathogens that they have successfully entered their host. Different mechanisms, including changes in DNA topology and thermoresponsive proteins, are known to be involved in temperature measurement [58]. In *Listeria monocytogenes*, an RNA thermometer controls translation of the virulence gene activator PrfA [59]. The *prfA* transcript is preceded by an RNA structure in which the SD sequence is poorly accessible. This structure is stable at 30 °C, but destabilized at 37 °C, ensuring that PrfA-dependent virulence genes are maximally expressed in the mammalian host.

The RNA thermometers described above do not function as molecular switches since melting of the RNA structure proceeds gradually as the temperature increases. This regulatory mode results in a graded response. There are a few exceptions, in which temperature controls the existence of two mutually exclusive structures that act in a switch-like fashion. Translation of the phage λ cIII gene is controlled by temperature [60]. The cIII gene product is involved in the lysis–lysogeny decision of the phage. High concentrations of cIII protein at optimal growth temperatures (37 °C) favor the lysogenic pathway. During severe heat stress (45 °C), phage λ enters the lytic pathway because the cIII amount is kept at a low level. This is accomplished by alternative RNA structures upstream of the cIII. High temperatures shift the equilibrium towards the energetically more stable conformation, in which the SD sequence is blocked. At 37 °C, the structure with the accessible ribosome-binding site is preferred.

Secondary structures in the 5′-UTR also have been predicted to influence the expression of cold shock genes. A 159-nucleotide-long sequence upstream of the *E. coli cspA* gene has been postulated to act as an RNA thermometer controlling the translation efficiency by alternative RNA structures [61]. The RNA structure might also affect stability of the *cspA* transcript [62]. *E. coli* contains nine paralogs of CspA, among them CspE. The *cspE* gene contains a much shorter 5′-UTR of 43 nucleotides. Alternative RNA structures in this region are thought to control differential stability of the *cspE* transcript, which is stabilized at 15 °C [63]. The evidence for this model is that a point mutation predicted to relax the structure resulted in reduced transcript stability.

Like riboswitches, RNA thermometers might be useful for practical applications. It is possible to generate sensitive synthetic RNA thermometers from random sequences [64]. Imitating the melting principle, artificial thermosensors have been created by introducing G-rich quadruplex-forming sequences close to the SD sequence [65]. Quadruplexes with moderate stability responded to changes in temperature. As well as applications in fine-tuning gene expression, RNA thermometers have been proposed to serve as biocompatible nanoscale temperature sensors for clinical sensing and imaging [66].

23.5
Conclusions

Riboswitches and RNA thermometers are widespread sensory and regulatory elements implemented in the mRNA to control gene expression in response to chemical or physical signals. It is believed that they represent an ancient control mechanism [26]. The widely accepted "RNA world" hypothesis states that RNA appeared before the event of DNA and proteins. It is easily conceivable that RNA-based control elements descend from such a setting. However, they are not just outdated control mechanisms left over from a protein-devoid period on Earth – apparently, they still have a place in modern cells.

What benefit might a cell have from using RNA instead of protein for regulatory purposes? The striking advantage of these in-built regulatory elements is that the sensory device is directly hooked up to the gene that is to be regulated. Riboswitches and RNA thermometers directly measure an input signal and convert it into an immediate response. All trans-acting mechanisms, regardless of whether they are protein or RNA-mediated, are delayed in comparison to mechanisms that directly control the outcome of an already existing transcript. Therefore, it is not surprising that cis-regulatory elements are often being used to rapidly respond to potentially harmful stress and starvation conditions.

Within only a few years, RNA biology has been established as an important research field. We are just beginning to understand some of the principles in RNA-mediated gene regulation. It is possible that the currently known collection of regulatory RNA elements reflects only a small proportion of the true contribution that RNA makes to gene regulation in prokaryotic and eukaryotic organisms.

Acknowledgments

I thank my current and previous coworkers for their valuable contributions to the RNA thermometer project. Funding from the Deutsche Forschungsgemeinschaft (DFG NA/240 and SPP 1258) is gratefully acknowledged.

References

1 Serganov, A. and Patel, D.J. (2007) Ribozymes, riboswitches and beyond: regulation of gene expression without proteins. *Nat. Rev. Genet.*, **8**, 776–790.

2 Winkler, W.C. and Breaker, R.R. (2005) Regulation of bacterial gene expression by riboswitches. *Annu. Rev. Microbiol.*, **59**, 487–517.

3 Majdalani, N., Vanderpool, C.K., and Gottesman, S. (2005) Bacterial small RNA regulators. *Crit. Rev. Biochem. Mol. Biol.*, **40**, 93–113.

4 Valentin-Hansen, P., Eriksen, M., and Udesen, C. (2004) The bacterial Sm-like protein Hfq: a key player in RNA transactions. *Mol. Microbiol.*, **51**, 1525–1533.

5 Leontis, N.B., Stombaugh, J., and Westhof, E. (2002) The non-Watson–Crick base pairs and their associated isostericity matrices. *Nucleic Acids Res.*, **30**, 3497–3531.

6 Guillier, M., Gottesman, S., and Storz, G. (2006) Modulating the outer membrane with small RNAs. *Genes Dev.*, **20**, 2338–2348.

7 Romby, P., Vandenesch, F., and Wagner, E.G. (2006) The role of RNAs in the regulation of virulence-gene expression. *Curr. Opin. Microbiol.*, **9**, 229–236.

8 Winkler, W., Nahvi, A., and Breaker, R.R. (2002) Thiamine derivatives bind messenger RNAs directly to regulate bacterial gene expression. *Nature*, **419**, 952–956.

9 Winkler, W.C. and Breaker, R.R. (2003) Genetic control by metabolite-binding riboswitches. *ChemBioChem*, **4**, 1024–1032.

10 Barrick, J.E. and Breaker, R.R. (2007) The distributions, mechanisms, and structures of metabolite-binding riboswitches. *Genome Biol.*, **8**, R239.

11 Dann, C.E. 3rd., Wakeman, C.A. Sieling, C.L. Baker, S.C. Irnov, I. Winkler, W.C. (2007) Structure and mechanism of a metal-sensing regulatory RNA. *Cell*, **130**, 878–892.

12 Sudarsan, N., Wickiser, J.K., Nakamura, S., Ebert, M.S., and Breaker, R.R. (2003) An mRNA structure in bacteria that controls

gene expression by binding lysine. *Genes Dev.*, **17**, 2688–2697.

13 Mandal, M., Lee, M., Barrick, J.E., Weinberg, Z., Emilsson, G.M., Ruzzo, W.L., and Breaker, R.R. (2004) A glycine-dependent riboswitch that uses cooperative binding to control gene expression. *Science*, **306**, 275–279.

14 Mandal, M. and Breaker, R.R. (2004) Adenine riboswitches and gene activation by disruption of a transcription terminator. *Nat. Struct. Mol. Biol.*, **11**, 29–35.

15 Mandal, M., Boese, B., Barrick, J.E., Winkler, W.C., and Breaker, R.R. (2003) Riboswitches control fundamental biochemical pathways in *Bacillus subtilis* and other bacteria. *Cell*, **113**, 577–586.

16 Sudarsan, N., Lee, E.R., Weinberg, Z., Moy, R.H., Kim, J.N., Link, K.H., and Breaker, R.R. (2008) Riboswitches in eubacteria sense the second messenger cyclic di-GMP. *Science*, **321**, 411–413.

17 Winkler, W.C., Nahvi, A., Roth, A., Collins, J.A., and Breaker, R.R. (2004) Control of gene expression by a natural metabolite-responsive ribozyme. *Nature*, **428**, 281–286.

18 Nou, X. and Kadner, R.J. (2000) Adenosylcobalamin inhibits ribosome binding to *btuB* RNA. *Proc. Natl. Acad. Sci. USA*, **97**, 7190–7195.

19 Mironov, A.S., Gusarov, I., Rafikov, R., Lopez, L.E., Shatalin, K., Kreneva, R.A., Perumov, D.A., and Nudler, E. (2002) Sensing small molecules by nascent RNA: a mechanism to control transcription in bacteria. *Cell*, **111**, 747–756.

20 Winkler, W.C., Cohen-Chalamish, S., and Breaker, R.R. (2002) An mRNA structure that controls gene expression by binding FMN. *Proc. Natl. Acad. Sci. USA*, **99**, 15908–15913.

21 Epshtein, V., Mironov, A.S., and Nudler, E. (2003) The riboswitch-mediated control of sulfur metabolism in bacteria. *Proc. Natl. Acad. Sci. USA*, **100**, 5052–5056.

22 Winkler, W.C., Nahvi, A., Sudarsan, N., Barrick, J.E., and Breaker, R.R. (2003) An mRNA structure that controls gene expression by binding S-adenosyl-methionine. *Nat. Struct. Biol.*, **10**, 701–707.

23 Wang, J.X., Lee, E.R., Morales, D.R., Lim, J., and Breaker, R.R. (2008) Riboswitches that sense S-adenosylhomocysteine and activate genes involved in coenzyme recycling. *Mol. Cell*, **29**, 691–702.

24 Regulski, E.E., Moy, R.H., Weinberg, Z., Barrick, J.E., Yao, Z., Ruzzo, W.L., and Breaker, R.R. (2008) A widespread riboswitch candidate that controls bacterial genes involved in molybdenum cofactor and tungsten cofactor metabolism. *Mol. Microbiol.*, **68**, 918–932.

25 Wang, J.X. and Breaker, R.R. (2008) Riboswitches that sense S-adenosyl-methionine and S-adenosylhomocysteine. *Biochem. Cell Biol.*, **86**, 157–168.

26 Vitreschak, A.G., Rodionov, D.A., Mironov, A.A., and Gelfand, M.S. (2004) Riboswitches: the oldest mechanism for the regulation of gene expression? *Trends Genet.*, **20**, 44–50.

27 Noeske, J., Richter, C., Grundl, M.A., Nasiri, H.R., Schwalbe, H., and Wöhnert, J. (2005) An intermolecular base triple as the basis of ligand specificity and affinity in the guanine- and adenine-sensing riboswitch RNAs. *Proc. Natl. Acad. Sci. USA*, **102**, 1372–1377.

28 Serganov, A., Yuan, Y.R., Pikovskaya, O., Polonskaia, A., Malinina, L., Phan, A.T., Hobartner, C., Micura, R., Breaker, R.R., and Patel, D.J. (2004) Structural basis for discriminative regulation of gene expression by adenine- and guanine-sensing mRNAs. *Chem. Biol.*, **11**, 1729–1741.

29 Sudarsan, N., Hammond, M.C., Block, K.F., Welz, R., Barrick, J.E., Roth, A., and Breaker, R.R. (2006) Tandem riboswitch architectures exhibit complex gene control functions. *Science*, **314**, 300–304.

30 Collins, J.A., Irnov, I., Baker, S., and Winkler, W.C. (2007) Mechanism of mRNA destabilization by the *glmS* ribozyme. *Genes Dev.*, **21**, 3356–3368.

31 Kubodera, T., Watanabe, M., Yoshiuchi, K., Yamashita, N., Nishimura, A., Nakai, S., Gomi, K., and Hanamoto, H. (2003) Thiamine-regulated gene expression of *Aspergillus oryzae thiA* requires splicing of the intron containing a riboswitch-like domain in the 5′-UTR. *FEBS Lett.*, **555**, 516–520.

32 Sudarsan, N., Barrick, J.E., and Breaker, R.R. (2003) Metabolite-binding RNA

domains are present in the genes of eukaryotes. *RNA*, **9**, 644–647.

33 Cheah, M.T., Wachter, A., Sudarsan, N., and Breaker, R.R. (2007) Control of alternative RNA splicing and gene expression by eukaryotic riboswitches. *Nature*, **447**, 497–500.

34 Wachter, A., Tunc-Ozdemir, M., Grove, B.C., Green, P.J., Shintani, D.K., and Breaker, R.R. (2007) Riboswitch control of gene expression in plants by splicing and alternative 3′ end processing of mRNAs. *Plant Cell*, **19**, 3437–3450.

35 Montange, R.K. and Batey, R.T. (2006) Structure of the *S*-adenosylmethionine riboswitch regulatory mRNA element. *Nature*, **441**, 1172–1175.

36 Batey, R.T., Gilbert, S.D., and Montange, R.K. (2004) Structure of a natural guanine-responsive riboswitch complexed with the metabolite hypoxanthine. *Nature*, **432**, 411–415.

37 Thore, S., Leibundgut, M., and Ban, N. (2006) Structure of the eukaryotic thiamine pyrophosphate riboswitch with its regulatory ligand. *Science*, **312**, 1208–1211.

38 Serganov, A., Polonskaia, A., Phan, A.T., Breaker, R.R., and Patel, D.J. (2006) Structural basis for gene regulation by a thiamine pyrophosphate-sensing riboswitch. *Nature*, **441**, 1167–1171.

39 Klein, D.J. and Ferre-D'Amare, A.R. (2006) Structural basis of *glmS* ribozyme activation by glucosamine-6-phosphate. *Science*, **313**, 1752–1756.

40 Wieland, M. and Hartig, J.S. (2008) Artificial riboswitches: synthetic mRNA-based regulators of gene expression. *Chembiochem*, **9**, 1873–1878.

41 Suess, B. and Weigand, J.E. (2008) Engineered riboswitches: overview, problems and trends. *RNA Biol.*, **5**, 1–6.

42 Ogawa, A. and Maeda, M. (2008) An artificial aptazyme-based riboswitch and its cascading system in *E. coli*. *ChemBiochem*, **9**, 206–209.

43 Suess, B., Hanson, S., Berens, C., Fink, B., Schroeder, R., and Hillen, W. (2003) Conditional gene expression by controlling translation with tetracycline-binding aptamers. *Nucleic Acids Res.*, **31**, 1853–1858.

44 Weigand, J.E., Sanchez, M., Gunnesch, E.B., Zeiher, S., Schroeder, R., and Suess, B. (2008) Screening for engineered neomycin riboswitches that control translation initiation. *RNA*, **14**, 89–97.

45 Weigand, J.E. and Suess, B. (2007) Tetracycline aptamer-controlled regulation of pre-mRNA splicing in yeast. *Nucleic Acids Res.*, **35**, 4179–4185.

46 Bocobza, S., Adato, A., Mandel, T., Shapira, M., Nudler, E., and Aharoni, A. (2007) Riboswitch-dependent gene regulation and its evolution in the plant kingdom. *Genes Dev.*, **21**, 2874–2879.

47 Soukup, G.A. and Breaker, R.R. (1999) Engineering precision RNA molecular switches. *Proc. Natl. Acad. Sci. USA*, **96**, 3584–3589.

48 Blount, K.F. and Breaker, R.R. (2006) Riboswitches as antibacterial drug targets. *Nat. Biotechnol.*, **24**, 1558–1564.

49 Narberhaus, F., Waldminghaus, T., and Chowdhury, S. (2006) RNA thermometers. *FEMS Microbiol. Rev.*, **30**, 3–16.

50 Morita, M., Kanemori, M., Yanagi, H., and Yura, T. (1999) Heat-induced synthesis of σ^{32} in *Escherichia coli*: structural and functional dissection of *rpoH* mRNA secondary structure. *J. Bacteriol.*, **181**, 401–410.

51 Morita, M.T., Tanaka, Y., Kodama, T.S., Kyogoku, Y., Yanagi, H., and Yura, T. (1999) Translational induction of heat shock transcription factor σ^{32}: evidence for a built-in RNA thermosensor. *Genes Dev.*, **13**, 655–665.

52 Nocker, A., Hausherr, T., Balsiger, S., Krstulovic, N.P., Hennecke, H., and Narberhaus, F. (2001) mRNA-based thermosensor controls expression of rhizobial heat shock genes. *Nucleic Acids Res.*, **29**, 4800–4807.

53 Waldminghaus, T., Fippinger, A., Alfsmann, J., and Narberhaus, F. (2005) RNA thermometers are common in α- and γ-proteobacteria. *Biol. Chem.*, **386**, 1279–1286.

54 Chowdhury, S., Ragaz, C., Kreuger, E., and Narberhaus, F. (2003) Temperature-controlled structural alterations of an RNA thermometer. *J. Biol. Chem.*, **278**, 47915–47921.

55 Chowdhury, S., Maris, C., Allain, F.H., and Narberhaus, F. (2006) Molecular basis for temperature sensing by an RNA thermometer. *EMBO J.*, **25**, 2487–2497.

56 Waldminghaus, T., Heidrich, N., Brantl, S., and Narberhaus, F. (2007) FourU: a novel type of RNA thermometer in *Salmonella*. *Mol. Microbiol.*, **65**, 413–424.

57 Hoe, N.P. and Goguen, J.D. (1993) Temperature sensing in *Yersinia pestis*: translation of the LcrF activator protein is thermally regulated. *J. Bacteriol.*, **175**, 7901–7909.

58 Hurme, R. and Rhen, M. (1998) Temperature sensing in bacterial gene regulation – what it all boils down to. *Mol. Microbiol.*, **30**, 1–6.

59 Johansson, J., Mandin, P., Renzoni, A., Chiaruttini, C., Springer, M., and Cossart, P. (2002) An RNA thermosensor controls expression of virulence genes in Listeria monocytogenes. *Cell*, **110**, 551–561.

60 Altuvia, S., Kornitzer, D., Teff, D., and Oppenheim, A.B. (1989) Alternative mRNA structures of the cIII gene of bacteriophage λ determine the rate of its translation initiation. *J. Mol. Biol.*, **210**, 265–280.

61 Yamanaka, K., Mitta, M., and Inouye, M. (1999) Mutation analysis of the 5′ untranslated region of the cold shock *cspA* mRNA of *Escherichia coli*. *J. Bacteriol.*, **181**, 6284–6291.

62 Fang, L., Jiang, W., Bae, W., and Inouye, M. (1997) Promoter-independent cold-shock induction of *cspA* and its derepression at 37 °C by mRNA stabilization. *Mol. Microbiol.*, **23**, 355–364.

63 Uppal, S., Akkipeddi, V.S., and Jawali, N. (2008) Posttranscriptional regulation of *cspE* in *Escherichia coli*: involvement of the short 5′-untranslated region. *FEMS Microbiol. Lett.*, **279**, 83–91.

64 Waldminghaus, T., Kortmann, J., Gesing, S., and Narberhaus, F. (2008) Generation of synthetic RNA-based thermosensors. *Biol. Chem.*, **389**, 1319–1326.

65 Wieland, M. and Hartig, J.S. (2007) RNA quadruplex-based modulation of gene expression. *Chem. Biol.*, **14**, 757–763.

66 Lee, J. and Kotow, N.A. (2007) Thermometer design at the nanoscale. *Nanotoday*, **2**, 48–51.

24
Signal Transduction by Serine/Threonine Protein Kinases in Bacteria
Michael Bott

24.1
Introduction

In order to survive, bacteria must be able to adapt to changing environmental conditions, such as nutrient availability, pH, osmolarity, temperature, toxins, and so on. This process includes the sensing of extracellular or intracellular changes and signal transduction to proteins that finally elicit a suitable response to these changes. Two-component systems consisting of a histidine kinase (also termed sensor kinase) and a response regulator are considered as the most widely distributed signal transduction systems in bacteria. As described in detail in Chapter 8, the histidine kinase responds to a certain stimulus by autophosphorylation of a conserved histidine residue and the phosphoryl group is subsequently transferred to a conserved aspartate residue of the response regulator, which then mediates changes in gene expression or cell behavior. In the past decade, it has become obvious, however, that serine/threonine protein kinases (STPKs) also play an important role in bacterial signal transduction.

24.2
Discovery and Distribution of STPKs in Prokaryotes

STPKs were previously regarded as exclusively eukaryotic proteins. However, in 1991 [1], the first gene for a "eukaryotic-like" STPK, *pkn1*, was identified in *Myxococcus xanthus*, a member of the δ-proteobacteria, by using the polymerase chain reaction with oligonucleotides derived from two of the highly conserved subdomains of the catalytic domain of STPKs (see Section 24.5) [2]. In the following years, particularly after genome sequencing had been established, it became evident that many bacteria harbor genes for STPKs (see, e.g., [3–10]).

Recently, Perez *et al.* [11] analyzed 626 prokaryotic genomes, of which 577 belonged to bacteria and 49 to archaea, for the presence of eukaryotic-like protein

Bacterial Signaling. Edited by Reinhard Krämer and Kirsten Jung
Copyright © 2010 WILEY-VCH Verlag GmbH & Co. KGaA, Weinheim
ISBN: 978-3-527-32365-4

kinases (ELKs) by using the Pfam matrix PF00069 [12]. ELKs were identified in 404 of the sequenced strains, whereas 222 strains were devoid of ELKs (Supplementary Table 4 in [11]). These numbers underline the fact that ELKs, the large majority of which are most likely STPKs, play an important role in signal transduction in prokaryotes. However, the distribution of ELKs is quite uneven. Of the 2697 ELKs identified in total, 892 were present in seven strains of myxobacteria, which belong to the δ-proteobacteria. The largest number of ELKs, namely 313, is found in *Sorangium cellulosum* So ce 56, which also contains the largest bacterial genome known to date (13.03 Mb). Apart from δ-proteobacteria, strains with large numbers of ELKs are found in actinobacteria such as *Frankia alni* ACN14a (53 ELKs), *Rhodococcus* sp. RHA1 (37 ELKs), or *Streptomyces coelicolor* A3(2) (31 ELKs), in chloroflexi such as *Herpetosiphon aurantiacus* ATCC 23779 (35 ELKs) or *Roseiflexus* sp. RS-1 (25 ELKs), in cyanobacteria such as *Anabaena variabilis* ATCC 29413 (55 ELKs) or *Nostoc* sp. PCC 7120 (48 ELKs), in planctomycetes such as *Rhodopirellula baltica* SH 1 (54 ELKs), and in acidobacteria such as *Solibacter usitatus* Ellin6076 (75 ELKs). Based on the distribution of ELKs, Perez *et al.* [11] suggested that multicellular behavior is the main evolutionary driving force for an extensive kinome (i.e., the set of protein kinases in an organism).

24.3
Serine/Threonine Phosphorylation versus Histidine/Aspartate Phosphorylation

A major difference between signal transduction by histidine kinase/response regulator systems and STPKs is the stability of the phosphorylated residues. The "high-energy" phosphohistidine and phosphoaspartate residues are much more labile than the phosphomonoesters. At neutral pH, the phosphoaspartate residue of response regulators has a half-life of seconds to several minutes, depending on the individual protein, while phosphoserine, phosphothreonine, and phosphotyrosine residues in proteins are stable for weeks [13, 14]. As a consequence, signal transduction by two-component systems is rapidly attenuated by chemical hydrolysis of phosphoaspartate. In addition, phosphorylated response regulators can be dephosphorylated by their cognate kinases, due to the fact that the phosphotransfer from histidine to aspartate is a reversible reaction (see Chapter 8). In contrast, phosphorylation of serine, threonine, and tyrosine residues allows the production of long-term signals that require phosphatases to be shut off.

24.4
Domain Architecture of STPKs

As described in Section 24.5, the common feature of all STPKs is the catalytic domain that comprises about 300 residues. This domain is described in more detail below. Apart from the kinase domain, most bacterial STPKs contain additional domains, most often in the C-terminal part of the protein. These additional regions are extremely variable – some showing no similarity to other proteins at all, some representing known protein domains [15]. In myxobacteria, more than 30 known protein domains were found to be associated with the kinase core domain in different combinations and arrangements [11]. Interestingly, there are several examples were STPKs are associated with the characteristic domains of histidine kinases and response regulators, showing that these two types of signal transduction systems can be combined. Other domains found in STPKs are second messenger signaling domains like the adenylate and guanylate cylase catalytic domain or the GGDEF domain which possesses diguanylate cyclase activity, or protein–protein interaction domains like the phosphothreonine-binding FHA (forkhead-associated) domains (see Section 24.8.2).

Another level of variation in STPKs is based on the presence or absence of transmembrane helices. Many bacterial STPKs possess a single transmembrane helix and thus are integral membrane proteins, with the C-terminal portion located in the extracytoplasmic space. In *Mycobacterium tuberculosis*, for example, nine of the 11 STPKs contain one transmembrane helix [8]. These "receptor" kinases are apparently able to sense extracellular or periplasmic stimuli and transmit them to the cytoplasm. One of the domains that has been identified in the extracytoplasmic portion of several bacterial STPKs is the PASTA (penicillin-binding protein and serine/threonine kinase-associated) domain (indicating that it is also found in high-molecular-weight penicillin-binding proteins) [16]. A PASTA domain consists of about 70–80 amino acids that form a $\beta_3\alpha$ structure. PASTA domains bind to peptidoglycan and the amino acid composition of the peptide stem plays an important role in recognition, as recently shown for the STPK PrkC from *Bacillus subtilis* [17]. In PrkC and many STPKs, more than one PASTA domain is present. According to the information given above, there exists a great diversity in the domain architecture of bacterial STPKs, which allows them to detect numerous different stimuli.

24.5
Structural Studies on STPKs

The catalytic domain of STPKs (and other kinases) is composed of 12 subdomains or regions (numbered I, II, III, IV, V, VIA, VIB, VII, VIII, IX, X, and XI) with specific consensus motifs that were identified by sequence alignments of eukaryotic STPKs [2, 18]. About 10 residues in these regions are invariant or nearly invariant in all members, and are known to play crucial roles in binding of the substrates and catalysis (see Figure 24.1A and B). The crystal structures of eukaryotic STPKs, the first one being the cAMP-dependent protein kinase PKA from mouse [19, 20], revealed that the conserved regions correspond quite well to units of higher-order structure, such as β-sheets and α-helices, as depicted in Figure 24.1(A).

The catalytic domain folds into a two-lobed structure. The smaller N-terminal lobe includes subdomains I–IV, has a predominantly antiparallel β-sheet structure, and has a long α-helix (αC). It is primarily involved in anchoring and orienting the nucleotide. The larger C-terminal lobe includes subdomains VIA–XI, is predominantly α-helical in content, and is largely responsible for binding the peptide substrate and initiation of phosphotransfer. Subdomain V forms the connection between the two lobes, which form a deep cleft that is the site of catalysis. Depending on the relative orientations of the two lobes to each other, STPKs can be found in two conformational states, referred to as "open" and "closed" conformations, the

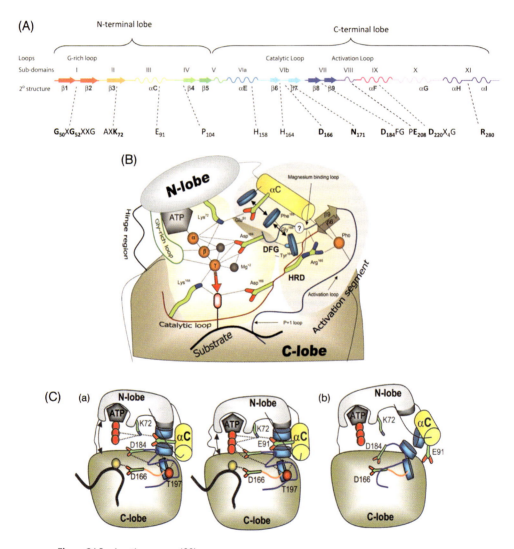

Figure 24.1 (caption see p. 431)

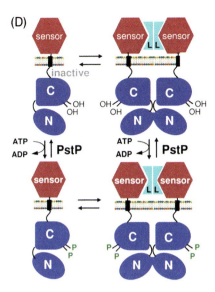

Figure 24.1 Characteristic features of the kinase domain of STPKs and modes of its activation. (A) Secondary structures, key motifs, and residues of the catalytic domain of STPKs and other protein kinases (modified from [63]). The numbers refer to eukaryotic PKA. (B) Diagram of known interactions between protein kinase catalytic core (PKA numbering), ATP, and a substrate (taken from [23]). Red arrow indicates catalyzed transfer of the ATP phosphate to a hydroxyl group of a protein substrate. Catalytically important residues that are in contact with ATP and/or substrate are shaded yellow. Secondary structures and residues that are known to be involved in regulation of the catalytic activity are shaded gray. Hydrophobic interactions between the HRD motif, the DFG motif, and the C helix are shown by black arrows. The important polar contacts are shown by dashed lines. (C) Model of protein kinase activation (taken from [23]). (a) Active conformation. Phosphothreonine T197 arranges the Mg^{2+}-binding loop positioning the DFG aspartate for interaction with the ATP and the DFG phenylalanine for building up a hydrophobic "spine". The αC helix flips to complete the spine formation and simultaneously secures it with the K72–E91 polar contact. The spine residues are shown as blue disks and the shaded gray portion of the N-lobe. The spine stabilizes the protein kinase molecule, which can perform coordinated motions going through open (left) and closed (right) conformations during the catalytic cycle. (b) Inactive conformation. The Mg^{2+}-binding loop and the spine are distorted, destabilizing the molecule. The lobes move independently. The unconstrained Mg^{2+}-binding loop becomes flexible and can attain different inactive configurations. (D) Model of activation of the *M. tuberculosis* receptor kinase PknD (taken from [30]). The monomeric, dephosphorylated kinase (dark blue) is the inactive form (upper left). Extracellular ligand (light blue) binding to the extracellular domain (maroon) dimerizes the kinase, forming an active complex (upper right). Phosphorylation of the kinase domain follows (lower right) further stimulating activity. The phosphorylated kinase monomers remain active in the absence of the extracellular ligand (lower left). The activity of the phosphorylated monomers implies that trans-phosphorylation by other STPKs may provide a distinct mechanism of regulation capable of activating the kinase in the absence of dimerizing signal. The *M. tuberculosis* serine/threonine phosphatase PstP (middle) dephosphorylates the kinase, returning the system to the inactive state only in the absence of the extracellular ligand.

latter corresponding to the active state [21, 22]. In Figure 24.1(C), a current model of kinase activation as proposed by Kornev *et al.* [23] is shown. In this model, autophosphorylation of a threonine residue in the so-called activation loop plays a critical role. In fact, most STPKs catalyze autophosphorylation as part of the signal transduction cascade [24]. Autophosphorylation can have two functions – activation of the enzymatic activity by inducing conformational changes in the protein and provision of a docking site for substrate proteins, such as FHA domains.

Structures of several mycobacterial STPKs have been determined, such as the cytoplasmic domains of the membrane-bound kinases PknB and PknE, and of the soluble kinase PknG [25–29]. All showed the classic STPK architecture described above, supporting a universal activation mechanism for these kinases. In contrast to most eukaryotic STPKs, the kinase domains of PknB and PknE formed dimers, and it was suggested that dimerization plays a role in the regulation of these membrane-integral STPKs. Studies on the receptor kinase PknD of *M. tuberculosis* led to the model of regulation shown in Figure 24.1(D), in which binding of a ligand to the extracellular domain triggers dimerization and activation by autophosphorylation [30]. Apart from the structures of kinase domains, the structure of the extracytoplasmic sensor domain of *M. tuberculosis* PknD, which forms a rigid, six-bladed β-propeller, has also been solved. However, the ligand(s) recognized by this structure are not yet known.

24.6
Signal Transduction by STPKs

A physiological understanding of the function of individual STPKs requires knowledge of (i) the stimuli recognized, (ii) the influence of stimuli on the kinase activity and/or the binding of target proteins to the kinase, (iii) the identity of the target proteins which become phosphorylated, (iv) the function of the target proteins, and (v) the influence of phosphorylation on the function of the target proteins. Numerous studies have been dedicated to these topics and it is not within the scope of this chapter to give an overview of these. Rather, a few selected examples taken from studies on STPKs in *M. tuberculosis* and *Corynebacterium glutamicum* are described. In particular, the 11 STPKs of *M. tuberculosis* [8] have been intensively analyzed in the last decade for their physiological function and potential application for drug development to combat tuberculosis.

In Table 24.1, proteins of *C. glutamicum* and *M. tuberculosis* are listed that were experimentally shown to be phosphorylated by STPKs, usually by *in vitro* studies, but sometimes also *in vivo*. These proteins function as enzymes, as enzyme inhibitors, as cell division proteins, as transcriptional regulators, as sigma factors, anti-sigma factors or anti-anti-sigma factors, as transporters, or are of unknown function. Accordingly, presumably all types of proteins irrespective of their function can be targeted by STPKs and consequently all kinds of physiological processes can be influenced by serine/threonine phosphorylation. In the following, some examples of signal transduction by STPKs are discussed in more detail.

Table 24.1 Examples of identified STPK substrate proteins in C. glutamicum und M. tuberculosis (order according to the locus tag number).

Species	Protein name	Locus tag	Annotated function	STPK(s)	References
C. glutamicum	OdhI	cg1630	inhibitor of 2-oxoglutarate dehydrogenase	PknG	[55, 56]
C. glutamicum	MurC	cg2368	UDP-N-acetylmuramate-L-alanine ligase	PknA	[64]
M. tuberculosis	PbpA	Rv0016c	penicillin-binding protein A	PknB	[65]
M. tuberculosis	—	Rv0020c	protein of unknown function with a C-terminal FHA domain	PknB, PknF	[66]
M. tuberculosis	—	Rv0516c	putative anti-anti-sigma factor	PknD (PknB, PknE)	[35]
M. tuberculosis	FabH	Rv0533c	β-ketoacyl-acyl carrier protein synthase	PknA, PknF	[47]
M. tuberculosis	—	Rv0681	transcriptional regulator of TetR family	PknH	[67]
M. tuberculosis	GlmU	Rv1018c	N-acetylglucosamine-1-phosphate uridyltransferase (peptidoglycan synthesis)	PknB	[68]
M. tuberculosis	EmbR	Rv1267c	transcriptional regulator of embCAB genes involved in arabinogalactan synthesis, contains FHA domain	PknH, PknA, PknB	[38, 44]
M. tuberculosis	RsfA	Rv1365c	anti-anti-SigF	PknE	[35]
M. tuberculosis	—	Rv1422	hypothetical protein	PknA, PknB	[69]
M. tuberculosis	—	Rv1747	putative ABC transporter with two FHA domains	PknB, PknD, PknE, PknF	[66, 70]
M. tuberculosis	GarA	Rv1827	homolog of C. glutamicum OdhI, inhibits ODH and glutamate dehydrogenase	PknG, PknB	[58, 59]
M. tuberculosis	—	Rv1904	anti-anti-sigma factor	PknE	[35]
M. tuberculosis	Wag31	Rv2145c	homolog of cell division protein DivIVA	PknA, PknB	[69]
M. tuberculosis	FtsZ	Rv2150c	cell division protein with GTPase activity	PknA	[71]
M. tuberculosis	MurD	Rv2155c	UDP-N-acetylmuramoyl-L-alanine: D-glutamate ligase	PknA	[72]

Table 24.1 (Continued)

Species	Protein name	Locus tag	Annotated function	STPK(s)	References
M. tuberculosis	FabD	Rv2243	malonyl-coenzyme A: AcpM transacylase, part of the FAS-II system involved in mycolic acid biosynthesis	PknA, PknB, PknE, PknF, PknH	[46]
M. tuberculosis	KasA	Rv2245	β-ketoacyl-AcpM synthase A, part of the FAS-II system involved in mycolic acid biosynthesis	PknA, PknB, PknE, PknF, PknH	[46]
M. tuberculosis	KasB	Rv2246	β-ketoacyl-AcpM synthase A, part of FAS-II system involved in mycolic acid biosynthesis	PknA, PknB, PknE, PknF, PknH	[46]
M. tuberculosis	MmpL7	Rv2942	transporter of the RND family	PknD	[73]
M. tuberculosis	RshA	Rv3221	anti-SigH	PknB, PknE	[34, 35]
M. tuberculosis	SigH	Rv3223c	sigma factor H	PknB	[34]
M. tuberculosis	DacB1	Rv3330	putative D-alanyl-D-alanine carboxypeptidase (peptidoglycan synthesis)	PknH	[67]

24.7
Control of Gene Expression by PknB via the Activity of Sigma Factors

The alternative sigma factor SigH of *M. tuberculosis* is a key regulator of the response to oxidative, nitrosative, and heat stress [31, 32]. Apart from autoregulation of *sigH* expression, SigH activity is regulated post-translationally via its anti-sigma factor RshA [33]. Binding of RshA to SigH inhibits SigH activity. Under oxidizing conditions, which are sensed by the redox-switch protein RshA, the interaction of the two proteins is disrupted, and SigH can activate transcription of stress response genes and additional transcriptional regulators. Both SigH (on Thr26 and/or Thr106) and RshA (at Thr94) were shown to be phosphorylated by PknB. Evidence was provided that phosphorylation of RshA, but not of SigH, leads to a partial disruption of the RshA–SigB complex and allowed increased transcription of some strong SigH-dependent promoters, such as *trxB2* [34]. Thus, phosphorylation of RshA provides an additional means of signal input besides the recognition of stress conditions by RshA itself. The influence of SigH phosphorylation on its activity is not yet clear. As indicated in Table 24.1, other sigma factors of *M. tuberculosis* besides SigH can presumably also be regulated by STPKs, as indicated by the phosphorylation of the anti-anti-sigma factors Rv0516c and Rv1365c [35].

24.8
Control of Gene Expression by PknH via the Transcriptional Regulator EmbR

24.8.1
Discovery of EmbR in Mycobacteria and its Phosphorylation by PknH

The *embR* gene was first discovered in *Mycobacterium avium* in a study aimed at identifying genes conferring resistance to ethambutol – a compound used to treat tuberculosis and infections caused by *M. avium* [36]. In *M. avium*, *embR* is located upstream and divergent to the *embAB* genes, which encode membrane-embedded arabinosyltransferases that are involved in arabinogalactan and lipoarabinomannan synthesis, and play an important role in ethambutol resistance [36]. In *M. tuberculosis*, the homologous *embR* gene (Rv1267c) is not located in the immediate vicinity of the *embCAB* genes (Rv3793–3795), but upstream of *pknH* (Rc1266c), which encodes a membrane-integral STPK with a single transmembrane helix [8]. EmbR shows homology to the transcriptional regulator AfsR from *S. coelicolor* (SCO4426) – the first protein in bacteria shown to be phosphorylated by a STPK (AfsK, SCO4423) [37]. *In vitro* phosphorylation experiments revealed that PknH phosphorylates EmbR on threonine, and that this reaction requires the prior autophosphorylation of PknH at Thr170 in the activation loop and the FHA domain of EmbR [38].

24.8.2
Structure of EmbR

The crystal structure of *M. tuberculosis* EmbR protein alone or in complex with a nonspecific phosphopeptide has been determined [39]. EmbR is composed of three major domains: an N-terminal winged helix–turn–helix domain, a central bacterial transcription activation domain, and a C-terminal FHA domain [38]. FHA domains are known to bind to phosphothreonine epitopes of proteins [40–42]. They fold into a β-sandwich of two β-sheets, and a conserved serine and a conserved arginine recognize the phosphorylation of the threonine residue targeted.

24.8.3
Effects of EmbR Phosphorylation by PknH

The effect of phosphorylation by PknH on the activity of EmbR was studied both *in vitro* and *in vivo* [43]. Purified EmbR binds to the upstream regions of *embC*, *embA*, and *embB*, and phosphorylation increases the DNA-binding affinity about 10-fold in all three cases. Expression of *M. tuberculosis pknH* in *Mycobacterium smegmatis*, which lacks a PknH homolog, but contains an EmbR homolog and *embCAB* genes, led to an increased phosphorylation of the endogenous EmbR protein and to an increased expression of the endogenous *embCAB* genes. These data are consistent with the view that phosphorylation of EmbR increases its activity as a transcriptional activator of *embCAB*. EmbR was shown to bind to holo-RNA polymerase of *M. smegmatis*, and to possess ATPase and GTPase activity, which was sixfold stimulated by phosphorylation of EmbR [44]. The ATPase activity might play a role in the conversion of the closed complex to the open complex. Apart from PknH, PknA and PknB of *M. tuberculosis* were also found to phosphorylate EmbR *in vitro*. Dephosphorylation of PknH, PknA, PknB, and EmbR was catalyzed *in vitro* by PstP – the only known phospho-serine/threonine protein phosphatase of *M. tuberculosis*. Dephosphorylation of PknH, PknA, and PknB inhibited their interaction with EmbR, and dephoshorylation of EmbR inhibited its interaction with the DNA targets [44].

Increased levels of EmbC, which catalyzes the arabinosylation of lipomannan (LM) to lipoarabinonmannan (LAM), led to a fourfold increased LAM: LM ratio, which presumably is correlated with an increased mycobacterial virulence. Increased levels of EmbA and EmbB, which are involved in the synthesis of arabinogalactan, led to a threefold increase in ethambutol resistance, in agreement with the observation that ethambutol resistance depends on the copy number of EmbA and EmbB [36].

24.8.4
Model of Signal Transduction by PknH and EmbR

In Figure 24.2(A), a simplified model of signal transduction by PknH and EmbR is shown, which does not include the fact that not only PknH, but also PknA and PknB are able to phosphorylate EmbR *in vitro* [44], and which neglects the observation than in some *M. tuberculosis* strains such as CDC1551 the EmbR homolog EmbR2 influences PknH activity and thus EmbR phosphorylation [45].

24.9
Direct Control of Enzyme Activities by STPKs

Several examples are known in which the activity of enzymes is influenced by STPK-dependent phosphorylation. The β-ketoacyl AcpM synthases KasA and KasB are part of the type II fatty acid synthase (FAS-II) system of *M. tuberculosis* and involved in mycolic acid synthesis. They were shown to be phosphorylated on threonine residues both *in vitro* (by PknA) and *in vivo*. Whereas phosphorylation decreased the activity of KasA, that of KasB was enhanced. It was suggested that this type of regulation may allow pathogenic mycobacteria to produce full-length mycolates required for adaptation and survival in macrophages [46]. Also β-ketoacyl-acyl carrier protein synthase III (FabH) of *M. tuberculosis* was shown to be phosphorylated on a single

Figure 24.2 Models of signal transduction by two selected STPKs in bacteria. (A) Signal transduction by the membrane-bound kinase PknH and its target protein EmbR in *M. tuberculosis*. The model is based on the studies by Molle *et al.* [38], Sharma *et al.* [43, 44], and Alderwick *et al.* [39]. HTH, helix–turn–helix; KD, kinase domain; Pro, proline-rich domain; PstP, phospho-serine/threonine protein phosphatase; TM, transmembrane helix. (B) Signal transduction by the soluble kinase PknG and its target protein OdhI/GarA in corynebacteria and mycobacteria. The model is based on the studies by Niebisch *et al.* [55], O'Hare *et al.* [59], and England *et al.* [62]. Ppp is the PstP homolog in *C. glutamicum*. For further details, see text.

24.9 Direct Control of Enzyme Activities by STPKs

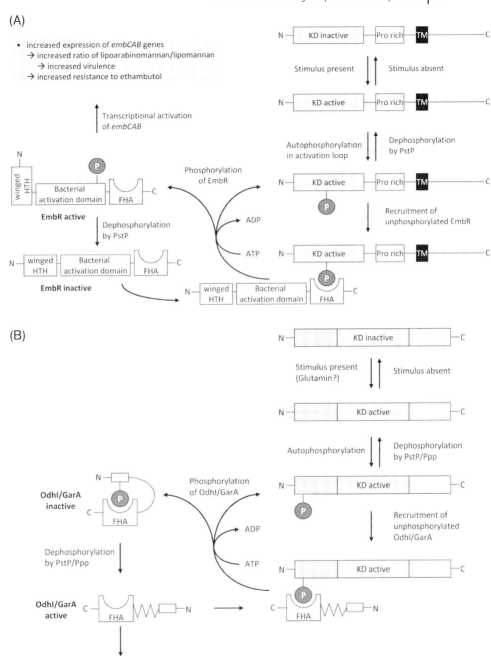

Figure 24.2 (caption see p. 436)

threonine residue (Thr45) and indirect evidence was provided that phosphorylation inhibits enzymatic activity [47].

Apart from enzymes involved in fatty acid biosynthesis, enzymes involved in peptidoglycan biosynthesis were shown to be regulated via phosphorylation by STPKs. The MurC ligase is responsible for converting UDP-N-acetylmuramic acid to UDP-N-acetylmuramoyl-L-alanine and thus adds the first amino acid of the pentapeptide chain. The MurC protein of C. glutamicum was shown to be phosphorylated in vitro by PknA on six threonine residues (Thr51, Thr120, Thr133, Thr167, Thr362, and Thr365) and evidence was provided that phosphorylation inhibits the activity of the enzyme [48]. The GlmU protein functions as N-acetylglucosamine-1-phosphate uridyltransferase, which carries out two enzymatic activities. The C-terminal domain catalyzes the transfer of an acetyl group from acetyl-coenzyme A to glucosamine-1-phosphate to produce N-acetyl-glucosamine-1-phosphate, which is converted by the N-terminal domain to UDP-N-acetylglucosamine-1-phosphate by transfer of the UMP moiety of UTP. Evidence was provided that phosphorylation of M. tuberculosis GlmU by PknB at a threonine residue in the region 414–439 inhibits the acetyltransferase activity, but not the uridyltransferase activity [49]. These examples strongly suggest that PknA and PknB, both of which are essential in M. tuberculosis, are involved in the regulation of peptidoglycan synthesis.

24.10
Indirect Control of Enzyme Activity by PknG and its Target Protein OdhI/GarA

24.10.1
Distribution of PknG

Signal transduction by the soluble STPK PknG is a particularly interesting case, as this protein plays roles both in central metabolism and virulence. PknG has primarily been studied in M. tuberculosis and C. glutamicum, but homologs are present in all members of the suborder Corynebacterineae and further members of the order Actinomycetales, such as Streptomyces species. The PknG protein levels can differ strongly between different mycobacterial strains and this difference appears to result from differences in translation rather from differences in transcription [50]. In corynebacteria and mycobacteria, the pknG gene (Rv0410c, cg3046) is the last gene of a putative tricistronic operon. The first gene of this cluster (Rv0412c, cg3044), designated glnX in C. glutamicum, encodes an integral membrane protein, whereas the second gene, designated glnH (Rv0411c, cg3045), encodes a putative glutamine-binding lipoprotein.

24.10.2
Structure of PknG

The structure of PknG from M. tuberculosis lacking the first 73 amino acid residues was solved and showed that the protein is composed of three distinct structural domains [27]. The N-terminal part contains two iron-binding Cys-X-X-Cys-Gly motifs characteristic of rubredoxins. Exchange of the four cysteines to serine residues led to an inactive protein that was no longer capable of autophosphorylation [27]. The kinase domain, which is sandwiched between the N- and C-terminal domains, shows the characteristic two-lobed structure typical for STPKs. The C-terminal half contains a tetratricopeptide repeat, a structural element that is known to mediate protein-protein interactions [51]. Interestingly, the two Cys-X-X-Cys-Gly motifs are absent in the corynebacterial PknG proteins, indicating variations in the function of these homologs.

24.10.3
Evidence for a Role of PknG in the Pathogenicity of Mycobacteria

A *pknG* deletion mutant of *M. tuberculosis* showed slower growth and a reduced final cell density in three different media [52], indicating that PknG plays a role for growth *"in vitro"*. Moreover, the mutant caused delayed mortality in immunodeficient mice and decreased viability in immunocompetent mice. The intracellular glutamine/glutamate levels were analyzed and found to be elevated in the *pknG* mutant. In contrast to *M. tuberculosis*, a *pknG* deletion mutant of *Mycobacterium bovis* BCG showed no growth defect and no increased intracellular levels of glutamine/glutamate [53]. The reason for these differences are not known yet.

Using *M. bovis* BCG, a crucial role for PknG in the survival of mycobacteria in macrophages by inhibition of phagosome–lysosome fusion was shown [54]. Deletion of *pknG* in *M. bovis* BCG resulted in lysosomal localization and mycobacterial cell death in infected macrophages. PknG was reported to be present in the cytosol of infected macrophages and was therefore suggested to interfere with as-yet unknown host signaling pathways [54].

24.10.4
Studies on PknG and its Target Protein OdhI in *C. glutamicum*

In *C. glutamicum*, a *pknG* deletion mutant had a severe growth defect on agar plates containing glutamine as sole carbon and nitrogen source, and contained twofold elevated levels of cytoplasmic glutamate, but not of glutamine [55]. The latter fact indicated that the catabolism of glutamate was disturbed in the mutant. By a comparative two-dimensional gel analysis of wild-type and *pknG* mutant, and subsequent biochemical studies, the 15-kDa OdhI protein (cg1630) was identified as the first target protein of PknG [55]. OdhI is composed of three parts: an N-terminal domain extending to amino acid residue 42, followed by an FHA domain comprising residues 43–138, and a C-terminal portion extending up to the C-terminal residue 143. Phosphorylation of OdhI by PknG was shown to occur at Thr14. In contrast to the *pknG* deletion mutant, an *odhI* deletion mutant of *C. glutamicum* had no obvious defect in glutamine utilization. However, deletion of *odhI* in the Δ*pknG* background abolished the growth defect on glutamine agar plates, showing that OdhI is part of the PknG signaling pathway and most probably has an inhibitory function. Consistent with this hypothesis, transformation of the Δ*odhI* mutant with a plasmid encoding an OdhI-T14A derivative led to the same growth defect on glutamine as the *pknG* deletion [55]. Consequently, the presence of an OdhI protein that cannot be phosphorylated on Thr14 by PknG inhibited growth on glutamine.

24.10.5
Inhibition of 2-Oxoglutarate Dehydrogenase by Corynebacterial OdhI

As OdhI itself has no obvious functional domains except for the FHA domain, its inhibitory action was suggested to be due to an interaction with other cellular proteins. Indeed, it could be shown that unphosphorylated OdhI interacts with the E1 subunit (OdhA, cg1280) of the 2-oxoglutarate dehydrogenase (ODH) complex in *C. glutamicum*. Studies with cell extracts and the purified ODH complex showed that unphosphorylated OdhI inhibits the activity of this enzyme (i.e., the oxidative decarboxylation of 2-oxoglutarate to succinyl-coenzyme A [55]). Half-maximal inhibition was observed at an OdhI concentration of about 4 nM. Kinetic studies indicated that OdhI acts as a partial mixed type, essentially noncompetitive inhibitor with respect to 2-oxoglutarate [55]. As the ODH complex is required for glutamate catabolism, the inhibition of its activity by unphosphorylated OdhI explains the defect in glutamine utilization caused by deletion of PknG or exchange of Thr14 by alanine. The stimulus that is sensed directly or indirectly by PknG is still unknown. It has been demonstrated that the dephosphorylation of phosphorylated OdhI is catalyzed by Ppp (cg0062) – the only phospho-serine/threonine protein phosphatase encoded by the *C. glutamicum* genome [56].

24.10.6
Identification of the OdhI Homolog GarA in Mycobacteria

The OdhI homolog in mycobacteria is designated GarA (*g*lycogen *a*ccumulation *r*egulator) [57]. The *garA* gene (Rv1827) was discovered as a multicopy suppressor of the temperature-sensitive *M. smegmatis* mutant SMEG53, which stops growing at 42 °C due to the accumulation of high glycogen levels. The mutant carries a His359Tyr exchange in the GlgE protein – a putative glycosyl transferase of the α-amylase family that presumably is involved in glycogen degradation. Overexpression of *garA* leads to a reduction of the glycogen levels that allows growth at 42 °C [57]. The molecular mechanism by which GarA reduces the glycogen level is still unknown. It has to be kept in mind, however, that suppression was only obtained when *garA* was present in multicopy, which presumably represents an unphysiological situation.

24.10.7
Identification of GarA as a Substrate of Mycobacterial PknB

In a study aimed to identify endogenous protein substrates of the essential STPK PknB, GarA of *M. tuberculosis* and *M. smegmatis* was found to be the best detectable substrate of the PknB kinase domain [58]. Studies with purified proteins revealed that $PknB_{1-279}$ phosphorylates GarA on a single threonine residue, Thr22. This residue corresponds to Thr15 in the OdhI protein of *C. glutamicum*. The interaction of GarA with PknB$_{1-279}$ was found to be dependent on the FHA domain of GarA and two phosphorylated threonine residues (Thr171 and Thr173) in the activation loop of PknB. Mutation of either one of the threonine residues to alanine reduced complex formation; mutation of both almost completely abolished binding of GarA and its phosphorylation. The N-terminal part of GarA containing the phosphorylation site was dispensable for binding to PknB [58]. Apart from PknB, also PknD, PknE, and PknF of *M. tuberculosis* were able to phosphorylate GarA *in vitro*, but with a significantly lower activity than PknB. Based on these data, a model was proposed in which phosphorylation of the activation loop is required for both substrate recruitment and enzyme activation [58].

24.10.8
Identification of GarA as Substrate of Mycobacterial PknG

Based on the studies in *C. glutamicum*, O'Hare *et al.* [59] analyzed whether PknG of *M. tuberculosis* and *M. smegmatis* perform similar functions. Indeed they showed that PknG phosphorylates the OdhI homolog GarA on a single threonine residue, Thr21, which corresponds to Thr14 in OdhI. Phosphorylations of GarA by PknG at Thr21 and by PknB at Thr22 were found to be mutually exclusive events *in vitro*. In contrast to PknB and other transmembrane mycobacterial STPKs, autophosphorylation of PknG does not occur in the activation loop, but rather in the N-terminal part of the protein. Four phosphorylation sites were identified, namely Thr23, Thr32, and two residues in the sequence Thr63–Thr64–Ser65 [59]. Phosphorylation of these N-terminal sites, in contrast to phosphorylations in the activation loop of PknB, does not seem to play a role in the activation of PknG. However, phosphorylation of these N-terminal residues is important for recruitment of unphosphorylated GarA via its FHA domain [59]. The interaction between PknG and unphosphorylated GarA was not only demonstrated *in vitro*, but also *in vivo* with *M. smegmatis* using the Split-Trp method. In addition, mass spectrometry analysis of phosphopeptides enriched from *M. tuberculosis* and *M. smegmatis* revealed that peptides from GarA were among the most abundant phosphopeptides detected and the site of phosphorylation was Thr21. This supports the assumption that under the chosen conditions GarA is predominantly phosphorylated by PknG rather than by PknB. Overexpression of GarA in *M. smegmatis* caused a growth defect that was enhanced by a Thr21Ala mutation, but not by a Thr22Ala mutation. Transformation of *M. smegmatis* with a plasmid encoding a GarA protein with Thr21 and Thr22 exchanged by alanine proved to be impossible, either because of a defect in metabolic regulation or because of a blockade of PknG and PknB [59].

24.10.9
Functions of GarA in Mycobacteria

Similar to the results obtained in *C. glutamicum*, unphosphorylated GarA was found to bind to the OdhA homolog of *M. tuberculosis* (ketoglutarate decarboxylase, Rv1248c) and *M. smegmatis* (ketoglutarate decarboxylase, MSMEG_5049). *M. tuberculosis* was reported to lack a functional ODH complex [60]. Instead, 2-oxoglutarate is decarboxylated by ketoglutarate decarboxylase to succinate semialdehyde, which then is oxidized by succinate semialdehyde dehydrogenase to succinate [61]. The activity of ketoglutarate decarboxylase was shown to be inhibited noncompetitively by unphosphorylated GarA, with half-maximal inhibition at 53 nM GarA. Apart from ketoglutarate decarboxylase (and PknG), another interaction partner of unphosphorylated GarA was identified in *M. tuberculosis* and *M. smegmatis*, namely an NAD^+-specific glutamate dehydrogenase (Rv2476c and MSMEG_4699), which catalyses the reversible reductive amination of 2-oxoglutarate to glutamate. In this case, GarA was found to increase the K_m values for glutamate and 2-oxoglutarate, and half-maximal inhibition of the glutamate-forming reaction was observed at 540 nM GarA. An important finding in the course of these studies was that a Ser95Ala mutation in the FHA domain of GarA resulted in a loss of binding to ketoglutarate decarboxylase and glutamate dehydrogenase. This indicated that the FHA domain rather than the N-terminal portion of GarA is responsible for the interaction and, consequently, this domain apparently can also bind to proteins that presumably do not contain a phosphorylated threonine residue.

24.10.10
Putative Mechanism of GarA/OdhI Function

Novel studies on the biophysical and biochemical properties of GarA suggested a molecular mechanism by which GarA and presumably also OdhI can act as a phosphorylation-dependent molecular switch [62]. GarA was found to be a monomer in solution, independent of the phosphorylation state. However, temperature-induced denaturation revealed that GarA phosphorylated on Thr21 or Thr22 is much more stable than unphosphorylated GarA, as the melting temperature was more than 15 K higher. This was confirmed by limited proteolysis studies, where phosphorylation protected GarA against proteolysis. The stabilization of phosphorylated GarA was suggested to result from binding of the phosphorylated N-terminal segment of GarA by its C-terminal FHA domain. Indeed, a GarA derivative lacking the first 43 residues was shown to bind a GarA-derived 17mer synthetic peptide centered on Thr22 in a 1 : 1 stoichiometry and with a K_d of 0.39 µM when the peptide was phosphorylated on Thr22. No binding was observed with the unphosphorylated peptide. Based on these data, a model was proposed in which unphosphorylated GarA contains a structurally flexible N-terminal segment and a structured C-terminal FHA domain with a free phosphothreonine-binding site, both of which are accessible for interaction with and modulation of other proteins, such as ketoglutarate decarboxylase or glutamate dehydrogenase. Phosphorylation of Thr21 or Thr22 by PknG and PknB, respectively, was suggested to trigger a tight intramolecular interaction between the N-terminal phosphothreonine-containing region and the C-terminal FHA domain. In this way, the regulatory functions of the protein are switched off.

The intramolecular interaction should be favored over interactions of the FHA domain with other phosphothreonine-containing proteins due to a higher effective local concentration of the two partners (as they are part of a single protein) and due to the high affinity [62].

24.10.11
Model of Signal Transduction by PknG and OdhI/GarA

In Figure 24.2(B), a simplified model of signal transduction by PknG and OdhI/GarA is presented, which does not include the fact that not only PknG, but also other STPKs are able to phosphorylate OdhI [55, 56] and GarA [58], and which neglects the fact that the binding site for OdhI within corynebacterial PknG has not yet been shown experimentally.

24.11
Conclusions and Outlook

STPKs, previously thought to be exclusively eukaryotic proteins, have been clearly established as key players also in bacterial signal transduction. Analysis of the Global Ocean Sampling dataset revealed that eukaryotic-like kinases are as prevalent as histidine kinases [63]. Whereas histidine kinase/response regulator systems in the vast majority of cases regulate gene expression by activating or repressing transcription, STPKs act in much more variable fashion, as many different types of proteins can be modulated in their activity by phosphorylation. Although phosphorylation on serine and threonine is much more easy to analyze compared to phosphorylation on aspartate or histidine, one has to be aware that many STPKs are promiscuous *in vitro*. Therefore, *in vivo* phosphorylation data should be obtained whenever possible. The studies in mycobacteria indicate that many of their STPKs have several target proteins and that many target proteins can be phosphorylated by several different STPKs. The latter feature offers the possibility to integrate different stimuli. A major deficiency in our current understanding of STPKs is the fact that the stimuli sensed by these proteins are almost completely unknown. Exceptions are the STPKs with PASTA domains in their extracytoplasmic part, which presumably bind to peptidoglycan moieties, as shown for PrkC of *B. subtilis* [17]. Thus, future studies need to be directed not only on the identification of novel substrates, but also on the identification of the stimuli, which certainly is a much more difficult task. A particular challenge is presented by the myxobacterial species containing up to 300 STPK-encoding genes in their genome.

Acknowledgment

The author thanks Dr. Christian Schultz for his great support in collecting the relevant literature.

References

1 Munoz-Dorado, J., Inouye, S., and Inouye, M. (1991) A gene encoding a protein serine/threonine kinase is required for normal development of *M. xanthus*, a Gram-negative bacterium. *Cell*, **67**, 995–1006.

2 Hanks, S.K. and Hunter, T. (1995) The eukaryotic protein kinase superfamily – kinase (catalytic) domain structure and classification. *FASEB J.*, **9**, 576–596.

3 Zhang, C.-C. (1996) Bacterial signalling involving eukaryotic-type protein kinases. *Mol. Microbiol.*, **20**, 9–15.

4 Bakal, C.J. and Davies, J.E. (2000) No longer an exclusive club: eukaryotic signalling domains in bacteria. *Trends Cell Biol.*, **10**, 32–38.

5 Kennelly, P.J. (2002) Protein kinases and protein phosphatases in prokaryotes: a genomic perspective. *FEMS Microbiol. Lett.*, **206**, 1–8.

6 Deutscher, J. and Saier, M.H. Jr. (2005) Ser/Thr/Tyr protein phosphorylation in bacteria – for long time neglected, now well established. *J. Mol. Microbiol. Biotechnol.*, **9**, 125–131.

7 Zhang, C.C., Gonzalez, L., and Phalip, V. (1998) Survey, analysis and genetic organization of genes encoding eukaryotic-like signaling proteins on a cyanobacterial genome. *Nucleic Acids Res.*, **26**, 3619–3625.

8 Av-Gay, Y. and Everett, M. (2000) The eukaryotic-like Ser/Thr protein kinases of *Mycobacterium tuberculosis*. *Trends Microbiol.*, **8**, 238–244.

9 Petrickova, K. and Petricek, M. (2003) Eukaryotic-type protein kinases in *Streptomyces coelicolor*: variations on a common theme. *Microbiology*, **149**, 1609–1621.

10 Zhang, C.C., Jang, J., Sakr, S., and Wang, L. (2005) Protein phosphorylation on Ser, Thr and Tyr residues in cyanobacteria. *J. Mol. Microbiol. Biotechnol.*, **9**, 154–166.

11 Perez, J., Castaneda-Garcia, A., Jenke-Kodama, H., Müller, R., and Munoz-Dorado, J. (2008) Eukaryotic-like protein kinases in the prokaryotes and the myxobacterial kinome. *Proc. Natl. Acad. Sci. USA*, **105**, 15950–15955.

12 Finn, R.D., Tate, J., Mistry, J., Coggill, P.C., Sammut, S.J., Hotz, H.R., Ceric, G., Forslund, K., Eddy, S.R., Sonnhammer, E.L.L. et al. (2008) The Pfam protein families database. *Nucleic Acids Res.*, **36**, D281–D288.

13 Sickmann, A. and Meyer, H.E. (2001) Phosphoamino acid analysis. *Proteomics*, **1**, 200–206.

14 Greenstein, A.E., Grundner, C., Echols, N., Gay, L.M., Lombana, T.N., Miecskowski, C.A., Pullen, K.E., Sung, P.Y., and Alber, T. (2005) Structure/function studies of Ser/Thr and Tyr protein phosphorylation in *Mycobacterium tuberculosis*. *J. Mol. Microbiol. Biotechnol.*, **9**, 167–181.

15 Krupa, A. and Srinivasan, N. (2005) Diversity in domain architectures of Ser/Thr kinases and their homologues in prokaryotes. *BMC Genomics*, **6**, 129.

16 Yeats, C., Finn, R.D., and Bateman, A. (2002) The PASTA domain: a beta-lactam-binding domain. *Trends Biochem. Sci.*, **27**, 438–440.

17 Shah, I.M., Laaberki, M.H., Popham, D.L., and Dworkin, J. (2008) A eukaryotic-like Ser/Thr kinase signals bacteria to exit dormancy in response to peptidoglycan fragments. *Cell*, **135**, 486–496.

18 Hanks, S.K., Quinn, A.M., and Hunter, T. (1988) The protein kinase family – conserved features and deduced phylogeny of the catalytic domains. *Science*, **241**, 42–52.

19 Knighton, D.R., Zheng, J.H., Teneyck, L.F., Ashford, V.A., Xuong, N.H., Taylor, S.S., and Sowadski, J.M. (1991) Crystal structure of the catalytic subunit of cyclic adenosine-monophosphate dependent protein kinase. *Science*, **253**, 407–414.

20 Knighton, D.R., Zheng, J.H., Teneyck, L.F., Xuong, N.H., Taylor, S.S., and Sowadski, J.M. (1991) Structure of a peptide inhibitor bound to the catalytic subunit of cyclic adenosine monophosphate-dependent protein kinase. *Science*, **253**, 414–420.

21 Huse, M. and Kuriyan, J. (2002) The conformational plasticity of protein kinases. *Cell*, **109**, 275–282.

22 Cox, S., Radzioandzelm, E., and Taylor, S.S. (1994) Domain movements in

protein-kinases. *Curr. Opin. Struct. Biol.*, **4**, 893–901.

23 Kornev, A.P., Haste, N.M., Taylor, S.S., and Ten Eyck, L. F. (2006) Surface comparison of active and inactive protein kinases identifies a conserved activation mechanism. *Proc. Natl. Acad. Sci. USA*, **103**, 17783–17788.

24 Smith, J.A., Francis, S.H., and Corbin, J.D. (1993) Autophosphorylation – a salient feature of protein kinases. *Mol. Cell. Biochem.*, **128**, 51–70.

25 Ortiz-Lombardia, M., Pompeo, F., Boitel, B., and Alzari, P.M. (2003) Crystal structure of the catalytic domain of the PknB serine/threonine kinase from *Mycobacterium tuberculosis*. *J. Biol. Chem.*, **278**, 13094–13100.

26 Young, T.A., Delagoutte, B., Endrizzi, J.A., Falick, A.M., and Alber, T. (2003) Structure of *Mycobacterium tuberculosis* PknB supports a universal activation mechanism for Ser/Thr protein kinases. *Nat. Struct. Biol.*, **10**, 168–174.

27 Scherr, N., Honnappa, S., Kunz, G., Mueller, P., Jayachandran, R., Winkler, F., Pieters, J., and Steinmetz, M.O. (2007) Structural basis for the specific inhibition of protein kinase G, a virulence factor of *Mycobacterium tuberculosis*. *Proc. Natl. Acad. Sci. USA*, **104**, 12151–12156.

28 Gay, L.M., Ng, H.L., and Alber, T. (2006) A conserved dimer and global conformational changes in the structure of apo-PknE Ser/Thr protein kinase from *Mycobacterium tuberculosis*. *J. Mol. Biol.*, **360**, 409–420.

29 Mieczkowski, C., Iavarone, A.T., and Alber, T. (2008) Auto-activation mechanism of the *Mycobacterium tuberculosis* PknB receptor Ser/Thr kinase. *EMBO J.*, **27**, 3186–3197.

30 Greenstein, A.E., Echols, N., Lombana, T.N., King, D.S., and Alber, T. (2007) Allosteric activation by dimerization of the PknD receptor Ser/Thr protein kinase from *Mycobacterium tuberculosis*. *J. Biol. Chem.*, **282**, 11427–11435.

31 Fernandes, N.D., Wu, Q.L., Kong, D.Q., Puyang, X.L., Garg, S., and Husson, R.N. (1999) A mycobacterial extracytoplasmic sigma factor involved in survival following heat shock and oxidative stress. *J. Bacteriol.*, **181**, 4266–4274.

32 Raman, S., Song, T.S., Puyang, X.L., Bardarov, S., Jacobs, W.R., and Husson, R.N. (2001) The alternative sigma factor SigH regulates major components of oxidative and heat stress responses in *Mycobacterium tuberculosis*. *J. Bacteriol.*, **183**, 6119–6125.

33 Song, T.S., Dove, S.L., Lee, K.H., and Husson, R.N. (2003) RshA, an anti-sigma factor that regulates the activity of the mycobacterial stress response sigma factor SigH. *Mol. Microbiol.*, **50**, 949–959.

34 Park, S.T., Kang, C.M., and Husson, R.N. (2008) Regulation of the SigH stress response regulon by an essential protein kinase in *Mycobacterium tuberculosis*. *Proc. Natl. Acad. Sci. USA*, **105**, 13105–13110.

35 Greenstein, A.E., MacGurn, J.A., Baer, C.E., Falick, A.M., Cox, J.S., and Alber, T. (2007) *M. tuberculosis* Ser/Thr protein kinase D phosphorylates an anti-anti-sigma factor homolog. *PLoS Pathog.*, **3**, e49.

36 Belanger, A.E., Besra, G.S., Ford, M.E., Mikusova, K., Belisle, J.T., Brennan, P.J., and Inamine, J.M. (1996) The *embAB* genes of *Mycobacterium avium* encode an arabinosyl transferase involved in cell wall arabinan biosynthesis that is the target for the antimycobacterial drug ethambutol. *Proc. Natl. Acad. Sci. USA*, **93**, 11919–11924.

37 Matsumoto, A., Hong, S.K., Ishizuka, H., Horinouchi, S., and Beppu, T. (1994) Phosphorylation of the AfsR protein involved in secondary metabolism in *Streptomyces* species by a eukaryotic-type protein kinase. *Gene*, **146**, 47–56.

38 Molle, V., Kremer, L., Girard-Blanc, C., Besra, G.S., Cozzone, A.J., and Prost, J.F. (2003) An FHA phosphoprotein recognition domain mediates protein EmbR phosphorylation by PknH, a Ser/Thr protein kinase from *Mycobacterium tuberculosis*. *Biochemistry*, **42**, 15300–15309.

39 Alderwick, L.J., Molle, V., Kremer, L., Cozzone, A.J., Dafforn, T.R., Besra, G.S., and Fütterer, K. (2006) Molecular structure of EmbR, a response element of Ser/Thr kinase signaling in *Mycobacterium tuberculosis*. *Proc. Natl. Acad. Sci. USA*, **103**, 2558–2563.

40 Pallen, M., Chuadhuri, R., and Khan, A. (2002) Bacterial FHA domains: neglected players in the phospho-threonine signalling game? *Trends Microbiol.*, **10**, 556–563.

41 Liang, X.Y. and Van Doren, S.R. (2008) Mechanistic insights into phosphoprotein-binding FHA domains. *Acc. Chem. Res.*, **41**, 991–999.

42 Durocher, D. and Jackson, S.P. (2002) The FHA domain. *FEBS Lett.*, **513**, 58–66.

43 Sharma, K., Gupta, M., Pathak, M., Gupta, N., Koul, A., Sarangi, S., Baweja, R., and Singh, Y. (2006) Transcriptional control of the mycobacterial *embCAB* operon by PknH through a regulatory protein, EmbR, in vivo. *J. Bacteriol.*, **188**, 2936–2944.

44 Sharma, K., Gupta, M., Krupa, A., Srinivasan, N., and Singh, Y. (2006) EmbR, a regulatory protein with ATPase activity, is a substrate of multiple serine/threonine kinases and phosphatase in *Mycobacterium tuberculosis*. *FEBS J.*, **273**, 2711–2721.

45 Molle, V., Reynolds, R.C., Alderwick, L.J., Besra, G.S., Cozzone, A.J., Fütterer, K., and Kremer, L. (2008) EmbR2, a structural homologue of EmbR, inhibits the *Mycobacterium tuberculosis* kinase/substrate pair PknH/EmbR. *Biochem J.*, **410**, 309–317.

46 Molle, V., Brown, A.K., Besra, G.S., Cozzone, A.J., and Kremer, L. (2006) The condensing activities of the *Mycobacterium tuberculosis* type II fatty acid synthase are differentially regulated by phosphorylation. *J. Biol. Chem.*, **281**, 30094–30103.

47 Veyron-Churlet, R., Molle, V., Taylor, R.C., Brown, A.K., Besra, G.S., Zanella-Cleon, I., Futterer, K., and Kremer, L. (2009) The *Mycobacterium tuberculosis* β-ketoacyl-acyl carrier protein synthase III activity is inhibited by phosphorylation on a single threonine residue. *J. Biol. Chem.*, **284**, 6414–6424.

48 Fiuza, M., Canova, M.J., Patin, D., Letek, M., Zanella-Cleon, I., Becchi, M., Mateos, L.M., Mengin-Lecreulx, D., Molle, V., and Gil, J.A. (2008) The MurC ligase essential for peptidoglycan biosynthesis is regulated by the serine/threonine protein kinase PknA in *Corynebacterium glutamicum*. *J. Biol. Chem.*, **283**, 36553–36563.

49 Parikh, A., Verma, S.K., Khan, S., Prakash, B., and Nandicoori, V.K. (2009) PknB-mediated phosphorylation of a novel substrate, *N*-acetylglucosamine-1-phosphate uridyltransferase, modulates its acetyltransferase activity. *J. Mol. Biol.*, **386**, 451–464.

50 Houben, E.N.G., Walburger, A., Ferrari, F., Nguyen, L., Thompson, C.J., Miess, C., Vogel, G., Mueller, B., and Pieters, J. (2009) Differential expression of a virulence factor in pathogenic and non-pathogenic mycobacteria. *Mol. Microbiol.*, **72**, 41–52.

51 D'Andrea, L.D. and Regan, L. (2003) TPR proteins: the versatile helix. *Trends Biochem. Sci.*, **28**, 655–662.

52 Cowley, S., Ko, M., Pick, N., Chow, R., Downing, K.J., Gordhan, B.G., Betts, J.C., Mizrahi, V., Smith, D.A., Stokes, R.W. *et al.* (2004) The *Mycobacterium tuberculosis* protein serine/threonine kinase PknG is linked to cellular glutamate/glutamine levels and is important for growth *in vivo*. *Mol. Microbiol.*, **52**, 1691–1702.

53 Nguyen, L., Walburger, A., Houben, E., Koul, A., Muller, S., Morbitzer, M., Klebl, B., Ferrari, G., and Pieters, J. (2005) Role of protein kinase G in growth and glutamine metabolism of *Mycobacterium bovis* BCG. *J. Bacteriol.*, **187**, 5852–5856.

54 Walburger, A., Koul, A., Ferrari, G., Nguyen, L., Prescianotto-Baschong, C., Huygen, K., Klebl, B., Thompson, C., Bacher, G., and Pieters, J. (2004) Protein kinase G from pathogenic mycobacteria promotes survival within macrophages. *Science*, **304**, 1800–1804.

55 Niebisch, A., Kabus, A., Schultz, C., Weil, B., and Bott, M. (2006) Coryne-bacterial protein kinase G controls 2-oxoglutarate dehydrogenase activity via the phosphorylation status of the OdhI protein. *J. Biol. Chem.*, **281**, 12300–12307.

56 Schultz, C., Niebisch, A., Gebel, L., and Bott, M. (2007) Glutamate production by *Corynebacterium glutamicum*: dependence on the oxoglutarate dehydrogenase inhibitor protein OdhI and protein kinase PknG. *Appl. Microbiol. Biotechnol.*, **76**, 691–700.

57 Belanger, A.E. and Hatfull, G.F. (1999) Exponential-phase glycogen recycling is

essential for growth of *Mycobacterium smegmatis*. *J. Bacteriol.*, **181**, 6670–6678.

58 Villarino, A., Duran, R., Wehenkel, A., Fernandez, P., England, P., Brodin, P., Cole, S.T., Zimny-Arndt, U., Jungblut, P.R., Cervenansky, C. *et al.* (2005) Proteomic identification of *M. tuberculosis* protein kinase substrates: PknB recruits GarA, a FHA domain-containing protein, through activation loop-mediated interactions. *J. Mol. Biol.*, **350**, 953–963.

59 O'Hare, H.M., Duran, R., Cervenansky, C., Bellinzoni, M., Wehenkel, A.M., Pritsch, O., Obal, G., Baumgartner, J., Vialaret, J., Johnsson, K. *et al.* (2008) Regulation of glutamate metabolism by protein kinases in mycobacteria. *Mol. Microbiol.*, **70**, 1408–1423.

60 Tian, J., Bryk, R., Shi, S.P., Erdjument-Bromage, H., Tempst, P., and Nathan, C. (2005) *Mycobacterium tuberculosis* appears to lack α-ketoglutarate dehydrogenase and encodes pyruvate dehydrogenase in widely separated genes. *Mol. Microbiol.*, **57**, 859–868.

61 Tian, J., Bryk, R., Itoh, M., Suematsu, M., and Nathan, C. (2005) Variant tricarboxylic acid cycle in *Mycobacterium tuberculosis*: identification of α-ketoglutarate decarboxylase. *Proc. Natl. Acad. Sci. USA*, **102**, 10670–10675.

62 England, P., Wehenkel, A., Martins, S., Hoos, S., Andre-Leroux, G., Villarino, A., and Alzari, P.M. (2009) The FHA-containing protein GarA acts as a phosphorylation-dependent molecular switch in mycobacterial signaling. *FEBS Lett.*, **583**, 301–307.

63 Kannan, N., Taylor, S.S., Zhai, Y.F., Venter, J.C., and Manning, G. (2007) Structural and functional diversity of the microbial kinome. *PLoS Biol.*, **5**, 467–478.

64 Fiuza, M., Canova, M.J., Patin, D., Letek, M., Zanella-Cleon, I., Becchi, M., Mateos, L.M., Mengin-Lecreulx, D., Molle, V., and Gil, J.A. (2008) The MurC ligase essential for peptidoglycan biosynthesis is regulated by the serine/threonine protein kinase PknA in *Corynebacterium glutamicum*. *J. Biol. Chem.*, **283**, 36553–36563.

65 Dasgupta, A., Datta, P., Kundu, M., and Basu, J. (2006) The serine/threonine kinase PknB of *Mycobacterium tuberculosis* phosphorylates PBPA, a penicillin-binding protein required for cell division. *Microbiology*, **152**, 493–504.

66 Grundner, C., Gay, L.M., and Alber, T. (2005) *Mycobacterium tuberculosis* serine/threonine kinases PknB, PknD, PknE, and PknF phosphorylate multiple FHA domains. *Protein Sci.*, **14**, 1918–1921.

67 Zheng, X., Papavinasasundaram, K.G., and Av-Gay, Y. (2007) Novel substrates of *Mycobacterium tuberculosis* PknH Ser/Thr kinase. *Biochem. Biophys. Res. Commun.*, **355**, 162–168.

68 Parikh, A., Verma, S.K., Khan, S., Prakash, B., and Nandicoori, V.K. (2009) PknB-mediated phosphorylation of a novel substrate, N-acetylglucosamine-1-phosphate uridyltransferase, modulates its acetyltransferase activity. *J. Mol. Biol.*, **386**, 451–464.

69 Kang, C.M., Abbott, D.W., Park, S.T., Dascher, C.C., Cantley, L.C., and Husson, R.N. (2005) The *Mycobacterium tuberculosis* serine/threonine kinases PknA and PknB: substrate identification and regulation of cell shape. *Genes Dev.*, **19**, 1692–1704.

70 Molle, V., Soulat, D., Jault, J.M., Grangeasse, C., Cozzone, A.J., and Prost, J.F. (2004) Two FHA domains on an ABC transporter, Rv1747, mediate its phosphorylation by PknF, a Ser/Thr protein kinase from *Mycobacterium tuberculosis*. *FEMS Microbiol. Lett.*, **234**, 215–223.

71 Thakur, M. and Chakraborti, P.K. (2006) GTPase activity of mycobacterial FtsZ is impaired due to its transphosphorylation by the eukaryotic-type Ser/Thr kinase PknA. *J. Biol. Chem.*, **281**, 40107–40113.

72 Thakur, M. and Chakraborti, P.K. (2008) Ability of PknA, a mycobacterial eukaryotic-type serine/threonine kinase, to transphosphorylate MurD, a ligase involved in the process of peptidoglycan biosynthesis. *Biochem J.*, **415**, 27–33.

73 Perez, J., Garcia, R., Bach, H., de Waard, J.H., Jacobs, W.R. Jr., Av-Gay, Y., Bubis, J., and Takiff, H.E. (2006) *Mycobacterium tuberculosis* transporter MmpL7 is a potential substrate for kinase PknD. *Biochem. Biophys. Res. Commun.*, **348**, 6–12.

25
Regulatory Proteolysis and Signal Transduction in Bacteria
Kürşad Turgay

25.1
Introduction

Protein degradation plays an important part in protein homeostasis and in protein quality control, ensuring the proper function of proteins in their cellular environment. Protein degradation is mediated by dedicated molecular machines, which are ATP-dependent proteases consisting of ring-forming hexameric Hsp100/Clp proteins of the AAA+ (*A*TPases *a*ssociated with diverse cellular *a*ctivities) family, associated with barrel-forming compartmentalized peptidases. The AAA+ proteins of the protease can recognize, unfold, and translocate target proteins into the proteolytic chamber of the interacting peptidase structure. An example of such proteases in eukaryotes is the proteasome. In prokaryotic cells, homologous and analogous functioning proteases consisting of AAA+ proteins like ClpA, ClpX, or ClpC interacting with ClpP fulfill this role [1–3].

In addition, the same proteolytic machines are used for regulatory purposes, such as regulated proteolysis of key transcription factors to control, for example, cell cycle, developmental, or adaptation processes [4–6].

In eukaryotic cells the recognition of regulatory and general substrates for proteolysis by the proteasome is quite well understood. An intricate system of ubiquitin-activating enzymes (E1), ubiquitin transferases (E2), and ubiquitin ligases (E3) was identified and characterized, which can specifically recognize and mark proteins by multiubiquitination for subsequent degradation by the proteasome (Figure 25.1) [3, 6]. In contrast, substrate recognition by the analogous prokaryotic Hsp100/Clp proteins is only partially and less well understood, although some intrinsic and specific recognition sequences and/or adaptor proteins have been implicated in substrate recognition and targeting by the different AAA+ proteins of these protease complexes.

Different aspects of the function, mechanism, and biological role of the Hsp100/Clp, FtsH, and Lon proteins have been described in recent reviews [2, 4, 5, 7–11].

In this chapter, I give a short overview of the proteins important for regulatory and general proteolysis, their mechanism, and the substrate recognition of these complex

---

Bacterial Signaling. Edited by Reinhard Krämer and Kirsten Jung
Copyright © 2010 WILEY-VCH Verlag GmbH & Co. KGaA, Weinheim
ISBN: 978-3-527-32365-4

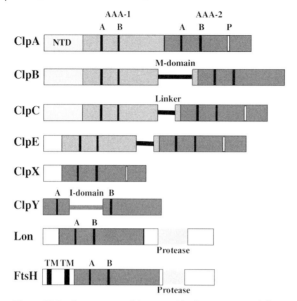

Figure 25.1 Comparison of the Hsp100/Clp proteins and the AAA+ proteases Lon and FtsH. "A" and "B" mark the position of the Walker A and B sites. "P" marks the position of the conserved loop necessary for interaction with ClpP.

molecular machines. To conclude, a more detailed description and comparison of two relatively well-characterized bacterial regulatory systems, namely competence development and heat shock adaptation in *Bacillus subtilis*, where proteolysis plays an important role in signal transduction in the cytosol, is discussed.

The transmembrane signal transduction system of regulated intramembrane proteolysis (RIP), which is also very important in bacterial signaling processes, such as the periplasmic stress response in *Escherichia coli* and sporulation in *B. subtilis*, is described in Chapter 13.

25.2
Hsp100/Clp and other AAA+ Protease Systems in Bacteria

Hsp100/Clp proteins are ATPases and members of the AAA+ superfamily of proteins that form oligomeric ring structures [12, 13]. Proteins of this superfamily are found in all kingdoms life. They participate in various, sometimes very different cellular functions, which all involve moving or changing the conformation of molecular complexes. A core domain, homologous to the P-loop ATPase domain, is the defining element of this superfamily. It consists of two subdomains – a larger ATP-binding domain with a RecA-like α/β-fold and a smaller C-terminal mostly α-helical sensor domain [14, 15]. Conformational changes upon ATP binding and hydrolysis in the oligomeric ring are believed to be the driving force for the AAA+ and the Hsp100/Clp proteins [2, 15–19].

As already mentioned, most members of the Hsp100/Clp family associate with an oligomeric peptidase to form an ATP-dependent protease. For some Hsp100/Clp proteins, like ClpA and ClpX, it was shown that they act as chaperones by promoting the recognition, ATP-dependent unfolding [20], and translocation of substrate proteins into the catalytically active sites of the associated ClpP peptidase ring [2, 21, 22]. These and other Hsp100/Clp proteins also have chaperone activities independent of their role in the protease complex [20, 23–26].

ClpB, another member of the Hsp100/Clp family, does not interact with a peptidase, but, in cooperation with the Hsp70 chaperone system (DnaK, DnaJ, and GrpE in bacteria), has the capacity to facilitate the efficient disaggregation and refolding of a large variety of protein aggregates *in vitro* and *in vivo* [27–30]. It has been demonstrated that the disaggregation ability of ClpB is intimately connected to the ability of ClpB to translocate protein, as previously observed for the other Hsp100/Clp proteins [31], but also needs its accessory middle domain for the disaggregation activity [32]. A variant of Hsp100/Clp proteins, ClpV, was recently identified as part of a new bacterial protein secretion system (type VI) [33, 34].

The AAA+ core domain of the Hsp100/Clp proteins can contain at different positions extensions like a C-terminal recognition loop necessary for the interaction with ClpP (P in Figure 25.1) [35, 36] or accessory domains, like the linker or middle domain, the different N-terminal domains (NTD), or the I domain found in ClpYQ (HslUV) proteins of *E. coli* and *B. subtilis* (depicted in Figure 25.1) [37–41].

Lon is an AAA+ protein with a NTD and a covalently linked C-terminal serine protease domain. Lon monomers also form hexameric barrel-like structures, which recognize substrate proteins, unfold them, and target them for degradation by the covalently linked protease domain. Lon from *E. coli* is the major protease for misfolded and unfolded damaged proteins [42], but has also a number of regulatory functions, such as in the SOS response [11]. *B. subtilis* encodes two Lon homologs, LonA and LonB, which have been inferred, for example, in the regulation of sigma factor activity [11, 43–46].

FtsH is a membrane-localized AAA+ protein with an additional metalloprotease domain. It is a general protease for membrane proteins with additional regulatory functions [10]. In *E. coli*, FtsH is responsible for the quality control-related protein degradation of membrane proteins, but is also involved, for example, in the degradation of the heat shock sigma factor σ^{32} in the heat shock regulation system [47]. *B. subtilis* cells with a deletion in *ftsH* exhibited a pleiotropic phenotype including filamentous growth, hypersensitivity towards heat and salt stress, and a failure to sporulate [48, 49]. A small protein, SpoVM, which modulates the activity of FtsH, has been identified in *B. subtilis* [50]. Finally, FtsH exerts an indirect influence on the gene expression of a membrane stress regulon controlled by σ^W [51].

25.3
Substrate Recognition and Adaptor Proteins

25.3.1
Substrate Recognition

In order to unfold and translocate substrate proteins through the pore formed by one or two AAA+ domains, the hexameric Hsp100/Clp ATPases have to recognize and specifically bind to their respective substrate protein [2, 52].

A number of different possibilities for substrate recognition by the Hsp100/Clp proteins have been identified. A direct recognition by (i) adaptor proteins which themselves interact with accessory domains of the AAA+ core domain, like the NTD or I domain [7, 53], (ii) the core AAA+ domain itself (e.g., by a stretch of amino acids located close to the pore of the first and even second AAA+ domain [52, 54]), or (iii) the accessory NTD, I, linker, or middle domain [2, 53, 55, 56].

The direct recognition by determinants on the AAA+ core domain could also reflect the necessary substrate interaction for basic translocation mechanism of these molecular machines [52, 54, 57]. Some adaptor proteins like SspB enhance the recognition of substrate by the AAA+ core domain by concurrent binding to the NTD of the AAA+ domain and the substrate.

On the substrate side there are some recognition sequences, such as the SsrA-tag or the N-end-rule sequences, whose presence targets the modified protein for degradation by Hsp100/Clp proteases; however, other peptide sequences can also be recognized. For some ClpXP substrates it has been demonstrated that their recognition sequences were only exposed after a primary cleavage event [2, 58–61]. Specific sequences rich in aromatic acids, normally hidden in the hydrophobic core of native proteins, have recently been characterized as targeting sequences for Lon [62]. One or two recognition sequences, which can be located at the termini or inside the protein sequence [63], can be sufficient for substrate recognition. It has been suggested that in addition to a specific recognition sequence, a second unfolded region has to be present for a successful unfolding and translocation of the target protein [64].

25.3.2
Adaptor Proteins

A number of adaptor proteins have been identified not only for bacterial Hsp100/Clp proteins, but also for related eukaryotic AAA+ proteins like p97/NSF [53]. Since these proteins not only target substrate proteins for degradation, but are also often regulated themselves, it was suggested to rename them recognition, targeting, or cofactors [5, 9]; however, it appears that the term is already too established in the field to be changed.

The adaptor proteins identified in *E. coli* include SspB and RssB for ClpX, ClpS for ClpA, and even inorganic polyphosphate for Lon [53, 65]. SspB was identified as a specificity factor enhancing the targeting and degradation of SsrA-tagged proteins and increasing the affinity of ClpXP for this specific substrate [2]. RssB is an unusual response regulator, which, depending on its phosphorylation state, targets the starvation sigma factor σ^S for degradation by ClpXP. σ^S is recognized and bound by the adaptor protein RssB, upon which a conformational change is induced and a second recognition sequence that is directly recognized by ClpXP is exposed [66–69]. It was demonstrated that the phosphorylation of RssB is partially controlled by the sensor kinase ArcB, which monitors the cellular energy state [70]. A more recent study identified a small signaling protein IraP that is synthesized upon phosphate starvation and interacts with RssB, which results in the release of σ^S from ClpXP dependent degradation [71]. ClpS interacting with the NTD of ClpA inhibits degradation of SsrA-tagged substrates, and enables ClpAP to recognize and degrade aggregated proteins [72, 73]. Moreover, ClpS specifically recognizes N-end-rule substrate proteins in *E. coli* [58, 74].

In the control of competence development, MecA of *B. subtilis* enables the recognition and targeting of the transcription factor ComK or the signaling peptide ComS to ClpCP and simultaneously induces the ATPase activity of ClpC [26, 75–77]. This proteolytic switch is described in Section 25.4.1 in more detail. YpbH, a paralog of MecA, has comparable, but not identical, activities to MecA. It displays chaperone activity similar to MecA [26], but is not involved in ComK degradation [78].

The mechanism of ClpC activation and substrate targeting by MecA was elucidated in more detail. It could be demonstrated that the presence of an adaptor protein, such as MecA, is essential for all the activities of ClpC because it facilitates ClpC oligomerization. In addition, it could be observed that ClpP of *B. subtilis* oligomerizes in its active $ClpP_{14}$ form only in the presence of the active ClpC hexamer. Both the NTD and the linker domain of ClpC were important determinants for the interaction with MecA [79]. This is a very effective mechanism to control AAA+ protein activity. The pre-existing ClpC will be rendered inactive, unless activated by an adaptor protein. Thereby, ClpCP is immediately available on demand and the adaptor, when not delivering substrate protein, becomes itself a target for degradation by ClpCP [26, 75] (Figure 25.2). *In vivo*, different adaptor proteins with different substrate affinities, targeting abilities, or localization could activate ClpC, thereby providing oligomeric ClpC complexes with distinct substrate specificities for different physiological or regulatory purposes at various times, conditions, and cellular locations [79]. This also suggests that more adaptor proteins that can compete for ClpC should exist. The identification and characterization of a new adaptor protein McsB and the implications for adaptor-mediated substrate recognition will be described in Section 25.4.2 on the regulation of heat shock adaptation.

It is interesting to note that two substrates of ClpCP, MurAA [80] and SpoIIAB [81, 82], were identified *in vivo*, but a direct *in vitro* degradation of these substrates by ClpCP could not be demonstrated in the presence or absence of any of the known adaptor proteins MecA, YpbH, or McsB (J. Kirstein and K. Turgay, unpublished results). This suggests the existence of more unknown adaptor proteins and/or targeting mechanisms for the ClpCP system.

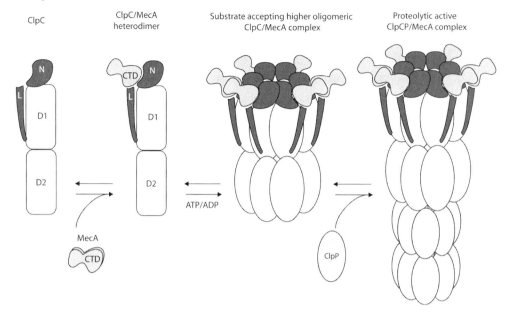

Figure 25.2 Activation of ClpCP by adaptor-mediated oligomerization of ClpC.

25.4
Examples of Regulatory Proteolysis in *B. subtilis*

ClpC of the Gram-positive bacterium *B. subtilis* is an Hsp100/Clp protein with unique properties. Through regulated proteolysis, ClpC controls not only key steps of developmental processes, like competence [75], sporulation [81], and regulation of stress response [83–85], but also participates in protein quality control [86, 87].

25.4.1
Competence Development and the Proteolytic Switch

Natural competence is a distinct physiological state of a subpopulation of stationary-phase *B. subtilis* cells [88] that are able to actively import DNA [89]. The development of competence is a highly regulated process controlled by the transcription factor ComK, which positively controls its own expression [90]. The activity and the expression of this transcription factor were shown to be regulated at both transcriptional and post-translational levels, thereby integrating signals from different signal transduction systems [91, 92]. ClpC was initially discovered together with the adaptor protein MecA in a genetic screen for repressors of competence development [93]. Subsequently, it could be demonstrated that the important post-translational control of competence development is mediated by a proteolytic switch [75, 76, 94]. In noncompetent cells the adaptor protein MecA inhibits ComK activity and targets ComK for degradation by ClpCP. Competence development is initiated by a pheromone-mediated quorum sensing (QS) mechanism [95–97], resulting in the expression of a signaling peptide ComS [98, 99]. After its synthesis ComS interacts with MecA, causing the release of ComK [77, 94] and a halt of proteolysis [75]. The release from the proteolytic complex as well as the positive autoregulation of ComK leads to a fast increase of the concentration of this transcription factor, resulting in the subsequent expression of the proteins necessary for competence, including the proteins forming the DNA receptor and translocation machinery encoded by the late competence genes [75, 89]. Cells can escape competence within 1–2 h after dilution in fresh medium. During this time lag, the very abundant ComK can be retargeted by MecA for degradation by ClpCP [75].

The adaptor protein MecA, due to its ability to interact and inhibit ComK and to interact with ComS, which results in the release and reactivation of ComK, integrates both signal reception and signal transduction of this developmental switch [75, 77, 94]. Another important aspect of this switch is the bistable nature of the positive autoregulatory activity of ComK, explaining the heterogeneous population structure of cells grown into competence [100–104].

25.4.2
Heat Shock Adaptation

The heat shock response in *B. subtilis* is controlled by at least five different mechanisms. The major regulatory proteins of four of these classes were identified and characterized. Class I genes are regulated primarily by the repressor HrcA, the alternative sigma factor σ^B controls the class II genes, the repressor CtsR class III genes, and the two component system CssRS controls the class V genes necessary for secretion stress response. Class IV genes include stress-induced genes like *clpX* and *htpG*, controlled by an unknown mechanism [105, 106]. The major regulator CtsR is a repressor that binds upstream of *clpP*, *clpE*, and the *clpC* operon [107, 108]. Several lines of evidence, including pulse–chase experiments, demonstrated that this repressor is stable at 37 °C, but rapidly degraded at elevated temperatures by ClpCP [84, 85]. All proteins encoded by the *clpC* operon, which consists of *ctsR*, *mcsA*, *mcsB* and *clpC*, are involved in the regulation of CtsR activity. McsB is a kinase whose activity is induced by McsA, repressed by ClpC, and antagonized by the phosphatase YwlE. McsB interacts with and phosphorylates CtsR, thereby inhibiting its DNA-binding activity. It could be demonstrated that both *mcsB* and *mcsA* were necessary for the additional heat-induced degradation of CtsR *in vivo* [84]. The mechanism by which the repressor of heat shock regulation CtsR is removed by regulatory proteolysis from the stressed cells includes the same proteins. Again, this is a mechanism that includes a regulated adaptor protein. In this case, it could be demonstrated that McsB in its phosphorylated form acts as an adaptor for ClpC, targeting CtsR for degradation by ClpCP. [83]. Like MecA, McsB is a regulated adaptor and part of a sensory switch. McsB, in its nonphosphorylated state, interacts with ClpC, which inhibits the kinase activity of McsB. This interaction may already occur when McsB is synthesized cotranslationally with McsA and ClpC. The release of unphosphorylated McsB from ClpC can be triggered by competition with other adaptor proteins, like MecA and YpbH, interacting with unfolded proteins and thereby sensing protein conformational stress. McsB, which is thereby released from the inhibition of its kinase activity, can be activated by McsA, allowing autophosphorylation to occur. The phosphorylated form of McsB inhibits the repressor activity of CtsR and acts as an adaptor protein targeting CtsR for degradation by ClpCP. The repressor CtsR is inhibited and constantly removed from the cell, thereby the heat shock response is turned on. When the stress situation ceases to continue, newly synthesized McsB is kept inactive by ClpC, which is not redirected to its protein quality control assignment anymore. The activity of the phosphatase YwlE, which dephosphorylates McsB, McsA, and CtsR, will also support this return to the repression of the class III heat shock genes in *B. subtilis* [83, 84].

25.5
Conclusions

Proteolysis can be an important part in intracellular signal transduction processes. Proteases, which are also involved in general tasks, such as protein quality control, are used to degrade specific regulatory proteins.

In the first example, the activator ComK is constantly removed from the cell to a level beyond the detection limit. Upon a QS signal, the degradation of ComK is halted and as a consequence the cellular level of ComK is raised significantly and competence development ensues. In the second, example CtsR a repressor, which inhibits its own synthesis, is present in the cell until a signal mediated by heat shock results in the constant removal of this repressor from the cell, resulting in the expression of the *clpC* operon, *clpP*, and *clpE*.

Somehow, opposite cellular strategies – the stabilization or the destabilization of a regulatory protein – are mediated by the same ClpCP protease, but by different regulated adaptor proteins. Both adaptor proteins can activate ClpC in a very similar manner and are themselves subject to degradation when active as an adaptor, but not presenting substrate to ClpCP. It is also interesting to note that both adaptors are inhibitors of the DNA-binding activity of their respective substrates as well.

Many different mechanisms of proteolysis acting in various regulatory circuits with different logic in their structure have been discovered and described [4, 5]. Even a relatively simple mechanism of constitutive degradation of a transcription factor in the context of a very specific response system, such as the *E. coli* acid stress response, can have an important influence. In this case, it was demonstrated that constitutive proteolysis is necessary for a fast response, timing, and fine-tuning of this specific regulatory circuit embedded in the σ^S regulon, which itself is also regulated by proteolysis [109, 110].

It was also demonstrated that in order to adjust the cellular response to DNA damage in *E. coli*, regulatory proteolysis results in the reshaping of a significant part of the proteome, in addition to the transcriptional response, to ensure a fast cellular adaptation on all levels [111].

It should also be noted that ClpP of Gram-negative and -positive bacteria was recently discovered as a target for antibiotics. These compounds seem to activate

ClpP to perform unrestricted intracellular proteolysis, resulting in inhibition of cell division and, finally, cell death [112]. This finding additionally demonstrates the importance of regulated and controlled proteolysis for the bacterial physiology.

Acknowledgments

I thank Janine Kirstein, Noel Moliere, Leendert Hamoen, Axel Mogk for discussions, Noel Moliere for critical reading of the manuscript, and Regine Hengge for discussion and support. The work in my laboratory is supported by the Deutsche Forschungsgemeinschaft.

References

1 Wickner, S., Maurizi, M.R., and Gottesman, S. (1999) Posttranslational quality control: folding, refolding, and degrading proteins. *Science*, **286**, 1888–1893.

2 Sauer, R.T., Bolon, D.N., Burton, B.M., Burton, R.E., Flynn, J.M., Grant, R.A., Hersch, G.L., Joshi, S.A., Kenniston, J.A., Levchenko, I. *et al.* (2004) Sculpting the proteome with AAA+ proteases and disassembly machines. *Cell*, **119**, 9–18.

3 Pickart, C.M. and Cohen, R.E. (2004) Proteasomes and their kin: proteases in the machine age. *Nat. Rev. Mol. Cell Biol.*, **5**, 177–187.

4 Gottesman, S. (2003) Proteolysis in bacterial regulatory circuits. *Annu. Rev. Cell Dev. Biol.*, **19**, 565–587.

5 Jenal, U. and Hengge-Aronis, R. (2003) Regulation by proteolysis in bacterial cells. *Curr. Opin. Microbiol.*, **6**, 163–172.

6 Ciechanover, A. (1998) The ubiquitin–proteasome pathway: on protein death and cell life. *EMBO J.*, **17**, 7151–7160.

7 Baker, T.A. and Sauer, R.T. (2006) ATP-dependent proteases of bacteria: recognition logic and operating principles. *Trends Biochem. Sci.*, **31**, 647–653.

8 Butler, S.M., Festa, R.A., Pearce, M.J., and Darwin, K.H. (2006) Self-compartmentalized bacterial proteases and pathogenesis. *Mol. Microbiol.*, **60**, 553–562.

9 Frees, D., Savijoki, K., Varmanen, P., and Ingmer, H. (2007) Clp ATPases and ClpP proteolytic complexes regulate vital biological processes in low GC, Gram-positive bacteria. *Mol. Microbiol.*, **63**, 1285–1295.

10 Ito, K. and Akiyama, Y. (2005) Cellular functions, mechanism of action, and regulation of FtsH protease. *Annu. Rev. Microbiol.*, **59**, 211–231.

11 Tsilibaris, V., Maenhaut-Michel, G., and Van Melderen, L. (2006) Biological roles of the Lon ATP-dependent protease. *Res. Microbiol.*, **157**, 701–713.

12 Lupas, A., Flanagan, J.M., Tamura, T., and Baumeister, W. (1997) Self-compartmentalizing proteases. *Trends Biochem. Sci.*, **22**, 399–404.

13 Neuwald, A.F., Aravind, L., Spouge, J.L., and Koonin, E.V. (1999) AAA+: a class of chaperone-like ATPases associated with the assembly, operation, and disassembly of protein complexes. *Genome Res.*, **9**, 27–43.

14 Erzberger, J.P. and Berger, J.M. (2006) Evolutionary relationships and structural mechanisms of AAA+ proteins. *Annu. Rev. Biophys. Biomol. Struct.*, **35**, 93–114.

15 Ogura, T. and Wilkinson, A.J. (2001) AAA+ superfamily ATPases: common structure–diverse function. *Genes Cells*, **6**, 575–597.

16 Martin, A., Baker, T.A., and Sauer, R.T. (2005) Rebuilt AAA+ motors reveal

operating principles for ATP-fuelled machines. *Nature*, **437**, 1115–1120.

17 Hersch, G.L., Burton, R.E., Bolon, D.N., Baker, T.A., and Sauer, R.T. (2005) Asymmetric interactions of ATP with the AAA+ ClpX$_6$ unfoldase: allosteric control of a protein machine. *Cell*, **121**, 1017–1027.

18 Horwich, A.L., Weber-Ban, E.U., and Finley, D. (1999) Chaperone rings in protein folding and degradation. *Proc. Natl. Acad. Sci. USA*, **96**, 11033–11040.

19 Vale, R.D. (2000) AAA proteins. Lords of the ring. *J. Cell Biol.*, **150**, 13–19.

20 Weber-Ban, E.U., Reid, B.G., Miranker, A.D., and Horwich, A.L. (1999) Global unfolding of a substrate protein by the Hsp100 chaperone ClpA. *Nature*, **401**, 90–93.

21 Kim, Y.I., Burton, R.E., Burton, B.M., Sauer, R.T., and Baker, T.A. (2000) Dynamics of substrate denaturation and translocation by the ClpXP degradation machine. *Mol. Cell*, **5**, 639–648.

22 Reid, B.G., Fenton, W.A., Horwich, A.L., and Weber-Ban, E.U. (2001) ClpA mediates directional translocation of substrate proteins into the ClpP protease. *Proc. Natl. Acad. Sci. USA*, **98**, 3768–3772.

23 Levchenko, I., Luo, L., and Baker, T.A. (1995) Disassembly of the Mu transposase tetramer by the ClpX chaperone. *Genes Dev.*, **9**, 2399–2408.

24 Burton, B.M., Williams, T.L., and Baker, T.A. (2001) ClpX-mediated remodeling of mu transpososomes: selective unfolding of subunits destabilizes the entire complex. *Mol Cell*, **8**, 449–454.

25 Wickner, S., Gottesman, S., Skowyra, D., Hoskins, J., McKenney, K., and Maurizi, M.R. (1994) A molecular chaperone, ClpA, functions like DnaK and DnaJ. *Proc. Natl. Acad. Sci. USA*, **91**, 12218–12222.

26 Schlothauer, T., Mogk, A., Dougan, D.A., Bukau, B., and Turgay, K. (2003) MecA, an adaptor protein necessary for ClpC chaperone activity. *Proc. Natl. Acad. Sci. USA*, **100**, 2306.

27 Mogk, A., Tomoyasu, T., Goloubinoff, P., Rudiger, S., Roder, D., Langen, H., and Bukau, B. (1999) Identification of thermolabile *Escherichia coli* proteins: prevention and reversion of aggregation by DnaK and ClpB. *EMBO J.*, **18**, 6934–6949.

28 Parsell, D.A., Kowal, A.S., Singer, M.A., and Lindquist, S. (1994) Protein disaggregation mediated by heat-shock protein Hsp104. *Nature*, **372**, 475–478.

29 Goloubinoff, P., Mogk, A., Zvi, A.P., Tomoyasu, T., and Bukau, B. (1999) Sequential mechanism of solubilization and refolding of stable protein aggregates by a bichaperone network. *Proc. Natl. Acad. Sci. USA*, **96**, 13732–13737.

30 Glover, J.R. and Lindquist, S. (1998) Hsp104, Hsp70, and Hsp40: a novel chaperone system that rescues previously aggregated proteins. *Cell*, **94**, 73–82.

31 Weibezahn, J., Tessarz, P., Schlieker, C., Zahn, R., Maglica, Z., Lee, S., Zentgraf, H., Weber-Ban, E.U., Dougan, D.A., Tsai, F.T. et al. (2004) Thermotolerance requires refolding of aggregated proteins by substrate translocation through the central pore of ClpB. *Cell*, **119**, 653–665.

32 Haslberger, T., Weibezahn, J., Zahn, R., Lee, S., Tsai, F.T., Bukau, B., and Mogk, A. (2007) M domains couple the ClpB threading motor with the DnaK chaperone activity. *Mol. Cell*, **25**, 247–260.

33 Schlieker, C., Zentgraf, H., Dersch, P., and Mogk, A. (2005) ClpV, a unique Hsp100/Clp member of pathogenic proteobacteria. *Biol. Chem.*, **386**, 1115–1127.

34 Mougous, J.D., Cuff, M.E., Raunser, S., Shen, A., Zhou, M., Gifford, C.A., Goodman, A.L., Joachimiak, G., Ordonez, C.L., Lory, S. et al. (2006) A virulence locus of *Pseudomonas aeruginosa* encodes a protein secretion apparatus. *Science*, **312**, 1526–1530.

35 Singh, S.K., Rozycki, J., Ortega, J., Ishikawa, T., Lo, J., Steven, A.C., and Maurizi, M.R. (2001) Functional domains of the ClpA and ClpX molecular chaperones identified by limited proteolysis and deletion analysis. *J. Biol. Chem.*, **276**, 29420–29429.

36 Kim, Y.I., Levchenko, I., Fraczkowska, K., Woodruff, R.V., Sauer, R.T., and Baker, T.A. (2001) Molecular determinants of complex formation between Clp/Hsp100 ATPases and the ClpP peptidase. *Nat. Struct. Biol.*, **8**, 230–233.

37 Lee, S., Sowa, M.E., Watanabe, Y., Sigler, P.B., Chiu, W., Yoshida, M., and Tsai, F.T.F. (2003) The structure of ClpB: a molecular chaperone that rescues proteins from an aggregated state. *Cell*, **115**, 229–240.

38 Sousa, M.C., Trame, C.B., Tsuruta, H., Wilbanks, S.M., Reddy, V.S., and McKay, D.B. (2000) Crystal and solution structures of an HslUV protease–chaperone complex. *Cell*, **103**, 633–643.

39 Guo, F., Maurizi, M.R., Esser, L., and Xia, D. (2002) Crystal structure of ClpA, an Hsp100 chaperone and regulator of ClpAP protease. *J. Biol. Chem.*, **277**, 46743–46752.

40 Bochtler, M., Hartmann, C., Song, H.K., Bourenkov, G.P., Bartunik, H.D., and Huber, R. (2000) The structures of HsIU and the ATP-dependent protease HsIU–HsIV. *Nature*, **403**, 800–805.

41 Song, H.K., Hartmann, C., Ramachandran, R., Bochtler, M., Behrendt, R., Moroder, L., and Huber, R. (2000) Mutational studies on HslU and its docking mode with HslV. *Proc. Natl. Acad. Sci. USA*, **97**, 14103–14108.

42 Tomoyasu, T., Mogk, A., Langen, H., Goloubinoff, P., and Bukau, B. (2001) Genetic dissection of the roles of chaperones and proteases in protein folding and degradation in the *Escherichia coli* cytosol. *Mol. Microbiol.*, **40**, 397–413.

43 Riethdorf, S., Volker, U., Gerth, U., Winkler, A., Engelmann, S., and Hecker, M. (1994) Cloning, nucleotide sequence, and expression of the *Bacillus subtilis lon* gene. *J. Bacteriol.*, **176**, 6518–6527.

44 Schmidt, R., Decatur, A.L., Rather, P.N., Moran, C.P. Jr., and Losick, R. (1994) *Bacillus subtilis lon* protease prevents inappropriate transcription of genes under the control of the sporulation transcription factor sigma G. *J. Bacteriol.*, **176**, 6528–6537.

45 Liu, J., Cosby, W.M., and Zuber, P. (1999) Role of lon and ClpX in the post-translational regulation of a sigma subunit of RNA polymerase required for cellular differentiation in *Bacillus subtilis*. *Mol. Microbiol.*, **33**, 415–428.

46 Serrano, M., Hovel, S., Moran, C.P. Jr., Henriques, A.O., and Volker, U. (2001) Forespore-specific transcription of the *lonB* gene during sporulation in *Bacillus subtilis*. *J. Bacteriol.*, **183**, 2995–3003.

47 Yura, T., Kanemori, M., and Morita, M.T. (2000) The heat shock response: regulation and function, in *Bacterial Stress Responses* (eds G. Storz and R. Hengge-Aronis), ASM Press, Washington, DC, pp. 3–18.

48 Deuerling, E., Mogk, A., Richter, C., Purucker, M., and Schumann, W. (1997) The *ftsH* gene of *Bacillus subtilis* is involved in major cellular processes such as sporulation, stress adaptation and secretion. *Mol. Microbiol.*, **23**, 921–933.

49 Lysenko, E., Ogura, T., and Cutting, S.M. (1997) Characterization of the *ftsH* gene of *Bacillus subtilis*. *Microbiology*, **143**, 971–978.

50 Cutting, S., Anderson, M., Lysenko, E., Page, A., Tomoyasu, T., Tatematsu, K., Tatsuta, T., Kroos, L., and Ogura, T. (1997) SpoVM, a small protein essential to development in *Bacillus subtilis*, interacts with the ATP-dependent protease FtsH. *J. Bacteriol.*, **179**, 5534–5542.

51 Zellmeier, S., Zuber, U., Schumann, W., and Wiegert, T. (2003) The absence of FtsH metalloprotease activity causes overexpression of the sigmaW-controlled *pbpE* gene, resulting in filamentous growth of *Bacillus subtilis*. *J. Bacteriol.*, **185**, 973–982.

52 Schlieker, C., Weibezahn, J., Patzelt, H., Tessarz, P., Strub, C., Zeth, K., Erbse, A., Schneider-Mergener, J., Chin, J.W., Schultz, P.G. et al. (2004) Substrate recognition by the AAA+ chaperone ClpB. *Nat. Struct. Mol. Biol.*, **11**, 607–615.

53 Dougan, D., Mogk, A., Zeth, K., Turgay, K., and Bukau, B. (2002) AAA+ proteins and substrate recognition, it all depends on their partner in crime. *FEBS Lett.*, **529**, 6.

54 Hinnerwisch, J., Fenton, W.A., Furtak, K.J., Farr, G.W., and Horwich, A.L. (2005) Loops in the central channel of ClpA chaperone mediate protein binding, unfolding, and translocation. *Cell*, **121**, 1029–1041.

55 Mogk, A., Dougan, D., Weibezahn, J., Schlieker, C., Turgay, K., and Bukau, B.

56 Bukau, B., Weissman, J., and Horwich, A.L. (2006) Molecular chaperones and protein quality control. *Cell*, **125**, 443–451.

57 Smith, C.K., Baker, T.A., and Sauer, R.T. (1999) Lon and Clp family proteases and chaperones share homologous substrate-recognition domains. *Proc. Natl. Acad. Sci. USA*, **96**, 6678–6682.

58 Erbse, A., Schmidt, R., Bornemann, T., Schneider-Mergener, J., Mogk, A., Zahn, R., Dougan, D.A., and Bukau, B. (2006) ClpS is an essential component of the N-end rule pathway in *Escherichia coli*. *Nature*, **439**, 753–756.

59 Neher, S.B., Sauer, R.T., and Baker, T.A. (2003) Distinct peptide signals in the UmuD and UmuD′ subunits of UmuD/D′ mediate tethering and substrate processing by the ClpXP protease. *Proc. Natl. Acad. Sci. USA*, **100**, 13219–13224.

60 Neher, S.B., Flynn, J.M., Sauer, R.T., and Baker, T.A. (2003) Latent ClpX-recognition signals ensure LexA destruction after DNA damage. *Genes Dev.*, **17**, 1084–1089.

61 Flynn, J.M., Neher, S.B., Kim, Y.I., Sauer, R.T., and Baker, T.A. (2003) Proteomic discovery of cellular substrates of the ClpXP protease reveals five classes of ClpX-recognition signals. *Mol. Cell*, **11**, 671–683.

62 Gur, E. and Sauer, R.T. (2008) Recognition of misfolded proteins by Lon, a AAA+ protease. *Genes Dev.*, **22**, 2267–2277.

63 Hoskins, J.R. and Wickner, S. (2006) Two peptide sequences can function cooperatively to facilitate binding and unfolding by ClpA and degradation by ClpAP. *Proc. Natl. Acad. Sci. USA*, **103**, 909–914.

64 Prakash, S., Tian, L., Ratliff, K.S., Lehotzky, R.E., and Matouschek, A. (2004) An unstructured initiation site is required for efficient proteasome-mediated degradation. *Nat. Struct. Mol. Biol.*, **11**, 830–837.

65 Kuroda, A., Nomura, K., Ohtomo, R., Kato, J., Ikeda, T., Takiguchi, N., Ohtake, H., and Kornberg, A. (2001) Role of inorganic polyphosphate in promoting ribosomal protein degradation by the Lon protease in *E. coli*. *Science*, **293**, 705–708.

66 Becker, G., Klauck, E., and Hengge-Aronis, R. (1999) Regulation of RpoS proteolysis in *Escherichia coli*: the response regulator RssB is a recognition factor that interacts with the turnover element in RpoS. *Proc. Natl. Acad. Sci. USA*, **96**, 6439–6444.

67 Zhou, Y., Gottesman, S., Hoskins, J.R., Maurizi, M.R., and Wickner, S. (2001) The RssB response regulator directly targets sigma^S for degradation by ClpXP. *Genes Dev.*, **15**, 627–637.

68 Becker, G., Klauck, E., and Hengge-Aronis, R. (2000) The response regulator RssB, a recognition factor for sigmaS proteolysis in *Escherichia coli*, can act like an anti-sigma^S factor. *Mol. Microbiol.*, **35**, 657–666.

69 Stüdemann, A., Noirclerc-Savoye, M., Klauck, E., Becker, G., Schneider, D., and Hengge, R. (2003) Sequential recognition of two distinct sites in sigma^S by the proteolytic targeting factor RssB and ClpX. *EMBO J.*, **22**, 4111–4120.

70 Mika, F. and Hengge, R. (2005) A two-component phosphotransfer network involving ArcB, ArcA, and RssB coordinates synthesis and proteolysis of sigma^S (RpoS) in *E. coli*. *Genes Dev.*, **19**, 2770–2781.

71 Bougdour, A., Wickner, S., and Gottesman, S. (2006) Modulating RssB activity: IraP, a novel regulator of sigma^S stability in *Escherichia coli*. *Genes Dev.*, **20**, 884–897.

72 Zeth, K., Ravelli, R.B., Paal, K., Cusack, S., Bukau, B., and Dougan, D.A. (2002) Structural analysis of the adaptor protein ClpS in complex with the N-terminal domain of ClpA. *Nat. Struct. Biol.*, **9**, 906–911.

73 Dougan, D.A., Reid, B.G., Horwich, A.L., and Bukau, B. (2002) ClpS, a substrate modulator of the ClpAP machine. *Mol. Cell*, **9**, 673–683.

74 Mogk, A., Schmidt, R., and Bukau, B. (2007) The N-end rule pathway for regulated proteolysis: prokaryotic and eukaryotic strategies. *Trends Cell Biol.*, **17**, 165–172.

75. Turgay, K., Hahn, J., Burghoorn, J., and Dubnau, D. (1998) Competence in *Bacillus subtilis* is controlled by regulated proteolysis of a transcription factor. *EMBO J.*, **17**, 6730–6738.

76. Persuh, M., Turgay, K., Mandic-Mulec, I., and Dubnau, D. (1999) The N- and C-terminal domains of MecA recognize different partners in the competence molecular switch. *Mol. Microbiol.*, **33**, 886–894.

77. Prepiak, P. and Dubnau, D. (2007) A peptide signal for adaptor protein-mediated degradation by the AAA+ protease ClpCP. *Mol. Cell*, **26**, 639–647.

78. Persuh, M., Mandic-Mulec, I., and Dubnau, D. (2002) A MecA paralog, YpbH, binds ClpC, affecting both competence and sporulation. *J. Bacteriol.*, **184**, 2310–2313.

79. Kirstein, J., Schlothauer, T., Dougan, D.A., Lilie, H., Tischendorf, G., Mogk, A., Bukau, B., and Turgay, K. (2006) Adaptor protein controlled oligomerization activates the AAA+ protein ClpC. *EMBO J.*, **25**, 1481–1491.

80. Kock, H., Gerth, U., and Hecker, M. (2004) MurAA, catalysing the first committed step in peptidoglycan biosynthesis, is a target of Clp-dependent proteolysis in *Bacillus subtilis*. *Mol. Microbiol.*, **51**, 1087–1102.

81. Pan, Q., Garsin, D.A., and Losick, R. (2001) Self-reinforcing activation of a cell-specific transcription factor by proteolysis of an anti-sigma factor in *B. subtilis*. *Mol. Cell*, **8**, 873–883.

82. Pan, Q. and Losick, R. (2003) Unique degradation signal for ClpCP in *Bacillus subtilis*. *J. Bacteriol.*, **185**, 5275–5278.

83. Kirstein, J., Dougan, D.A., Gerth, U., Hecker, M., and Turgay, K. (2007) The tyrosine kinase McsB is a regulated adaptor protein for ClpCP. *EMBO J.*, **26**, 2061–2070.

84. Kirstein, J., Zühlke, D., Gerth, U., Turgay, K., and Hecker, M. (2005) A tyrosine kinase and its activator control the activity of the CtsR heat shock repressor in *B. subtilis*. *EMBO J.*, **24**, 3435–3445.

85. Krüger, E., Zühlke, D., Witt, E., Ludwig, H., and Hecker, M. (2001) Clp-mediated proteolysis in Gram-positive bacteria is autoregulated by the stability of a repressor. *EMBO J.*, **20**, 852–863.

86. Kramer, G., Patzelt, H., Rauch, T., Kurz, T.A., Vorderwülbecke, S., Bukau, B., and Deuerling, E. (2004) Trigger factor peptidyl-prolyl *cis/trans* isomerase activity is not essential for the folding of cytosolic proteins in *Escherichia coli*. *J. Biol. Chem.*, **279**, 14165–14170.

87. Krüger, E., Witt, E., Ohlmeier, S., Hanschke, R., and Hecker, M. (2000) The Clp proteases of *Bacillus subtilis* are directly involved in degradation of misfolded proteins. *J. Bacteriol.*, **182**, 3259–3265.

88. Berka, R.M., Hahn, J., Albano, M., Draskovic, I., Persuh, M., Cui, X., Sloma, A., Widner, W., and Dubnau, D. (2002) Microarray analysis of the *Bacillus subtilis* K-state: genome-wide expression changes dependent on ComK. *Mol. Microbiol.*, **43**, 1331–1345.

89. Chen, I., Christie, P.J., and Dubnau, D. (2005) The ins and outs of DNA transfer in bacteria. *Science*, **310**, 1456–1460.

90. van Sinderen, D., Luttinger, A., Kong, L., Dubnau, D., Venema, G., and Hamoen, L. (1995) *comK* encodes the competence transcription factor, the key regulatory protein for competence development in *Bacillus subtilis*. *Mol. Microbiol.*, **15**, 455–462.

91. Dubnau, D. and Turgay, K. (2000) The regulation of competence in *Bacillus subtilis* and its relation to stress response, in *Bacterial Stress Responses* (eds G. Storz and R. Hengge-Aronis), ASM Press, Washington, DC, pp. 249–260.

92. Hamoen, L.W., Venema, G., and Kuipers, O.P. (2003) Controlling competence in *Bacillus subtilis*: shared use of regulators. *Microbiology*, **149**, 9–17.

93. Dubnau, D. and Roggiani, M. (1990) Growth medium-independent genetic competence mutants of *Bacillus subtilis*. *J. Bacteriol*, **172**, 4048–4055.

94. Turgay, K., Hamoen, L.W., Venema, G., and Dubnau, D. (1997) Biochemical characterization of a molecular switch involving the heat shock protein ClpC, which controls the activity of ComK, the competence transcription factor of *Bacillus subtilis*. *Genes Dev.*, **11**, 119–128.

95 Magnuson, R., Solomon, J., and Grossman, A.D. (1994) Biochemical and genetic characterization of a competence pheromone from *B. subtilis*. *Cell*, **77**, 207–216.

96 Okada, M., Sato, I., Cho, S.J., Iwata, H., Nishio, T., Dubnau, D., and Sakagami, Y. (2005) Structure of the *Bacillus subtilis* quorum-sensing peptide pheromone ComX. *Nat. Chem. Biol.*, **1**, 23–24.

97 Tortosa, P. and Dubnau, D. (1999) Competence for transformation: a matter of taste. *Curr. Opin. Microbiol.*, **2**, 588–592.

98 D'Souza, C., Nakano, M.M., and Zuber, P. (1994) Identification of *comS*, a gene of the *srfA* operon that regulates the establishment of genetic competence in *Bacillus subtilis*. *Proc. Natl. Acad. Sci. USA*, **91**, 9397–9401.

99 Hamoen, L.W., Eshuis, H., Jongbloed, J., Venema, G., and van Sinderen, D. (1995) A small gene, designated *comS*, located within the coding region of the fourth amino acid-activation domain of *srfA*, is required for competence development in *Bacillus subtilis*. *Mol. Microbiol.*, **15**, 55–63.

100 Dubnau, D. and Losick, R. (2006) Bistability in bacteria. *Mol. Microbiol.*, **61**, 564–572.

101 Smits, W.K., Kuipers, O.P., and Veening, J.W. (2006) Phenotypic variation in bacteria: the role of feedback regulation. *Nat. Rev. Microbiol.*, **4**, 259–271.

102 Maamar, H. and Dubnau, D. (2005) Bistability in the *Bacillus subtilis* K-state (competence) system requires a positive feedback loop. *Mol. Microbiol.*, **56**, 615–624.

103 Maamar, H., Raj, A., and Dubnau, D. (2007) Noise in gene expression determines cell fate in *Bacillus subtilis*. *Science*, **317**, 526–529.

104 Smits, W.K., Eschevins, C.C., Susanna, K.A., Bron, S., Kuipers, O.P., and Hamoen, L.W. (2005) Stripping *Bacillus*: ComK auto-stimulation is responsible for the bistable response in competence development. *Mol. Microbiol.*, **56**, 604–614.

105 Darmon, E., Noone, D., Masson, A., Bron, S., Kuipers, O.P., Devine, K.M., and van Dijl, J.M. (2002) A novel class of heat and secretion stress-responsive genes is controlled by the autoregulated CssRS two-component system of *Bacillus subtilis*. *J. Bacteriol.*, **184**, 5661–5671.

106 Schumann, W., Hecker, M., and Msadek, T. (2002) Regulation and function of heat-inducible genes in *Bacillus subtilis*, in *Bacillus subtilis and its Closest Relatives: From Genes to Cells* (eds A.L. Sonenshein, J.A. Hoch and R. Losick), ASM Press, Washington, DC.

107 Derre, I., Rapoport, G., and Msadek, T. (1999) CtsR, a novel regulator of stress and heat shock response, controls Clp and molecular chaperone gene expression in Gram-positive bacteria. *Mol. Microbiol.*, **31**, 117–131.

108 Krüger, E. and Hecker, M. (1998) The first gene of the *Bacillus subtilis clpC* operon, *ctsR*, encodes a negative regulator of its own operon and other class III heat shock genes. *J. Bacteriol.*, **180**, 6681–6688.

109 Heuveling, J., Possling, A., and Hengge, R. (2008) A role for Lon protease in the control of the acid resistance genes of *Escherichia coli*. *Mol. Microbiol.*, **69**, 534–547.

110 Weber, H., Polen, T., Heuveling, J., Wendisch, V.F., and Hengge, R. (2005) Genome-wide analysis of the general stress response network in *Escherichia coli*: sigma^S-dependent genes, promoters, and sigma factor selectivity. *J. Bacteriol.*, **187**, 1591–1603.

111 Neher, S.B., Villen, J., Oakes, E.C., Bakalarski, C.E., Sauer, R.T., Gygi, S.P., and Baker, T.A. (2006) Proteomic profiling of ClpXP substrates after DNA damage reveals extensive instability within SOS regulon. *Mol. Cell*, **22**, 193–204.

112 Brötz-Oesterhelt, H., Beyer, D., Kroll, H.P., Endermann, R., Ladel, C., Schroeder, W., Hinzen, B., Raddatz, S., Paulsen, H., Henninger, K. *et al.* Dysregulation of bacterial proteolytic machinery by a new class of antibiotics. *Nat. Med.*, **11**, 1082–1087.

26
Intracellular Signaling and Gene Target Analysis – *Methods Chapter*
Jörn Kalinowski

26.1
Introduction

Deciphering regulatory networks in bacterial cells is an important goal in molecular biology. The identification of regulatory interactions as well as the kinetics between regulatory proteins and their DNA or RNA targets under a variety of conditions is a fundamental step towards this goal. The daunting task of reconstructing a complete transcriptional regulatory network for a bacterial organism is far from complete even in model organisms like *Escherichia coli* or *Bacillus subtilis*. However, the recent developments in high-throughput experimental techniques such as DNA microarrays have gathered comprehensive data sets for a number of these organisms and, in conjunction with other experimental approaches and bioinformatics methods, this enables us to build up regulatory networks on a genome-wide scale [7].

The regulatory interactions in such a regulatory network can have different mechanisms. Transcription regulator proteins (sometimes simply termed transcription factors) can act as repressors or activators of transcription and a number of them have both functions. Regulator proteins themselves might be negatively or positively influenced by small molecules, mostly metabolites. This chapter does not go into much detail concerning transcriptional regulation, but does describe some of the modern high-throughput and single gene-oriented experimental approaches that have proved useful in describing regulatory interactions and networks in bacteria.

26.2
Genome-Wide Expression Analysis

Genome-wide expression in bacteria and other organisms is mostly determined by the use of DNA microarrays in a process called transcriptional profiling. With this method, the abundance of transcripts derived from any gene of a bacterium

Bacterial Signaling. Edited by Reinhard Krämer and Kirsten Jung
Copyright © 2010 WILEY-VCH Verlag GmbH & Co. KGaA, Weinheim
ISBN: 978-3-527-32365-4

(transcriptome) is determined, either in a relative fashion applying competitive hybridization of a sample and a reference or (semi)quantitatively by applying Affymetrix gene chips. In the following, the focus is on the standard microarray used for transcriptional profiling by competitive hybridization.

The most important prerequisite for microarray analysis is a completely sequenced genome. For transcriptional profiling of all genes in this genome, bacterial RNA is harvested, purified, and transcribed into cDNA by reverse transcriptase. The cDNA is then labeled with fluorescent dyes and hybridized to an array of short double-stranded DNA (e.g., polymerase chain reaction (PCR) products) or single-stranded oligonucleotides immobilized on a glass surface. Such arrays contain probes representing all genes of a bacterium in several replicates. After hybridization, fluorescence is read from the arrays with the help of laser excitation. In case of competitive hybridization, the sample and the reference cDNAs are labeled with different fluorophores and mixed. After laser scanning of these arrays, signals from both fluorophores are recorded and the data compared.

Microarray experiments deliver a huge amount of data that cannot be handled without computational automation. The data must be filtered for noise and normalized before transcript ratios can be determined. Although microarrays deliver a comprehensive view on all known genes of a given organism, there are some limitations.

A significant limitation of standard microarrays is their dynamic range of transcript abundance, and very high and very low amounts of transcripts might cause problems. Second, the standard microarray addresses each known gene of an organism by only one probe on the array. These probes have differences in composition leading to differing hybridization kinetics. Although bioinformatics methods used for array construction helped to limit this compositional bias, this leads to a situation where some probes address the 5′-end of a gene, some the 3′-end, and others the middle. Since bacterial RNA is highly unstable and easily degraded by endo- or by exonucleolytic activities (see [1] for review) the position of the probe might not be representative for an active transcript.

These limitations can be overcome at least in part by validating candidate genes derived from the microarray experiment by real-time reverse transcriptase PCR. This method also needs cDNA, but in this case a PCR reaction with gene-specific primers is run on the cDNA and the online monitoring of the exponential amplification of the gene-specific transcript allows the determination of the initial amount of gene-specific RNA in the sample. In addition, the method has a very wide dynamic range and delivers highly reproducible data. Furthermore, it can be used for an absolute quantification of a specific RNA if an appropriate calibration is performed.

In conclusion, microarray analysis of the whole bacterial transcriptome under specific conditions or by using regulatory mutants has proven extremely useful for finding candidate genes involved in a certain signal transduction pathway or regulatory network. It can best be used as a screening system and candidate genes have to be confirmed by other complementary methods, some of which are described in the following.

26.3
Finding Unknown Target Genes

If a transcriptional regulator is known and the DNA or RNA targets within a genome are unknown, there are besides the use of microarrays and a regulator mutant, two principal methods for the analysis of regulator-operator binding: the SELEX (systematic evolution of ligands by exponential enrichment) method for finding an interaction partner *in vitro* or the ChIP (chromatin immunoprecipitation) method for such interactions *in vivo*. Both methods require a purified regulator protein. In SELEX, this protein is coupled to a matrix and used for DNA or RNA affinity purification. In ChIP, a specific antiserum is raised against this protein and later used for immunoaffinity purification of the regulator–DNA/RNA complex. This chapter does not deal with the plethora of tools and methods for the expression of epitope-tagged proteins, but gives an overview of SELEX and ChIP techniques applied to analyze regulatory interactions in bacteria.

26.3.1
Systematic Evolution of Ligands by Exponential Enrichment (SELEX)

SELEX is a method for the extraction and selective enrichment of sequences from a random pool of oligonucleotides for those that have the highest binding affinity for a given molecular target (Figure 26.1). These targets can be either proteins or small

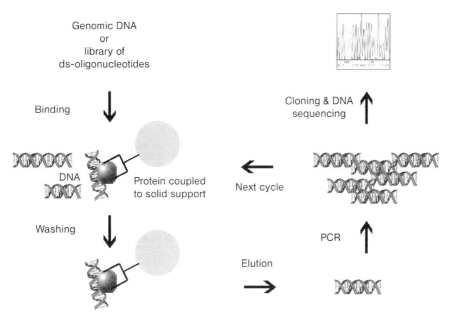

Figure 26.1 SELEX procedure. The starting DNA material is applied to a cyclic procedure after which the final library is sequenced. ds, double-stranded.

molecules. An initial oligonucleotide library can consist of single- or double-stranded DNA as well as RNA. It can be obtained by chemical synthesis, from a genome, or from a clone library. From these libraries, strong binding partners are selected by repeated cycles of binding, selection, and amplification. It has been used successfully in the characterization of protein-binding sites as well as in the selection of optimized binding partners for nucleic acids and construction of specific ribozymes. Excellent reviews on the different applications of the SELEX procedure are available [2, 3]. In the following, the application of SELEX for the characterization of bacterial operator sites is described exemplarily by a derivative – the genomic SELEX procedure.

Genomic SELEX is used to find the DNA targets for a given DNA-binding protein, mostly a transcriptional regulator. In genomic SELEX, a library of small (200–300 base pairs) DNA fragments is constructed, normally by shearing chromosomal DNA, size selection in agarose gels, repair of ends, and cloning into a blunt-end restriction site of a plasmid vector. From the resulting clone library, plasmid DNA is isolated from the pool of clones and the cloned DNA fragments are amplified by PCR using standard primers flanking the cloned chromosomal fragments. This pool of fragments is then applied to a column in which the binding protein is immobilized, mostly by using a specific epitope-tag (e.g., His-tag, Strep-tag). After binding to the protein, several washing steps are performed to remove DNA binding unspecifically or weakly to the immobilized protein. After washing, the stronger binding DNA is eluted from the column and used as template in a second PCR reaction with the same flanking primers. This PCR product mixture is again applied to the column and specifically binding DNA fragments are enriched further.

After several steps of enrichment, the fragments can be cloned into a sequencing plasmid and the sequences of enriched clones determined by DNA sequencing. In this case, the abundance of specific sequences among the fragments roughly represents the strength of binding. As an alternative, the enriched PCR product can be used in microarray analysis. For this purpose, special microarrays are applied, since these microarrays must represent the 5′-regions of genes or, better, the intergenic regions of a genome. In such microarray experiments, the strength of the hybridization signal to a specific region represented on the array is a rough measure of binding site strength.

Although genomic SELEX delivers the regions where binding occurs, and identifies the genes that are close by and might be regulated, it is a slight disadvantage that larger fragments are selected in this procedure. Therefore, many overlapping clones have to be sequenced and sequence alignment by bioinformatics methods has to be applied to narrow down the true binding site. Another way is the selective cloning of smaller DNA fragments from the region of interest and further rounds of SELEX. Alternatively, the electrophoretic mobility shift assay (EMSA) procedure described in Section 26.4.2 might be applied.

SELEX is also possible with an artificial library of double-stranded oligonucleotides, in order to find an optimal binding partner for a given protein. In this case, oligonucleotides must be synthesized that carry a randomly generated sequence in between constant parts that are used for PCR amplification of enriched oligonucleotides. Also in this procedure, cloned DNA fragments from the final round of enrichment have to be sequenced or hybridized to microarrays. This procedure can be used to locate the binding site and, in combination with sound statistics, to provide a comprehensive description of the specific contribution of single nucleotide positions to the protein–DNA binding – at least in an *in vitro* situation.

26.3.2
Chromatin Immunoprecipitation (ChIP)

ChIP is a powerful method for the *in vivo* determination of both the position and the strength of protein–DNA or protein–RNA interactions. It is therefore superior to genetic methods that require mutants and suffer from indirect effects or the inability to study essential proteins.

In brief, the method involves treatment of cells with formaldehyde, an agent that cross-links DNA–protein complexes in the cells, followed by cell lysis and sonication to fragment the DNA to 300–400 base pairs on average (Figure 26.2). The DNA-binding protein of interest is then immunoprecipitated and cross-linking is reversed by heat treatment. By this, the formerly cross-linked DNA is released and, after purification, the enriched binding sites can be analyzed by microarray analysis (ChIP-chip or ChIP-on-chip) or by sequencing (ChIP-seq).

In order to minimize false-positive signals, either total genomic DNA or DNA from a mock immunoprecipitation experiment (e.g., using preimmune serum) is used as

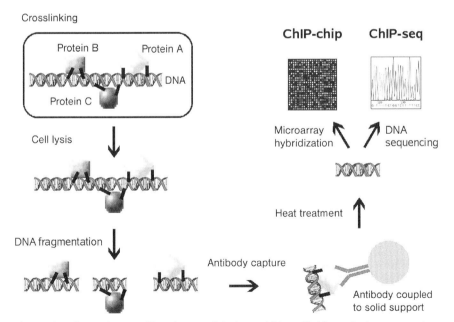

Figure 26.2 ChIP techniques. Here the cross-linked material is purified by antibody capture and finally analyzed by microarray hybridization or DNA sequencing.

hybridization control in microarray analysis. The latter is also used as control in ChIP-seq experiments.

The ChIP-chip method has been developed for eukaryotes and has been applied to bacteria on a broader scale only recently. This is despite the fact that bacteria with their small genome sizes and their high-quality genome sequences represent ideal systems for comprehensive microarray analyses. With the advances in bacterial genome engineering, epitope tagging of proteins within the cell is now a routine task. Therefore, the lack of a specific antibody can be overcome by using epitope-specific antibodies for the immunoaffinity step.

ChIP-chip studies on bacterial transcription regulators, RNA polymerase, and nucleoid-folding proteins in different bacteria yielded fascinating results that could not be obtained by other methods (for a recent review, see [4]). However, the method obviously yields false-positives and false-negatives since all the technical limitations of microarrays regarding sensitivity and different binding kinetics apply also to these experiments. In addition, these analyses showed that binding of a transcriptional regulator does not necessarily mean that the neighboring gene is regulated upon binding of the factor. Hence, the possibility of analyzing protein–DNA interactions *in vivo* revealed the current lack of knowledge on the full complexity of the components and the many interactions within cellular regulatory networks.

The ChIP-chip analyses that have been published to date open exciting new fields of research, not only in transcriptional processes, but also in chromosome folding and replication. It can be expected, at least for several bacterial model organisms, that the interactions of all RNA polymerase sigma factors with their promoters will be determined, as well as the binding sites for many transcriptional regulators. In addition, the complex structure of the bacterial nucleoid, together with the specific chromosome-organizing proteins and the interplay of chromosome structure and gene expression, will be analyzed on a large scale.

26.4
Analyzing Known Targets

If a target gene is known or suspected, there are different methods for the identification and characterization of (unknown) proteins binding to its operator region. Since it is almost impossible to describe all of them, this chapter focuses on two methods. The first is DNA affinity purification chromatography (DAC) sometimes also named "GDAC" (genomic DAC) or "regulator fishing". As with the ChIP methods, it was first developed for eukaryotic transcription factors and later transferred successfully to bacteria. It serves to isolate a DNA-binding protein in an amount sufficient for identification.

The second method is electrophoretic mobility shift assay (EMSA) – a method routinely applied to characterization and quantification of protein–DNA interactions. The method is chosen here because of its simplicity and its frequent use in the literature. It is obvious that it does not compare to other kinds of interaction analyses that require expensive instrumentation.

26.4.1
DNA Affinity Chromatography (DAC)

The DAC method is used to find proteins binding to a specific region of the genome, most often an operator site upstream of a regulated gene. The method was pioneered by the work of Gabrielsen *et al.* [5] and Ghislain and Fish [8] in eukaryotic systems, and later modified and applied to bacteria. The method is known also as DNA affinity chromatography assay (DACA [9]). Like other gene targeting methods, it requires a sequenced genome, since the proteins binding to the DNA fragment used for affinity chromatography are identified by a tryptic fingerprinting technique using mass spectrometry. For this, the whole genome sequence is used to generate a database of theoretical mass spectra from all proteins of an organism to identify the binding protein by pattern matching techniques.

The DAC procedure involves the following steps (Figure 26.3). First, a biotinylated DNA fragment of interest is prepared by a PCR reaction in which one of the primers is biotinylated. Then, the resulting DNA fragment is purified and incubated with a cytosolic protein extract. During several rounds of washing and recapture of the biotinylated fragment with streptavidin-coated magnetic beads, weakly binding proteins are removed and strongly binding proteins retained. After elution from the DNA, the strongly binding proteins are separated by polyacrylamide gel electrophoresis in the presence of sodium dodecylsulfate (SDS). Most often, the DNA fragments used are 100–500 base pairs long and several proteins are eluted from the biotinylated DNA. Among these are proteins that bind specifically to the fragment and those that bind unspecifically. It is therefore recommended to run control experiments (e.g., with a DNA fragment of comparable size that is derived from an intragenic region). The experiment and the control patterns can be compared in order to find proteins that only bind to the region of interest. In addition, the elution of proteins from the biotinylated DNA can be performed by using a salt gradient, giving some information on the strength of binding for individual proteins.

As most other methods, DAC has limitations. The most important is that it is an *in vitro* procedure. It might therefore yield some false-positive and miss true-positive proteins. However, several proven transcriptional regulators, among them repressors and activators, have been purified by this method – indicating at least that a lot of such interactions are robust enough to occur also in the *in vitro* experiment. Another limitation of DAC can be overcome – in standard DAC experiments, proteins are isolated that bind the DNA in the absence of an effector; however, the whole procedure might also be performed in the presence of a specific effector molecule in order to detect such proteins.

26.4.2
Electrophoretic Mobility Shift Assay (EMSA)

The EMSA technique to demonstrate binding of a protein to DNA has been known for decades, but was significantly improved by the availability of fluorescent dyes for DNA labeling that mostly replaced radioactive or other labels. The principle of the

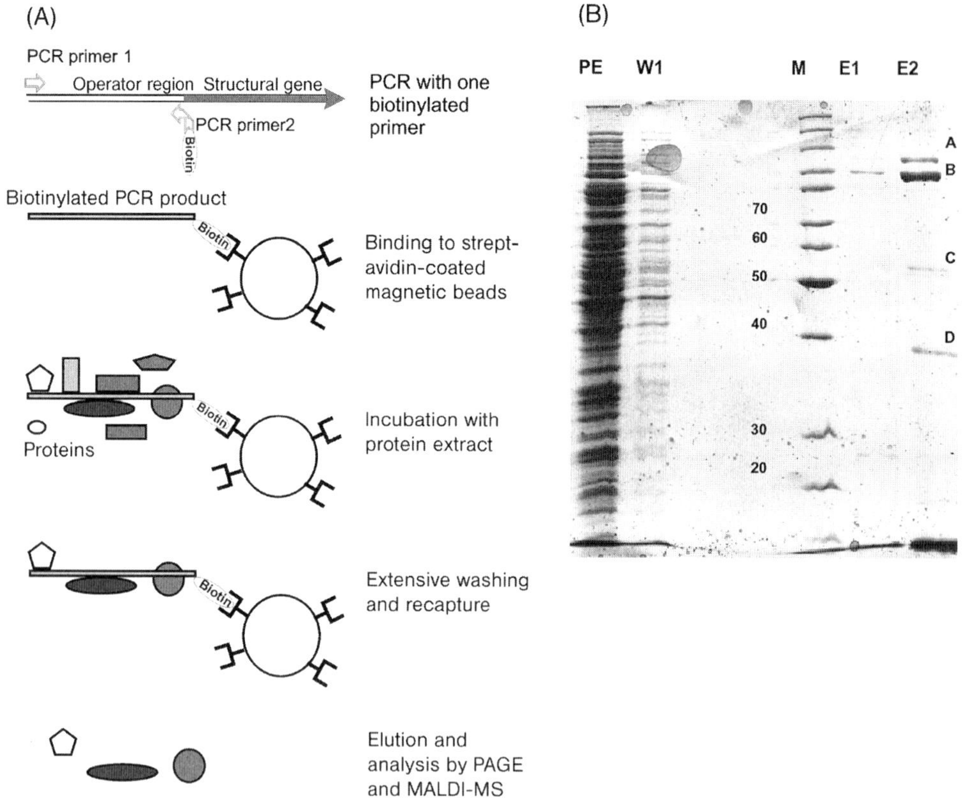

Figure 26.3 Genomic DNA affinity chromatography. (A) Experimental steps (PAGE, polyacrylamide gel electrophoresis; MALDI-MS, matrix-assisted laser desorption ionization-mass spectrometry) and an example result (B). In the example, a raw protein extract from *Corynebacterium glutamicum* was used for affinity purification of proteins binding to the upstream region of the ribokinase gene *rbsK2* [6]. The Coomassie Blue-stained SDS–polyacrylamide gel shows proteins from different steps of the protocol. PE: protein extract obtained after cell lysis; W1: first wash fraction; E1: elution fraction obtained after a single washing step; E2: elution fraction obtained after a second washing step. M: protein size marker with numbers indicating the molecular weight of the marker proteins (kDa). The eluted proteins were identified by tryptic fingerprinting. A: excinuclease subunit UvrA; B: DNA polymerase I; C: DNA polymerase III subunit; D: transcriptional regulator UriR.

method is that a protein–DNA complex adopts a conformation that migrates slower in gel electrophoresis. If this binding is specific, the retarded DNA forms a distinct band that is shifted relative to the DNA without the binding protein (band-shift). As already noted, the band-shift procedure (Figure 26.4) is most often performed with fluorescently labeled PCR products or double-stranded oligonucleotides of at least 50 base pairs length in order to show band-shifting in agarose gels. For shorter DNA fragments it is also possible to use polyacrylamide gels, which provide improved sensitivity and resolution.

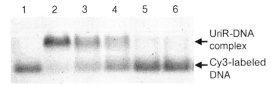

Figure 26.4 EMSA. The image shows an agarose gel in which different mixtures of the UriR repressor protein, a 59-base-pair double-stranded DNA target upstream of the *uri* operon in *C. glutamicum* [6], and competing double-stranded oligonucleotides of length 59 base pairs were separated. The DNA target was labeled by the fluorescent dye Cy3 and the image was obtained by a fluorescence scanner. The lane numbers denote: (1) 0.1 pmol of the DNA target alone; (2) the DNA target and 19 pmol UriR protein; (3–6) DNA target, UriR protein, and increasing amounts of unlabeled competing double-stranded oligonucleotides: (3) 10-, (4) 25-, (5) 50-, and (6) 75-fold molar excess.

EMSA is used in multiple ways – the protein concentration might be varied in order to assess the binding stoichiometry or to find intermediate complexes with different composition. With such studies, multiple binding sites for a protein can be demonstrated. A band-shift caused by specific binding can also be validated by adding an antibody specific for the binding protein. In this case, a stronger retardation (super-shift) is observed. Another form of validation is the addition of unlabeled competitor DNA, either nonspecific competitor DNA (polydI·dC) able to resolve nonspecific binding or specific subfragments of the band-shifted DNA. With the addition of such subfragments, which can be short double-stranded oligonucleotides (down to 20mers) only carrying the presumed binding site and added in a molar excess, the specific binding is resolved and the band-shift is reversed (Figure 26.4). Such short competitors can either contain the native or mutated sequences, allowing us to determine the contribution of every single base pair to the binding.

Due to the ease of use and broad applicability, the EMSA method has been used in many studies to demonstrate specific binding of proteins to DNA. Although this is also an *in vitro* procedure, it has specific advantages with respect to higher throughput and the possibility to apply small-molecule effectors in order to characterize their effects on binding.

26.5
Conclusions and Outlook

This chapter presents an overview on some, but by far not all, methods currently used in the identification and characterization of binding partners involved in DNA–protein interactions. However, all of these methods have advantages and disadvantages. It is therefore highly recommended to use *in vivo* and *in vitro* methods in combination in order to draw coherent conclusions.

In addition, the field is rapidly developing with the introduction of novel or improved techniques. A striking example is the development in the field of DNA sequencing. With current ultra-high-throughput sequencers, the sequencing of complete transcriptomes is now not only possible, but will soon be economic. This will also have a direct impact on procedures like SELEX and ChIP-seq.

Although sequencing of transcriptomes is a special challenge to DNA microarray analysis, this field is also evolving rapidly by increasing the number of features that can be tested on a single microarray. It is to be expected that microarrays will soon comprise high-density coverage of complete bacterial genomes at reasonable prices. Since a considerable number of genes in bacteria do not encode proteins, but specify noncoding (regulatory) RNA, this will allow us for the first time to identify all genes in a bacterial organism.

This knowledge will then be the basis for the description of all regulatory interactions in a bacterial cell and, together with input from bioinformatics and mathematical modeling, for significant contributions to the emerging field of Systems Biology.

Acknowledgments

Dr. Karina Brinkrolf (Bielefeld University) is acknowledged for the supply of unpublished information and figures.

References

1 Condon, C. (2007) Maturation and degradation of RNA in bacteria. *Curr. Opin. Microbiol.*, **10**, 271–278.

2 Djordjevic, M. (2007) SELEX experiments: new prospects, applications and data analysis in inferring regulatory pathways. *Biomol. Eng.*, **24**, 179–189.

3 Gopinath, S.C. (2007) Methods developed for SELEX. *Anal. Bioanal. Chem.*, **387**, 171–182.

4 Wade, J.T., Struhl, K., Busby, S.J.W., and Grainger, D.C. (2007) Genomic analysis of protein–DNA interactions in bacteria: insights into transcription and chromosome organization. *Mol. Microbiol.*, **65**, 21–26.

5 Gabrielsen, O.S., Hornes, E., Korsnes, L., Ruet, A., and Oyen, T.B. (1989) Magnetic DNA affinity purification of yeast transcriptional factor: a new purification principle for the ultrarapid isolation of near homogeneous factor. *Nucleic Acids Res.*, **17**, 6253–6267.

6 Brinkrolf, K., Plöger, S., Solle, S., Brune, I., Nentwich, S.S., Huser, A.T., Kalinowski, J., Puhler, A., and Tauch, A. (2008) The LacI/GalR family transcriptional regulator UriR negatively controls uridine utilization of *Corynebacterium glutamicum* by binding to catabolite-responsive element (*cre*)-like sequences. *Microbiology*, **154**, 1068–1081.

7 Balaji, S., Iyer, L.M., Babu, M.M., and Aravind, L. (2008) Comparison of transcription regulatory interactions inferred from high-throughput methods: what do they reveal? *Trends Genet.*, **24**, 319–323.

8 Ghislain, J.J. and Fish, E.N. (1996) Application of genomic DNA affinity chromatography identifies multiple interferon-alpha-regulated Stat2 complexes. *J. Biol. Chem.* **271**, 12408–12413.

9 Park, S.S., Ko, B.J., and Kim, B.G. (2005) Mass spectrometric screening of transcriptional regulators using DNA affinity capture assay. *Anal. Biochem.* **344**, 152–154.

Index

a

AAA⁺ protein   449, 451, 452
Acetobacter xylinum   380
N-acetyl-glucosaminyl transferase   11
activator protein-1 (AP-1)   79
acyl carrier protein (ACP)   396
N-acyl-homoserine lactones (AHLs)
  systems   7, 24, 95, 105
– AHL synthase   24
– antagonists   33
– gene expression   33
– mediated cell-cell communication   32
– mediated cross-talk   34
– polar nature   105
– quorum sensing (QS) system   24, 85
– – role   26
– receptor protein   24
adaptor proteins   453–456
adenine riboswitch, aptamer domain   419
adenyl cyclase (AC)   361
– C-terminal regulatory domain   361
– negative mutants   365
– N-terminal catalytic domain   361
Aeromonas hydrophila, homologs   279
Aeschynomene sensitiva   66
Agrobacterium tumefaciens   143
– VirA/VirG system   143
– – transmembrane domains   143
aldonic acids, *Lotus albus*   61
alkylation reaction   248
allosteric receptor behavior   171
– mathematical models   171
– – Ising-type models   171
– – Monod–Wyman–Changeux (MWC)-type models   171
alternative sigma factors   408
– PpGpp-dependent activation   408
ammonium signaling   218, 219

antigen-presenting cells (APCs)   86
– major histocompatibility complex (MHC) class II receptor   86
antimicrobial defense program   77
Archaeoglobus fulgidus protein   206
ATP-binding cassette (ABC)-type transport systems   212
– export proteins   144
– transporter   118, 212
ATP-dependent proteases   449
attractant-specific receptors   166
autoinducer peptide (AIP) pheromone   145
autoinducers (AIs)   3, 7, 95–99, 103, 104, 203
– acid-base association constant   104
– analytical tools   97
– biofilm formation   95
– biosensor construction   98
– biosynthesis gene I   95
– identification strategy   95, 97
– – steps   96
– molecular structure   96
– negative mutant   98
– physicochemical properties   96
– quantification techniques   103–107
– – AHL-based QS signals analysis   105
– – HAQ-based QS signals analysis   106
– – principles   103
– sample preparation   99
– – liquid chromatography, principles   99–102
– – liquid-liquid extractions   99
– screening   98
– secondary metabolite production   95
– structural analysis techniques   102
– – mass spectrometry   102
– – NMR Spectroscopy   103
– types   7
– virulence/bioluminescence production   95

Bacterial Signaling. Edited by Reinhard Krämer and Kirsten Jung
Copyright © 2010 WILEY-VCH Verlag GmbH & Co. KGaA, Weinheim
ISBN: 978-3-527-32365-4

autoinducing peptide (AIP) 14
Azotobacter vinelandii 297
– NifL/NifA system 297
– – FADH2 297
– – function 297

b

Bacillus anthracis 352
Bacillus clausii 419
Bacillus subtilis 232, 235, 237, 294, 296, 334, 336, 349, 352, 397, 402, 406, 417, 450, 453, 454
– alternative sporulation sigma factor σ^K activation 237, 238
– CcpA-mediated CCR pathway 349
– ClpP 453
– FNR 295
– – TnrA, DNA-binding activity 336
– glutamate biosynthesis control 337
– heat shock adaptation 450
– MecA 453
– nitrogen metabolism control, by TnrA 336
– proteolysis, role 450
– regulatory proteolysis 454–456
– – competence development 454
– – heat shock adaptation 455
– – proteolytic switch 454
– σ^W regulon, regulation 235
– rRNA regulation mechanism 406
– thiamine biosynthesis 417
bacteria 3, 310, 450
– AAA$^+$ protease systems 450
– adaptor proteins 452
– cells, regulatory networks 463
– control systems 310
– genome analysis, goals 270
– H$_2$ metabolism 310
– Hsp100/Clp systems 450
– I-CLiPs 231, 232
– – substrate recognition mechanism 231
– PTS, features 344, 345
– receptors, sensory domains 271
– regulatory proteolysis 449
– signal transduction 450
– singled-cell living organisms 3
– substrate recognition 452
– surface colonies, oxygen gradients 291
bacteria-host interactions 400
– suicide function 401
bacterial adhesins 88, 89
– fimbrial adhesins 89
– nonfimbrial adhesins 89
bacterial autoinducers, *see* autoinducers (AIs)
bacterial blue-light photoreceptors 178–185

– BLUF domain proteins 183, 184
– cryptochromes 181–183
– microbial rhodopsins 178–180
– photoactive yellow proteins 183
– phototropin-like microbial photoreceptors 184, 185
bacterial cell-cell communication, *see* quorum sensing (QS) system
bacterial model organisms 468
bacterial pathogens 76, 88, 399
– *Chlamydia* 399
– infection process 76
– PpGpp 399, 400
bacterial photoreceptors 182, 275
– light-induced signal transduction 182
– – schematic drawing 182
– PHY (phytochrome) domain 275
bacterial regulatory systems 450
bacterial RNA polymerase 329
– sigma factors, activity 329
bacterial RNA, transcriptional profiling 464
bacterial sensors 292, 298
bacterial signal transduction systems 269, 270, 280, 450
– complexity 269
– computational analysis 270
– organization 280
– property 280
bacterial strains 398
– *Bacillus subtilis* 398
– *Corynebacterium glutamicum* 398
– transcriptional profiling 398, 399
– *Vibrio cholerae* 398
bacterial toxins 86
– Bordetella modulins 87
– cholera toxin 86
– *Clostridium difficile* toxins 87, 88
– *Helicobacter*, VacA 87
– superantigens 86
bacterial transcriptome 464
– microarray analysis 464
bacterial transmembrane signaling processes 230
bacterial two-hybrid system 121, 122
bacterial UV/blue-light photoreceptors 180
– domain organization 180
bacteriophytochromes 189
– spectroscopic analyses 189
bacteriorhodopsin 179, 258
bacteriostasis 401
BarA/UvrY system 143
– PH-responsive system 143
BetP 221–223
– activation 223

– C-terminal domain   222
BglG/SacY family   350, 351
– antiterminator proteins, activity control   351
– RNA-binding antiterminator proteins   350
bifunctional enzymes, *see* histidine kinases
biofilm accumulation   12
– surface proteins   12
– – accumulation-associated protein Aap   12
– – *S. aureus* surface protein G (SasG)   12
biofilm-associated protein (Bap)   9
biofilm formation   24–29
– cycle, QS role   27–29
– QS-regulated factors   29–32
– – biosurfactants   30, 31
– – dissimilatory nitrate reduction   32
– – DNA release   31
– – EPSs   29, 30
biofilms eradication, target   32, 33
biological marker genes   98
biological processes   381
biosurfactants   30, 31
blue light using flavin (BLUF)/EAL domain protein   183, 184, 178, 380
– blue-light-sensing minimal modules   183
Bordetella bronchiseptica   126, 235
– putative ferric siderophore outer membrane transport protein (BfrZ)   126
– Fec type transcriptional regulation   126
Bordetella pertussis   87, 151, 219
– BvgS/BvgA system   151
– whooping cough causative agent   87
bovine serum albumin (BSA)   248
Bradyrhizobium japonicum   59, 143, 311
– FixL/FixJ histidine kinase/response regulator system   143
– Nod factor   63
Burkholderia xenovorans   321
N-butanoyl-l-homoserine lactone   24

c
CadC system   218, 237
– ToxR-like family   237
– transcriptional activator   218
Calvin–Benson–Bassham reductive pentose phosphate cycle   320
cAMP–cAMP receptor protein complex   347, 358, 364–366
– function   364–366
– regulatory network   386
– – binding site   365
– – dependent promoters   359, 365
– – dependent regulation   366
– – mediated catabolite repression modeling   370
– – modulon   366, 367
– – regulated genes   367
cAMP receptor protein (CRP)   294, 318, 347, 358, 364
– binding sites   366
– C-reactive protein complex   214
– dependent protein kinase   429
– DNA complex   364
– excretion   363
– modulon   366, 367
– – systematic approaches   366
– properties   364, 365
– proteins   364, 365
– transcription factor, MrpC   49
capillary electrophoresis (CE)   103
carbohydrate-specific enzyme II (EII) transporters   345
carbohydrate uptake system   344
carbohydrate utilization systems   348
carbon and energy metabolism   362
– pyruvate ratio   362
carbon catabolite repression (CCR)   343, 345–347, 357
– cAMP excretion/phosphodiesterase activity   363
– historical account   357–363
– intracellular cAMP levels, regulation   358–362
– mechanism   347, 348
– non-PTS substrates, catabolite repression   362
carbon monoxide sensing   315–324
– hypothetical/secondary sensors systems   321
– – detecting multiple diatomic gases   323
– – eukaryotic-style NO sensing in prokaryotes   321
– – NO sensing by Fur/SoxR/OxyR   322
– primary sensors for CO   319–321
– – CooA   320
– – RcoM   321
carcinoembryonic antigen cell adhesion molecules (CEACAMs)   77, 90
– surface-exposed domain organization   90
catabolite control protein A (CcpA)   348
– transcriptome analysis   348
catabolic processes   290
catabolic systems, CCR   348
catabolite activator protein, *see* cAMP receptor protein (CRP)
catabolite repression   369

– mathematical/computer-assisted
  modeling 369.
catalytic and ATP-binding (CA) domain 136
Caulobacter crescentus 147, 151, 167, 238, 379,
  380, 387, 388
– c-di-GMP signaling 387, 388
– – role 388
– cell cycle 387
– cell polarity determinant, PodJ
  regulation 238, 239
– EAL protein 379
– GGDEF protein 379, 381
– LovK/LovR system 147
– PleC/DivJ, autokinase activities 151
c-di-GMP signaling system 377, 386, 387
– degrading enzymes, PDEs 377
– module 382
– principles 378
– protein domains 377–381
– – c-di-GMP-binding effectors 380
– – c-di-GMP-unrelated functions,
    GGDEF/EAL domains recruitment 379
– – composite GGDEF, EAL, and HD-GYP
    proteins 379
– – DGC and PDE activity and expression,
    regulation of 380
– – making and breaking 377–379
– synthesizing enzyme, DGCs 377
cell-cell communication, *see* quorum sensing
  (QS) system
cell differentiation program 238
– cell types 238
cell receptors, integrins 90
cellular functional unit (CFU) 362
cellular receptor proteins 270
central signaling proteins 163
– histidine kinase, CheA 163
– response regulator, CheY 163
cGMP molecule 272, 273, 317
cGMP-specific phosphodiesterases 186,
  273, 318
chemiluminescence spectrophotometry
  101
chemoreceptor(s) 165, 197, 300
– clusters 167–169
– – assembly 168
– – positioning 168, 169
– – stability 169
– function 165, 166
– structure 165, 166
chemoreceptor domain 199, 201
– architecture 199, 201
– extracellular domain 205
– motions 200

– periplasmic sensory domains 201
chemotaxis 352
– PTS-dependent regulation 352
– receptor 275
– – transmembrane signaling 204, 205
chemotaxis receptor-transducer proteins,
  see methyl-accepting chemotaxis proteins
  (MCPs)
chemotaxis response regulators 197
ChIP-chip method 468
ChIP (chromatin immunoprecipitation)
  method 465, 467, 468
chromatography-based methods 106
Chromobacterium violaceum 98
– violacein production 98
chromophore (s) 177–179, 183, 184,
  186–190, 254, 273
chromosomal DNA 466
CitA 139
– autophosphorylation 139
– periplasmic domain 139
– transmembrane signaling 202, 203
c-Jun N-terminal kinase (JNK) 80
Clostridium botulinum 321
ClpC activation mechanism 453
ClpCP activation 454
– by adaptor-mediated oligomerization 454
coiled-coil cytoplasmic domains 201
colonization factor 13, 15, 16, 88
ComK, autoregulatory activity 455
communication systems 75, 77
– long-distance communication 77–88
– *b/w* pathogens and eukaryotic cells 75
– short-distance communication 88–91
competitive hybridization 464
complementary chromatic adaptation
  (CCA) 188
continuous-wave (cw) EPR spectrum 255
Corynebacterium glutamicum 146, 221, 222,
  432, 433, 438–441
– MurC protein 438
– OdhI deletion mutant 439
– osmoresponsive MtrB/MtrA system 146
– osmosensing 222
– 2-oxoglutarate dehydrogenase (ODH)
    complex 440
– proteins 432
– STPK substrate proteins 433
Crp gene 364
– transcriptional regulation 364
CRY-DASH proteins 181, 182
cryptochromes 181–183
– domains 182
– – C-terminal region 182

– – N-terminal photolyase-related domain (PHR) 182
– photoreceptors 181
CtrA-mediated inhibition 388
– PpGpp 395
– PppGpp 395
curli fimbriae, expression 383
curli synthesis 384
– activator gene 384
– regulator 384
– YedQ 385
Cya gene 359
– expression 359
– transcriptional regulation 359
cyanobacterial adenylate cyclases 186, 273
– GAF domains 273
cyanobacterial phytochromes 187, 188
cyclic oligopeptides 7
cyclobutane pyrimidine dimer (CPD) photolyases 183
cysteic acid 249
cysteine 248
– accessibility approach 249–251
– – schematic presentation 249
– amino acid 248
– applications of 250
– – cysteine accessibility analysis 250–252
– – proteins, proximity relationships 252, 253
– cross-linking 252
– labeling 248, 249
– modification 248
– nucleophilic addition 248
– residue 249, 252
– specific biotin derivative 250
cystic fibrosis transmembrane conductance regulator (CFTR) chloride channel 86
cytoplasmic membrane 121, 145
– sensor proteins 136
– signal transfer 121
– transmembrane domains 145

d

damage-associated molecular-pattern (DAMP) molecules/components 78
DcuS 139
– autophosphorylation 139
– cytoplasmic PAS domain 139
– periplasmic domain 139
Deinococcus radiodurans photosensory core 189
– crystal structure 189
diazotrophic bacteria 291
– O_2-sensitive nitrogenase 291

differential interference contrast microscopy 43
diffusible signal factor (DSF) 27
diguanylate cyclases (DGC) 380
– expression 380
– PDE activity 380
– regulation 380
dihydroxy acetone (DHA) metabolism 215
dimerization and histidine phosphotransfer (DHp) domain 198
diphosphoryl transfer protein (DTP) 344
direct oxygen sensor (Dos) 274, 291–298, 319
– FAD-containing sensors, NifL 297
– $[4Fe-4S]^{2+}$-containing sensors 294
– – cofactor for O_2 sensing 296
– – $FNRB_{Bs}$ 296
– – $FNRE_{Ec}$ 294, 295
– – NreB 296
– – WhiB3 296
– heme B-containing sensors 291–294
– – Dos 294
– – FixL 291–294
– – HemAT 294
DNA 319, 320, 323, 467, 468
– binding domain 319, 320, 338
– binding regions 296, 323
– double-stranded 464
– fragments 466, 469, 470
– labeling, fluorescent dyes 469
– receptor 455
– recognition motifs 332
– sequences 15, 322
DNA affinity chromatography assay 469
DNA affinity purification chromatography (DAC) 468, 469
– limitation 469
– steps 469
DNA microarrays 369, 463
– experiments 369
DNA-protein complexes 467
DNA-protein interactions 471
double-stranded oligonucleotides 466
– artificial library 466
double electron-electron resonance (DEER) 258
Drosophila melanogaster, Toll receptor 78

e

EAL proteins 383, 384, 386, 387
– PDE YhjH 384
electron paramagnetic resonance (EPR) spectroscopy 247, 253, 312
– continuous-wave 255
– double spin labeling 258

– – intra/intermolecular distances 258, 259
– interspin distance measurements 258
– – intra/intermolecular distances 258, 259
– spectral line shape 255
– spectrum 312
– spin-labeled proteins 255, 256
– – protein structure 256, 257
– – site-directed 253–255
– spin label microenvironment 257
– – polarity/proticity 257, 258
electrophoretic mobility shift assay (EMSA) 466, 469, 471
– advantages 471
– analysis 468
Ellman's assay 250
Ellman's reagent 250
enhancer binding proteins (EBPs) 50
– transcription factors 50
enteric bacteria 139, 345, 357
– carbon catabolite repression (CCR) 357
– IIAGlc subunit, phosphorylation 345–347
– PhoQ/PhoP systems 139
Enterococcus faecalis 144, 239
– peptide sex pheromones, production 239
– virulence factor gelatinase biosynthesis 144
Escherichia coli 85, 88, 126, 128, 138, 143–146, 163–166, 168, 170, 213–218, 272, 289, 290, 294, 296, 298, 300, 318, 319, 322, 329, 334, 335, 344, 346, 352, 368, 382, 383, 385, 397, 406, 417, 420, 451, 452, 456
– aconitases 334
– adaptor proteins 452
– anaerobic respiratory gene expression control 143
– ArcB/ArcA two-component system, role 298
– artificial ribozyme-based ON switch 420
– attractant-specific receptors 166
– cAMP receptor protein 329
– causing meningitis 89
– CCR, mechanisms 346
– c-di-GMP signaling 382–385
– chemosensory clusters 168
– chemotaxis system 163–165
– – proteins, homologs 294
– – response 170
– – receptor, flavin-binding domain 272
– Cpx system 144
– σ^E-dependent envelope stress response, regulation 233, 234
– dicarboxylic acid uptake 216–218
– direct oxygen sensor protein 294
– DNA damage, cellular response 456
– enterohemorrhagic 83, 145

– – noradrenaline effect 83
– enterotoxic 89
– EnvZ/OmpR 146
– FecI/FecR interactions 126
– fumarate/nitrate reductase regulator (FNR) 296, 319
– – protein 319
– GGDEF/EAL genes 382
– glucose transport system, regulation 368
– β-glucoside transport 335
– high-affinity K^+ uptake system, role 146
– histidine kinases 138
– LysP/CadC system 218
– maltose regulon, transcription factors 213
– maltose system 213–215
– – regulation 213
– – transport sensing 213
– Mlc transcription regulator 335
– motility, c-di-GMP-mediated coordination 383
– O_2 sensor, FNR of 294
– peroxide/superoxide sensors 322
– PpGpp, induction scheme 397
– rRNA regulation mechanism 406
– transport sensing 217
– Uhp system 215, 216
Escherichia coli K-12 117, 118, 357, 368
– ferric citrate transport 117, 118
– genome 269
– regulatory system 118
eubacteria 181, 399
– *Cytophaga hutchinsonii* 181
– *Helicobacter pylori* 399
– *Synechocystis* sp. PCC6803 181
– *Vibrio cholerae* 181
eukaryotic-like protein kinases (ELKs) 428
eukaryotic translation initiation factor, phosphorylation 85
extracellular matrix (ECM) proteins 10, 90
extracytoplasmic function (ECF) sigma factors 122–124, 129, 233
– activity 123
– FecI, deletion analysis 122
– regulation 233–236
extrapolymeric substances 28
– production, regulation 28

f

fatty acid biosynthesis 438
FecA 119–121, 124, 128
– crystal structure 119, 124
– electron paramagnetic resonance spectroscopy 120
– mutants 120

– signaling domain   121
FecCDE encode cytoplasmic membrane proteins   118
Fec genes   121, 122
FecR   118, 121, 123
– cytoplasmic transmembrane protein   118
– gene fragment   121
Fec transport system   119
Fec-type signaling cascade   123
ferric citrate transcription regulation mechanism   123
fibronectin-binding proteins   10
Fick's first law   251
FixL/FixJ histidine kinase/response regulator system   143
flavin adenine dinucleotide (FAD)   272
flavin mononucleotide (FMN)   184, 272
– chromophore   184
flavonoid induced nodulation genes   59–61
flavonoid stimulons   65
– microarray studies   65, 66
fluorescence analysis   254
N-formyl-methionyl-leucyl-phenylalanine (fMLF), GPCR scheme   82, 84
FMN-containing PAS domains, see LOV (light, oxygen, and voltage) domains
formyl-peptide receptors (FPR)   81
– receptor family   82
Fourier transform infrared spectroscopy   310
Fourier transform ion cyclotron resonance (FTICR)-MS   102
fumarate/nitrate reductase regulator (FNR), form   295

g
GAF domains   186, 324
gas chromatography (GC)   103, 104, 404
– phases   104
– retention time   104
– rich sequence element   404
gas-sensing systems   307–309
– proteins, schematic overview   309
gene expression control   434, 435
– C-signal control   49–51
– by PknB via sigma factors activity   434
– by PknH via transcriptional regulator EmbR   435
– – EmbR discovery in mycobacteria   435
– – EmbR phosphorylation by PknH effects   435
– – EmbR structure   435
– – phosphorylation by PknH   435
– – signal transduction model by PknH and EmbR   436

gene regulation   95
gene target analysis   463
genome-based microbiology, paradigm   270
genome-sequencing projects   269
genome-wide expression analysis   463
genomic DNA affinity chromatography   470
GGDEF protein   383–386, 388
– expression   383
– *Vibrio cholerae*   385
– YdaM   384
globin-coupled sensor (GCS) domains   274
Gluconacetobacter xylinus   377
G-protein   84, 86, 397
– $G_{\alpha s}$ subunit   86
– hormones language   83–86
G-protein CgtA   397
– cell-cell signaling mechanisms, mediator   400
– promoter-specific effects   404, 405
– SpoT-mediated synthesis   397
G-protein-coupled receptor (GPCR)   84
gram-negative bacteria   7, 23, 83, 95, 117, 128, 234, 236, 382
– *Actinobacillus actinomycetemcomitans*   12
– N-acyl-homoserine lactones (AHLs) production   95
– *Escherichia coli*   382
– *Myxococcus xanthus*   236
– outer membrane signaling   117
– proteobacterium, *V. cholerae*   385
– QS   23, 24, 83
– – bacteria, communication   4
– – lipid-based molecules   83
– SE homologous systems   234, 235
gram-positive organisms   7, 83, 95, 144, 301, 338, 382, 398, 417
– autoinducing cyclic peptides   83
– *Bacillus subtilis*   348, 382, 398
– *Corynebacterium glutamicum*   398
– cyclic oligopeptides   95
– transcriptional control   417
– *Vibrio cholerae*   398
GTP pool size   406
– sigma factor competition   407, 408

h
Halobacterium salinarum   178
HAMP domain   166, 205
– linker domain, structure   205, 206
hand shaking communication   88
HD-GYP domain   379, 382
– PDE activity   379
– proteins   379, 382
Helicobacter pylori   87

– vacuolating toxin, VacA secretion 87
heme-containing biosensors, FixL 320
heterotrophic bacteria 185, 189
– phytochromes 189, 190
hidden Markov model, application 123
high-performance liquid chromatography (HPLC) 100, 103–106
– retention time 104
high-throughput experimental techniques 463
– DNA microarrays 463
histidine/aspartate signaling pathways 139
histidine kinase(s) 135–137, 139, 143, 148, 149, 184, 199, 200
– accessory proteins 148–151
– architecture. 199
– class I 136
– class II 136, 138
– domains 137, 184
– extracellular sensory domains 198
– family 135–139
– periplasmic/extracellular sensing/input domain 136
– – physical stimuli 145–148
– – vs. serine/threonine/tyrosine kinases 138
– signal integration 149
– stimulus perception/signaling 135, 139–148
– – chemical stimuli 139–145
– transmitter domain, parts 136
– transmembrane helices 136
histidine kinase/response regulator systems 135, 138, 140, 148, 150, 151, 428, 443
– membrane-integrated histidine kinase 135
– molecular signals 140–142
– vs. STPKs 428
– use 138
H-NOX (heme nitric oxide and oxygen binding) domain 274
HOG pathway 147
homeostasis 75
Homo bacteriens 75
hormone-like effector molecules 395
– PpGpp 395
– PppGpp 395
hormone-like signaling molecules 83
host-adapted metabolism 76
host cell receptors 77, 91
– CEACAMs 77
– integrins 77, 91
host complement system 81
host-pathogen communication 91

host receptors 78, 89
HoxA 313, 315
– nonphosphorylated form 315
– phosphorylation status 315
Hsp100/Clp proteins 449–452
– AAA$^+$ core domain 451
– AAA$^+$ proteases 450
– biological role 449
hybrid sensor kinase/response regulator system 145
– QseC/QseB 145
hydrogenases 310
– biosynthesis, regulation 310
– classes 310
hydrophilic cavity 249, 251
4-hydroxy-2-alkylquinolones (HAQs) 106
– derivatives 106
– GC-MS-based method 106
hypoxia-inducible factor (HIF) 85, 319
– DNA binding activity 319
– HIF-1α subunit regulation 85

i

immunoreceptor tyrosine-based inhibitory motif (ITIM) 90
indirect O_2 sensors 297–301
– electron transport-linked sensors 298
– – Aer 300
– – ArcB/ArcA 298, 299
– – PrrB/PrrA and RegB/RegA 300
– – ResE/ResD and SrrA/SrrB 300, 301
– NADH-linked systems 301
– – Rex 299, 301
inducer exclusion mechanism 347, 349, 367, 368
innate immune system 75, 77, 79
interleukin (IL)-1 receptor 79
intracellular regulatory network, reversal 45, 46
intracellular sensory domains 271–276
– BLUF domain 273
– GAF domain, roles 273
– GCS domain 274
– HNOB domain 274
– Hr domain, role 275
– KdpD domain 275
– PAS domain, ligand-binding pocket 272
– PHY domain 275
intracellular signaling pathways, diversity 280
intracellular signal-transducing/output domains 276–280
– c-di-GMP-mediated signaling 279
– proteolysis 456

- serine/threonine protein phosphorylation signaling system   279
- sugar, PTS   279
- types   278, 279
intramembrane cleaving proteases (I-CLiPs)   229, 231
ion-exchange chromatography (IEC)   100
iron-responsive elements (IREs)   333
- bound aconitase, structure determination   333
iron-sulfur cluster   333
iron transport systems   126, 129

k

ketoglutarate decarboxylase   442
- 2-oxoglutarate decarboxylase activity   442
kinase domains   293, 432
- structure   432
Klebsiella pneumoniae   202

l

Lactobacillus brevis   349
Lactococcus lactis   221
lipopolysaccharide (LPS), binding proteins   89
liquid chromatography (LC) techniques   99–102
- mobile phase, composition   100
- principles   99–102
- (semi)preparative liquid chromatography   101
- - advantage/disadvantage   101
- set-up, role   100
- SPE   100
- TLC   101
liquid-liquid extraction (LLE)   99
Listeria monocytogenes   90, 421
- internalins   90
- PrfA translation   421
long-distance communication   77–88
- pathogen-associated molecular patterns language   77–82
- - cytosolic PAMP receptors   80
- - PAMPs as chemoattractants   81–82
- - TLRs   78–80
- pathogen-host cell communication   86
- - extracellular bacterial toxins   86–88
LOV (light, oxygen, and voltage) domain   147, 184
LuxPQ transmembrane signaling   203, 204
LuxQ   203, 204
- cytoplasmic domains   203, 204
- periplasmic domains   203
LuxS/AI-2 system   7, 17

- gene   17
- QS system   17

m

Macroptilium atropurpureum   65
maltose transporter   347
- MalK subunit   347
MAPK signaling pathways   81
marker gene, expression   99
mass spectrometry, set-ups   102
MBH operon   313
- structural/regulatory hydrogenase genes   313
- - arrangement   313
mechanosensitive channels   220, 221
- MscL   220
- MscS   220
membrane-bound [NiFe]-hydrogenase (MBH)   313
membrane proteins   247, 248, 251, 336
- CadC   335
- cysteine   248
- topography   251
Mesorhizobium loti   64
- type IV secretion system (T4SS)   64
methane thiosulfonate (MTS)   248
methyl-accepting chemotaxis proteins (MCPs)   179, 269, 276
- classes   276
methylation system   170
- allosteric models   170, 171
- role   170, 171
methyl methane thiosulfonate (MMTS)   251
micellar electrokinetic chromatography   106
microarray analysis   464, 466, 467, 472
- experiments   464, 466
microbial-associated molecular pattern (MAMP)   78
microbial iron chelators, siderophores   85
microbial metabolites, advantage   75
microbial rhodopsins   178–180
- haloarchaeal types   178
microbial sensor systems   307, 308
- for carbon monoxide   307
- for dihydrogen   307
- for nitric oxide   307
microbial signaling molecules   76, 82
microbial surface components recognizing adhesive matrix molecules (MSCRAMMs)   9, 10, 16
microbiota-host consortium   75
mitogen-activated protein kinases (MAPKs)   80
molecular hydrogen (H_2)   309

– oxidative cleavage 309
– sensing hydrogenases, characteristic 311
– sensing model 316
– sensing proteins, Ni-Fe-active site 312
– signaling regulatory hydrogenases 311
– signal transduction, regulatory components 313
molecular hydrogen sensing 309–315
– direct hydrogenase control, environmental signals 310
– H_2 sensor complex in action 314
– H_2 signaling cascade 312–314
– H_2 signaling hydrogenase *vs.* energy-providing hydrogenase 311
– hydrogen-activating proteins in nature 310
molecular light-switches, *see* phytochromes
molecular oxygen (O_2) 289, 290
– chemical versatility 290
– nitrogenase formation 323
– sensing 289
– – by bacteria 289
– – O_2 as signal 290
– – proteins 316
– sensitive enzyme 289, 323
– sensors 289, 290
molecular probes 315
moonlighting proteins 330
mRNA motifs 381
– genes for the environment, membranes and motility (GEMM) 381
Mycobacterium bovis BCG 439
– PknG deletion mutant 439
Mycobacterium tuberculosis 324, 429, 432–436, 438, 439, 441
– crystal structure 435
– hypoxic response 324
– phosphorylation 438, 441
– phosphoserine/threonine protein phosphatase 436
– PknF 441
– PknG deletion mutant 439
– PknG structure 438
– receptor kinase PknD 432
– sigma factors 434
– STPKs 429, 433
– target protein OdhI in *C. glutamicum* studies 439
– type II fatty acid synthase (FAS-II) system 436
Mycobacterium xanthus 42, 46, 49, 50
– elastic wall 49
– forkhead-associated (FHA) domain 50
– fruiting body development 46–48
– genes, inventory 41
– lifecycle 42
Myxobacteria 41, 44, 48
– motility 41
– multicellular organisms 41
– reversal frequency management 48, 49
– swarming 44
Myxococcus stipitatis 47

n

National Center for Biotechnology Information (NCBI) 271
– protein database 271, 272
– RefSeq database 271
Natronomonas pharaonis 205
– transmembrane domains 205
neuronal PAS domain protein 2 (NPAS2) 319
– eukaryotic CO receptor 319
nitric oxide (NO) 316, 317
– biological functions 316
– detoxification 317
– sensing 315–324
– – detecting multiple diatomic gases 323
– – DosT/DosS/DosR system 324
– – eukaryotic-style sensing in prokaryotes 321
– – by FNR 323
– – FixL 323
– – by Fur/SoxR/OxyR 322
– – NorR-type regulators 317
– – NsrR-type regulators 318
– – primary sensors 316–319
– – regulators containing CAP domains 318
– toxicity 317
nodulation genes 60–64, 66
– accessory components 66
– – rhizobial core genome 66
– activation 61–64
– ATP-binding cassette (ABC) transporter proteins 63
– functions 62
– LysR family, regulator 60
– sequences 66
nonphotosynthetic bacterium 183
– *Salinibacter ruber* 183
normal-phase liquid chromatography (NPLC) 100
N-terminal chromophore-binding domain, architecture 189
N-terminal DNA-binding domain 322
N-terminal domains (NTD) 81, 221, 366, 451
nuclear factor-κB (NF-κB) 79

nuclear magnetic resonance (NMR) spectroscopy 103, 120
nucleotide-binding and oligomerization domain (NOD)-like receptor 80
nucleotide-binding domain and leucine-rich repeat (NLR) 80, 81
– family gene I 80
– inflammasome 81
– proteins, classification 81
– role 81
– tripartite domain structure 81

o

obligate intracellular bacterial pathogens 399
– *Helicobacter pylori* 399
– (p)ppGpp signaling 399
OdhI protein 439, 440
one-component system 318
osmosensitive histidine kinase response/regulator system 146
– paradigm for 146
osmosensory uptake systems 221–223
outer membrane signaling mechanism 117–120, 127, 128
– biofilm formation 128
– cell-cell communication 128
– ferric citrate transport genes 117
– – transcriptional regulation 117
– transporters 127
2-oxoglutarate dehydrogenase (ODH) complex 440

p

Pachyrhizus tuberosus 65
passive regulation 407
– sigma factor competition 407–408
PASTA (penicillin-binding protein and serine/threonine kinase-associated) domain 429
– role 429
pathogen 77, 91
– translocated intimin receptor (Tir) 91
pathogen-associated molecular patterns (PAMPs) 78–80
– diversity 79
– signaling molecules 78
pathogen-host cell communication 91
pathogen-host interaction 400
– (p)ppGpp signaling 400, 401
pathogenic *Neisseria* species 89
– language 77
– short-range adhesins 89
pattern-recognition receptors (PRRs) 78

– cell/non-cell-associated 78
Per/Arnt/Sim (PAS) domain 139, 183, 202, 293, 300, 314, 315
– protein-ligand interaction domain 202
– protein-protein interaction domain 202
– roles 315
persister cells 401, 402
– DNA-dependent RNA synthesizing enzyme 402
Peyer's patches 90
phenol-soluble modulins (PSMs), classes 13, 16
phenylpropanoid and acetate-malonate pathway 59
– secondary metabolites 59
phosphodiesterase activity 363
phosphoenolpyruvate (PEP)-dependent sugar phosphotransferase system 343, 358
– dependent protein kinase 362
phosphotransferase system (PTS) 270, 334, 335, 343, 345, 359
– chemotaxis sensors 270
– dependent phosphorylation cascade 369
– *E. coli* glucose permease 335
– energy-coupling proteins 359
– – enzyme I, histidine-containing protein 359
– features 345
– multiprotein phosphorelay system 343
– nitrogen-related 352
– paralogous, regulatory functions 352, 353
– phosphoenolpyruvate-dependent 334
– phosphorylation cascade 360
– proteins, signal transfer mechanism 345
– regulation domain 335
– transport systems 215
phosphotransferase system regulation domain (PRDs) 350
– containing regulators 350, 351
– catabolite control 351
– HPr-mediated phosphorylation 351
photoactive yellow protein (PYP) 183
photoautotrophic organisms, cyanobacteria 177
photoreceptors 177
– chromoproteins 177
photosensory domain 186
– GAF 186
– PAS 186
– PHY 186
photosensory receptors 178
phototaxis receptors 179
– SRI/SRI I 179
phototrophic bacteria 188

– phytochromes 188, 189
phototropin-like microbial photoreceptors 184, 185
phytochromes 186
– action, principle 185, 186
– BV adducts 189
– domain organization 186, 187
– like photoreceptors 188
– like proteins 187, 188
PknG 438, 439
– distribution 438
– enzyme activity, indirect control 438
– GarA identification 440, 441
– – as substrate of mycobacterial PknB 440
– – as substrate of mycobacterial PknG 441
– GarA/OdhI function, putative mechanism 442
– – in Mycobacteria 441
– OdhI homolog identification, GarA in Mycobacteria 440
– 2-oxoglutarate dehydrogenase inhibition 440
– – by corynebacterial OdhI 440
– role in pathogenicity of mycobacteria 439
– signal transduction model 443
– – by PknG and OdhI/GarA 443
– structure 438
– target protein 438
plant photoreceptors 275
– PHY (phytochrome) domain 275
poly-N-acetyl-glucosamine 11
polysaccharide-associated biofilm accumulation 11
polysaccharide intercellular adhesin, see poly-N-acetyl-glucosamine
positive stringent control mechanism 407
ppGpp 398–402, 408
– (p)ppGpp signaling 399
– – lack of, in bacterial pathogens/archaea 399
– – in virulence/pathogen–host interaction 400
– as mediator of bacterial social behavior 400
– as mediator of cell–cell signaling mechanisms 400
– as regulator for toxin–antitoxin systems 401
– – in bacterial programmed cell death 401
– dependent inhibition 405
– dependent transcription initiation, rate-limiting step 405, 406
– global effects 398
– global regulator 398–402

– in plants 399
– persister cells/enhanced mutation frequency 401
– signaling 401
– transcriptional profiling 398, 399
(p)ppGpp 397, 402, 404
– effects 402
– promoter-specific effects 404, 405
– signaling 396, 404
– – induction 396, 397
– – role 404
– SpoT-mediated synthesis 397
primase enzyme 402
– DNA-dependent RNA synthesizing enzyme 402
– role 404
programmed cell death, see bacteriostasis
proline dehydrogenase (PDH) 333
– domain 333
proline metabolism 332
– PutA localization 332
– two-step degradation 332
– – bifunctional enzyme 332
proteasome 449
– proteolysis, substrates for 449
protein 253, 255, 415, 427, 442
– degradation, role 449
– domain 269, 270, 332, 429
– – definition 270
– – identification 270
– – involvement in intracellular signal transduction 269
– extracellular/intracellular changes 427
– labeling techniques 247
– – site-directed cysteine modification 247
– localization, role 170
– paramagnetic center 255
– purification 250
– regulatory functions 442
– role 415
– sequence databases, UniProt 271
– signal transduction 427
– site-directed spin labeling 253
– topology 250, 255
protein-DNA complex 467, 470
– interactions 468
protein domain databases 270
– COG 270
– Pfam 270
– SMART 270
protein-encoding mRNAs 416
– 5′-untranslated region (5′-UTR) 416
protein interactions by cysteine modification (PICM) 252

protein-mediated control mechanisms   415
protein-protein interaction   150, 249, 314
α-proteobacteria   185
– phototrophic purple bacteria   185
– – *Rhodobacter sphaeroides*   185
– – *Rhodopseudomonas palustris*   185
proteolysis mechanisms   456
Providencia stuartii   231, 240
– rhomboid-mediated QS   240
Pseudomonas aeruginosa   24, 26, 29–32, 85, 124, 126, 234, 235, 380
– dissimilatory nitrate reduction   32
– FecR homologs   126
– Fec type transcription regulation   124–126
– lipid-like AHL molecules, production   85
– PA14 mutant   30
– quinolone signal   26
– – 2-heptyl-3-hydroxy-4-quinolone   26
– rhamnolipid synthesis operon, expression   31
Pseudomonas putida   30, 123
– Fec type transcription regulation   123
– PCL1445, lipodepsipeptides   30
Pseudomonas quinolone signal (PQS)   85, 106
pulse-chase experiments   455
PutA   332
– domains   332
– membrane-bound form   332

q

quorum quenching approach   26, 32
quorum sensing (QS) system   3, 23, 28, 83, 95, 144, 385, 400
– biofilm formation   83
– cascade   32
– inhibition targets   32
– inhibitors   33
– pathogenic bacteria, virulence genes expression   83
– regulated factors   25
– regulated proteome   26
– regulation system   98

r

Ralstonia eutropha   312, 313, 315, 317
– H$_2$ sensors   312
– RH samples, characteristic features   312
Ralstonia solanacearum   127
– ECF signaling   127
– regulatory cascade   127
rapid sampling method   363
reactive nitrogen species (RNS)   322
receptors   166, 167, 172
– allosteric interactions   172
– CheW-CheA complexes   165, 166
– kinase complexes   167, 170, 171
– NFR1/NFR5   64
– operational modules   166
red-light sensing, phytochromes   185–190
– action principle   185, 186
– cyanobacterial   187, 188
– domain organization   186, 187
– proteins   187, 188
redox-switch protein, RshA   434
regulated intramembrane proteolysis (RIP)   229, 232, 234, 236, 237, 239, 450
– aspects   234
– *Bacillus subtilis* cell division   237
– bacterial cell division regulation   237–239
– cell-cell communication   239, 240
– differentiation regulation   237–239
– ECF sigma factors regulation   236
– role   229
– transmembrane signal transduction system   450
regulator-operator binding methods   465
– ChIP   465
– SELEX   465
regulatory proteins   338, 456
regulatory RNAs, types   416
repression of heat shock gene expression (ROSE) element   421
resistance-nodulation-cell division (RND) family   66
retinoic acid-inducible gene I (RIG-I)-like receptors   80
reversed-phase liquid chromatography (RPLC)   100, 105, 107
rhizobia   57, 64, 216
– dicarboxylic acid uptake   216–218
– gram-negative soil bacteria   57
– legume interaction   65
– *Rhizobium leguminosarum*   216
– *Sinorhizobium meliloti*   216
– strains   60
Rhizobium leguminosarum   59, 216
– nod genes   59
Rhizobium meliloti   143, 216
RH mutants   314
– phenotypic characterization   314
Rhodobacter capsulatus   311, 312
– H$_2$ sensors   312
Rhodobacter sphaeroides   28, 273, 379
– BLUF domain   273
rhodopsins   178–180, 205
Rho proteins, multifunctionality   87
RH proteins   311
– structural comparison   311

ribbon-helix-helix (RHH) motif   333
– characteristic   333
riboswitch(s)   381, 416–420, 422
– classes   417
– fork-like architecture   420
– glycine-dependent   419
– metabolite-loaded   417
– ON/OFF switches   417
– regulatory principles, schematic illustration   418
– three-dimensional structure   420
riboswitch-mediated control mechanisms   419
RNA   14, 333, 334, 415, 416
– regulators   420
– as regulatory molecule   415
– sensory/regulatory potential   416
– structures   334
– world hypothesis   422
– transcription   14
RNA polymerase   404, 405
– core enzyme   123, 125
– holoenzyme   366, 403, 408
– promoter complexes   405
– role   404
– secondary channel-binding proteins   406, 407
RNA thermometers   420–422
– applications   422
– *vs.* riboswitches   420
– melting mechanism   420
ROK (repressor, open reading frame, kinase) family   338
– glucose kinase   338
RseA transmembrane regulatory protein   122, 233, 234
– anti-sigma factor   233
– RIP, aspects   234

s

Saccharomyces cerevisiae   147
– osmoresponsive hybrid sensor kinase   147
S-adenosylmethionine (SAM)   417
Salmonella enterica   143, 144
– anaerobic respiratory gene expression control   143
– PmrB/PmrA system   144
Salmonella species   385
– c-di-GMP signaling   385–387
Salmonella typhimurium   357, 359, 400
scaffolding proteins   150
SELEX (systematic evolution of ligands by exponential enrichment) method   465–467

– application   466
– procedure   465
sensing protein, function   312
sensor kinase   142, 300, 453
– BarA   144
– diversity   280
– transmembrane domains   144
sensor kinase/response regulator systems, *see* histidine kinase/response regulator systems
sensor proteins   279, 316
– primary sensors   316
– secondary sensors   316
– type   279
sensory complex   165, 167
– architecture   165–167
– clustering   167–169
sensory rhodopsin transducer protein, HtrII   201
sensory transporters, classes   211
serine/threonine phosphorylation   428
– *vs.* histidine/aspartate phosphorylation   428
serine/threonine protein kinases (STPKs)   279, 427, 428–430, 432, 436, 443
– catalytic domain   427, 429
– conformational states   430
– discovery and distribution in prokaryotes   427
– domain architecture   428, 429
– enzyme activities, direct control   436
– feature   428
– function   432
– histidine kinases, characteristic domains   429
– *in vivo* phosphorylation data   443
– kinase domain, characteristic features   431
– level of variation   429
– mycobacterial, structures   432
– response regulators, characteristic domains   429
– role   427, 428
– signal transduction models   427, 432, 436
– structural studies   429–432
serine/threonine/tyrosine phosphorylation-dependent signaling pathways   138
Serratia marcescens   27, 29, 127
– ECF signaling   127
– high-affinity hemin transport system   127
Shine–Dalgarno (SD) sequence   417
short-distance communication   77, 88–91
– bacterial adhesins   88–91
– host cell receptors   88–91
sigma factors   233, 368
signal processing   170, 172
– allosteric receptor units   172

– amplification 170
– clustering, role 170–173
– integration 172
– molecules, classes 23
signal transduction systems 77, 135, 269, 358
– domain 280
– histidine kinase/response regulator systems 135
– organization 269
signaling domains 125, 270
– computational analysis 270
– structural characterization 270
signaling protein 270, 453
– computational analysis 270
signaling specificity 381, 382
signaling systems 76, 197, 269
– functional components 76
– two-component signaling system 197
Sinorhizobium meliloti 60, 63, 66, 292, 323
– FixL proteins 323
– O_2-responsive regulation 292
site-directed spin labeling (SDSL) 253
– proteins 253–255
site-1 proteases (S1Ps) 229
site-2 proteases (S2Ps) 229, 231
– HExxH motif 231
small regulatory RNAs (sRNAs) 416
– expression 416
– roles 416
sodium dodecylsulfate-polyacrylamide gel electrophoresis 252
solid-phase extraction (SPE) 100, 101, 105
soluble bactericidal proteins 77
– advantage/disadvantage 101
– application 105
– defensins 77
– lysozyme 77
spin-labeled proteins 255
– EPR spectra of 255
standard [NiFe]-hydrogenase, structure 311
staphylococcal biofilms 13
– formation phases, model 9
– QS 13–17
– – Agr QS Locus 13–16
– – LuxS/AI-2 system 17
staphylococci 8
– abiotic surfaces attachment 8–10
– accumulation process 11, 12
– – extracellular DNA 12
– – polysaccharide-associated biofilm accumulation 11
– – protein-associated biofilm accumulation 12

– agr QS system model 14
– biofilm formation 8–13
– – escape factors 12
– – molecular basis 8–13
– biotic surfaces attachment 10
– host-factor-binding proteins 10
– infections 8
– surface-associated proteins 10
Staphylococcus aureus 8, 9, 144, 150
– ApsS/ApsR/ApsX system, scaffolding protein 150
– colonization 9
– DltA mutant 9
– untranslated RNA molecule synthesis 144
Staphylococcus epidermidis 8, 11, 13, 150
– agr mutant 15, 16
– – biofilm formation 16
– ApsS/ApsR/ApsX system, scaffolding protein 150
– IcaADBC operon 11
– luxS mutant 17
Stigmatella aurantiaca 47
stimulus perception mechanism 136
stochastic nucleation mechanism 168
streptococci 349
– lactose transport proteins 349
streptococcus mutans 397
– *Bona fide* global regulator 398–402
– global effects 398
– role 402, 403
Streptococcus pneumoniae 3
Streptococcus pyogenes 352
Streptomyces coelicolor 338
stress response genes, transcription 434
stress sensing, transport proteins 219–223
stringent control phenomenon 395
sugar-specific PTS EII components 279
– roles 279
sulfate transporter and anti-sigma antagonist (STAS) domain 185

t
target cell 77
– surface-exposed receptors 77
target gene 465, 468
– analysis 468–471
– – DNA affinity purification chromatography (DAC) 468, 469
– – EMSA analysis 468, 469
– finding 465–468
– – ChIP 467
– – SELEX 465–467

temperature-sensing histidine kinase/
  response regulator system 148
– CorS/CorR 148
Thermus aquaticus 407
– Gre factor homolog (Gfh1) 407
Thermus thermophilus 335, 338, 403, 404
– Mlc repressor 335
– rate-limiting step 405, 406
thiamine pyrophosphate (TPP)-responsive
  RNA 416
thin-layer chromatography (TLC) 100, 101,
  105
– advantage/disadvantage 101
– AIs isolation 101
threonine residue, autophosphorylation
  432
toll-like receptor (TLR) 78–80
– heterodimers, crystal structures 79
TonB box 119, 120, 124
– flexibility 120
Ton system 119
toxin-antitoxin systems 401
– regulator 401
– suicide function 401
ToxR-like transcriptional regulators 236
– regulation 236, 237
transcription, regulation 403–408
transcription factors 329
– ComK 454
transcription regulators 350
– control 350
– proteins 463
transcriptional profiling 463
transcriptomes 472
transgenic plants 32
translation, inhibition 403
transmembrane domain 257
transmembrane helix 1 (TM1) 198
transmembrane histidine kinase domain
  architecture 198–201
transmembrane protein, AgrB 145
transmembrane signaling pathways 201,
  229
– characterization 229
– domains, architecture 198–201
– structural analysis 201–206
transmembrane signal transduction 122
– molecular mechanism 122
transmitter molecules 307
– inter/intracellular signaling 307
transport activity 212
– sensing 212–219
transport proteins 212

tricarboxylic acid (TCA) cycle 333
trigger enzymes 329–335, 338
– activity 335
– as DNA-binding transcription factors
  332, 333
– compilation 330, 331
– evolution 338, 339
– – importance 338
– functions 338
– gene expression control 334
– – by signal-dependent phosphorylation
    of transcription regulators 334
– post-transcriptional regulation via
    protein-RNA interaction 333
– role 330
– signal transduction 329
– transcription factors activity control
  335–337
– – by protein-protein interactions 335
tris(2-carboxymethyl)phosphine (TCEP)
  249
turgor sensor, intracellular parameters
  147
two-component receptor 307
– ethylene receptor ETR1 (ethylene
  resistant 1) 307
two-component regulatory systems 314
two-component signal transduction
  system 83, 276–278
– chemotaxis 276–278
– histidine kinase-type ATPase catalytic
  (HATPase_c) domain 276
– QseE/QseF 84
two-component system 148–150, 300,
  314, 315, 318, 427, 428
– BceS/BceR 149
– histidine kinase 427
– intriguing questions 314
– NorR 318
– signal transduction 428
type I secretion system (T1SS) 87
type II secretion system (T2SS) 86
type III secretion system (T3SS) 64, 77,
  128
– protein transport machinery 64
type IV secretion systems 77

u
UhpABC regulatory system 148, 215
– two-component system 215
ultra-performance liquid chromatography
  (UPLC) 106
– advantage 106

ultraviolet/visible (UV/Vis) spectrophotometry 101
universal stress protein (USP) family 275
– amino-acid domain 275
unphosphorylated EIIAGlc 367
– inducer exclusion 367, 368
upregulated genes, classes 49

v

van der Waals interactions 9, 256
Vibrio cholerae 86, 380, 385, 386
– c-di-GMP signaling 385–387
– phage-encoded cholera toxin (CTX) secretion 86
Vibrio harveyi 145, 203
– complex QS regulation 145
virulence genes 400
– associated genes 17
– (p)ppGpp signaling 400, 401
– toxin-antitoxin systems, regulator 401

w

Watson–Crick base pair 419
– base pairing 416
– formation 419
Western blot analysis 250
Wood–Ljungdahl pathway 320

x

X-ray diffraction data 310
Xanthomonas campestris 27
Xanthopsins 183
Xenorhabdus nematophila 146
– EnvZ 146
Xylella fastidiosa 27

y

Yersinia enterocolitica 90
– invasin interaction 90
Yersinia pseudotuberculosis 28, 90
– invasin interaction 90